DEVELOPMENTAL MATHEMATICS

SECOND EDITION

MARTHA M. WOOD

DeKalb Community College

PEGGY CAPELL

Clayton Junior College

JAMES W. HALL

Parkland College

PRINDLE, WEBER & SCHMIDT
BOSTON, MASSACHUSETTS

Library of Congress Cataloging in Publication Data

Wood, Martha M
    Developmental mathematics
    Includes index
    1. Algebra   2. Arithmetic
 I. Capell, Peggy, joint author
 II. Hall, James W., joint author
III. Title
QA 152.2 W654   1980   512.9   80-11571
ISBN 0-87150-287-9

Fourth printing: July, 1981

Text design by Elizabeth Thomson.
Cover art: ''Pulse'' by Frank Rowland. © 1976 by Circle Fine Art Corporation. Used by permission of the publisher.
Text composition by Omegatype.
Text and cover printing: The Alpine Press
Binding: The Alpine Press

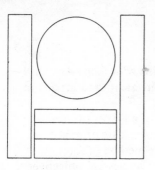

PREFACE

This text, like the successful first edition, is intended for college students who have not previously studied algebra or who need a review of elementary algebra. An optional arithmetic review is provided for those students who need to refresh their basic arithmetic skills.

Developmental Mathematics is an informal, non-rigorous, readable approach to elementary algebra. A student who successfully completes this text will be able to enter a subsequent algebra course, a technical mathematics course, or a liberal arts mathematics course.

Developmental Mathematics has several important features:

1. It covers both arithmetic and algebra. While the arithmetic review can be omitted, some students will benefit from reviewing these basic procedures before beginning a study of algebra. This is particularly true since the arithmetic topics are developed to facilitate the learning of corresponding algebra topics.

2. Although the arithmetic review is optional, the first ten chapters are sequential. After Chapter 10, the remaining chapters can be selected independently according to the needs of the student and the time available. The flowchart that follows partially illustrates this flexibility, as well as chapter dependencies. However, the individual instructor is best equipped to draw up a course outline as determined by the needs and abilities of the students.

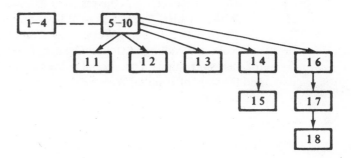

3. The arithmetic review starts with decimals instead of fractions. This facilitates a natural review of whole number arithmetic using a form that is more familiar to most students.

4. Verbal statements of problems are emphasized throughout the text. By experiencing success in solving verbally-stated problems such as "increase two by five," students are able to develop and strengthen the skills needed to solve algebraic word problems. Students gain confidence in translating mathematics into English, and back again, and by solving practical "useful" problems.

5. Most topics are completely developed when they are introduced rather than treated intermittently throughout the text. For example, the student who wishes to review linear equations can refer to one complete chapter, instead of sections in several different chapters.

6. Each chapter contains the following:
   a. A list of objectives referenced by section.
   b. A pretest with problems referenced to the objective being tested.
   c. A multitude of examples, each immediately following the introduction of a procedure or rule.

d. An average of 345 exercises. Each exercise set begins with many easy drill type problems. The even and odd exercises are equivalent, and graded in difficulty.

e. A set of review exercises. These provide a sampling of all types of problems presented in the chapter. These various types of problems are arranged randomly so that the student must select the appropriate method of solution. A star is used to denote optional exercises that may cover several objectives, or require insight or the extension of previous skills. They may also require the use of skills from previous chapters.

7. There are more than 6200 exercises in the text.

Developmental Mathematics has been used successfully by both teacher and student in a traditional classroom setting, an individualized setting, and in a mathematics laboratory. Its success is due to its readability, clearly stated rules, carefully outlined objectives, numerous examples, and abundant exercises. These strengths have been retained and improved in the light of our own experience and that of others who used the first edition. We would like to thank:

Helen Atkins, John C. Calhoun Community College, Alabama
Joseph Collinson, Baruch College—C.U.N.Y.
Terry Czerwinski, University of Illinois—Chicago Circle
Carlyle Flemming, Loma Linda University, California
L.C. Glaser, University of Utah
Allan Gottlieb, York College—C.U.N.Y.
Howard E. Hudson, Theodore Lawson State Community College
John C. Miller, City College—C.U.N.Y.
W. Kent Moore, Gardner-Webb College, North Carolina
Paul Pontius, Pan American University, Texas
Joan Richardson, University of Northern Colorado
Douglas Robertson, University of Minnesota—Twin Cities
Walter Sadler, University of Wisconsin—Waukesha Center
Constance Saulsbery, Westchester Community College, New York
Robert Wendling, Ashland College, Ohio

for their suggestions. Our special thanks go to mathematics editor Connie Caldwell and the staff of Prindle, Weber & Schmidt for their assistance in the preparation of this edition.

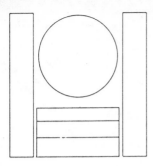

CONTENTS

REVIEW ARITHMETIC

1

DECIMALS

OBJECTIVES

Upon completion of this chapter you should be able to:

1. Identify the place value of a digit in a numeral.	1.1
2. Write numerals for verbal expressions.	1.1
3. Write verbal expressions for numerals.	1.1
4. Round decimals.	1.2
5. Add two or more decimals.	1.3
6. Subtract decimals.	1.4
7. Multiply two or more decimals.	1.5
8. Divide decimals.	1.6
9. Work word problems involving addition, subtraction, multiplication, and/or division of decimals.	1.7
10. Identify the operation implied by each of the following words or phrases.	1.3–1.6

decrease	less	subtract from
difference	less than	sum
divide by	minus	take away
divided by	more than	times
divide into	plus	triple
double	product	twice
increase	quotient	

PRETEST

1. Identify the place value of the digit 7 in each of the following numerals (Objective 1).
   a. 743   b. 347
   c. 4.37   d. 0.04037
   e. 70,430   f. 407,030

2. Write a numeral for each of the following verbal expressions (Objective 2).
   a. one thousand thirty-nine
   b. three hundred and seven-tenths
   c. five million three hundred thousand seven hundred forty
   d. nine thousand and three millionths

3. Write a verbal expression for each of the following numerals (Objective 3).
   a. 307   b. 4009.08
   c. 0.0037   d. 4,000,710.009

4. Round 583.547 to each of the following places (Objective 4).
   a. the nearest tenth   b. the nearest hundred
   c. the nearest hundredth   d. the nearest unit

5. Add the following decimals (Objective 5).
   a. 74.031
      + 9.0508
   b. 408.9 + 17.48 + 9.031 + 8.09 + 79

6. Perform the following subtractions (Objective 6).

   a. 119.46
      $-\ 95.08$

   b. $43.51 - 27.567$

7. Perform the following multiplications (Objective 7).

   a. 35.718
      $\times\ \ \ 9.05$

   b. $7.43 \times 8.01 \times 1.001$

8. Perform the following divisions (Objective 8).

   a. $3.5061 \div 4.03$
   b. $0.108823 \div 0.00761$

9. Solve the following problems (Objective 9).

   a. An employee worked thirty-five hours at four dollars and fifteen cents per hour. What was his gross pay for the week?

   b. A product is thirty-eight and seven tenths centimeters wide. When the product is shipped, a padding one and seventeen hundredths centimeters thick is placed on each side of the product. How wide must the shipping carton be?

   c. At the beginning of a trip, a car's odometer read eighteen thousand four hundred thirty-three and five tenths miles. At the end of the trip, the reading was eighteen thousand nine hundred ninety-seven and one tenth miles. How long was the trip?

   d. A club contributed $233.91 to a civic project. If there were twenty-three members in the club, what was the average contribution per member?

10. Identify the operation implied, and then work the problem (Objective 10).

   a. 117.8 more than 83.409

   b. 0.0002862 divided by 14.31

   c. 12.9 less than 182.4

   d. 43.082 less 41.23

   e. 1007.04 decreased by 909.8

   f. 0.0179 subtracted from 0.179

   g. 0.0782 divided into 0.008602

   h. 4.85 multiplied by 3.002

It may seem logical to assume that college students are able to perform basic arithmetic computations. However, it is not unusual to find students who have forgotten some of the basic procedures because they have not had occasion to use the required skills regularly. It should not be a difficult task to review or to relearn the important procedures. Computational skills are becoming almost as important as language skills in day-to-day activities; so students should exert whatever effort is needed to obtain competency in arithmetic computations and their applications.

1.1 Reading Numerals

A study of the historical development of ways numbers have been recorded and of procedures that were devised for performing computations is an interesting but time-consuming endeavor. We will concentrate in this text on the numeration system used in most of the world today. This is the decimal system of numeration, using a base of ten. It is known as a **place value** or positional system. The symbols used to represent numbers in this system are the ten Hindu-Arabic symbols 0, 1, 2, 3, 4, 5, 6, 7, 8, 9. Any real number can be represented by using combinations of these symbols, which are known as **digits**.

In a place value system every digit in a numeral has associated with it the place value of the digit and the value of the digit. The **place value** is determined by the base of ten, which means that as we move from right to left in a numeral, the place value of each position is ten times the place value of the position to its right. Some of the place values of the positions in a base ten numeral are given in the following figure.

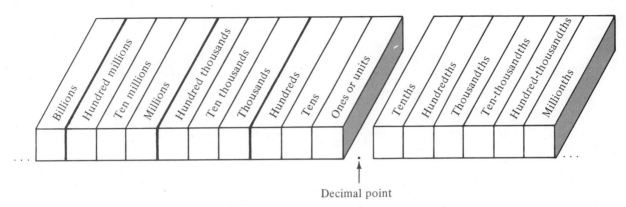

Decimal point

The **value** of each digit is found by multiplying the digit by its place value. Consider the numeral 37.9, as shown in the next figure.

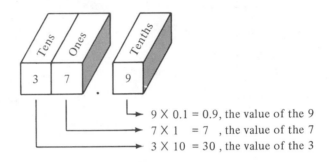

9 × 0.1 = 0.9, the value of the 9
7 × 1 = 7 , the value of the 7
3 × 10 = 30 , the value of the 3

Example 1   Give the value of each digit in the numeral 5243.917.

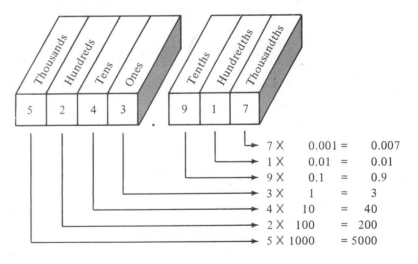

7 × 0.001 = 0.007
1 × 0.01 = 0.01
9 × 0.1 = 0.9
3 × 1 = 3
4 × 10 = 40
2 × 100 = 200
5 × 1000 = 5000

Example 2   Identify the place value of the digit 7 in each of the following numerals.

a. 7049.82   thousands
b. 12.6479   thousandths
c. 17.43   ones
d. 31.74   tenths
e. 7,462,013.85   millions

Example 3   Give the place value of each digit in the numeral 4278.391.

4   thousands   3   tenths

2   hundreds   9   hundredths

7   tens   1   thousandths

8   ones

It is important to note that the numerals 5.0, 5.00, 005, 5., and 05.0 all name the same number. Even though these names are not the standard or simplest form of the numeral 5, you need to be familiar with such forms because they will occur in computational problems.

In order to read or write the verbal name for a numeral correctly, you must consider the digits and the position they occupy.

Example 4   Write the verbal name for 5,203,456.

Recall that the digits in a whole number are grouped in threes, starting with the units digit, as shown in the following figure.

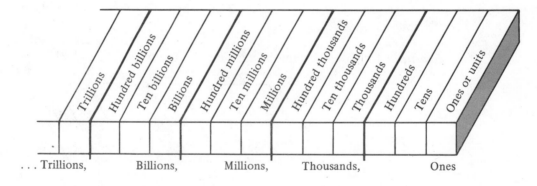

Note that each place in the second group ends with thousands, the third group with millions, the fourth group with billions, etc. Thus, a whole number is read by reading the number in each group and then giving it the group ending. The first group is understood to be ones but the group ending is omitted in reading. Thus, 5,203,456 is read "five million, two hundred three thousand, four hundred fifty-six."

To read a decimal numeral, follow these steps:

1. Read the whole number to the left of the decimal point.

2. Read "and" for the decimal point.

3. Read the numeral to the right of the decimal point as if it were a whole number and follow it by the place value of the last digit on the right.

Example 5   Write the verbal name for 5243.917.

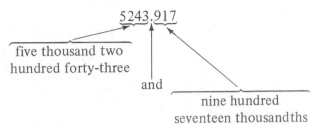

Example 6   Write the verbal name for each of the following numerals.

    a. 429.7   four hundred twenty-nine and seven tenths

    b. 1,004,000.03 one million four thousand and three hundredths

    c. 1,484.2876   one thousand four hundred eighty-four and two thousand eight hundred seventy-six ten-thousandths

Example 7   a. 429 is read "four hundred twenty-nine" (not "four hundred and twenty-nine"). Note that the decimal point does not appear in this numeral. However, since the digit 9 is in the units position, the decimal point is understood to follow the digit 9, so that 429 and 429. name the same number.

    b. 3948 is read "three thousand nine hundred forty-eight." Where is the decimal point in this numeral?

Example 8   a. 0.4 is read "four tenths." Even though a decimal point appears in the numeral, "and" is not read.

    b. 0.803 is read "eight hundred three thousandths."

       If you need to write the standard name for a numeral given in verbal form, the above process can be reversed.

Example 9   Write the numeral for each of the following verbal expressions.

    a. Forty-five and six tenths

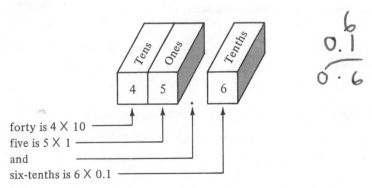

b. Fifteen thousand four hundred six and seven thousandths

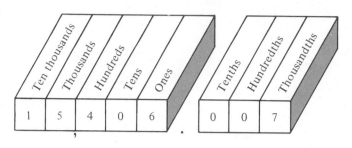

1.1   Exercises

Identify the place value of the digit 5 in each of the numerals in Exercises 1–8.

1. 351
2. 4025.1
3. 95,200.01
4. 14.2456
5. 1,047,394.5
6. 0.00156
7. 0.40135
8. 5,001,203,000.7

Identify the place value of the digit 1 in each of the numerals in Exercises 9–16.

9. 351
10. 4025.1
11. 95,200.01
12. 14.2456
13. 1,047,394.5
14. 0.00156
15. 0.40135
16. 5,001,203,000.7

17. Identify the place value of each digit in the numeral 27,398.614.
18. Identify the place value of each digit in the numeral 1,234.5678.

Write each of the numerals in Exercises 19–22 so that the numeral contains a decimal point.

19. 5740
20. 4
21. 23
22. 1,471,305

23. Which of the following are equal to 42?

    042,   42.,   4.2,   0.42,   42.000

24. Which of the following are equal to 0.7?

    00.7,   .7,   .70,   0.700,   7,   0.69

Write a numeral for each of the verbal expressions in Exercises 25–36.

25. Five hundred sixty-one
26. Eight thousand twenty-five
27. Four hundred twenty-nine ten-thousandths
28. Seven hundred twenty-eight million four hundred six thousand six hundred eighty-four and three hundred thirty-five thousandths
29. One thousand eight hundred and eighteen thousandths
30. Two hundred thirty-five ten-thousandths
31. Five thousand six hundred two
32. Three hundred forty-eight
33. Five hundred and seventy-two ten-thousandths
34. Eighty-two million three hundred ninety-four thousand one hundred twenty-three and four hundred twenty-three thousandths
35. Five thousand and five hundredths
36. Ninety-three thousandths

Write a verbal expression for each of the numerals in Exercises 37–50.

37. 582
38. 6302.9
39. 72.3492
40. 53,213,051.89
41. 0.00105
42. 8600.09
43. 351
44. 1025.6
45. 14.2456
46. 1,047,394.52
47. 0.00056
48. 95,000.05
49. 5,000,713,400
50. 892

1.2   Rounding Decimals

Numbers are used for many purposes: to count, to measure, to order or rank, to index, to identify. Numbers may be exact or approximate, depending upon the purpose for which they are used. For example, if two cities are 547 miles apart, you would probably say that the distance between them is about 550 miles. The number 547 is said to be **rounded** to the nearest ten when expressed as 550. Similarly, a measurement of 5.47 feet rounded to the nearest tenth of a foot is 5.5 feet. It is customary to say that 5.47 is approximately equal to 5.5. Symbolically, $5.47 \doteq 5.5$. Note that the symbol $\doteq$ means "is approximately equal to."

RULE FOR ROUNDING DECIMALS	*EXAMPLE A*	*EXAMPLE B*
	Round 52167.7364 to the nearest hundredth. *Solution:*	Round 52167.7364 to the nearest thousand. *Solution:*
1. Underline the digit which appears in the position to which the number is to be rounded.	52,167.7<u>3</u>64	5<u>2</u>,167.7364
2. Examine the *first* digit to the right of the underlined position.	6 is the first digit to the right of the underlined position.	1 is the first digit to the right of the underlined position.
a. If this digit is 0, 1, 2, 3, or 4, replace all digits to the right of the underlined position by zeros.		Since 1 is a 0, 1, 2, 3, or 4, all digits to the right of 2 are replaced by zeros. Thus we have 5<u>2</u>,000.0000
b. If this digit is 5, 6, 7, 8, or 9, add 1 to the digit in the underlined position and replace all digits to the right of the underlined position by zeros.	Since 6 is a 5, 6, 7, 8 or 9, add 1 to the 3 and replace all digits to the right of 3 by zeros. Thus we have 52,167.7<u>4</u>00	
3. If any of the zeros from step 2 are to the right of the decimal, omit them.	or 52,167.74	or 52,000
	Thus 52,167.7364 rounded to the nearest hundredth is 52,167.74.	Thus 52,167.7364 rounded to the nearest thousand is 52,000.

Example 1   Round 41.736 to the hundredths position.

$$41.7\underline{3}6 \longrightarrow 41.740 \longrightarrow 41.74$$

Thus 41.736 rounded to the hundredths position is 41.74.

Example 2   Round 41.736 to the nearest unit.

$$4\underline{1}.736 \longrightarrow 42.000 \longrightarrow 42$$

Thus $41.736 \doteq 42$.

Example 3   Round 7406.2 to the nearest hundred.

$$7\underline{4}06.2 \longrightarrow 7400.0 \longrightarrow 7400$$

Thus, $7406.2 \doteq 7400$.

Example 4   Round each of the following to the nearest thousandth.

a. $11.03\underline{43} \longrightarrow 11.0340 \longrightarrow 11.034$

Thus $11.0343 \doteq 11.034$.

b. $15.96\underline{95} \longrightarrow 15.9700 \longrightarrow 15.970$

Thus, 15.9695 rounded to the nearest thousandth is 15.970.

Example 5   Round each of the following to the nearest thousand.

a. $1,40\underline{5},789 \longrightarrow 1,406,000$

Thus, $1,405,789 \doteq 1,406,000$.

b. $752 = \underline{0}752 \longrightarrow 1000$

Thus, 752 rounded to the nearest thousand is 1000.

Example 6   Round 0.05 to one decimal place.

This means to round the number so that there is one place to the right of the decimal point.

$0.\underline{05} \longrightarrow 0.10 \longrightarrow 0.1$

Thus, $0.05 \doteq 0.1$.

Example 7   Round each of the following to two decimal places.

a. $172.5\underline{082} \longrightarrow 172.5100 \longrightarrow 172.51$

Thus, $172.5082 \doteq 172.51$.

b. $0.0\underline{03} \longrightarrow 0.000 \longrightarrow 0.00$

Thus, 0.003 rounded to two decimal places is 0.00.

c. $999.9\underline{99} \longrightarrow 1000.000 \longrightarrow 1000.00$

Thus, $999.999 \doteq 1000.00$.

With practice, you should be able to write the answer without showing any intermediate steps.

1.2   Exercises

Round each of the numbers in Exercises 1–8 to the nearest tenth.

1. 37.84

2. 215.96

3. 0.06

4. 122.925

5. 9.999

6. 22.09

7. 4814.55

8. 0.09

Round each of the numbers in Exercises 9–16 to the nearest unit.

9. 134.7899

10. 122.925

11. 9.993

12. 19.95967

13. 754.2

14. 0.5

15. 0.91

16. 1,784,320.499

Round each of the numbers in Exercises 17–24 to two decimal places.

17. 19.9596

18. 134.7894

19. 9.993

20. 457.004

21. 122.925

22. 0.809

23. 1,784,320.499

24. 4.395

Express each of the numbers in Exercises 25–32 accurate to the nearest thousand.

25. 2346.9

26. 789

27. 800.0095

28. 10,500.0459

29. 1,784,320.4991

30. 4814.5564

31. 52,671

32. 9004

33. Express 2346 to the nearest hundred.

34. Round 0.02357 to the nearest ten-thousandth.

35. Round 0.12345 to four decimal places.

36. Round 505.75 to the nearest ten.

37. Round 19.9596 to the nearest thousandth.

38. Express 489,701 to the nearest thousand.

39. Round 26,489,701 to the nearest million.

40. Round 6403.8051 to three decimal places.

1.3 Addition of Decimals

The purpose of this section is to refresh and perfect your addition skills. In the process, you should begin to recall many facts which you already know and to establish good work habits which will aid you in later sections.

To master this material, it will be necessary to memorize the basic addition facts. Since it would be impossible to memorize all the possible addition combinations, the purpose here is to help you learn rules and procedures that will enable you to solve the variety of new problems that you may encounter.

Try to solve as many of the following six problems as rapidly as you can.

1. $2 + 5 = ?$

2. $7 + 4 = ?$

3. $3 + 9 = ?$

4. $6 + 5 = ?$

5. $12 + 37 = ?$

6. $193,078 + 85,319 = ?$

If you didn't answer the first four problems in ten seconds, then you need to practice and memorize the basic addition facts which are given in Appendix C. It is unlikely that you have memorized the sum of 12 and 37. Instead, you should recall the following procedure for calculating that sum:

$$\begin{array}{r} 12 \\ +37 \\ \hline 49 \end{array}$$

(5)

Align the numbers so that the digits having the same place value are in columns, and then add the columns from right to left. Even though problem 6 is considerably longer, the same procedure applies.

$$\begin{array}{r} 193{,}078 \\ +\ 85{,}319 \\ \hline 278{,}397 \end{array}$$

(6)

First align the digits and then perform the addition. This problem involves a process called **carrying** from one position to another when the sum of any single column is equal to 10 or more. The carrying process can be done mentally, or notations can be made above the columns to indicate the numbers which are carried.

$$\begin{array}{r} {\scriptstyle 1\qquad 1} \\ 193{,}078 \\ +\ 85{,}319 \\ \hline 278{,}397 \end{array}$$

Note that in the two problems we have solved for you, finding the answer requires knowledge of certain facts. For example, to solve (5), you should already know that $2 + 7 = 9$ and that $1 + 3 = 4$. These are basic facts that need to be memorized. You will see that learning one procedure in mathematics will enable you to work many different problems without memorizing the answers to all of them.

Addition of decimals is very similar to addition of **whole numbers** (the numbers 0, 1, 2, 3, 4, etc.). Compare the following examples:

Addition of Whole Numbers	*Addition of Decimals*
2917	29.17
+ 146	+ 1.46
3063	30.63

Note that in order to align the digits having the same place value, it is necessary to align the decimal points. This is specified in the following *rule for adding decimals.*

RULE FOR ADDING DECIMALS

1. Align the digits by first aligning the decimal points.
2. Add as if working with whole numbers.
3. Align the decimal point in the sum with the other decimal points.

EXAMPLE $10.207 + 3.49 = ?$
Solution:

10.207	addend	
+ 3.490	addend	Note that
13.697	sum	$3.49 = 3.490$

Thus, $10.207 + 3.49 = 13.697$.

Example 1 Perform the following addition.

579.430   Add column by column
8.900   starting from the right.
45.604
+ 0.940
634.874

Example 2 Find the sum of 85.073 and 11.

85.073
+11.000   Recall that $11 = 11.000$.
96.073

The sum of 85.073 and 11 is 96.073.

Example 3 Find the total of $845.95, $72.02, $552.34, $50.08, and $9.15.

$$\begin{array}{r} \$\ 845.95 \\ 72.02 \\ 552.34 \\ 50.08 \\ +\quad 9.15 \\ \hline \$1529.54 \end{array}$$

The total is $1529.54.

Recall that the order in which you add numbers is immaterial. In the first column, $5 + 2 + 4 + 8 + 5$ is the same as $5 + 5 + 2 + 8 + 4$. Choose the most convenient order.

In Examples 2 and 3, the words "sum" and "total" were used to indicate that addition was to be performed. Other terms which are used to indicate addition are "plus," "add," "increased by," and "more than." Terms which indicate subtraction, multiplication, and division will be covered in later sections.

1.3 Exercises

Add in each of Exercises 1–24.

1.
552
+841

2.
87
92
5
57
+80

3.
46
82
99
82
+41

4.
75
+930

5.
841
4
1452
92
100
+2009

6.
50000
72
5649
+ 7000

7.
8.9
+243.6

8.
8.481
+2.42

9.
7.53
+94.00

10.
142.9
+ 3.78

11.
7.43
.9
.05
27.99
+ 5.0

12.
$75.22
83.04
+ 28.17

13.
$25.72
48.30
+ 17.28

14.
1.1
1.11
11.11
+ .111

15.
7019.304
+ 5.070

16.
489.8643
+ 0.1050

17.
$ 458.92
4.95
1752.00
14.05
+ .92

18.
5789.
7.1432
+ 450.9

19.
89.4444
0.5
2000.
+ 45.

20.
$57905
4525
12005
975
85
+ 2505

21.
25.0000
85.1049
+72.3051

22.
345.839
765.271
200.000
+555.555

23.
25.906
0.2734
200.
4.2
5241.14
+ 0.9

24.
78942.1
405.
9.3
.0004
+ 93.897

Each of Exercises 25–45 involves addition. First identify the word(s) or symbol which indicates addition, and then work the exercise.

25. 1257.063 + 8

26. 47.523 + 7.47 + 578

27. Find the total of $215.42, $60.78, and $789.90.

28. Find the sum of 135.789, 0.0005, and 52.

29. Find the result if 952.8 is increased by 1500.42.

30.   23.059
     5.6
  + 0.08

31. What is 12.01 more than 19.47?

32. Find the total of 49, 7084.9, 0.5704, and 92.

33. 5222.1975 + 2.36 + 12 + 0.8

34. 53.4256 + 1.0421 + 0.2956 + 100

35. Find the total of 5.4, 9.6, 4.3, 7.9, and 8.2.

36. Find the result if 34.52 is increased by 15.

37. 95.6508 plus 35.03

38. 649.3 + 47.95

39. 53.0206 + 32.4050 + 232.0000 + 0.3200

40. Find the sum of 95.02 and 7.48.

41. Add seventy-six thousand and four hundredths to five hundred thirty-five and eighty-two thousandths.

42. What is nine and four tenths increased by ninety-four hundredths?

43. Find the sum of 87.5, 7.58, 578, and 0.785.

44. Add 50.2, 17, 43, and 17.43.

45. What is the sum of 92, 9.2, 902, and 0.902?

1.4 Subtraction of Decimals

Before considering subtraction of decimals, it will be helpful to review the vocabulary associated with the subtraction of whole numbers.

Subtraction can be indicated using a vertical or a horizontal notation.

$$9786 \quad \text{minuend}$$
$$\underline{-4320} \quad \text{subtrahend}$$

$$\underline{9786} \quad - \quad \underline{4320}$$
$$\text{minuend} \qquad \text{subtrahend}$$

Verbally, this problem can be stated in any of the following ways:

9786 minus 4320	4320 subtracted from 9786
9786 take away 4320	9786 less 4320
subtract 4320 from 9786	4320 less than 9786
9786 subtract 4320	9786 decreased by 4320

Note that in the verbal subtraction statements, the minuend is given first in some forms while the subtrahend is mentioned first in other forms. You should be able to determine from the statement of the problem which number is the subtrahend and which is the minuend. (In this chapter we will consider only those subtraction problems where the minuend is larger than the subtrahend.) Remember that the subtrahend is written beneath the minuend in the vertical format and is the second number in the horizontal format.

In order to subtract, align the numerals so that the digits having the same place value are in columns, and then subtract the columns from right to left. Consider the following examples with whole numbers.

Example 1 Perform the following subtractions.

a.   9786
    −4320
     5466

b.   2 10 6 18
      3̸ 0̸ 7̸ 8̸
    − 3 1 9
     2 7 5 9

The problem in Example 1b involves a process called **borrowing**. You borrow from one position and add to another position when the value of a digit in the subtrahend is greater than the value of the corresponding digit in the minuend. The borrowing process can be done mentally, or notations can be made above the columns.

Subtraction of decimals is very similar to subtraction of whole numbers. Align the digits by aligning the decimal points and then perform the subtraction. The procedure is given in the following *Rule for Subtracting Decimals.*

RULE FOR SUBTRACTING DECIMALS	EXAMPLE Find $97.067 - 43.2$.

1. Identify the minuend and subtrahend.
 	
2. Place the subtrahend beneath the minuend and align the digits by first aligning the decimal points.
 	
3. Subtract as if working with whole numbers.
 	
4. Align the decimal point in the difference with the other decimal points.
	Solution:
	
	  97.067 minuend
	  $\underline{-43.2}$ subtrahend
	
	  97.067
	  $\underline{-43.200}$
	  53.867 difference
	
	Thus, $97.067 - 43.2 = 53.867$

Example 2 $436.72 - 49.46 = ?$

$$
\begin{array}{r}
{\scriptstyle 3\ \ 12\ 16\ \ \ \ \ 6\ \ 12} \\
\cancel{4}\ \cancel{3}\cancel{6}\ .\ \cancel{7}\ \cancel{2} \\
-\ \ \ 4\ 9\ .\ 4\ 6 \\
\hline
3\ 8\ 7\ .\ 2\ 6
\end{array}
$$

So, $436.72 - 49.46 = 387.26$.

Example 3 $\underbrace{4.6}$ minus $\underbrace{3.942}$ $= ?$

  minuend   subtrahend

$$
\begin{array}{ll}
4.6 \ \longrightarrow & 4.600 \\
\underline{-3.942} & \underline{-3.942} \\
 & 0.658
\end{array}
$$
   Recall that $4.6 = 4.600$.

Therefore, 4.6 minus 3.942 is 0.658.

Note that a subtraction problem may be checked by adding the difference and the subtrahend. The resulting sum should be the minuend. $0.658 + 3.942 = 4.600$.

In subtraction we may, if necessary, annex zeros so that both numerals have the same number of digits to the right of the decimal point. Also, remember that the order in which you subtract *is* important. You always subtract the numbers in the subtrahend from those in the minuend.

Example 4   Find the difference if 18.05 is subtracted from 49.

subtrahend   minuend

49. → 49.00   Recall that 49 = 49.00
−18.05   −18.05
          30.95

Check: 30.95
         +18.05
          49.00

1.4 Exercises

Perform the subtractions in Exercises 1–25.

1. 87 −54	2. 34 −19	3. 4897 −3958
4. 518 − 47	5. 10000 − 4320	6. 80567 − 9489
7. $25.34 − 7.25	8. $5.73 − 1.24	9. 575.2 −234.7
10. 7520.4 − 435.6	11. 83.096 − 2.73	12. 15.1 − 4.4
13. 45.92 −15.45	14. 17.894 − 8.1	15. 244. − 17.1
16. 0.704 −0.052	17. 257.1 −213.1	18. 235. − 92.4
19. 0.8111 −0.5555	20. 10. − 0.444	21. 147.8 − 56.
22. 1576.95 − 357.	23. 584. − 29.504	24. 0.5124 − .1065
25. 752. − 0.999		

In Exercises 26–45, identify the word or symbol which indicates the operation of subtraction and then perform the subtraction and check your answer.

26. Subtract $125.33 from $483.17.

27. 747.26 − 601.07

28. 29.06 take away 21.96

29. 87.0468 minus 72.3007

30. Find the difference if 43.95 is subtracted from 801.6.

31. Subtract 724 from 8402.93.

32. Decrease $576 by $123.13.

33.   154.3   34. 103 less 9.82
     − 44.367

35. Subtract thirty-five and seven hundred eighty-nine thousandths from three hundred seventy-six and one hundred eleven thousandths.

36. 7.072 − 6.830   37. 46 less 0.891

38. Subtract seven and forty-five hundredths from four thousand forty-five.

39. 117.4 diminished by 87.4

40. Subtract 0.481 from 42.7.

41. 2.589 − 1.9

42. 7.01 less than 25

43. Find the difference if 189.35 is subtracted from 5999.78.

44. 11.56 minus 8

45. Subtract 92,576.473 from 142,439.111.

1.5 Multiplication of Decimals

Multiplying decimals is very much like multiplying whole numbers, as the following example points out. The only distinction is that you must determine the position of the decimal point in the product.

Multiplying Whole Numbers   *Multiplying Decimals*

317	3.17
X 85	X 8.5
1585	1585
2536	2536
26945	26.945

Why is this row shifted over one column?

The procedure is given below in the *Rule for Multiplying Decimals*. Note that the decimal points need not be aligned for multiplication.

RULE FOR MULTIPLYING DECIMALS

1. Multiply the numbers as though they are whole numbers.

2. Count the number of digits to the right of the decimal point in the factors.

3. Position the decimal point in the product so that the number of digits to the right of the decimal point is the sum of the number of digits to the right of the decimal points in the factors.

EXAMPLE   29.36 X 47.4 = ?
Solution:

```
    29.3 6   factor
  X  47.4   factor
   11 74 4
  205 52
  1174 4
  1391.66 4   product
```

In this example there are two digits to the right of the decimal point in 29.36 and one digit to the right of the decimal point in 47.4.

Since 2 + 1 = 3, there should be 3 digits to the right of the decimal point in the product. The product is 1391.664.

Example 1

```
    44.72  ←— 2 decimal places
  X   1.3  ←— 1 decimal place
   13416
   4472
   58.136  ←— 3 decimal places
```

Example 2   Multiply 23 by 17.5.

```
    2 3     or    17.5
   17.5           2 3
   11 5           52 5
  161            350
  23             402.5
  402.5
```

The order in which you multiply is not important. Note that 23 has 0 decimal places while 17.5 has 1 decimal place. Therefore, the product should contain 1 decimal place.

There are several different ways to indicate multiplication. Each of the following is a way of indicating 23 × 17.5.

$$23 \cdot 17.5 \qquad\qquad 23(17.5)$$
$$(23)(17.5) \qquad\qquad (23)17.5$$

Also, terms such as "product," "times," "multiply," "double" (which means multiply by 2), and "triple" (multiply by 3) are words which indicate that multiplication is to be performed.

Example 3   25.103 times 11.09 = ?

$$
\begin{array}{r}
25.103 \\
\times\ 11.09 \\
\hline
2\ 259\ 27 \\
25\ 103\ 0 \\
251\ 03 \\
\hline
278.392\ 27
\end{array}
$$
      (Why is this row shifted over two columns instead of one?)

Therefore, 25.103 times 11.09 equals 278.39227.

Example 4   Find the product of 1.073 and 0.0498.

$$
\begin{array}{r}
1.073 \\
\times 0.0498 \\
\hline
8584 \\
9657 \\
4292 \\
\hline
0.0534354
\end{array}
$$

1.073 ← three digits to right of decimal point
×0.0498 ← four digits to right of decimal point
0.0534354 ← seven digits to right of decimal point

Therefore, the product of 1.073 and 0.0498 is 0.0534354.

Note: If there are not enough digits in the product, then you should insert zeros to the left of the product until you have the correct number of digits.

Example 5   (4.3)(0.07)(231) = ?

$$
\begin{array}{rr}
23\ 1 & 9\ 9\ 3.3 \\
4.3 & 0.0\ 7 \\
\hline
69\ 3 & 69.5\ 3\ 1 \\
924 & \\
\hline
993.3 &
\end{array}
$$

Thus, (4.3)(0.07)(231) = 69.531.

1.5 Exercises

Find the products in Exercises 1–20.

1. 75	2. 43	3. 23		7. 23.1	8. 17.5	9. 50.09
× 6	× 9	×84		× 4.3	× 9.4	× 7.4
4. 57	5. 173	6. 52		10. 92	11. 3.24	12. 0.078
×18	× 20	× 0		×4.5	×0.045	× 3.3

13. 0.012
          ×0.706

14. 0.045 ·
          ×   52

15.   94
          ×.067

16. 0.104
          ×0.444

17. 4.307
          × 1149

18. 0.079
          × 348

19. 70.01
          × 5.09

20. 9.999
          × 1.72

In Exercises 21–24, compare each product with its factors and determine a quick way to multiply by numbers such as 10, 100, 1000, etc. Then write the answers to Exercises 25–30 by merely relocating the decimal point.

21. 43.1067
          ×   10

22. 43.1067
          ×   100

23. 43.1067
          × 1000

24. 43.1067
          × 10,000

25. 123.82 × 10

26. 0.2382 × 100

27. 123 × 1000

28. 123.82 × 10,000

29. 12.3 × 100

30. 0.2382 × 10

In Exercises 31–34, compare each product with its factors and determine a quick way to multiply by numbers such as 0.1, 0.01, 0.001, etc. Then write the answers to Exercises 35–40 by merely relocating the decimal point.

31. 43.1067
          ×   0.1

32. 43.1067
          ×   0.01

33. 43.1067
          × 0.001

34. 43.1067
          × 0.0001

35. 123.82 × 0.1

36. 0.2382 × 0.01

37. 123 × 0.001

38. 123.82 × 0.0001

39. 0.123 × 0.01

40. 123 × 0.1

In Exercises 41–60, identify the word, symbol, or phrase which indicates the operation of multiplication. Perform the multiplication.

41. Triple 17.5431.

42. 230.5 × 25

43. (43.7)(15)

44. (14)(7.06)(3.8)

45. 56 · 100

46. Find the product of 2001.51, 2.4, and 5.

47. Multiply five thousand six hundred by seventy-two and nine thousandths.

48. 31.09 times 25.403

49. (0.0432)(40)(9.7)

50. Find the product of 17.08 and 0.3.

51. 8.9 × 1.03 × 7

52. Find the product of seven hundred ten-thousandths and seven hundred ten.

53. Double 3.25.

54. Multiply 3.5 times itself.

55. 8907 · 768

56. (567.04)(0.00009)

57. Multiply 782.4 by 100.

58. Triple 0.04.

59. Multiply 0.016 times itself.

60. Double 0.016.

1.6 Division of Decimals

In this section you will need to recall some of the techniques used in dividing whole numbers. To work a long division problem, set it up as indicated below.

$$19\overline{)39786}$$

divisor dividend

Then divide.

```
      2094
19/39786
  38
  178
  171
   76
   76
```

Since 19 is larger than 3, divide 19 into 39. Thus the first nonzero digit in the quotient is placed above the 9 in the dividend.

Next multiply 2 by 19 and place the result beneath the dividend. Then subtract and bring down the next digit. Since 19 will not divide into 17, a 0 is placed above the 7 in the dividend and the next digit is brought down.

Divide 19 into 178 and place this digit above the 8 in the dividend.

Continue in this manner until the division is complete.

Compare the following examples.

Division of Whole Numbers	*Division of Decimals*
4635 ÷ 103	4.635 ÷ 1.03

$$\begin{array}{r} 45 \\ 103\overline{)4635} \\ \underline{412} \\ 515 \\ \underline{515} \end{array} \qquad \begin{array}{r} 4.5 \\ 1.03\overline{)4.635} \\ \underline{412} \\ 515 \\ \underline{515} \end{array}$$

Division of decimals is quite similar to division of whole numbers. The only distinction comes in the placement of the decimal point. You should not start the division until the divisor is a whole number. The procedure to follow is given in the *Rule for Dividing Decimals.*

RULE FOR DIVIDING DECIMALS

1. Identify the dividend and divisor.

2. Position the divisor and the dividend as though you were preparing to divide one whole number by another.

3. Reposition the decimal point in the divisor to the right of the last given digit, so that the divisor is a whole number.

4. Count the number of places the decimal point was moved in the divisor and reposition the decimal point in the dividend the same number of places to the right.

5. Place the decimal point for the quotient directly above the repositioned decimal point in the dividend.

6. Divide as though you were dividing by a whole number. Be sure to position each digit in the quotient directly above the last digit in the dividend used to obtain this part of the quotient.

EXAMPLE   262.629 ÷ 21.3
Solution:

$$\text{divisor} \longrightarrow 21.3\overline{)262.629} \longleftarrow \text{dividend}$$

$$\begin{array}{r} 1\,2\,.33 \quad \text{quotient} \\ 21.3_\wedge\overline{)262.6_\wedge 29} \\ \underline{213} \\ 496 \\ \underline{426} \\ 70\ 2 \\ \underline{63\ 9} \\ 6\ 39 \\ \underline{6\ 39} \end{array}$$

Study the following examples carefully.

Example 1   Position the decimal point and place the first nonzero digit in the quotient in each of the following problems.

a. $8.6_\wedge\overline{)97.1_\wedge 4}$ with quotient $\underset{}{1}\,.$

b. $3\overline{)89.4}$ with quotient $\underset{}{2}\,.$

c. $2.35_\wedge\overline{)704.60_\wedge}$ with quotient $\underset{}{2}\,.$

d. $43.2_\wedge\overline{)0.9_\wedge 07}$ with quotient $.02$

e. $5.2_\wedge\overline{)2.6_\wedge 804}$ with quotient $.5$

Example 2   Perform the following divisions.

a.
$$
\begin{array}{r}
3\,7. \\
2.3_{\wedge}\overline{)85.1_{\wedge}} \\
\underline{69} \\
16\ 1 \\
\underline{16\ 1}
\end{array}
\qquad
\begin{array}{r}
Check: \quad 37 \\
\times 2.3 \\
\hline
111 \\
\underline{74} \\
85.1
\end{array}
$$

b.
$$
\begin{array}{r}
.005 \\
99.9_{\wedge}\overline{)0.4_{\wedge}995} \\
\underline{4\ 995}
\end{array}
$$
After relocating the decimal point and placing the 5 in the quotient, you must place zeros between the decimal point and the 5.

$$
\begin{array}{r}
Check: \quad 99.9 \\
\times 0.005 \\
\hline
0.4995
\end{array}
$$

The problem $19\overline{)39786}$ can be written in several different ways.

$39786 \div 19$

39786 divided by 19

19 divided into 39786

Since $39786 \div 19$ is not the same as $19 \div 39786$, you must be able to determine which number is the divisor so that the problem can be set up correctly.

Example 3   a. Divide 8 into 0.0016.

$$
\begin{array}{r}
.0002 \\
8\overline{)0.0016} \\
\underline{16}
\end{array}
$$
Since the divisor is a whole number, the decimal point does not need to be repositioned.

b. Find the quotient if eight is divided by sixteen ten-thousandths.

$$
\begin{array}{r}
5000. \\
.0016_{\wedge}\overline{)8.0000_{\wedge}} \\
\underline{8\ 0} \\
0000
\end{array}
$$
Zeros must be annexed to the dividend so that the decimal point can be moved 4 places.

In each of the examples we have just seen, the division process was continued until the remainder was zero. However, it is not always possible or practical to continue the process until a zero remainder is obtained. You can discontinue the division when you have the desired number of decimal places in the quotient.

Example 4   Find the quotient, accurate to hundredths, when 1.587 is divided by 20.7.

$$
\begin{array}{r}
0.076 \\
20.7_\wedge\overline{)1.5_\wedge870} \\
1\ 4\ 49 \\
\hline
1\ 380 \\
1\ 242 \\
\hline
138 \quad \text{remainder}
\end{array}
$$

Thus, $1.587 \div 20.7 \doteq 0.08$.

Note: The digit in the thousandths position in the quotient is found so that the rounding rule can be applied. Therefore, the quotient accurate to hundredths is 0.08.

1.6   Exercises

Perform the divisions in Exercises 1–20.

1. $9\overline{)2781}$

2. $6\overline{)2934}$

3. $32\overline{)16000}$

4. $47\overline{)33276}$

5. $145\overline{)43645}$

6. $781\overline{)27335}$

7. $7\overline{)34.23}$

8. $4\overline{)225.6}$

9. $24\overline{)1200.72}$

10. $75\overline{)300.675}$

11. $230\overline{)323.380}$

12. $103\overline{)57813.9}$

13. $4.5\overline{)16.56}$

14. $0.0028\overline{)0.168}$

15. $0.014\overline{)9.8}$

16. $0.00007\overline{).0000385}$

17. $540.2\overline{)32.9522}$

18. $0.9\overline{)4.5882}$

19. $6.218\overline{)351.317}$

20. $13.41\overline{)577.971}$

In Exercises 21–24, compare each quotient with its dividend and determine a quick way to divide by numbers such as 10, 100, 1000, etc. Then write the answers to Exercises 25–30 by merely relocating the decimal point.

21. $10\overline{)118.728}$

22. $100\overline{)118.728}$

23. $1000\overline{)118.728}$

24. $10000\overline{)118.728}$

25. $1000\overline{)578.41}$

26. $100\overline{)6.66}$

27. $10\overline{)118}$

28. $10000\overline{)118.728}$

29. $100\overline{)0.892}$

30. $10\overline{)6.66}$

In Exercises 31–34, compare each quotient with its dividend and determine a quick way to divide by numbers such as 0.1, 0.01, 0.001, etc. Then write the answers to Exercises 35–40 by merely relocating the decimal point.

31. $0.1\overline{)118.728}$

32. $0.01\overline{)289.2}$

33. $0.001\overline{)118.728}$

34. $0.0001\overline{)118.728}$

35. $0.01\overline{)578.41}$

36. $0.01\overline{)6.66}$

37. $0.1\overline{)118}$

38. $0.001\overline{)118}$

39. $0.001\overline{)6.66}$

40. $0.1\overline{)0.892}$

In Exercises 41–60, give the word, symbol, or phrase which indicates the operation of division, and then perform that division. Give the exact answer unless otherwise specified.

41. Divide 5.2 into 19.76.

42. Divide 25.3 by 0.55.

43. Find the quotient when 1.112184 is divided by 0.00216.

44. $111.2184 \div 514.9$

45. Round the quotient $425.78 \div 204.5$ to two decimal places.

46. Find the quotient, to the nearest ten-thousandth, when 0.3897 is divided by 100.2.

47. Round the quotient $250 \div 1.91$ to the nearest unit.

48. Round the quotient 48.93 divided by 0.087 to the nearest tenth.

49. $118.728 \div 58.2$

50. Round the quotient $48 \div 237$ to the nearest thousandth.

51. Divide 88.038 by 0.876.

52. Divide sixty-four by sixteen hundred-thousandths.

53. Divide 100 into 568.4.

54. Divide 3 into 347.7.

55. Divide 237 by 481 and round the quotient to the nearest thousandth.

56. $0.0064 \div 16$

57. Divide 250 into 477.25. Round the quotient to two decimal places.

58. Find the quotient if 9.1506 is divided by 3.02.

59. Divide 7.05 into 0.49773.

60. Round the quotient $98704 \div 0.06$ to the nearest unit.

1.7   Word Problems

In this section we will consider word problems involving addition, subtraction, multiplication, and division of decimals. The difficulty in solving these problems will be in determining which operation or combination of operations to use. You should avoid trying first one operation and then another until you obtain the correct answer. Instead you should read the problem and carefully consider your choice of operations. It may be helpful to recall the different key words which you have studied in this chapter that indicate that addition, subtraction, multiplication, or division are to be performed. Other key words will be pointed out in the examples. Sometimes it is helpful to estimate the answer you would obtain by using a particular operation to see if the result seems sensible. If the result is unreasonable, then you have selected the wrong operation and you should try another.

Perhaps the best way to learn to work word problems is to study worked examples and then to work some similar problems by yourself. Your skill in solving word problems will improve with additional practice.

You may find the following suggestions helpful while working the problems in this section.

1. Read the problem until you understand exactly what you are asked to find.

2. Look for phrases in the problem which identify the operation or operations that should be used.

3. Perform the calculations.

4. Check to make sure that you have answered the question asked, that your answer is in the correct form, and that your answer is reasonable.

Example 1   **a.** A student had two part-time jobs. His take-home pay from these jobs was $52.35 and $78.92 per week. In addition to these jobs, he also tutored students in mathematics. If he made $25 tutoring one week, what was his total income for that week?

   1. What are you asked to find?   *The student's total income.*

   2. What words indicate the operation to be used?   *"Total" indicates that addition is to be used.*

3. Perform the calculations.

$$\begin{array}{r} \$52.35 \\ 78.92 \\ + \ 25.00 \\ \hline \$156.27 \end{array}$$

Thus, his total income was $156.27.

4. Check to see if you answered the question and if the answer is reasonable. *We have found the total income and $156.27 is a reasonable answer.*

b. If your bank balance is $483.17 and you write a check for $125.33, what is the new balance?

1. You are asked to find the new bank balance.

2. You have written a check. This implies that you have withdrawn or taken money from your checking account, so you should subtract.

3.
$$\begin{array}{r} \$483.17 \\ - \ 125.33 \\ \hline \$357.84 \end{array}$$ Thus, your balance is $357.84.

4. The question has been answered and $357.84 is a reasonable answer since it is less than your initial balance.

c. What is the cost of a set of four tires if each tire costs $47.95?

1. You are asked to find the cost of four tires when you are given the cost of one tire.

2. There is no key word or phrase used in this problem. However, whenever you are given the cost of one item and need to find the cost of several identical items, you multiply.

3.
$$\begin{array}{r} \$47.95 \\ \times \qquad 4 \\ \hline \$191.80 \end{array}$$ Thus, the cost of four tires is $191.80.

4. The question has been answered and $191.80 is a reasonable answer.

d. If 17 records cost $169.15, what would be the cost of 1 record if the price of each record is the same?

1. You are asked to find the cost of 1 record if you know the cost of 17.

2. Again, there is no key word or phrase which indicates the operation to be used. However, in general if you are given the total cost of a number of equally-priced items, the cost of one item can be found by dividing.

3.
$$\begin{array}{r} 9.95 \\ 17\overline{)169.15} \\ \underline{153} \\ 16\ 1 \\ \underline{15\ 3} \\ 85 \\ \underline{85} \end{array}$$ Thus, each record costs $9.95.

4. The question has been answered and $9.95 is a reasonable answer.

Example 2   In each of the following, first determine whether you should multiply or divide to obtain the answer. Then work the problem.

 a. If twenty-four cans of pineapple cost $17.04, what is the cost of one can?

  Since you are given the cost of 24 cans and wish to find the cost of one can, you divide.

$$\$17.04 \div 24 = \$.71$$

 b. Find the cost of 9 pounds of franks if one pound cost $1.79.

  Since you are given the cost of one pound and wish to find the cost of 9 pounds, you multiply.

$$\$1.79 \times 9 = \$16.11$$

 c. An electrician is paid $22.50 for each hour she works. What would her pay be for 6 hours work?

  Multiply: $\$22.50 \cdot 6 = \$135$

 d. A doctor gives a patient 51 ounces of medicine which must be consumed in doses of 3 ounces each. How many times will the patient have to take the medicine?

  Divide: 51 oz $\div$ 3 oz = 17 times

 e. A driver travels a distance of 7200 miles in 12 days. If he traveled the same distance each day, how many miles did he travel each day?

  Divide: 7200 mi $\div$ 12 = 600 miles

 f. A bus driver's route covers 233 miles each day. How many miles does he travel over his route in 31 working days?

  Multiply: 233 miles $\cdot$ 31 = 7223 miles

Example 3   At the beginning of the month, Peggy had a bank balance of $1091.46. During the month she wrote checks for $12, $27.93, $119.20, and $88.93. She made a deposit of $500. If the bank charges a service charge of $2.25 plus $.10 for each check, what would her balance be at the end of the month?

  Several operations will be involved in solving this problem. First, writing checks implies that we are to subtract from the balance. However, since more than one check was written, we will have to add all the checks before we subtract. Next, the word "deposit" implies that we add to the balance. Then since a service charge is made, we subtract from the balance. To calculate the service charge, we must multiply the number of checks by the charge per check and add this amount to the base rate.

$1091.46	Beginning balance	Checks written:	$12.00
+ 500.00	Deposit		27.93
$1591.46			119.20
− 248.06	Checks written		+ 88.93
$1343.40			$248.06
− 2.65	Service charge	Service charge:	
$1340.75	New balance	$2.25	
		+ .40	(4 checks at a charge
		$2.65	of $.10 each)

Thus, the new balance is $1340.75.

Example 4   A salesman calculated that he had traveled 243.5 miles since his last gasoline refill. When he had the tank filled, it required seventeen and two-tenths gallons. Find, to the nearest tenth, the number of miles per gallon he averaged.

The word "per" will occur in many word problems. Here we are asked to find miles per gallon. You may be asked to find cost per ounce, miles per hour, etc. To find these, you must divide the first quantity by the second quantity. Thus, in this problem 243.5 miles must be divided by 17.2 gallons.

$$
\begin{array}{r}
1\,4.\,15 \doteq 14.2 \\
17.2_\wedge\overline{)243.5_\wedge00} \\
\underline{172} \\
71\,5 \\
\underline{68\,8} \\
2\,7\,0 \\
\underline{1\,7\,2} \\
9\,80 \\
\underline{8\,60} \\
1\,20
\end{array}
$$

Thus, 243.5 miles ÷ 17.2 gallons ≐ 14.2 miles per gallon. Is this a reasonable answer?

This answer was rounded since we are asked to find the answer to the nearest tenth.

Example 5   During the quarter a student made grades of 58, 93, 74, 85, and 79 on tests. What is his average on these tests?

The average of a set of numbers is their sum divided by the total number of numbers in the set.

$$\frac{58 + 93 + 74 + 85 + 79}{5} = \frac{389}{5} = 77.8$$

Thus, his average is 77.8. Is this reasonable?

Example 6   A ten-pound box of Brand A laundry detergent is priced at $4.39 while 3 three-pound boxes of Brand B sell for $4.00. Which brand is the better buy?

To determine the better buy, we need to compare the costs per pound. Thus, we should find the cost per pound of Brand A and the cost per pound of Brand B. To find cost per pound, we divide the cost by the number of pounds.

Brand A

10 lb at $4.39

$$
\begin{array}{r}
.439 \\
10\overline{)4.390} \\
\underline{4\,0} \\
39 \\
\underline{30} \\
90 \\
\underline{90}
\end{array}
$$

$.439 per pound

Brand B

3 × 3 lb = 9 lb at $4.00

$$
\begin{array}{r}
.4444 \doteq .444 \\
9\overline{)4.000} \\
\underline{3\,6} \\
40 \\
\underline{36} \\
40 \\
\underline{36} \\
4
\end{array}
$$

$.444 per pound

Therefore, Brand A is the better buy since the cost per pound is less.

Example 7   **a.** A vacant rectangular lot measures 90 feet by 150 feet. If one pound of grass seed covers 250 square feet, how many pounds of seed will be needed to cover the lot?

We must find the area of the lot, that is, the number of square feet contained in the lot. The area of a rectangular region is found by multiplying the length by the width.

Area = 90 feet X 150 feet = 13,500 square feet

To determine the number of pounds of seed needed to cover 13,500 square feet, we divide 13,500 sq ft by 250 sq ft per pound.

13500 sq ft ÷ 250 sq ft per pound = 54 pounds

b. A rectangular lot is 125.74 feet long and 100.5 feet wide. How many feet of fencing would be required to enclose the lot?

We must find the perimeter of the lot, that is, the distance around the lot.

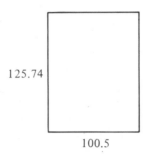

125.74

100.5

Since a rectangle has two lengths and two widths, the perimeter is

125.74 + 125.74 + 100.5 + 100.5 = 452.48

Thus, the amount of fencing required is 452.48 feet.

1.7   Exercises

1. A worker augmented his regular weekly salary of $215.33 with a $60.49 check from a part-time job. What was his total weekly income?

2. A Miami Dolphin player gained 746 yards in 196 carries. What was his average gain, to the nearest tenth of a yard, on each carry?

3. During September a commuter bought gas four times. She bought 15.3 gallons, 14.9 gallons, 11.2 gallons, and 14.7 gallons. How much gas did she buy?

4. If the gas in Exercise 3 cost $1.45 per gallon, how much did she pay for gas in September?

5. Seven checks in the amounts of $88.97, $50, $12, $32.43, $100, $5.14, and $235.01 are written on a checking account containing $1089.32. What will the balance be after these checks have been cashed?

6. If you buy an item which costs $35.92, how much change will you receive if you give the cashier two twenty-dollar bills?

7. A student tutor earns $2.95 an hour. If she worked 19.5 hours during the week, how much did she earn?

8. If you travel at a rate of 50 miles per hour, how many miles will you cover in 3.5 hours?

9. A mechanic saved $876 over a twelve-month period. What was his average monthly savings?

10. If a dozen apples cost $2.94, what is the cost per apple?

11. The list price of a set of radial tires was $191.80. Federal tax was $11.56 extra while state tax was an additional $6.10. In order to "drive out" with the tires on the car, the front end had to be aligned and the tires had to be balanced for a cost of $9.95 and $10, respectively. Find the total "drive out" price for a set of four radial tires.

12. An 8-meter length of cloth is cut into pieces each of length 0.4 meters. How many 0.4-meter pieces are formed?

13. If a car goes 208.2 miles on 13.3 gallons of gas, find to the nearest tenth the number of miles per gallon that the car averages.

14. If it takes 10 hours to travel 472.4 miles, what is the average number of miles traveled per hour?

15. A minibus has a load capacity of 2531 pounds. On one trip, there were five passengers weighing 251.5 pounds, 132 pounds, 98 pounds, 340 pounds, and 152.5 pounds. Their luggage weighed 562 pounds. Find the total load in the bus if the driver's weight was 192.3 pounds. Is this more or less than the load capacity?

16. In Exercise 15 find the average weight of the five passengers.

17. A soft drink is sold in containers of several different sizes. Which of the following is the best buy?

72 ounces for $1.85
32 ounces for $ .48
64 ounces for $1.19

18. A salesman's total income for last year was $42,374.82. He had deductions of $2389.93, $374.04, and $1576. Find his total deductions. If taxable income is the total income minus the deductions, what was his taxable income for the year?

19. Find the area of a rectangle which measures 16 feet by 21.5 feet.

20. In 1968 the suggested retail price for a new automobile was two thousand eight hundred thirteen dollars and sixty cents. The destination charges were one hundred forty-eight dollars and the accessories on the car were priced as follows: air-conditioning, three hundred ninety-seven dollars; rear and front bumper guards, twenty dollars; deluxe wheel covers, thirty-one dollars and fifty-two cents; AM radio, sixty-five dollars. Find the total price of the automobile.

21. Find the perimeter of a rectangle which measures 16 feet by 21.5 feet.

22. How much fencing would be needed to enclose the following lot?

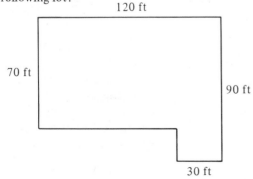

23. A businessman rents heavy equipment. Three pieces of his equipment rent for $32, $21.75, and $15.50 per hour. If all three pieces of this equipment are rented for five days, nine hours per day, how much rental does he receive for the five days?

24. A company has seventeen thousand one hundred cubic centimeters of a drug. It will be placed in bottles which hold 3.8 cubic centimeters. How many bottles are necessary?

25. A student worked 52 hours during one week. If he earned $2.95 an hour for the first 40 hours and $4.45 an hour for overtime, how much did he earn that week?

26. In 3.5 hours a man walked 9.8 miles. What was his average rate per hour?

27. The cost of building an addition to a house is $5.76 per square foot. What will it cost to add a room which measures 18 feet by 21 feet?

28. The wholesale cost of a bolt was reduced from $.027 to $.023. How much will be saved if 1000 are bought?

29. A set of four steel-belted radial tires which cost $190.17 was driven for 32,000 miles. Find, to the nearest cent, the cost per mile of the set of tires.

30. If a car averages 17.8 miles per gallon of gas, how many gallons, to the nearest tenth, will be needed to travel 500 miles?

31. A botanist recorded the growth of a plant each week. The growths were 2 inches, 2.5 inches. 1.75 inches, 2.3 inches, 0.75 inches, and 3 inches. What was the average growth each week?

32. Soap powder is regularly priced at 64 ounces for $3.83. On special, a 100-ounce box cost $6.00. Which is the better buy?

33. A hotel clerk earns $952.19 per month. The F.I.C.A. tax (social security) and federal income tax are found by multiplying this amount by 0.059 and 0.17 respectively. What is the net pay, to the nearest cent, after these taxes are deducted?

34. The annual estimated kilowatt-hour use for operating a color TV is 460, a dishwasher is 363, a microwave oven is 190, a range is 1205, and a refrigerator/freezer is 1829. If the cost per kilowatt-hour is four cents, what would it cost to operate these appliances for one year?

35. During October and November, an employee worked the following number of hours each week: 42.5, 65, 40, 90, 30.25, 40, 38.75, 17.5. What was the average number of hours he worked each week?

36. A plane leaves Chicago at 12 noon. At 2.15 P.M. the plane is 199.5 miles south of Chicago. What is the average number of miles, to the nearest tenth, that the plane covered per minute?

37. Find the average of the following test scores: 59, 67, 49, 66, 89, 45.

38. John pays $97.12 a month on a 30-year mortgage. How much does he pay over the 30-year period?

39. A manufacturer produces hubs for automobiles. The diameter of each hub is supposed to measure 8.32 centimeters. However, a tolerance of 0.004 centimeters is allowed. That is, the diameter may be 0.004 centimeters more or less than 8.32 centimeters. Which of the following measurements are in this range?

    8.32 centimeters   7.32 centimeters
    8.004 centimeters   8.323 centimeters
    9.00 centimeters   8.28 centimeters

40. In a math contest students took a multiple-choice test of 35 questions. Each correct answer counted 3.5 points, but there was a penalty of 1 point for each wrong answer. What was the score of a student who attempted 32 questions but answered only 27 correctly?

41. On a trip a family traveled 430 miles on Tuesday, 232 miles on Wednesday, 95 miles on Thursday, 396 miles on Friday, 179 miles on Saturday, and 285 miles on Sunday. What was the average number of miles traveled each day?

42. Find the average of the following test scores: 84, 56, 89, 89, 98, 100, 100.

43. A department store charges $3.50 per square yard to install new carpet. What would the installation charge be for a rectangular room which measures 7 yards by 8 yards?

44. A syrup manufacturer mixed 250 quarts of maple syrup worth $3.50 per quart with 390 quarts of corn syrup worth $0.95 per quart. What will be the cost, to the nearest cent, of one quart of the mixture?

45. Over a six-year period in the 1960s, a person's income each year was as follows: $4600, $4900, $6500, $7200, $8600, and $1500. Find her average income per year for this six-year period.

Important Terms Used in this Chapter	add	minuend
	addend	multiply
	borrowing	place value
	carrying	product
	difference	quotient
	digit	remainder
	divide	rounding decimals
	dividend	subtract
	division	subtrahend
	divisor	sum
	factor	whole numbers

Important Symbols Used in this Chapter	$+$ add
	$-$ subtract
	$\times$ multiply
	$\div$ divide
	$\doteq$ is approximately equal to
	$=$ equals

REVIEW EXERCISES

Answer true (T) or false (F) in Exercises 1–15.

1. The verbal expression for 504.3 is five hundred and four and three tenths.

2. The number 784.927 rounded to the nearest hundred is 800.

3. The place value of the digit 5 in the numeral 5437.29 is thousands.

4. The numerals 73 and 73.0 name the same number.

5. The product 7.5 × 0.0007 will contain five places to the right of the decimal point.

6. The place value of each of the underlined digits in the numeral 842.78 is tenths.

7. The numeral for five thousand and two hundredths is 500.02.

8. To work the problem 7.82$\overline{)568.4}$, the decimal point in 7.82 must be moved two places to the right.

9. The numerals 45 and .45 name the same number.

10. The digit 1 in the numeral 8400.69014 has a place value of thousandths.

11. The number 784.927 rounded to the nearest tenth is 784.93.

12. The sum 842 + 4.1 + 3.7 + 9 will contain three places to the right of the decimal point.

13. The sum 842 + 4.1 + 3.7 + 9 will contain three places to the left of the decimal point.

14. To work the problem 72$\overline{)5.894}$, the decimal point in 5.894 must be moved three places to the right.

15. The two problems 45 ÷ 15 and 15 ÷ 45 have the same answer.

Perform the indicated operations in Exercises 16–24.

16. 154 · 100   17. 85.41 + 104
18. 32.456 ÷ 0.001   19. 32.456 × 0.001
20. 32.456 ÷ 1000   21. 32.456 × 10000
22. 104 − 85.41   23. (0.132)(0.0401)
24. 4.86012 ÷ 0.404

In each of Exercises 25–44, translate the expression to a symbolic statement. Do not perform the computation.

25. 18 less 9
26. 18 divided by 6
27. The product of 7, 10, and 40
28. Twice 32
29. Divide 42 into 21
30. 17 minus 13
31. 10 plus 4
32. 17 subtracted from 22
33. 15 decreased by 4
34. Subtract 52 from 100
35. Decrease 14 by 3
36. 2 less than 10
37. Triple 17
38. Divide 42 by 7
39. The sum of 4, 5, and 10
40. Double 84
41. 15 increased by 4
42. 45 take away 15
43. 15 times 100
44. 18 more than 7

Give the exact answer to each of Exercises 45–73, unless otherwise specified.

45. Write the sum of 0.14093, 107.8, 1.4097, and 236 as a verbal expression.

46. Divide 7.05 into 57034.5.

47. What is 112.8 less 90.7?

48. 154.561 minus 71.99 = ?

49. What is 0.18045 divided by 12.03?

50. What is twice the sum of 18.3 and 9.05?

51. Find the difference if 22.349 is subtracted from 51.

52. Divide 12 by 39.2. Round the quotient to three decimal places.

53. Calculate fifteen plus twenty-seven and three tenths. Write the answer as a verbal expression.

54. How much is 3600.7 decreased by 8.39?

55. Increase 83.091 by 71.5.

56. Find the product of 5.08, 11.8, and 1.01.

57. What is 18.03 more than 125.2?

58. Find the quotient to the nearest tenth if 477.5 is divided by 250.

59. Subtract 47.002 from the sum of 92.04 and 30.9.

60. Multiply the difference 28.9 − 14.03 by 3.4.

61. If a check for one thousand thirty-three and fourteen hundredths dollars is deposited in an account containing $135.47, what is the new balance?

62. The total collected at the gate at a football game was $477,500. If each person paid $9.55 for admission, how many people attended the game?

63. Thirty-one thousand six hundred twenty-eight people paid ten dollars and fifty cents each to attend a rock concert. Write the total amount paid as a verbal expression.

64. At the end of February a utility meter read 378.49 units. By the end of March it read 405.34 units. How many units were used in March?

65. A stock on the New York Exchange sold for $22.50 per share at the beginning of the week. During the week the price dropped $8.25 and then doubled. What was the price of a share at the end of the week?

66. The treasurer of a club collected dues of $17.50 from each of the twelve members of the club. If the club's account originally contained $148.17, what will the balance be when the dues have been deposited?

67. If gas costs $1.45 a gallon, find the cost, to the nearest cent, for a car which averages 24 miles per gallon to travel 972 miles.

68. A lot measures 100 feet by 95.5 feet. Find the perimeter of the lot and the area of the lot.

69. The number of kilowatt-hours used by a family each day during a six-day period were 20, 24, 28, 21, 16, and 14. Find the average number of kilowatt-hours used each day and the average daily cost, to the nearest cent, if the cost per kilowatt-hour is $0.05.

70. Max kept the following record of expenses for his car for the first 10,000 miles:

Gasoline: 522 gallons, 59.9¢ per gallon
Oil and lubrication (3 times): $13.95 per change
General maintenance costs: $192.35
Insurance premium for this period: $96.95
Cost of tag and taxes for this period: $39.42

Find the average cost per mile (to the nearest cent) of operating the car.

*71. The following table gives the sign-in and sign-out time of a part-time worker. Determine the average number of hours worked each day.

Sign-in	Sign-out
8:30 A.M.	11:45 A.M.
8:15 A.M.	10:15 A.M.
8:22 A.M.	11:37 A.M.
8:35 A.M.	9:40 A.M.
8:19 A.M.	1:44 P.M.

*72. A man plans to spend no more than $5 on two one-pound boxes of candy. The store has candy selling for $1.75, $1.99, $2.99, and $3.50 for one-pound boxes. If he buys two one-pound boxes and pays with a five-dollar bill, what is the smallest amount of change he could receive? What is the largest amount of change he could receive?

*73. A student has grades of 74, 82, 93, and 70 on four tests. What must he make on a fifth test in order to have an average of 80 on the five tests?

*These exercises are optional.

2 FRACTIONS

PRETEST

1. a. Give all factors of 42 (Objective 1).
b. Give four multiples of 42 (Objective 1).

2. Factor each of the following natural numbers into a product of primes (Objective 2).
a. 111   **b.** 143
c. 252   **d.** 690
e. 451   **f.** 351

3. Reduce the following fractions to lowest terms (Objective 3).
a. $\dfrac{88}{104}$ **b.** $\dfrac{88}{209}$ **c.** $\dfrac{210}{510}$

Fill in the missing numerators (Objective 3).
d. $\dfrac{7}{19} = \dfrac{?}{95}$ **e.** $\dfrac{71}{133} = \dfrac{?}{798}$
f. $\dfrac{5}{37} = \dfrac{?}{777}$

4. Multiply the following fractions (Objective 4).
a. $\dfrac{35}{88} \cdot \dfrac{99}{49}$ **b.** $\dfrac{36}{65} \cdot \dfrac{143}{108}$
c. $\dfrac{74}{38} \cdot \dfrac{30}{185} \cdot \dfrac{133}{16}$

5. Divide the following fractions (Objective 5).
a. $\dfrac{56}{77} \div \dfrac{40}{33}$ **b.** $\dfrac{26}{85} \div \dfrac{39}{119}$
c. $\dfrac{206}{46} \div \dfrac{721}{23}$

6. Find the least common denominator of the following sets of fractions (Objective 6).
a. $\dfrac{3}{40}$ and $\dfrac{7}{72}$ **b.** $\dfrac{13}{45}$ and $\dfrac{17}{60}$

7. Add the following fractions (Objective 7).
a. $\dfrac{11}{28} + \dfrac{3}{28}$ **b.** $\dfrac{1}{3} + \dfrac{1}{9} + \dfrac{7}{18}$
c. $\dfrac{1}{35} + \dfrac{1}{15} + \dfrac{23}{105}$

8. Subtract the following fractions (Objective 8).
a. $\dfrac{17}{18} - \dfrac{5}{18}$ **b.** $\dfrac{1}{2} - \dfrac{1}{3}$
c. $\dfrac{5}{6} - \dfrac{4}{15}$

9. Solve the following problems (Objective 9).

   a. Each symbol on a typed page occupies $\frac{1}{12}$ inch. How many symbols can be typed in a space $\frac{2}{3}$ inch wide?

   b. A recipe calls for $\frac{3}{4}$ cup of sugar. If you have only $\frac{2}{3}$ cup, how much more sugar do you need?

   c. Agnes owns $\frac{2}{3}$ of the stock in a manufacturing company. If there are 1122 shares of stock, how many shares does Agnes own?

   d. One item in a trading stamp catalog requires two-thirds of a book of stamps, and a second item requires one-fourth of a book. How many books are required for both items?

   e. In a shipment of 400 TV sets, 20 of them were damaged. What fractional part of the TV sets were damaged?

Leopold Kronecker, a famous mathematician, once stated that God gave man the **natural numbers** (the numbers 1, 2, 3, 4, 5, etc.) and that man invented all other numbers. It is easy to understand why the natural numbers were the first used by man. They are the easiest to understand because they come from the counting process. One can imagine an early shepherd looking out at his flock and telling his son that he has seven lambs. It would not be appropriate for the shepherd to look out on the hill and say, "I see eight and a third sheep grazing today."

Fractions such as $\frac{1}{2}, \frac{2}{3}, \frac{3}{4}, \frac{2}{7}$, etc., are very important numbers and were invented by man to describe measures more accurately than was possible with natural numbers. For example, with fractions man could measure $\frac{1}{2}$ a cubit, or relate the eating of $\frac{3}{4}$ of a pie or the planting of $\frac{1}{3}$ of a field.

One way of describing what is meant by a **fraction** is to think of the fraction bar as indicating division. The fraction $\frac{3}{4}$ would mean $3 \div 4$. However, a more familiar interpretation might be that $\frac{3}{4}$ represents 3 out of 4 equal parts of some whole quantity. For example, Figure 2.1 shows a rectangle divided

Figure 2.1

into 4 equal parts of which 3 parts are shaded. Thus, $\frac{3}{4}$ of the figure is shaded. Similarly, if a pound of cheese contains 16 equal slices and 11 slices are eaten, then $\frac{11}{16}$ of the pound has been eaten. Since 5 slices remain, $\frac{5}{16}$ of the pound of cheese has not been eaten. In any fraction, the number above the bar is called the **numerator** and the number below the bar is called the **denominator**.

Today we all need to be able to work with fractions. In particular, think how fractions are used by carpenters, engineers, cooks, nurses, and those taking prescription drugs. Even though we sometimes try to avoid fractions, they are needed so frequently in many areas that you should be able to deal with them whenever they occur.

2.1 Factors, Multiples, and Primes

In order to work effectively with fractions, it will be helpful to consider some topics related to the natural numbers, that is, the numbers 1, 2, 3, 4, 5, etc.

A **factor**, or **divisor**, of a natural number is any natural number whose product with some natural number is the given natural number. For example, 3 is a factor of 15 because $3 \cdot 5 = 15$. Other factors, or divisors, of 15 are 1 (since $1 \cdot 15 = 15$), 5 (since $5 \cdot 3 = 15$), and 15 (since $15 \cdot 1 = 15$).

Example 1 a. Since 12 can be written as $1 \cdot 12$, $2 \cdot 6$, or $3 \cdot 4$, the factors of 12 are 1, 2, 3, 4, 6, and 12.

b. Since the only way to write 7 as a product is $1 \cdot 7$, the divisors of 7 are 1 and 7.

Note that the factors, or divisors, of a natural number divide into the number exactly.

A **multiple** of a natural number is a number which is the product of the given natural number and another natural number. For example, if we wish to find a multiple of 15, it would be a number which is equal to 15 *times* some other natural number. Since $15 \cdot 1 = 15$, $15 \cdot 2 = 30$, $15 \cdot 3 = 45$, $15 \cdot 4 = 60$, $15 \cdot 5 = 75$, $15 \cdot 6 = 90$, and $15 \cdot 7 = 105$, then 15, 30, 45, 60, 75, 90, and 105 are multiples of 15.

Note that a given number is a factor, or divisor, of each of its multiples.

Example 2 a. The multiples of 2 are the natural numbers which have 2 as a factor. They are 2, 4, 6, 8, 10, 12, 14, ...

b. The multiples of 7 are:

$7 \cdot 1 = 7$ $7 \cdot 5 = 35$
$7 \cdot 2 = 14$ $7 \cdot 6 = 42$
$7 \cdot 3 = 21$ $7 \cdot 7 = 49$
$7 \cdot 4 = 28$ $7 \cdot 8 = 56$
                        etc.

Example 3 a. Is 10 a factor of 5?

No, because 10 will not divide into 5 exactly.

b. Is 10 a multiple of 5?

Yes, because $10 = 5 \cdot 2$.

c. Is 3 a divisor of 18?

Yes, since 3 divides into 18 exactly 6 times.

Consider the following chart which lists the factors and multiples of the natural numbers from 1 through 23.

Natural number	Factors or divisors	Multiples
1	1	1, 2, 3, 4, 5, 6, . . .
2	**1, 2**	2, 4, 6, 8, 10, 12, . . .
3	**1, 3**	3, 6, 9, 12, 15, . . .
4	1, 2, 4	4, 8, 12, 16, . . .
5	**1, 5**	5, 10, 15, 20, . . .
6	1, 2, 3, 6	6, 12, 18, 24, . . .
7	**1, 7**	7, 14, 21, 28, 35, . . .
8	1, 2, 4, 8	8, 16, 24, . . .
9	1, 3, 9	9, 18, 27, 36, . . .
10	1, 2, 5, 10	10, 20, 30, 40, 50, . . .
11	**1, 11**	11, 22, 33, 44, . . .
12	1, 2, 3, 4, 6, 12	12, 24, 36, . . .
13	**1, 13**	13, 26, 39, 52, . . .
14	1, 2, 7, 14	14, 28, 42, . . .
15	1, 3, 5, 15	15, 30, 45, . . .
16	1, 2, 4, 8, 16	16, 32, 48, . . .
17	**1, 17**	17, 34, 51, . . .
18	1, 2, 3, 6, 9, 18	18, 36, 54, . . .
19	**1, 19**	19, 38, 57, . . .
20	1, 2, 4, 5, 10, 20	20, 40, 60, 80, 100, . . .
21	1, 3, 7, 21	21, 42, 63, 84, . . .
22	1, 2, 11, 22	22, 44, 66, . . .
23	**1, 23**	23, 46, 69, . . .

From this chart, you can make several observations. The divisors of a number are less than or equal to the number, while the multiples are greater than or equal to the number. Each number has a finite number of divisors but an infinite number of multiples. Also, any natural number greater than 1 has at least two divisors, 1 and itself.

The natural numbers which have exactly two divisors are called **prime numbers**. Those which have more than two divisors are **composite numbers**. Thus, considering the numbers in the chart, 2, 3, 5, 7, 11, 13, 17, 19, and 23 are prime numbers while 4, 6, 8, 9, 10, 12, 14, 15, 16, 18, 20, 21, and 22 are composite. Note that 1 is neither prime nor composite. If you are given a number that is not listed in the chart, how do you determine whether it is prime or composite? You must determine if the number has more than two divisors.

Example 4   Determine whether each of the following numbers is prime or composite.

a. 51

To determine if 51 has divisors other than 1 and 51, start with 2 and continue through the natural numbers until you find a divisor or until you reach 51.

$$
\begin{array}{c}
25 \\
2 \text{ does not divide } 51: \quad 2\overline{)51} \\
\underline{4} \\
11 \\
\underline{10} \\
1
\end{array}
\qquad
\begin{array}{c}
17 \\
3\overline{)51} \quad 3 \text{ divides } 51. \\
\underline{3} \\
21 \\
\underline{21}
\end{array}
$$

Thus, 51 has more than two divisors; so it is composite.

b. 41

None of the numbers 2, 3, 4 , . . ., 40 will divide 41 exactly. Therefore, 41 is prime.

In Example 1b, it is not actually necessary to check all the numbers between 1 and 41 for possible divisors of 41. In fact, if you begin with 2 and proceed through the *prime numbers* in increasing order in search of a divisor of 41, you may stop your search for a divisor when you reach a prime number whose product with itself is greater than 41. Thus, you need try only 2, 3, 5, and 7 (since $7 \cdot 7 = 49$ which is greater than 41) before determining that 41 is prime.

The following tests for divisibility are very useful shortcuts. They enable you to determine whether a number is divisible by 2 or 3 or 5 without actually having to do the division. For other numbers, however, you must carry out the division.

DIVISIBILITY TESTS FOR 2, 3, AND 5:

1. A natural number is divisible by 2 if and only if the units digit is even.
2. A natural number is divisible by 3 if and only if the sum of the digits is divisible by 3.
3. A natural number is divisible by 5 if and only if the units digit is a 0 or 5.

Example 5   a. 1,247,986 is divisible by 2 since the units digit (6) is even.

b. 1,489,793 is not divisible by 2 since the units digit (3) is odd.

c. 486 is divisible by 3 since the sum of the digits ($4 + 8 + 6 = 18$) is divisible by 3.

d. 1145 is not divisible by 3 since the sum of the digits ($1 + 1 + 4 + 5 = 11$) is not divisible by 3.

e. 486 is not divisible by 5 since the units digit is not a 0 or a 5.

f. 1145 is divisible by 5 since the units digit is 5.

Example 6   Determine whether the following numbers are prime or composite.

a. 153

153 is not divisible by 2 since the units digit is not even.

153 is divisible by 3 since the sum of the digits ($1 + 5 + 3 = 9$) is divisible by 3.

Thus, 153 is composite because it is divisible by at least 3 numbers, 1, 153, and 3.

b. 143

143 is not divisible by 2.
143 is not divisible by 3.
143 is not divisible by 5.

We must continue the search for a divisor until we find a divisor or reach a prime whose product with itself is greater than 143. Since $5 \cdot 5 = 25$ and this is less than 143, we must continue. The next prime is 7.

$$
\begin{array}{r}
20 \\
7{\overline{\smash{\big)}\,143}} \\
\underline{14} \\
3 \\
\underline{0} \\
3
\end{array}
$$

143 is not divisible by 7.

$7 \cdot 7 = 49$ which is less than 143 so we must continue.

$$
\begin{array}{r}
13 \\
11{\overline{\smash{\big)}\,143}} \\
\underline{11} \\
33 \\
\underline{33}
\end{array}
$$

143 is divisible by 11.

Thus 143 is not prime; it is composite.

c. 251

By using the divisibility tests, we find that neither 2 nor 3 nor 5 divides 251. So, we continue our search for a divisor.

$$
\begin{array}{r}
35 \\
7{\overline{\smash{\big)}\,251}} \\
\underline{21} \\
41 \\
\underline{35} \\
6
\end{array}
\qquad
\begin{array}{r}
22 \\
11{\overline{\smash{\big)}\,251}} \\
\underline{22} \\
31 \\
\underline{22} \\
9
\end{array}
\qquad
\begin{array}{r}
19 \\
13{\overline{\smash{\big)}\,251}} \\
\underline{13} \\
121 \\
\underline{117} \\
4
\end{array}
\qquad
\begin{array}{r}
14 \\
17{\overline{\smash{\big)}\,251}} \\
\underline{17} \\
81 \\
\underline{68} \\
13
\end{array}
$$

Note $13 \cdot 13$ = 169, so we continue.

Note $17 \cdot 17$ = 289, so we are through.

Since we did not find a divisor of 251 other than 1 and 251, 251 is a prime number.

In the next section, it will be helpful for you to be familiar with at least the first 12 primes. They are 2, 3, 5, 7, 11, 13, 17, 19, 23, 29, 31, and 37. All of the prime numbers cannot be listed since the number of primes is unlimited.

2.1 Exercises

Use the divisibility tests to determine if 2, 3, and 5 are factors of each of the numbers in Exercises 1-6.

1. 213

2. 615

3. 111,111

4. 1,000,000,002

5. 606

6. 12,345

7. Which of the following numbers is divisible by 3? 13, 49, 97, 153, 679.

8. Which of the following numbers is divisible by 5? 13, 49, 97, 153, 679.

9. List the factors of 24.   10. List 5 multiples of 24.

11. List 6 multiples of 10.

12. List the divisors of 18.

13. Give 4 factors of 72.

14. List 2 multiples of 100.

15. Give 3 divisors of 45.

16. List 2 factors of 70.

Answer the following true (T) or false (F).

17. 50 is a factor of 10.

18. 13 is a multiple of 1.

19. 50 is a multiple of 10.

20. 15 is a factor of 15.

21. 50 is a divisor of 10.

22. 24 is a factor of 820.

23. 50 is divisible by 10.

24. 125 is a divisor of 25.

25. 15 is a multiple of 15.

26. 10 is a factor of 50.

27. 820 is divisible by 25.

28. 10 is a multiple of 50.

29. 1 is a factor of 13.

30. 10 is a divisor of 50.

31. 125 is divisible by 12.

32. 10 is divisible by 50.

33. Determine which of the following numbers is (are) prime. 13, 49, 97, 153, 679.

34. Determine which of the following numbers is (are) composite. 13, 49, 97, 153, 679.

35. List the numbers from the following array which are not divisible by 2, 3, 5, or 7. (Note that your list will be a list of the prime numbers greater than 7 but less than 100.)

	11	21	31	41	51	61	71	81	91
2	12	22	32	42	52	62	72	82	92
3	13	23	33	43	53	63	73	83	93
4	14	24	34	44	54	64	74	84	94
5	15	25	35	45	55	65	75	85	95
6	16	26	36	46	56	66	76	86	96
7	17	27	37	47	57	67	77	87	97
8	18	28	38	48	58	68	78	88	98
9	19	29	39	49	59	69	79	89	99
10	20	30	40	50	60	70	80	90	100

2.2 Prime Factorization

A number has been factored if it is written as a product of numbers. Throughout the following material, you will be required to perform numerous factorizations. Suppose a classmate asked you, "How do you factor 24? Is it 3 · 8, or 2 · 12, or 4 · 6?" What would you reply? Of course each of these factorizations is correct since the product is 24. However, each could be factored further as indicated below.

$$
\begin{aligned}
24 &= 3 \cdot 8 \\
&= 3 \cdot 2 \cdot 4 \\
&= 3 \cdot 2 \cdot 2 \cdot 2 \\
&= 2 \cdot 2 \cdot 2 \cdot 3
\end{aligned}
\qquad
\begin{aligned}
24 &= 4 \cdot 6 \\
&= 2 \cdot 2 \cdot 6 \\
&= 2 \cdot 2 \cdot 3 \cdot 2 \\
&= 2 \cdot 2 \cdot 2 \cdot 3
\end{aligned}
\qquad
\begin{aligned}
24 &= 2 \cdot 12 \\
&= 2 \cdot 3 \cdot 4 \\
&= 2 \cdot 3 \cdot 2 \cdot 2 \\
&= 2 \cdot 2 \cdot 2 \cdot 3
\end{aligned}
$$

Note that the factorizations were continued until each number in the product was a prime number. In fact, the same factors were obtained in each case. The final factorization, 24 = 2 · 2 · 2 · 3, is the prime factorization of 24, and it is unique except for the order of the factors. In general, the **prime factorization** of a natural number is the factorization that contains only prime numbers as factors. To obtain the prime factorization of a number, start with any nontrivial factorization of the number and continue breaking down the factors until you have only prime factors. (A nontrivial factorization of a number is any factorization other than 1 times the number.) This is illustrated in Example 1.

Example 1   Factor each of the following numbers into a product of prime numbers.

a. 36

We can start with any factorization of 36 except 36 · 1. If we start with 6 · 6 we get

$$
\begin{aligned}
36 &= 6 \cdot 6 \\
&= 2 \cdot 3 \cdot 2 \cdot 3 \quad \text{(since } 6 = 2 \cdot 3)
\end{aligned}
$$

If we start with $9 \cdot 4$, we get

$$36 = 9 \cdot 4$$
$$= 3 \cdot 3 \cdot 2 \cdot 2 \quad (\text{since } 9 = 3 \cdot 3 \text{ and } 4 = 2 \cdot 2)$$

Thus, we obtain the same prime factors regardless of the factorization with which we start, so the prime factorization of 36 is $2 \cdot 2 \cdot 3 \cdot 3$. *Note that we usually list the factors in increasing order.*

b. 45

$$45 = 5 \cdot 9$$
$$= 5 \cdot 3 \cdot 3$$

All the numbers in the factorization are prime, so the prime factorization of 45 is $3 \cdot 3 \cdot 5$.

c. 210

$$210 = 10 \cdot 21$$
$$= 2 \cdot 5 \cdot 7 \cdot 3$$

Thus, $210 = 2 \cdot 3 \cdot 5 \cdot 7$

d. 43

43 is prime since it is not divisible by 2, 3, 5, or 7. Thus, the only possible factorization of 43 is the trivial one: $43 = 1 \cdot 43$.

In these examples, the initial factorization you choose is not important. However, it is essential that you begin by finding a factor or divisor of the number other than 1 or itself. This may become more difficult when you are dealing with larger numbers. The following *prime factorization procedure* gives you a technique for finding the factors or divisors of a number. It is usually the most efficient method for determining the prime factorization of a number.

A PRIME FACTORIZATION PROCEDURE	*EXAMPLE* Obtain the prime factorization of 1599. *Solution:*
1. Begin with 2 (the smallest prime number) and proceed through the prime numbers *in increasing order* in search of a divisor of the given number. You may stop looking for a divisor when you reach a prime whose product with itself is larger than the number you are trying to factor.	2 doesn't divide 1599 3 divides 1599
2. If a divisor is found, express the number as a product of this divisor and the quotient. (If a divisor is not found, then the number is prime.)	$1599 = 3 \cdot 533$
3. Repeat this procedure in order to factor the quotient. Begin your search for a divisor of the quotient with the prime divisor found in step 1. (The procedure is completed when the quotient is prime.)	3 doesn't divide 533 5 doesn't divide 533 7 doesn't divide 533 11 doesn't divide 533 13 divides 533
4. Express the original number as the product of all the prime divisors found and the final quotient (which is also a prime number).	Therefore, $$1599 = 3 \cdot 533$$ $$= 3 \cdot 13 \cdot 41$$
	41 is prime; therefore, the factorization $1599 = 3 \cdot 13 \cdot 41$ is the prime factorization.

Example 2   Factor each of the following numbers completely. That is, give the prime factorization of each of the numbers.

    a. 2100

$2100 = 2 \cdot 1050$	2 divides 2100 so replace 2100 with $2 \cdot 1050$.
$= 2 \cdot 2 \cdot 525$	Continue with 1050. 2 divides 1050 so replace 1050 with $2 \cdot 525$.
$= 2 \cdot 2 \cdot 3 \cdot 175$	Continue with 525. 2 does not divide 525. 3 divides 525 so replace 525 with $3 \cdot 175$.
$= 2 \cdot 2 \cdot 3 \cdot 5 \cdot 35$	Continue with 175. 3 does not divide 175. 5 divides 175 so replace 175 with $5 \cdot 35$.
$= 2 \cdot 2 \cdot 3 \cdot 5 \cdot 5 \cdot 7$	Continue with 35. 5 divides 35 so replace 35 with $5 \cdot 7$.
	Since 7 is prime, we are finished.

Thus, $2100 = 2 \cdot 2 \cdot 3 \cdot 5 \cdot 5 \cdot 7$.

    b. 429

$429 = 3 \cdot 143$	2 does not divide 429, so try the next prime 3. 3 does divide 429 so replace 429 with $3 \cdot 143$.
$= 3 \cdot 11 \cdot 13$	Continue with 143. Neither 3, 5, nor 7 will divide 143. However, 11 divides 143. Replace 143 with $11 \cdot 13$.

So, $429 = 3 \cdot 11 \cdot 13$.

    c. 3300

$3300 = 2 \cdot 1650$	3300 is divisible by 2.
$= 2 \cdot 2 \cdot 825$	1650 is divisible by 2.
$= 2 \cdot 2 \cdot 3 \cdot 275$	825 is not divisible by 2 but is divisible by 3.
$= 2 \cdot 2 \cdot 3 \cdot 5 \cdot 55$	275 is not divisible by 3 but is divisible by 5.
$= 2 \cdot 2 \cdot 3 \cdot 5 \cdot 5 \cdot 11$	55 is divisible by 5.
$3300 = 2 \cdot 2 \cdot 3 \cdot 5 \cdot 5 \cdot 11$	11 is prime so the factorization is complete.

The form which we use when obtaining a prime factorization may be abbreviated as in the following example where the prime factors are written along the left side and where the quotients are written beneath the dividends. The problems from Example 2 are reworked in Example 3.

Example 3 **a.**

$$
\begin{array}{r|l}
2 & 2100 \\
2 & 1050 \\
3 & 525 \\
5 & 175 \\
5 & 35 \\
  & 7
\end{array}
$$

$2100 = 2 \cdot 2 \cdot 3 \cdot 5 \cdot 5 \cdot 7$

  b.

$$
\begin{array}{r|l}
3 & 429 \\
11 & 143 \\
   & 13
\end{array}
$$

$429 = 3 \cdot 11 \cdot 13$

c. 2|3300
    2|1650
    3|825
    5|275
    5|55
      11

$$3300 = 2 \cdot 2 \cdot 3 \cdot 5 \cdot 5 \cdot 11$$

2.2   Exercises

Find the smallest prime divisor of each of the numbers in Exercises 1-10.

1. 16	**2.** 95
3. 65	**4.** 38
5. 79	**6.** 63
7. 51	**8.** 77
9. 91	**10.** 98

Find the prime factorization of each of the numbers in Exercises 11-40.

11. 72	**12.** 39
13. 143	**14.** 165
15. 119	**16.** 104
17. 170	**18.** 101
19. 187	**20.** 560
21. 462	**22.** 215
23. 873	**24.** 468
25. 805	**26.** 582
27. 444	**28.** 705
29. 1200	**30.** 605
31. 1598	**32.** 1000
33. 2107	**34.** 2391
35. 5355	**36.** 1155
37. 4641	**38.** 1599
39. 5400	**40.** 7735

2.3   Equivalent Fractions

Since we communicate with each other using fractions, it would be helpful if we did this in the simplest way possible. As you may know, $\frac{2}{4}$, $\frac{3}{6}$, $\frac{4}{8}$, and $\frac{222}{444}$ can all be expressed more simply as $\frac{1}{2}$.

The rectangles in Figure 2.2 illustrate that $\frac{2}{4}$, $\frac{3}{6}$, and $\frac{4}{8}$ are each equal to $\frac{1}{2}$.

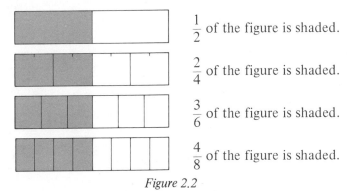

$\frac{1}{2}$ of the figure is shaded.

$\frac{2}{4}$ of the figure is shaded.

$\frac{3}{6}$ of the figure is shaded.

$\frac{4}{8}$ of the figure is shaded.

Figure 2.2

In each case, it is clear that $\frac{1}{2}$ of the figure is shaded. If the figures were divided into 444 equal parts and 222 of the parts were shaded, again exactly $\frac{1}{2}$ of the figure would be shaded. Thus $\frac{2}{4}$, $\frac{3}{6}$, $\frac{4}{8}$, and $\frac{222}{444}$ are different names for the number $\frac{1}{2}$. Such fractions are called **equivalent fractions**. There are two procedures for obtaining equivalent fractions. The first procedure we will discuss is **reducing fractions** to lowest terms.

A fraction is said to be in **lowest terms** if the numerator and denominator have no common factor greater than one. When it is possible to recognize factors common to both the numerator and the denominator, the fraction may be reduced by dividing both the numerator and denominator by the **common factor**. This common factor is referred to as a **reducing factor**.

Example 1   Reduce each of the following fractions.

a. $\frac{12}{15}$

Clearly, 3 is a reducing factor. If we divide the numerator and denominator by 3 we obtain

$$\frac{12}{15} = \frac{4}{5}$$

and since 4 and 5 have no common factor greater than 1, $\frac{4}{5}$ is the answer.

b. $\frac{2}{9}$

Since 2 and 9 have no common factor greater than 1, $\frac{2}{9}$ is already in lowest terms.

c. $\frac{77}{99}$

Since 11 divides both 77 and 99,

$$\frac{77}{99} = \frac{7}{9}.$$

d. $\frac{105}{231}$

This is more difficult. If you use the divisibility tests, you see that 3 divides both 105 and 231. Thus,

$$\frac{105}{231} = \frac{35}{77} = \frac{5}{11} \quad \text{(since 7 divides 35 and 77)}.$$

If you use the technique illustrated in Example 1, you must be good at recognizing common factors. If the reducing factor is not obvious, then *the rule for producing equivalent fractions by using reducing factors* should be used. The technique given in this rule *always* gives the fraction in lowest terms, and it does not depend upon your ability to recognize common factors.

RULE FOR PRODUCING EQUIVALENT FRACTIONS BY USING REDUCING FACTORS	EXAMPLE Reduce $\dfrac{42}{147}$.
1. Obtain the prime factorization of both the numerator and the denominator of the fraction.	$\dfrac{42}{147} = \dfrac{2 \cdot 3 \cdot 7}{3 \cdot 7 \cdot 7}$
2. Examine these factorizations for reducing factors (divisors common to both the numerator and the denominator).	3 is a reducing factor 7 is a reducing factor
3. Divide both the numerator and the denominator by each of these reducing factors to obtain an equivalent fraction in lowest terms.	$\dfrac{42}{147} = \dfrac{2 \cdot \not{3} \cdot \not{7}}{\not{3} \cdot \not{7} \cdot 7} = \dfrac{2}{7}$

Example 2 Reduce each of the following fractions using the rule for reducing fractions.

a. $\dfrac{12}{15}$

$\dfrac{12}{15} = \dfrac{2 \cdot 2 \cdot \not{3}}{\not{3} \cdot 5} = \dfrac{4}{5}$

Prime factorizations:

$2\underline{|12} \quad 3\underline{|15}$
$\;\;2\underline{|6} \qquad\; 5$
$\qquad 3$

b. $\dfrac{77}{99}$

$\dfrac{77}{99} = \dfrac{7 \cdot \not{11}}{3 \cdot 3 \cdot \not{11}} = \dfrac{7}{9}$

Prime factorizations:

$7\underline{|77} \quad 3\underline{|99}$
$\quad 11 \qquad 3\underline{|33}$
$\qquad\qquad\quad 11$

c. $\dfrac{105}{231}$

$\dfrac{105}{231} = \dfrac{\not{3} \cdot 5 \cdot \not{7}}{\not{3} \cdot \not{7} \cdot 11} = \dfrac{5}{11}$

Prime factorizations:

$3\underline{|105} \quad 3\underline{|231}$
$5\underline{|35} \qquad 7\underline{|77}$
$\quad 7 \qquad\qquad 11$

Example 3 a. $\dfrac{26}{65} = \dfrac{2 \cdot \not{13}}{5 \cdot \not{13}} = \dfrac{2}{5}$

$2\underline{|26} \quad 5\underline{|65}$
$\quad 13 \qquad 13$

b. $\dfrac{170}{187} = \dfrac{2 \cdot 5 \cdot \not{17}}{11 \cdot \not{17}} = \dfrac{2 \cdot 5}{11} = \dfrac{10}{11}$

$2\underline{|170} \quad 11\underline{|187}$
$5\underline{|85} \qquad\;\; 17$
$\quad 17$

c. $\dfrac{170}{130} = \dfrac{\not{2} \cdot \not{5} \cdot 17}{\not{2} \cdot \not{5} \cdot 13} = \dfrac{17}{13}$

$2\underline{|130}$
$5\underline{|65}$
$\quad 13$

d. $\dfrac{4641}{7735} = \dfrac{3 \cdot \not{7} \cdot \not{13} \cdot \not{17}}{5 \cdot \not{7} \cdot \not{13} \cdot \not{17}} = \dfrac{3}{5}$

$3\underline{|4641} \quad 5\underline{|7735}$
$7\underline{|1547} \qquad 7\underline{|1547}$
$13\underline{|221} \qquad 13\underline{|221}$
$\qquad 17 \qquad\qquad 17$

Example 4 $\dfrac{560}{215} = \dfrac{2 \cdot 2 \cdot 2 \cdot 2 \cdot \not{5} \cdot 7}{\not{5} \cdot 43} = \dfrac{112}{43}$

$2\underline{|560} \quad 5\underline{|215}$
$2\underline{|280} \qquad 43$
$2\underline{|140}$
$2\underline{|70}$
$5\underline{|35}$
$\quad 7$

Example 4 could also be done by recognizing that 5 is a common factor. Since 43 is prime and does not divide 112, the fraction is in lowest terms.

$$\frac{560}{215} = \frac{112}{43}$$

Example 5 $\frac{41}{97}$ is already in lowest terms since both 41 and 97 are prime.

Example 6 $\frac{63}{8} = \frac{3 \cdot 3 \cdot 7}{2 \cdot 2 \cdot 2}$

Since 63 and 8 have no common factors, $\frac{63}{8}$ is in lowest terms.

Example 7   a. $\frac{150}{130} = \frac{\cancel{10} \cdot 15}{\cancel{10} \cdot 13} = \frac{15}{13}$

Note: This problem illustrates that it is not necessary to obtain prime factorizations of the numerator and denominator if you can recognize common composite factors. Be careful, however, that you do not overlook common factors by failing to factor the numerator and the denominator.

b. $\frac{816}{272} = \frac{3 \cdot 272}{1 \cdot 272} = \frac{3}{1} = 3$ Recall that 1 is a factor of every number.

A familiar application of fractions involves considering a fractional part of a quantity. This application uses the interpretation that a fraction, say $\frac{3}{4}$, represents 3 out of 4 parts of a quantity. Also, 3 is $\frac{3}{4}$ of 4.

Example 8   a. Twenty-four is what fractional part of 48?

Since $\frac{24}{48} = \frac{1}{2}$, 24 is $\frac{1}{2}$ of 48.

b. What fractional part of 20 is 17?

$$\frac{17}{20}$$

Example 9   A pound of cheese contains 16 slices. If 14 slices have been eaten, what fractional part of the cheese has been eaten?

Since 14 out of 16 slices have been eaten, $\frac{14}{16}$ of the cheese has been eaten. But $\frac{14}{16} = \frac{7}{8}$, so $\frac{7}{8}$ of the cheese has been eaten.

Example 10   Fill in the missing numerators and denominators. Identify the reducing factor in each problem.

a. $\dfrac{15}{55} = \dfrac{?}{11}$ Since 55 must be divided by 5 to obtain 11, the numerator is divided by 5 to obtain 3. Note that the reducing factor 5 is equal to $55 \div 11$.

$$\frac{15 \div 5}{55 \div 5} = \frac{3}{11}$$

Thus, $\dfrac{15}{55} = \dfrac{3}{11}$

b. $\dfrac{15}{153} = \dfrac{5}{?}$ Note that the reducing factor 3 is equal to $15 \div 5$.

$$\frac{15 \div 3}{153 \div 3} = \frac{5}{51}$$

So, $\dfrac{15}{153} = \dfrac{5}{51}$

c. $\dfrac{88}{112} = \dfrac{?}{14}$

To find the reducing factor, we must determine what number divided into 112 is 14. Since $112 \div 14 = 8$, we know that $112 \div 8 = 14$. Thus 8 is the reducing factor.

$$\frac{88 \div 8}{112 \div 8} = \frac{11}{14}$$

Equivalent fractions may also be obtained by multiplying both the numerator and denominator by the same natural number. This procedure (which is equivalent to multiplying the fraction by 1) produces a fraction with a larger numerator and denominator. The number by which the numerator and denominator are multiplied is called the **building factor**. This procedure is summarized in the *rule for producing equivalent fractions by using a building factor*.

RULE FOR PRODUCING EQUIVALENT FRACTIONS BY USING A BUILDING FACTOR	*EXAMPLE* Convert $\dfrac{4}{5}$ to an equivalent fraction by using a building factor of 3. *Solution:*
Multiply both the numerator and the denominator by the same natural number to obtain an equivalent fraction.	Multiply both the numerator and the denominator by 3. $$\frac{4 \cdot 3}{5 \cdot 3} = \frac{12}{15}$$

Example 11   Build each of the following to an equivalent fraction using the indicated building factor.

a. $\dfrac{4}{5}$, building factor of 7

$$\frac{4}{5} = \frac{4 \times 7}{5 \times 7} = \frac{28}{35}$$

b. $\dfrac{17}{50}$, building factor of 12

$$\frac{17}{51} = \frac{17 \times 12}{50 \times 12} = \frac{204}{600}$$

c. $\dfrac{5}{2}$, building factor of 4

$$\frac{5}{2} = \frac{5 \times 4}{2 \times 4} = \frac{20}{8}$$

This procedure is used frequently when an equivalent fraction with a specified denominator is desired. In this case, you must determine the appropriate building factor.

Example 12   Determine the missing numerators.

a. $\dfrac{5}{7} = \dfrac{?}{77}$

What must you multiply 7 by to obtain 77? In other words what is $77 \div 7$? Since the answer is 11, this is the building factor.

$$\frac{5 \cdot 11}{7 \cdot 11} = \frac{55}{77}$$

b. $\dfrac{2}{3} = \dfrac{?}{12}$

The building factor is 4. Thus, $\dfrac{2}{3} = \dfrac{2 \times 4}{3 \times 4} = \dfrac{8}{12}$.

c. $\dfrac{2}{3} = \dfrac{?}{36}$

The building factor is 12. So, $\dfrac{2}{3} = \dfrac{24}{36}$.

d. $\dfrac{11}{13} = \dfrac{?}{91}$

The building factor is 7. Therefore, $\dfrac{11}{13} = \dfrac{77}{91}$.

Example 13   Fill in the missing denominator.

$$\frac{23}{47} = \frac{299}{?}$$

Since $299 \div 23 = 13$, 13 is the building factor.

$$\frac{23 \cdot 13}{47 \cdot 13} = \frac{299}{611}$$

2.3 Exercises

Reduce each of the fractions in Exercises 1–50 to lowest terms.

1. $\frac{10}{15}$
2. $\frac{34}{122}$
3. $\frac{99}{121}$

4. $\frac{290}{70}$
5. $\frac{14}{22}$
6. $\frac{35}{7}$

7. $\frac{3}{123}$
8. $\frac{20}{28}$
9. $\frac{21}{7}$

10. $\frac{11}{99}$
11. $\frac{130}{110}$
12. $\frac{12}{36}$

13. $\frac{14}{119}$
14. $\frac{146}{106}$
15. $\frac{15}{45}$

16. $\frac{8}{22}$
17. $\frac{200}{300}$
18. $\frac{1000}{15000}$

19. $\frac{7}{11}$
20. $\frac{3100}{4300}$
21. $\frac{1500}{700}$

22. $\frac{20}{230}$
23. $\frac{35000}{6000}$
24. $\frac{7000}{10000}$

25. $\frac{14300}{2000}$
26. $\frac{13}{23}$
27. $\frac{46}{194}$

28. $\frac{444}{111}$
29. $\frac{50}{790}$
30. $\frac{52}{56}$

31. $\frac{54}{270}$
32. $\frac{210}{66}$
33. $\frac{462}{468}$

34. $\frac{1155}{1617}$
35. $\frac{81}{405}$
36. $\frac{88}{520}$

37. $\frac{250}{50}$
38. $\frac{364}{819}$
39. $\frac{255}{97}$

40. $\frac{220}{132}$
41. $\frac{287}{123}$
42. $\frac{72}{252}$

43. $\frac{16}{240}$
44. $\frac{119}{45}$
45. $\frac{410}{630}$

46. $\frac{100}{4500}$
47. $\frac{805}{138}$
48. $\frac{161}{240}$

49. $\frac{873}{582}$
50. $\frac{82}{195}$

51. Thirty is what fractional part of forty-two?

52. What fractional part of 82 is 52?

53. What fractional part of 2 is 3?

54. Four is what fractional part of 3?

55. Seventeen is what fractional part of 80?

56. In a book of 500 pages, illustrations appeared on 235 of the pages. What fractional part of the book contained illustrations?

57. Two students in a class of 170 students made an A on the final exam. What fractional part of the students made A's? What fractional part of the students did not make A's?

58. During the winter 250 students were absent on one day because of illness. If the school had 890 students enrolled, what fractional part of the students were absent?

59. A baseball player got 68 hits in 272 times at bat. What fractional part of his times at bat did he get hits?

60. An old mansion was divided into 8 apartments. Only 6 of the apartments were rented. What fractional part of the apartments are rented? What fractional part of the apartments are not rented?

Determine the missing numerators and denominators, and identify the reducing factor for each fraction in Exercises 61–70.

61. $\frac{8}{16} = \frac{?}{2}$
62. $\frac{26}{39} = \frac{?}{3}$
63. $\frac{64}{56} = \frac{?}{7}$

64. $\frac{49}{70} = \frac{?}{10}$
65. $\frac{61}{183} = \frac{?}{3}$
66. $\frac{86}{129} = \frac{2}{?}$

67. $\frac{459}{792} = \frac{153}{?}$
68. $\frac{143}{198} = \frac{?}{18}$
69. $\frac{462}{210} = \frac{11}{?}$

70. $\frac{41}{82} = \frac{?}{2}$

Build each of the fractions in Exercises 71–80 to an equivalent fraction using the building factor indicated.

71. $\dfrac{3}{7}$ use 3 as a building factor.

72. $\dfrac{3}{7}$ use 7 as a building factor.

73. $\dfrac{5}{9}$ use 11 as a building factor.

74. $\dfrac{5}{9}$ use 2 as a building factor.

75. $\dfrac{11}{13}$ use 23 as a building factor.

76. $\dfrac{11}{23}$ use 7 as a building factor.

77. $\dfrac{7}{16}$ use 9 as a building factor.

78. $\dfrac{9}{24}$ use 6 as a building factor.

79. $\dfrac{13}{18}$ use 8 as a building factor.

80. $\dfrac{2}{49}$ use 12 as a building factor.

Determine the missing numerators or denominators, and identify the building factor for each fraction in Exercises 81–100.

81. $\dfrac{2}{3} = \dfrac{?}{12}$ **82.** $\dfrac{2}{3} = \dfrac{?}{30}$ **83.** $\dfrac{2}{3} = \dfrac{?}{300}$

84. $\dfrac{2}{3} = \dfrac{?}{96}$ **85.** $\dfrac{5}{24} = \dfrac{?}{48}$ **86.** $\dfrac{5}{24} = \dfrac{?}{120}$

87. $\dfrac{7}{18} = \dfrac{?}{18}$ **88.** $\dfrac{7}{18} = \dfrac{?}{54}$ **89.** $\dfrac{2}{11} = \dfrac{?}{121}$

90. $\dfrac{2}{11} = \dfrac{?}{99}$ **91.** $\dfrac{11}{13} = \dfrac{?}{26}$ **92.** $\dfrac{11}{13} = \dfrac{44}{?}$

93. $\dfrac{17}{19} = \dfrac{?}{437}$ **94.** $\dfrac{5}{7} = \dfrac{390}{?}$ **95.** $\dfrac{17}{35} = \dfrac{68}{?}$

96. $\dfrac{117}{228} = \dfrac{?}{684}$ **97.** $\dfrac{111}{123} = \dfrac{1221}{?}$ **98.** $\dfrac{5}{6} = \dfrac{785}{?}$

99. $\dfrac{1{,}000{,}001}{496} = \dfrac{3{,}000{,}003}{?}$

100. $\dfrac{5}{817} = \dfrac{?}{25{,}327}$

2.4   Multiplication of Fractions

Just as an ancient shepherd probably would not understand the necessity for fractions, students wonder why we add, subtract, multiply, and divide fractions. One reason is that these operations on fractions allow modern man to perform tasks that the shepherd never even dreamed of. Notice the variety of problems involving fractions as you work through the exercises and the chapter review.

In order to multiply fractions, you may recall that the numerators are multiplied and the denominators are multiplied. For example,

$$\frac{1}{3} \cdot \frac{2}{5} = \frac{2}{15}$$

However if the answer is not in lowest terms, it must be reduced. For example,

$$\frac{2}{3} \cdot \frac{3}{5} = \frac{6}{15} = \frac{2}{5}$$

This procedure can become cumbersome in problems involving large numbers. The *rule for multiplying fractions* describes a procedure which will eliminate the need for reducing, since the common factors are eliminated before multiplying.

RULE FOR MULTIPLYING FRACTIONS	EXAMPLE $\dfrac{2}{3} \cdot \dfrac{7}{50} = ?$
1. Indicate the product of the numerators and denominators. 2. Write the prime factorization of each factor in these indicated products. 3. Divide the numerator and denominator by any common factors. 4. Multiply the remaining factors in the numerator to obtain the numerator, and then multiply the remaining factors in the denominator to obtain the denominator.	*Solution:* $\dfrac{2}{3} \cdot \dfrac{7}{50} = \dfrac{2 \cdot 7}{3 \cdot 50}$ $= \dfrac{2 \cdot 7}{3 \cdot 2 \cdot 5 \cdot 5}$ $= \dfrac{\not2 \cdot 7}{3 \cdot \not2 \cdot 5 \cdot 5} = \dfrac{7}{75}$ Thus, $\dfrac{2}{3} \cdot \dfrac{7}{50} = \dfrac{7}{75}.$

The procedure given in the rule is used in the following examples.

Example 1   Find the following products.

a. $\dfrac{2}{3} \cdot \dfrac{3}{5} = \dfrac{2 \cdot \not3}{\not3 \cdot 5} = \dfrac{2}{5}$

Thus, $\dfrac{2}{3} \cdot \dfrac{3}{5} = \dfrac{2}{5}$

b. $2 \cdot \dfrac{5}{7} = \dfrac{2}{1} \cdot \dfrac{5}{7}$ Recall that $2 = \dfrac{2}{1}.$

$= \dfrac{2 \cdot 5}{1 \cdot 7} = \dfrac{10}{7}$

Thus, $2 \cdot \dfrac{5}{7} = \dfrac{10}{7}$

c. $\dfrac{2}{3} \cdot \dfrac{3}{4} = \dfrac{2 \cdot 3}{3 \cdot 4}$

$= \dfrac{\not2 \cdot \not3}{\not3 \cdot \not2 \cdot 2} = \dfrac{1}{2}$

Note that when all common prime factors are eliminated, no factors remain in the numerator. However, since 1 is a factor of every number, it is understood to remain when all other factors have been eliminated. Thus,

$\dfrac{2}{3} \cdot \dfrac{3}{4} = \dfrac{1}{2}$

Example 2   **a.** Find the product of $\frac{15}{7}, \frac{33}{36}$, and $\frac{14}{3}$.

$$\frac{15}{7} \cdot \frac{33}{36} \cdot \frac{14}{3} = \frac{15 \cdot 33 \cdot 14}{7 \cdot 36 \cdot 3}$$

$$= \frac{\cancel{3} \cdot 5 \cdot \cancel{3} \cdot 11 \cdot \cancel{2} \cdot \cancel{7}}{\cancel{7} \cdot \cancel{2} \cdot 2 \cdot \cancel{3} \cdot \cancel{3} \cdot 3}$$

$$= \frac{55}{6}$$

Thus, $\frac{15}{7} \cdot \frac{33}{36} \cdot \frac{14}{3} = \frac{55}{6}$.

b. Multiply $\frac{6}{70}$ by $\frac{105}{9}$.

$$\frac{6}{70} \cdot \frac{105}{9} = \frac{6 \cdot 105}{70 \cdot 9}$$

$$= \frac{\cancel{2} \cdot \cancel{3} \cdot \cancel{3} \cdot \cancel{5} \cdot \cancel{7}}{\cancel{2} \cdot \cancel{5} \cdot \cancel{7} \cdot \cancel{3} \cdot \cancel{3}}$$

$$= 1$$

Prime factorizations:

$$\begin{array}{cc} 3\underline{|105} & 2\underline{|70} \\ 5\underline{|35} & 5\underline{|35} \\ 7 & 7 \end{array}$$

Thus, $\frac{6}{70} \cdot \frac{105}{9} = 1$.

Example 3   Find the following products.

a. $\frac{56}{39} \cdot \frac{26}{49} = \frac{56 \cdot 26}{39 \cdot 49}$

$$= \frac{2 \cdot 2 \cdot 2 \cdot \cancel{7} \cdot 2 \cdot \cancel{13}}{3 \cdot \cancel{13} \cdot \cancel{7} \cdot 7}$$

$$= \frac{16}{21}$$

Prime factorization:

$$\begin{array}{c} 2\underline{|56} \\ 2\underline{|28} \\ 2\underline{|14} \\ 7 \end{array}$$

Thus, $\frac{56}{39} \cdot \frac{26}{49} = \frac{16}{21}$.

A more concise way to work this problem is to recognize, without using the prime factorizations, that 7 and 13 are the only factors common to the numerators and denominators and then divide by 7 and 13. This could be shown as follows:

$$\frac{\overset{8}{\cancel{56}}}{\underset{3}{\cancel{39}}} \cdot \frac{\overset{2}{\cancel{26}}}{\underset{7}{\cancel{49}}} = \frac{16}{21} \qquad \begin{array}{l} \text{Divide 56 and 49 by 7.} \\ \text{Divide 26 and 29 by 13.} \end{array}$$

b. $\frac{8}{21} \cdot \frac{9}{13} \cdot \frac{5}{12} = \frac{8 \cdot 9 \cdot 5}{21 \cdot 13 \cdot 12}$

$$= \frac{2 \cdot 2 \cdot 2 \cdot 3 \cdot 3 \cdot 5}{3 \cdot 7 \cdot 13 \cdot 2 \cdot 2 \cdot 3}$$

$$= \frac{10}{91}$$

Thus, $\dfrac{8}{21} \cdot \dfrac{9}{13} \cdot \dfrac{5}{12} = \dfrac{10}{91}$.

The condensed form would be:

$$\frac{\overset{2}{\cancel{8}}}{\underset{7}{\cancel{21}}} \cdot \frac{\overset{\overset{1}{\cancel{3}}}{\cancel{9}}}{13} \cdot \frac{5}{\underset{\underset{1}{\cancel{3}}}{\cancel{12}}} = \frac{10}{91}$$
Divide 8 and 12 by 4.
Divide 21 and 9 by 3.
Divide 3 and 3 by 3.

Note that success in using the condensed form depends on your ability to recognize common factors. If you overlook common factors, then your answer will not be in lowest terms. However, the procedure given in the rule will always give the answer in lowest terms.

In Example 4, a combination of the two techniques is used.

Example 4 $\dfrac{210}{429} \cdot \dfrac{117}{310} \cdot \dfrac{605}{805} = ?$

$$\frac{\overset{21}{\cancel{210}}}{\underset{143}{\cancel{429}}} \cdot \frac{\overset{39}{\cancel{117}}}{\underset{31}{\cancel{310}}} \cdot \frac{\overset{121}{\cancel{605}}}{\underset{161}{\cancel{805}}}$$
Divide 210 and 310 by 10.
Divide 605 and 805 by 5.
Divide 117 and 429 by 3.

At this point it may be difficult to recognize any other common factor. However, this does not mean there are none. Prime factorization can be used to see if there are other common factors.

$$\frac{21 \cdot 39 \cdot 121}{143 \cdot 31 \cdot 161} = \frac{3 \cdot \cancel{7} \cdot 3 \cdot \cancel{13} \cdot \cancel{11} \cdot 11}{\cancel{11} \cdot \cancel{13} \cdot 31 \cdot \cancel{7} \cdot 23}$$

$$= \frac{99}{713}$$

Prime factorizations:
$$11\underline{|121} \quad 11\underline{|143} \quad 7\underline{|161}$$
$$11 \quad 13 \quad 23$$

Example 5   A butcher is slicing a chunk of cheese into pieces $\dfrac{3}{32}$ of an inch thick. How thick a chunk will she need to get 10 slices?

We could add $\dfrac{3}{32}$ ten times, but it is easier to multiply $\dfrac{3}{32}$ by 10.

$$\frac{3}{32} \cdot \frac{10}{1} = \frac{3 \cdot \cancel{2} \cdot 5}{\cancel{2} \cdot 2 \cdot 2 \cdot 2 \cdot 2} = \frac{15}{16}$$

Example 6   Find $\frac{1}{3}$ of 27.

$$\frac{1}{\cancel{3}} \times \frac{\overset{9}{\cancel{27}}}{1} = 9$$

Note: Because multiplying 27 by $\frac{1}{3}$ involves dividing 27 by 3, students sometimes think that *of* indicates division. However, you should notice that we multiplied $\frac{1}{3}$ times 27. When "of" is used as in this problem, it indicates multiplication.

Example 7   Three-fourths of the students in the freshman class at Clayton University own an automobile. If there are 560 students in the freshman class, how many own an automobile?

$$\frac{3}{4} \text{ of } 560 = \frac{3}{\cancel{4}} \times \frac{\overset{140}{\cancel{560}}}{1} = 420$$

Therefore, 420 students own an automobile.

2.4   Exercises

In Exercises 1–30, multiply.

1. $\frac{3}{8} \cdot \frac{5}{7}$

2. $11 \cdot \frac{7}{10}$

3. $\left(\frac{2}{3}\right)\left(\frac{7}{5}\right)$

4. $\left(\frac{8}{9}\right)\left(\frac{5}{3}\right)$

5. $\frac{1}{3} \cdot 5$

6. $\frac{5}{7} \cdot \frac{1}{2} \cdot 13$

7. $3 \cdot \frac{2}{5} \cdot \frac{2}{7}$

8. $\left(\frac{1}{2}\right)\left(\frac{1}{3}\right)\left(\frac{1}{5}\right)$

9. $\frac{2}{5} \cdot \frac{2}{5} \cdot \frac{2}{5}$

10. $\frac{5}{3} \cdot \frac{2}{3} \cdot \frac{5}{3}$

11. $\frac{8}{9} \cdot \frac{15}{7}$

12. $\frac{7}{16} \cdot 2$

13. $\frac{4}{7} \cdot \frac{3}{8}$

14. $\frac{3}{8} \cdot \frac{7}{6}$

15. $6 \cdot \frac{3}{2}$

16. $\frac{10}{19} \cdot \frac{14}{25}$

17. $\frac{18}{15} \cdot \frac{4}{27}$

18. $\left(\frac{22}{7}\right)\left(\frac{5}{11}\right)$

19. $\frac{2}{3} \cdot \frac{5}{11} \cdot \frac{9}{7}$

20. $\frac{5}{8} \cdot \frac{13}{3} \cdot \frac{10}{7}$

21. $\left(\frac{4}{23}\right)\left(\frac{69}{8}\right)$

22. $\frac{10}{25} \cdot \frac{18}{27}$

23. $\frac{16}{27} \cdot \frac{18}{24}$

24. $8 \cdot \frac{5}{28}$

25. $\frac{15}{77} \cdot \frac{91}{20}$

26. $\frac{15}{27} \cdot \frac{18}{14} \cdot \frac{1}{25}$

27. $\frac{14}{25} \cdot 10 \cdot \frac{5}{28}$

28. $\frac{57}{18} \cdot \frac{18}{57}$

29. $\frac{1}{2} \cdot \frac{2}{3} \cdot \frac{3}{4} \cdot \frac{4}{5} \cdot \frac{5}{6} \cdot \frac{6}{7}$

30. $42 \cdot \frac{2}{31}$

31. Find $\frac{7}{8}$ of $\frac{64}{77}$.

32. Find $\frac{5}{9}$ of $\frac{21}{25}$.

33. What is $\frac{5}{12}$ of 120?

34. Find $\frac{1}{2}$ of $\frac{1}{5}$.

35. Find $\frac{2}{3}$ of $\frac{1}{2}$.

36. Multiply $\frac{11}{15}$ by $\frac{3}{2}$.

37. What is the total length of 1,216 skateboards each $\frac{7}{8}$ of a yard long?

38. What is the product of $\frac{1}{119}$ and 5355?

39. $\frac{123}{77} \cdot \frac{49}{82} = ?$

40. A dress originally priced at $150 was marked $\frac{1}{3}$ off.

What was the amount of the reduction? What was the sale price of the dress?

41. A recipe for 4 people called for $\frac{2}{3}$ cup of milk. How much milk will be necessary if the recipe is increased to serve 12 people?

42. It takes a typist $\frac{1}{2}$ hour to type each page of a technical paper. How long will it take him to type a paper 31 pages long?

43. A prescription drug has a suggested dosage of $\frac{1}{2}$ tsp every four hours. How many teaspoons are taken if the dosage is administered 6 times?

44. $\frac{16}{26} \cdot \frac{39}{240} = ?$

45. Find the product of $\frac{54}{143}$ and $\frac{11}{270}$.

46. In a shipment of 900 radios, $\frac{1}{20}$ of the radios were defective. How many of the radios were defective?

47. $\frac{52}{56} \cdot \frac{77}{55} \cdot \frac{20}{26} = ?$

48. $\frac{20}{65} \cdot \frac{91}{230} \cdot \frac{75}{2} = ?$

49. Muscles normally account for approximately $\frac{2}{5}$ of a person's body weight. Approximately how many pounds of muscle does a 155-pound person have?

50. $\frac{3}{143} \cdot \frac{4}{23} \cdot \frac{286}{12} = ?$

51. One of the most expensive films ever made was *War and Peace*. The total cost was said to be more than \$96,000,000. If two-thirds of the cost was for personnel, how much was the cost for personnel?

52. Multiply $\frac{27}{21}$ by $\frac{99}{15}$.

53. $\frac{33}{105} \cdot \frac{1155}{6} \cdot \frac{23}{22} = ?$

54. If it takes $\frac{2}{3}$ of a cup of sugar to make one pie, how many cups of sugar will be needed for 45 pies?

55. Find the product of $\frac{1}{7}$ and $\frac{4}{10}$ and $\frac{21}{10}$.

56. If $\frac{2}{21}$ of the 1239 students at a college are majoring in mathematics, how many math majors are there?

57. Multiply $\frac{4}{10}$ by 72.

58. Find the product of $\frac{4}{7}$ and $\frac{210}{105}$.

59. A printer estimated that he was losing $\frac{2}{11}$ of his materials due to waste. On an order requiring 330 lbs of paper, what was the waste?

60. Multiply $\frac{85}{175}$ by $\frac{91}{170}$.

2.5   Division of Fractions

The usual rule associated with the division of fractions is *invert the divisor and multiply*. The following example illustrates this by using two approaches to solving the same problem.

Example 1   A father's will provided that each of three children should share equally in his \$21,000 estate. What is the share of each child?

Divide the \$21,000 into three equal shares:

$$\$21,000 \div 3 = \$7,000$$

Alternate Solution:

Each child's share is one-third of the estate.

$$\frac{\$21,000}{1} \cdot \frac{1}{3} = \frac{\$21,000}{3} = \$7,000$$

Thus, we see that dividing by 3 is equivalent to multiplying by $\frac{1}{3}$. The number $\frac{1}{3}$ is the multiplicative inverse of 3. The **multiplicative inverse** of any nonzero fraction is formed by interchanging or reciprocating the numerator and denominator. For example, the multiplicative inverse of $\frac{2}{5}$ is $\frac{5}{2}$. The multiplicative inverse is sometimes referred to as the **reciprocal**.

Since dividing by a number is the same as multiplying by its reciprocal or multiplicative inverse, we divide fractions by converting to an equivalent multiplication problem. This is specified in the *rule for dividing fractions.*

RULE FOR DIVIDING FRACTIONS	EXAMPLE $\quad \frac{2}{3} \div \frac{5}{7} = ?$
	Solution:
1. Form the multiplicative inverse of the divisor.	The multiplicative inverse of $\frac{5}{7}$ is $\frac{7}{5}$.
2. Multiply the dividend by this multiplicative inverse, using the rule for multiplying fractions.	$\frac{2}{3} \cdot \frac{7}{5} = \frac{14}{15}$

Example 2   a. $\frac{5}{6} \div \frac{1}{2} = ?$

$$\frac{5}{6} \div \frac{1}{2} = \frac{5}{\cancel{6}_3} \cdot \frac{\cancel{2}^1}{1} = \frac{5}{3}$$

Thus, $\frac{5}{6} \div \frac{1}{2} = \frac{5}{3}$

b. $\frac{4}{9} \div \frac{2}{3} = ?$

$$\frac{4}{9} \div \frac{2}{3} = \frac{\cancel{4}^2}{\cancel{9}_3} \cdot \frac{\cancel{3}^1}{\cancel{2}_1} = \frac{2}{3}$$

Thus, $\frac{4}{9} \div \frac{2}{3} = \frac{2}{3}$.

Example 3   Perform the following divisions.

a. $2 \div \frac{5}{6} = \frac{2}{1} \div \frac{5}{6}$ *Note:* $2 = \frac{2}{1}$

$$= \frac{2}{1} \cdot \frac{6}{5} = \frac{12}{5}$$

Thus, $2 \div \frac{5}{6} = \frac{12}{5}$.

b. $\dfrac{5}{6} \div 2 = \dfrac{5}{6} \div \dfrac{2}{1}$

$\qquad\quad\ = \dfrac{5}{6} \cdot \dfrac{1}{2} = \dfrac{5}{12}$

Thus, $\dfrac{5}{6} \div 2 = \dfrac{5}{12}$.

c. $18 \div \dfrac{2}{3} = \dfrac{18}{1} \div \dfrac{2}{3}$

$\qquad\quad\ = \dfrac{\overset{9}{\cancel{18}}}{1} \cdot \dfrac{3}{\underset{1}{\cancel{2}}}$

$\qquad\quad\ = \dfrac{27}{1} = 27$

Thus, $18 \div \dfrac{2}{3} = 27$.

d. $\dfrac{154}{155} \div 77 = \dfrac{154}{155} \div \dfrac{77}{1}$

$\qquad\quad\ = \dfrac{154}{155} \cdot \dfrac{1}{77}$

$\qquad\quad\ = \dfrac{154 \cdot 1}{155 \cdot 77}$

$\qquad\quad\ = \dfrac{2 \cdot \cancel{7} \cdot \cancel{11} \cdot 1}{5 \cdot 31 \cdot \cancel{7} \cdot \cancel{11}}$

$\qquad\quad\ = \dfrac{2}{155}$

Prime factorizations:

$\begin{array}{r}2\,\rvert\underline{154}\\ 7\,\rvert\underline{77}\\ 11\end{array}$ $\begin{array}{r}5\,\rvert\underline{155}\\ 31\end{array}$

Thus, $\dfrac{154}{155} \div 77 = \dfrac{2}{155}$.

Example 4   a. Divide $\dfrac{15}{28}$ by $\dfrac{9}{14}$.

$$\dfrac{15}{28} \div \dfrac{9}{14} = \dfrac{\overset{5}{\cancel{15}}}{\underset{2}{\cancel{28}}} \cdot \dfrac{\overset{1}{\cancel{14}}}{\underset{3}{\cancel{9}}}$$

$$= \dfrac{5}{6}$$

b. Divide $\dfrac{82}{115}$ into $\dfrac{123}{46}$.

$$\dfrac{123}{46} \div \dfrac{82}{115} = \dfrac{123}{46} \cdot \dfrac{115}{82}$$

$$= \dfrac{3 \cdot \overset{1}{\cancel{41}} \cdot 5 \cdot \overset{1}{\cancel{23}}}{2 \cdot \underset{1}{\cancel{23}} \cdot 2 \cdot \underset{1}{\cancel{41}}}$$

$$= \dfrac{15}{4}$$

c. Find the quotient if $\frac{28}{25}$ is divided by $\frac{35}{15}$.

$$\frac{28}{25} \div \frac{35}{15} = \frac{\overset{4}{\cancel{28}}}{\underset{5}{\cancel{25}}} \cdot \frac{\overset{3}{\cancel{15}}}{\underset{1}{\cancel{35}}}$$

$$= \frac{12}{25}$$

2.5 Exercises

In Exercises 1–30, divide.

1. $\frac{4}{5} \div \frac{7}{11}$ 2. $\frac{2}{3} \div \frac{7}{5}$ 3. $\frac{7}{8} \div \frac{4}{5}$

4. $\frac{2}{3} \div 5$ 5. $\frac{4}{5} \div \frac{4}{5}$ 6. $\frac{2}{3} \div \frac{3}{2}$

7. $\frac{4}{5} \div 3$ 8. $\frac{3}{8} \div \frac{6}{7}$ 9. $\frac{17}{8} \div \frac{17}{8}$

10. $\frac{5}{6} \div \frac{3}{42}$ 11. $\frac{1}{6} \div \frac{5}{3}$ 12. $\frac{6}{7} \div \frac{1}{14}$

13. $15 \div \frac{3}{5}$ 14. $\frac{10}{19} \div \frac{14}{25}$ 15. $\frac{18}{5} \div \frac{1}{5}$

16. $35 \div \frac{10}{21}$ 17. $\frac{16}{27} \div \frac{18}{24}$ 18. $\frac{4}{23} \div \frac{8}{69}$

19. $\frac{15}{27} \div 25$ 20. $\frac{5}{18} \div \frac{10}{12}$ 21. $\frac{4}{9} \div \frac{20}{36}$

22. $\frac{5}{6} \div \frac{13}{6}$ 23. $\frac{17}{35} \div 1$ 24. $1 \div \frac{52}{18}$

25. $\frac{17}{8} \div \frac{34}{8}$ 26. $\frac{16}{18} \div \frac{22}{24}$ 27. $\frac{16}{18} \div \frac{24}{22}$

28. $250 \div \frac{2}{5}$ 29. $\frac{2}{5} \div 250$ 30. $\frac{849}{3} \div 3$

31. What is $\frac{33}{26}$ divided by $\frac{22}{91}$?

32. $\frac{36}{35} \div \frac{66}{65} = ?$

33. If a patient is to consume 15 oz of a liquid by taking $\frac{1}{2}$ oz at a time, how many times will the patient take the liquid?

34. A recipe for caramel frosting calls for 3 C sugar, 1 C milk, $\frac{3}{4}$ C butter, $\frac{1}{8}$ tsp soda, and 1 tsp vanilla. If $\frac{1}{2}$ of the recipe is to be made, how much butter and soda will be needed?

35. $\frac{420}{65} \div 110 = ?$

36. If a piece of lead is $\frac{1}{15}$ inch wide, how many pieces would be needed to form a line $\frac{2}{3}$ of an inch wide?

37. $\frac{86}{115} \div \frac{34}{46} = ?$

38. What is $\frac{117}{53}$ divided by $\frac{9}{2}$?

39. A man and wife inherited $\frac{3}{5}$ interest in a pine straw company. For tax purposes, they divided their interest equally among their four children. What fractional part of the business does each child own?

40. $\frac{143}{429} \div \frac{156}{385} = ?$

41. Mrs. Green had 12 cups of azalea plant food. She was supposed to use $\frac{3}{4}$ cup on each plant. How many plants could she fertilize?

42. What is $\frac{204}{75}$ divided by $\frac{3}{4}$?

43. $\frac{85}{175} \div \frac{340}{25} = ?$

44. $\frac{345}{72} \div \frac{161}{8} = ?$

45. Three-fourths of a pound of candy is to be divided equally among 5 persons. How much will each person receive?

46. A seamstress had 15 yards of material. She wanted to make vests for the school's drill team. If each vest required $\frac{5}{8}$ yard, how many vests could she make from the 15 yards?

47. Find the quotient if 52 is divided by $\frac{1}{2}$.

48. A recipe calls for $\frac{1}{4}$ cup of butter. If $\frac{7}{4}$ cups are available, how many times can the recipe be made?

49. $\dfrac{25}{38} \div \dfrac{51}{209} = ?$

50. A 12-inch piece of paper is to be cut into strips $\dfrac{3}{8}$ of an inch wide. How many strips can be cut from the paper?

51. Divide $\dfrac{18}{5}$ into 90.

52. If the product of $\dfrac{429}{46}$ and $\dfrac{82}{65}$ is divided by $\dfrac{902}{115}$, what is the result?

53. A 60-acre plot of land is being divided into building sites. If each site is to contain $\dfrac{2}{3}$ of an acre, how many sites are possible?

54. Divide the sum $\dfrac{5}{6} + \dfrac{7}{8}$ by $\dfrac{5}{6}$.

55. A butcher is slicing a slab of bacon which is six inches thick into slices $\dfrac{3}{16}$ inch thick. How many slices will he get from the slab of bacon?

2.6 Least Common Denominator

In Section 2.3 we converted fractions to equivalent fractions. In order to add and subtract fractions in the next section, it will be necessary to convert fractions with different denominators to equivalent fractions with the same (or a common) denominator. Just as the reduced fraction $\dfrac{1}{2}$ has many equivalent forms $\left(\dfrac{2}{4}, \dfrac{3}{6}, \dfrac{111}{222}, \text{etc.}\right)$, fractions with different denominators have many common denominators. In particular, the fractions $\dfrac{3}{4}$ and $\dfrac{5}{6}$ have common denominators 12, 24, 36, 48, 24,000, etc. Of course, we would prefer to work with the simplest of these common denominators if possible. In the previous case, this would be 12. Fortunately this is always possible, and the common denominator which is the easiest to work with will be called the *least common denominator.*

To prepare us to calculate a least common denominator, we will first examine the concept of a least common multiple (LCM). The **least common multiple** of a set of natural numbers is the smallest natural number that is a multiple of each of the given numbers. For example, the multiples of 2 are 2, 4, 6, 8, 10, etc., and the multiples of 3 are 3, 6, 9, 12, etc. Clearly, the smallest number which is a multiple of both 3 and 2 is 6.

Example 1　a. Find the LCM of 4 and 6.

The multiples of 4 are 4, 8, 12, 16, etc.
The multiples of 6 are 6, 12, 18, 24, etc.
Therefore the LCM of 4 and 6 is 12.

b. Find the LCM of 24 and 40.

The multiples of 24 are 24, 48, 72, 96, 120, etc.
The multiples of 40 are 40, 80, 120, 160, 200, etc.
Therefore the LCM of 24 and 40 is 120.

You can see that this procedure could be cumbersome. The *rule for finding the LCM of a set of natural numbers* is much more systematic and is appropriate for any set of natural numbers.

RULE FOR FINDING THE LCM OF A SET OF NATURAL NUMBERS	EXAMPLE Find the LCM of 14, 40, and 50.
	Solution
1. Obtain the prime factorizations of each of the numbers.	$14 = 2 \cdot 7$ $40 = 2 \cdot 2 \cdot 2 \cdot 5$ $50 = 2 \cdot 5 \cdot 5$
2. Examine each factorization and list each factor the maximum number of times it occurs in any single factorization.	$2, 2, 2$ The prime 2 is listed 3 times since it occurs at most 3 times in any single factorization. $5, 5$ The prime 5 occurs at most twice. 7 The prime 7 occurs at most once.
3. The LCM is the product of the factors listed in step 2.	The LCM is $2 \cdot 2 \cdot 2 \cdot 5 \cdot 5 \cdot 7 = 1400$. *Note* that 1400 is a multiple of 14, 40, and 50.

Example 2   a. Find the LCM of 4 and 6.

$$4 = 2 \cdot 2$$
$$6 = 2 \cdot 3$$
$$LCM = 2 \cdot 2 \cdot 3 = 12$$

Note that 4 and 6 both divide 12. Perhaps you could recognize that 12 is the LCM by inspection. Inspection is an acceptable technique, but it will be necessary to use the rule for some of the more difficult problems.

b. Find the LCM of 24 and 40.

$$24 = 2 \cdot 2 \cdot 2 \cdot 3$$
$$40 = 2 \cdot 2 \cdot 2 \cdot 5$$
$$LCM = 2 \cdot 2 \cdot 2 \cdot 3 \cdot 5 = 120$$

Is 120 a multiple of 24 and 40?

c. Find the LCM of 14, 20, 35, and 98.

$$14 = 2 \cdot 7$$
$$20 = 2 \cdot 2 \cdot 5$$
$$35 = 5 \cdot 7$$
$$98 = 2 \cdot 7 \cdot 7$$
$$LCM = 2 \cdot 2 \cdot 5 \cdot 7 \cdot 7 = 980$$

Is 980 a multiple of 14, 20, 35, and 98?

In order to add fractions, you must be able to find the least common denominator (LCD) of a set of fractions. Since the **least common denominator** of a set of fractions is the least common multiple of the denominators, we can use the preceding rule to find the LCD.

Example 3   Find the LCD of the following sets of fractions.

a. $\dfrac{3}{4}, \dfrac{5}{6}$

$$4 = 2 \cdot 2$$
$$6 = 2 \cdot 3$$
$$LCD = 2 \cdot 2 \cdot 3 = 12$$

b. $\dfrac{7}{16}, \dfrac{9}{24}, \dfrac{13}{18}$

$$16 = 2 \cdot 2 \cdot 2 \cdot 2$$
$$24 = 2 \cdot 2 \cdot 2 \cdot 3$$
$$18 = 2 \cdot 3 \cdot 3$$
$$LCD = 2 \cdot 2 \cdot 2 \cdot 2 \cdot 3 \cdot 3 = 144$$

2.6   Exercises

Find the LCM for the sets of whole numbers in Exercises 1–20.

1. 3 and 4

2. 5 and 9

3. 6 and 9

4. 4 and 6

5. 10 and 15

6. 4 and 8

7. 3 and 27

8. 12 and 20

9. 2, 3, and 5

10. 2, 7, and 11

11. 45 and 30

12. 121 and 33

13. 2, 4, and 8

14. 98 and 56

15. 30 and 75

16. 4, 6, and 15

17. 21 and 42

18. 21 and 43

19. 24, 45, and 50

20. 128 and 288

Find the LCD of the sets of fractions in Exercises 21–30.

21. $\dfrac{5}{6}, \dfrac{1}{18}$

22. $\dfrac{8}{49}, \dfrac{5}{21}$

23. $\dfrac{1}{3}, \dfrac{1}{4}, \dfrac{1}{6}$

24. $\dfrac{4}{13}, \dfrac{8}{39}$

25. $\dfrac{2}{33}, \dfrac{5}{77}$

26. $\dfrac{1}{2}, \dfrac{2}{3}, \dfrac{2}{9}$

27. $\dfrac{3}{54}, \dfrac{5}{36}$

28. $\dfrac{19}{34}, \dfrac{5}{51}$

29. $\dfrac{5}{40}, \dfrac{7}{50}$

30. $\dfrac{4}{27}, \dfrac{8}{45}, \dfrac{9}{50}$

2.7   Addition of Fractions

It is easy to add fractions which have the same denominator. The rectangles in Figure 2.3 illustrate the problem $\dfrac{2}{7} + \dfrac{4}{7} = \dfrac{6}{7}$.

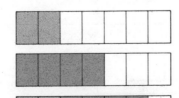

$\dfrac{2}{7}$ of the figure is shaded.

$\dfrac{4}{7}$ of the figure is shaded.

The two shaded regions are shown on the same figure and we see that $\dfrac{6}{7}$ of the figure is shaded.

Figure 2.3

When adding fractions with the same denominator, retain this common denominator as the denominator of the result and add the numerators to obtain the numerator of the result. If the resulting fraction is not in lowest terms, it should be reduced.

Example 1   Find the following sums.

a. $\dfrac{3}{7} + \dfrac{3}{7} = \dfrac{3+3}{7}$

$\qquad\qquad = \dfrac{6}{7}$

b. $\dfrac{2}{3} + \dfrac{1}{3} = \dfrac{2+1}{3}$

$\qquad\qquad = \dfrac{3}{3}$

$\qquad\qquad = 1$

c. $\dfrac{7}{16} + \dfrac{2}{16} + \dfrac{1}{16} = \dfrac{7+2+1}{16}$

$\qquad\qquad\qquad = \dfrac{10}{16}$

$\qquad\qquad\qquad = \dfrac{5}{8}$

d. $\dfrac{4}{153} + \dfrac{11}{153} = \dfrac{4+11}{153}$ Prime factorization:

$\qquad\qquad\qquad = \dfrac{15}{153}$ $\begin{array}{r} 3\,\underline{|\,153} \\ 3\,\underline{|\,51} \\ 17 \end{array}$

$\qquad\qquad\qquad = \dfrac{\cancel{3} \cdot 5}{\cancel{3} \cdot 3 \cdot 17}$

$\qquad\qquad\qquad = \dfrac{5}{51}$

To add fractions which do not have the same denominator, we must convert them to equivalent fractions which do have the same denominator. Consider the following example.

Example 2 $\dfrac{1}{4} + \dfrac{1}{6} = ?$

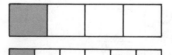 $\dfrac{1}{4}$ of the figure is shaded.

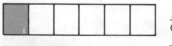 $\dfrac{1}{6}$ of the figure is shaded.

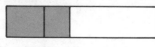   The parts indicated in this figure are not equal so we cannot represent the sum at this point.

However, since $\dfrac{1}{4} = \dfrac{3}{12}$ and $\dfrac{1}{6} = \dfrac{2}{12}$ we have

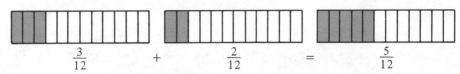

$$\dfrac{3}{12} \qquad + \qquad \dfrac{2}{12} \qquad = \qquad \dfrac{5}{12}$$

Thus, $\dfrac{1}{4} + \dfrac{1}{6} = \dfrac{3}{12} + \dfrac{2}{12} = \dfrac{5}{12}$.

In Example 2, we could have expressed $\dfrac{1}{4}$ as $\dfrac{6}{24}$ and $\dfrac{1}{6}$ as $\dfrac{4}{24}$ and added as follows:

$$\dfrac{1}{4} + \dfrac{1}{6} = \dfrac{6}{24} + \dfrac{4}{24}$$

$$= \dfrac{10}{24}$$

$$= \dfrac{5}{12}$$

However, it is usually more convenient to express the fractions in terms of their LCD, which in the example above was 12. Thus, if we are adding fractions which do not have the same denominator, we must find the lowest common denominator (see Section 2.6), and then convert each fraction to an equivalent fraction having the LCD as its denominator (see Section 2.3). Recall that we obtain these equivalent fractions by multiplying the numerator and denominator by the appropriate building factor.

Example 3 $\dfrac{5}{6} + \dfrac{4}{15} = ?$

First, find the LCD. By inspection, we see that the LCD is 30. Next rewrite each fraction as an equivalent fraction with a denominator of 30 and then complete the addition.

$$\dfrac{5}{6} + \dfrac{4}{15} = \dfrac{25}{30} + \dfrac{8}{30} \qquad \text{Building factor of 5 for } \dfrac{5}{6}. \text{ Building factor of 2 for } \dfrac{4}{15}.$$

$$= \dfrac{33}{30}$$

$$= \dfrac{11}{10}$$

Thus, $\dfrac{5}{6} + \dfrac{4}{15} = \dfrac{11}{10}$.

The procedure for adding fractions illustrated in Example 3 is given in the *rule for adding fractions.*

RULE FOR ADDING FRACTIONS	EXAMPLE $\dfrac{13}{18}+\dfrac{3}{14}=?$
	Solution:
1. If the fractions do not have the same denominator, calculate their LCD.	$18 = 2 \cdot 3 \cdot 3$ $14 = 2 \cdot 7$ $LCD = 2 \cdot 3 \cdot 3 \cdot 7 = 126$
2. Convert each fraction to an equivalent fraction with the LCD as the denominator.	$\dfrac{13}{18}=\dfrac{13}{18}\cdot\dfrac{7}{7}=\dfrac{91}{126}$ $\dfrac{3}{14}=\dfrac{3}{14}\cdot\dfrac{9}{9}=\dfrac{27}{126}$
3. The denominator of the result is the LCD. The numerator of the result is the sum of the numerators calculated in step 2.	$\dfrac{13}{18}+\dfrac{3}{14}=\dfrac{91}{126}+\dfrac{27}{126}$ $=\dfrac{91+27}{126}$ $=\dfrac{118}{126}$
4. If necessary, reduce the sum to lowest terms.	$\dfrac{118}{126}=\dfrac{59}{63}$
	Therefore, $\dfrac{13}{18}+\dfrac{3}{14}=\dfrac{59}{63}$.

Example 4   Add $\dfrac{5}{6}$ to $\dfrac{3}{4}$.

$$\dfrac{3}{4}+\dfrac{5}{6}=\dfrac{3\cdot 3}{4\cdot 3}+\dfrac{5\cdot 2}{6\cdot 2} \qquad \text{Find the LCD by inspection or as follows:}$$

$$\qquad\qquad\quad 6 = 2 \cdot 3$$
$$=\dfrac{9}{12}+\dfrac{10}{12} \qquad\qquad 4 = 2 \cdot 2$$
$$LCD = 2 \cdot 2 \cdot 3$$
$$=\dfrac{19}{12}$$

Example 5   Find the sum of $\dfrac{7}{16}$ and $\dfrac{9}{24}$.

$$\dfrac{7}{16}+\dfrac{9}{24}=\dfrac{7\cdot 3}{16\cdot 3}+\dfrac{9\cdot 2}{24\cdot 2} \qquad \begin{array}{l} 16 = 2 \cdot 2 \cdot 2 \cdot 2 \\ 24 = 2 \cdot 2 \cdot 2 \cdot 3 \\ LCD = 2 \cdot 2 \cdot 2 \cdot 2 \cdot 3 \\ \qquad = 48 \end{array}$$

$$=\dfrac{21}{48}+\dfrac{18}{48}$$

$$=\dfrac{39}{48}$$

Example 6 $\dfrac{3}{14}+\dfrac{2}{49}+\dfrac{5}{12}=?$ $\begin{array}{l} 14 = 2 \cdot 7 \\ 49 = 7 \cdot 7 \\ 12 = 2 \cdot 2 \cdot 3 \\ LCD = 2 \cdot 2 \cdot 3 \cdot 7 \cdot 7 = 588 \end{array}$

$$\frac{3}{14} = \frac{3 \cdot 42}{14 \cdot 42} = \frac{126}{588} \qquad \text{(building factor = 42)}$$

$$\frac{2}{49} = \frac{2 \cdot 12}{49 \cdot 12} = \frac{24}{588} \qquad \text{(building factor = 12)}$$

$$\frac{5}{12} = \frac{5 \cdot 49}{12 \cdot 49} = \frac{245}{588} \qquad \text{(building factor = 49)}$$

Thus,

$$\frac{3}{14} + \frac{2}{49} + \frac{5}{12} = \frac{126}{588} + \frac{24}{588} + \frac{245}{588}$$

$$= \frac{395}{588}$$

The factorizations of the denominators which are made in order to determine the LCD are also very useful in determining the building factors. In Example 6, to convert $\frac{3}{14}$ to an equivalent fraction with denominator 588, you must divide 588 by 14 in order to determine the building factor. This can be accomplished by comparing the prime factorizations of the two numbers.

$$588 \div 14 = \frac{588}{14} = \frac{2 \cdot \cancel{2} \cdot 3 \cdot \cancel{7} \cdot 7}{\cancel{2} \cdot \cancel{7}} = 42$$

The building factor is the product of the factors of 588 which remain after the factors of 14 are deleted; thus the building factor is $2 \cdot 3 \cdot 7 = 42$. To convert $\frac{2}{49}$ to an equivalent fraction with denominator 588, divide 49 into 588 by deleting the factors 7 and 7 from the factors of 588. The building factor is therefore $2 \cdot 2 \cdot 3 = 12$. In the last case, the building factor of $7 \cdot 7 = 49$ can be obtained by a similar process.

Addition problems can be worked in a horizontal or vertical format. The vertical format is illustrated in Example 7.

Example 7 $\dfrac{3}{4} + \dfrac{5}{6} + \dfrac{2}{15} = ?$

$$
\begin{array}{ll}
\dfrac{3}{4} = \dfrac{45}{60} & 4 = 2 \cdot 2 \\[2mm]
\dfrac{5}{6} = \dfrac{50}{60} & 6 = 2 \cdot 3 \\[2mm]
& 15 = 3 \cdot 5 \\[1mm]
+\dfrac{2}{15} = \dfrac{8}{60} & \text{LCD} = 2 \cdot 2 \cdot 3 \cdot 5 \\[1mm]
\overline{\phantom{+\dfrac{2}{15} =}\ \dfrac{103}{60}} & \phantom{\text{LCD} } = 60
\end{array}
$$

Thus, $\dfrac{3}{4} + \dfrac{5}{6} + \dfrac{2}{15} = \dfrac{103}{60}$.

2.7 Exercises

In Exercises 1–34, add.

1. $\dfrac{5}{8}+\dfrac{1}{8}$

2. $\dfrac{2}{9}+\dfrac{5}{9}$

3. $\dfrac{1}{5}+\dfrac{2}{5}+\dfrac{3}{5}$

4. $\dfrac{4}{13}+\dfrac{11}{13}$

5. $\dfrac{19}{15}+\dfrac{2}{15}$

6. $\dfrac{5}{8}+\dfrac{3}{8}$

7. $\dfrac{2}{3}+\dfrac{2}{3}+\dfrac{2}{3}$

8. $\dfrac{8}{21}+\dfrac{6}{21}$

9. $\dfrac{4}{7}+\dfrac{2}{7}+\dfrac{1}{7}$

10. $\dfrac{7}{10}+\dfrac{3}{10}$

11. $\dfrac{1}{2}+\dfrac{1}{3}$

12. $\dfrac{5}{3}+\dfrac{3}{5}$

13. $\dfrac{2}{5}+\dfrac{6}{7}$

14. $\dfrac{2}{7}+\dfrac{2}{11}$

15. $\dfrac{10}{11}+\dfrac{1}{2}$

16. $\dfrac{2}{3}+\dfrac{3}{5}+\dfrac{4}{7}$

17. $\dfrac{6}{7}+\dfrac{1}{2}$

18. $\dfrac{1}{2}+\dfrac{1}{5}$

19. $\dfrac{1}{2}+\dfrac{2}{3}+\dfrac{3}{5}$

20. $\dfrac{3}{4}+\dfrac{3}{7}$

21. $\dfrac{3}{8}+\dfrac{7}{3}$

22. $\dfrac{3}{4}+\dfrac{2}{3}+\dfrac{5}{6}$

23. $\dfrac{1}{6}+\dfrac{8}{5}$

24. $\dfrac{2}{3}+\dfrac{1}{2}$

25. $\dfrac{3}{2}+\dfrac{2}{9}+\dfrac{5}{6}$

26. $\dfrac{2}{5}+\dfrac{1}{9}$

27. $\dfrac{11}{10}+\dfrac{10}{11}$

28. $\dfrac{3}{4}+\dfrac{8}{11}$

29. $\dfrac{11}{15}+\dfrac{1}{4}$

30. $\dfrac{7}{10}+\dfrac{4}{9}$

31. $\dfrac{1}{3}+\dfrac{3}{4}$

32. $\dfrac{6}{5}+\dfrac{1}{9}$

33. $\dfrac{5}{6}+\dfrac{7}{9}$

34. $\dfrac{3}{4}+\dfrac{5}{6}$

35. What is $\dfrac{7}{10}$ increased by $\dfrac{13}{15}$?

36. $\dfrac{1}{4}+\dfrac{3}{8}=?$

37. Find $\dfrac{1}{3}$ plus $\dfrac{1}{27}$.

38. $\dfrac{5}{12}+\dfrac{7}{20}=?$

39. $\dfrac{1}{2}+\dfrac{1}{3}+\dfrac{1}{5}=?$

40. $\dfrac{3}{2}+\dfrac{4}{7}+\dfrac{2}{11}=?$

41. Find $\dfrac{49}{30}$ more than $\dfrac{4}{45}$.

42. $\dfrac{13}{121}+\dfrac{7}{33}=?$

43. $\dfrac{1}{2}+\dfrac{3}{4}+\dfrac{7}{8}=?$

44. What is $\dfrac{4}{98}$ more than $\dfrac{5}{56}$?

45. $\dfrac{1}{30}+\dfrac{1}{75}=?$

46. $\dfrac{7}{4}+\dfrac{23}{6}+\dfrac{17}{15}=?$

47. $\dfrac{4}{21}+\dfrac{1}{42}=?$

48. Increase $\dfrac{43}{21}$ by $\dfrac{21}{43}$.

49. $\dfrac{25}{24}+\dfrac{44}{45}+\dfrac{41}{50}=?$

50. What is $\dfrac{81}{128}$ plus $\dfrac{101}{288}$?

51. What is the sum of $\dfrac{3}{7}$ and $\dfrac{5}{7}$ and $\dfrac{1}{7}$?

52. A recipe calls for $\dfrac{1}{4}$ teaspoon basil, $\dfrac{1}{4}$ teaspoon oregano, and $\dfrac{1}{8}$ teaspoon garlic powder. What is the combined amount of these spices?

53. Approximately $\dfrac{1}{5}$ of a person's salary is withheld each month for federal taxes, $\dfrac{1}{25}$ for state taxes, $\dfrac{3}{50}$ for social security, and $\dfrac{3}{50}$ for a company retirement fund. What fractional part of the salary is withheld?

54. A rectangular piece of paper measures $\dfrac{1}{2}$ foot by $\dfrac{3}{4}$ foot. What is the total distance around the outside of this piece of paper?

55. A dollmaker was buying lace trim for a dress for an antique doll. Since the trim was $16.00 a yard, she wanted to buy the exact amount needed. If she needs $\dfrac{1}{8}$ yard for each sleeve, $\dfrac{3}{8}$ yard for the skirt, and $\dfrac{1}{8}$ yard for the neck, how much should she buy? How much will the trim cost?

56. Find $\dfrac{5}{6}$ of the sum of $\dfrac{7}{11}+\dfrac{1}{3}$.

57. A general store has three partially-filled barrels of apples. One is $\dfrac{3}{8}$ full, another $\dfrac{1}{3}$ full, and the third is $\dfrac{1}{4}$ full. If all the apples were placed in one barrel, what fractional part of the barrel would be filled?

58. A camper ate $\dfrac{1}{8}$ pound of fish for breakfast, $\dfrac{1}{4}$ pound for lunch, and $\dfrac{3}{8}$ pound for dinner. What is the total amount of fish he consumed at these three meals?

59. A cake was cut into 12 slices. Mary ate $\dfrac{1}{4}$ of the cake, Jim ate $\dfrac{1}{6}$ of it, and John ate $\dfrac{1}{2}$ of it. What part of the cake has been eaten?

60. The All American High School Band Booster Club decided to have a bake sale to raise money. One member volunteered to furnish a spice cake and a dozen cupcakes. The cake required $\dfrac{3}{4}$ cup of brown sugar, and the icing for the cupcakes called for $\dfrac{1}{3}$ cup of brown sugar. She had a 2-cup box of brown sugar, but $\dfrac{2}{3}$ of it had been used. How much brown sugar did she need? Did she have enough?

2.8 Subtraction of Fractions

Fractions that have the same denominator may be subtracted in much the same way that they are added. *To subtract fractions with the same denominator, find the difference in the numerators, and place this difference over the common denominator.*

Example 1 Subtract.

a.
$$\frac{9}{14}$$
$$-\frac{3}{14}$$
$$\frac{6}{14}=\frac{3}{7}$$

b. $\dfrac{11}{153}-\dfrac{4}{153}=\dfrac{11-4}{153}$

$$=\frac{7}{153}$$

If the fractions do not have the same denominator, then they must be converted to equivalent fractions having a common denominator and then subtracted.

Example 2 a. $\dfrac{2}{3}-\dfrac{3}{5}=\,?$

$$\frac{2}{3}-\frac{3}{5}=\frac{2\cdot 5}{3\cdot 5}-\frac{3\cdot 3}{5\cdot 3}$$

$$=\frac{10}{15}-\frac{9}{15}$$

$$=\frac{1}{15}$$

b.
$$\frac{7}{8}=\frac{7}{8}$$
$$-\frac{1}{4}=-\frac{2}{8}$$
$$\frac{5}{8}$$

The procedure is summarized in the *rule for subtracting fractions.*

RULE FOR SUBTRACTING FRACTIONS	*EXAMPLE* $\dfrac{13}{18} - \dfrac{3}{14} = ?$

Solution:

1. If the fractions do not have the same denominator, calculate their LCD.

$$18 = 2 \cdot 3 \cdot 3$$
$$14 = 2 \cdot 7$$
$$\text{LCD} = 2 \cdot 3 \cdot 3 \cdot 7 = 126$$

2. Convert each fraction to an equivalent fraction with the LCD as the denominator.

$$\frac{13}{18} = \frac{13}{18} \cdot \frac{7}{7} = \frac{91}{126}$$

$$\frac{3}{14} = \frac{3}{14} \cdot \frac{9}{9} = \frac{27}{126}$$

3. The denominator of the result is the LCD. The numerator of the result is the difference of the numerators calculated in step 2.

$$\frac{13}{18} - \frac{3}{14} = \frac{91}{126} - \frac{27}{126}$$

$$= \frac{91 - 27}{126}$$

$$= \frac{64}{126}$$

4. If necessary, reduce the difference to lowest terms.

$$= \frac{32}{63}$$

Therefore, $\dfrac{13}{18} - \dfrac{3}{14} = \dfrac{32}{63}$.

Example 3   Subtract $\dfrac{4}{39}$ from $\dfrac{5}{26}$.

$$\frac{5}{26} = \frac{5 \cdot 3}{26 \cdot 3} = \frac{15}{78} \qquad 26 = 2 \cdot 13$$
$$-\frac{4}{39} = -\frac{4 \cdot 2}{39 \cdot 2} = \frac{8}{78} \qquad 39 = 3 \cdot 13$$
$$\text{LCD} = 2 \cdot 3 \cdot 13 = 78$$
$$\frac{7}{78}$$

Example 4   a. What is 6 minus $\dfrac{15}{7}$?

$$6 - \frac{15}{7} = \frac{6}{1} - \frac{15}{7}$$

$$= \frac{42}{7} - \frac{15}{7}$$

$$= \frac{27}{7}$$

b. Find the difference if 2 is subtracted from $\dfrac{15}{7}$.

$$\frac{15}{7} - 2 = \frac{15}{7} - \frac{2}{1}$$

$$= \frac{15}{7} - \frac{14}{7}$$

$$= \frac{1}{7}$$

Example 5 $\dfrac{5}{6} - \dfrac{3}{10} + \dfrac{4}{15} = ?$

$$\dfrac{25}{30} - \dfrac{9}{30} + \dfrac{8}{30} = \dfrac{25 - 9 + 8}{30} \qquad 6 = 2 \cdot 3$$
$$10 = 2 \cdot 5$$
$$15 = 3 \cdot 5$$
$$= \dfrac{24}{30} = \dfrac{4}{5} \qquad \text{LCD} = 2 \cdot 3 \cdot 5 = 30$$

2.8 Exercises

In Exercises 1–31, subtract.

1. $\dfrac{5}{7} - \dfrac{2}{7}$

2. $\dfrac{18}{11} - \dfrac{14}{11}$

3. $\dfrac{8}{9} - \dfrac{7}{9}$

4. $\dfrac{5}{8} - \dfrac{3}{8}$

5. $\dfrac{13}{15} - \dfrac{10}{15}$

6. $\dfrac{7}{10} - \dfrac{3}{10}$

7. $\dfrac{19}{15} - \dfrac{4}{15}$

8. $\dfrac{24}{24} - \dfrac{8}{24}$

9. $\dfrac{7}{100} - \dfrac{6}{100}$

10. $\dfrac{43}{75} - \dfrac{11}{75}$

11. $\dfrac{3}{4} - \dfrac{2}{3}$

12. $\dfrac{10}{11} - \dfrac{1}{2}$

13. $\dfrac{5}{3} - \dfrac{3}{5}$

14. $7 - \dfrac{10}{11}$

15. $\dfrac{4}{5} - \dfrac{1}{7}$

16. $\dfrac{5}{6} - \dfrac{3}{5}$

17. $8 - \dfrac{2}{3}$

18. $\dfrac{1}{5} - \dfrac{1}{7}$

19. $\dfrac{3}{8} - \dfrac{1}{7}$

20. $\dfrac{1}{2} - \dfrac{2}{5}$

21. $\dfrac{17}{4} - 3$

22. $\dfrac{8}{5} - \dfrac{1}{6}$

23. $\dfrac{1}{9} - \dfrac{1}{10}$

24. $\dfrac{15}{2} - 7$

25. $\dfrac{7}{11} - \dfrac{2}{7}$

26. $\dfrac{7}{9} - \dfrac{1}{2}$

27. $\dfrac{3}{4} - \dfrac{5}{9}$

28. $\dfrac{3}{4} - \dfrac{3}{7}$

29. $\dfrac{1}{2} - \dfrac{1}{3}$

30. $\dfrac{3}{4} - \dfrac{8}{11}$

31. $\dfrac{7}{4} - \dfrac{7}{6} = ?$

32. If $\dfrac{1}{3}$ is subtracted from $\dfrac{11}{27}$, what is the result?

33. $\dfrac{5}{6}$ less $\dfrac{3}{10} = ?$

34. $\dfrac{8}{9} - \dfrac{2}{3} = ?$

35. Suzie wanted to lose 2 inches in her waist in 3 weeks. She lost $\dfrac{1}{2}$ inch the first week, $\dfrac{2}{3}$ inch the second week, and $\dfrac{3}{4}$ inch the third week. How much did she lose in all? Did she reach her goal?

36. $\dfrac{5}{12} - \dfrac{7}{18} = ?$

37. The wholesale price of gas went up $\dfrac{2}{7}$ of a cent and then went down $\dfrac{1}{9}$ of a cent. What was the net increase?

38. What is 5 less than $\dfrac{26}{5}$?

39. $\dfrac{5}{6} - \dfrac{3}{16} = ?$

40. $\dfrac{19}{15} - \dfrac{2}{9} = ?$

41. $\dfrac{41}{42} - \dfrac{7}{21} = ?$

42. $\dfrac{71}{63} - \dfrac{2}{7} = ?$

43. A part used in a manufacturing plant is supposed to measure $\dfrac{7}{8}$ of an inch in diameter. If a tolerance of $\dfrac{1}{1000}$ of an inch was allowed by the manufacturer, what is the shortest length the diameter could measure? What is the longest length it could measure?

44. $\dfrac{7}{44}$ less $\dfrac{1}{8} = ?$

45. What is $\dfrac{6}{41}$ minus $\dfrac{1}{7}$?

46. A recipe calls for $\dfrac{5}{8}$ cup of sugar. If Mary has $\dfrac{1}{3}$ cup on hand, how much more does she need in order to make the recipe?

47. What is 6 less than $\dfrac{41}{6}$?

48. $\dfrac{7}{16} - \dfrac{9}{24} = ?$

49. If three-fifths of the students in a class made A's, what fractional part of the class did not make A's?

50. $\dfrac{5}{14} - \dfrac{5}{18} = ?$

51. $\dfrac{16}{33}$ less $\dfrac{4}{15} = ?$

52. What is $\dfrac{5}{13}$ minus $\dfrac{2}{11}$?

53. $\dfrac{25}{24} - \dfrac{44}{45} = ?$

54. The length of a rectangle is $\dfrac{7}{8}$ of a yard while the width is $\dfrac{9}{10}$ of a yard. How much longer is the width than the length?

55. $\dfrac{13}{18} - \dfrac{9}{24} = ?$

56. If $\dfrac{7}{16}$ of the graduates of Monroe High School go to college, what fractional part do not go to college?

57. A doctor has ordered a patient to walk a mile a day.

At different times during the day, she has walked $\frac{1}{4}$ mile, $\frac{3}{8}$ mile, and $\frac{1}{8}$ mile. How much farther must she walk in order to cover one mile?

58. $\frac{7}{30} - \frac{4}{105} = ?$

59. If $\frac{5}{42}$ is subtracted from the sum of $\frac{8}{7}$ and $\frac{5}{6}$, what is the result?

60. If $\frac{3}{11}$ is subtracted from the sum of $\frac{5}{9}$ and $\frac{8}{33}$, what is the result?

Important Terms Used in this Chapter		
building factor	least common multiple
common factor	lowest terms
composite number	multiple
denominator	multiplicative inverse
divisible	natural numbers
divisibility	numerator
divisor	prime number
equivalent fractions	prime factorization
factor	reciprocal
fraction	reducing factor
least common denominator	reducing fractions

REVIEW EXERCISES

1. What is the least common multiple of 46 and 69?

2. What is the prime factorization of 259?

3. Which of the following fractions is equivalent to $\frac{5}{7}$?

$$\frac{25}{30} \qquad \frac{65}{150} \qquad \frac{85}{119}$$

4. What is the least common multiple of 10, 12, and 14?

5. Factor 345 into a product of primes.

6. Which of the following fractions is equivalent to $\frac{330}{390}$?

$$\frac{110}{120} \qquad \frac{30}{65} \qquad \frac{11}{13}$$

Replace the question marks with the correct values in Exercises 7-14.

7. $\frac{7}{13} = \frac{?}{143}$

8. $\frac{16}{56} = \frac{8}{?}$

9. $\frac{5}{11} = \frac{45}{?}$

10. $\frac{49}{301} = \frac{?}{43}$

11. $\frac{108}{177} = \frac{?}{59}$

12. $\frac{11}{13} = \frac{121}{?}$

13. $\frac{46}{69} = \frac{?}{3}$

14. $\frac{7}{17} = \frac{287}{?}$

15. Give 3 factors of 176.

16. Give 2 multiples of 176.

17. Give 4 factors of 75.

18. Give 3 multiples of 75.

In Exercises 19-23, perform the indicated operations and express all fractional answers in lowest terms.

19. $\frac{1}{2} \cdot \frac{2}{3}$

20. $\frac{2}{3} + \frac{1}{2}$

21. $\frac{2}{3} \div \frac{1}{2}$

22. $\frac{2}{3} - \frac{1}{2}$

23. $\frac{7}{6} + \frac{4}{39} - \frac{5}{26}$

24. What fractional part of 82 is 44?

25. $\frac{31}{65}$ increased by $\frac{7}{15} = ?$

26. A butcher has a piece of cheese $\frac{7}{8}$ inch thick, and from it he cuts a slice $\frac{3}{16}$ inch thick. What is the thickness of the remaining piece?

27. A $\frac{3}{8}$ share of a business is divided equally among five people. What fractional part does each of these five people receive?

28. $\dfrac{5}{6} \cdot \dfrac{3}{10} \cdot \dfrac{4}{15} = ?$ 29. $\dfrac{4}{3}$ less $\dfrac{2}{5} = ?$

30. $\dfrac{123}{63}$ more than $\dfrac{23}{42} = ?$

31. Add $\dfrac{5}{33}$ to $\dfrac{2}{3}$.

32. Four is what fractional part of ten?

33. John, Marie, Sue, and Jerry drank $\dfrac{1}{2}$ of a fifth of a gallon of tequila. What fractional part of a gallon did they drink?

34. When $\dfrac{51}{115}$ is divided by $\dfrac{34}{46}$, what is the quotient?

35. $\dfrac{5}{6} - \dfrac{3}{10} + \dfrac{4}{15} = ?$

36. Find two-thirds of four-sevenths.

37. A nurse administered a twelve-year-old child three-fourths of an adult dosage of a drug. If the adult dosage is 12 cubic centimeters, how much did she give the child?

38. $\dfrac{2}{7} \cdot \dfrac{5}{13} \cdot \dfrac{14}{25} \cdot \dfrac{13}{11} \cdot \dfrac{10}{4} = ?$ 39. $\dfrac{7}{9}$ less than $\dfrac{9}{7} = ?$

40. A recipe for six adults calls for $\dfrac{3}{4}$ tsp of salt. If this recipe is used to cook for four people, how much salt is necessary?

41. A man went to Las Vegas with $36,000 and lost five-sixths of it. How much does he still have?

42. A bartender has a 32-oz liquor bottle from which he dispenses drinks containing $\dfrac{4}{5}$ oz of liquor. How many drinks can be made from this bottle? If each drink costs $\dfrac{3}{4}$ of a dollar, what is the income produced by this bottle?

43. If $\dfrac{2}{7}$ of the price of gasoline selling for 56 cents per gallon is applied to taxes, how much tax per gallon is there?

44. Find the quotient $\dfrac{75}{175} \div \dfrac{121}{308}$.

45. A family is making a trip which will cover 5200 miles. After traveling for three days, they have covered $\dfrac{4}{13}$ of the distance. They have ten more days in which to complete the trip. How many miles per day should they average for each of the remaining ten days?

46. Find the sum of $\dfrac{21}{26}$ and $\dfrac{5}{39}$.

47. $\dfrac{17}{33}$ diminished by $\dfrac{5}{22} = ?$

48. A recipe for six people is to be reduced to serve four people. What fractional part of the original ingredients should be used?

*49. Subtract the sum $\dfrac{5}{7} + \dfrac{7}{8}$ from $\dfrac{15}{8}$, then multiply by $\dfrac{2}{3}$.

*50. Divide the product $\dfrac{4}{7} \cdot \dfrac{2}{3}$ by the sum $\dfrac{4}{7} + \dfrac{2}{3}$.

*51. $\dfrac{85}{86} \cdot \dfrac{86}{115} \div \dfrac{34}{46} = ?$

*52. Subtract the sum $\dfrac{5}{7} + \dfrac{7}{8}$ from the product $\dfrac{2}{3} \cdot \dfrac{15}{4}$.

*53. Divide $\dfrac{2}{3}$ by the difference obtained when $\dfrac{5}{9} \div \dfrac{5}{8}$ is subtracted from $\dfrac{15}{8}$.

*54. Divide the sum $\dfrac{4}{7} + \dfrac{2}{3}$ by the difference $\dfrac{3}{4} - \dfrac{5}{12}$.

*55. Multiply $\dfrac{9}{21}$ by the difference obtained when $\dfrac{7}{9}$ is subtracted from $\dfrac{7}{4} \div \dfrac{9}{16}$.

*These exercises are optional.

3

MIXED NUMBERS

OBJECTIVES

Upon completion of this chapter you should be able to:

1. Express a mixed number as an improper fraction.	3.1
2. Express an improper fraction as a mixed number.	3.1
3. Add two or more mixed numbers.	3.2
4. Subtract mixed numbers.	3.2
5. Multiply two or more mixed numbers.	3.3
6. Divide mixed numbers.	3.3
7. Work word problems involving addition, subtraction, multiplication, and division of mixed numbers.	3.2, 3.3

PRETEST

1. Express the following mixed numbers as improper fractions (Objective 1).

   a. $15\frac{3}{7}$ **b.** $11\frac{3}{26}$ **c.** $204\frac{2}{3}$

2. Express the following improper fractions as mixed numbers (Objective 2).

   a. $\frac{19}{5}$ **b.** $\frac{2046}{13}$ **c.** $\frac{508}{23}$

3. Add the following mixed numbers (Objective 3).

   a. $4\frac{5}{8} + 17\frac{7}{8}$ **b.** $5\frac{1}{2} + 3\frac{2}{3}$

   c. $13\frac{7}{10} + 4\frac{8}{15}$ **d.** $2\frac{5}{12} + 3\frac{7}{18} + 4\frac{11}{30}$

4. Perform the following subtractions (Objective 4).

   a. $22\frac{9}{17} - 8\frac{5}{17}$ **b.** $16\frac{5}{16} - 9\frac{9}{16}$

   c. $5\frac{3}{6} - 2\frac{7}{15}$ **d.** $4\frac{3}{26} - 1\frac{2}{39}$

5. Multiply the following mixed numbers (Objective 5).

   a. $9\frac{1}{5} \cdot 1\frac{22}{23}$ **b.** $1\frac{5}{6} \cdot 23\frac{47}{77}$

   c. $1\frac{1}{90} \cdot 3\frac{33}{49}$ **d.** $2\frac{4}{7} \cdot 5\frac{4}{9}$

6. Perform the following divisions (Objective 6).

   a. $3\frac{1}{5} \div 1\frac{3}{5}$ **b.** $6\frac{2}{7} \div 2\frac{5}{14}$

   c. $5\frac{1}{10} \div 7\frac{2}{7}$ **d.** $3\frac{1}{12} \div 8\frac{2}{9}$

7. Solve the following problems (Objective 7).

   a. A nurse planned to give each patient $1\frac{1}{2}$ ounces of a cold medicine. How many patients could be treated with a 36-ounce bottle?

   b. A new engine delivers one and three-sixteenths of the power of an old engine with the same input. If the old engine was rated at 320 horsepower, what is the rating of the new engine?

   c. A stone set in a ring weighs one and three-fourths carats. A companion setting in a necklace weighs five-eighths of a carat. How many carats are in these pieces of jewelry?

   d. It takes a carpenter $4\frac{3}{4}$ hours to install a window. How much longer must he work if he has already worked for $2\frac{7}{8}$ hours?

70

In Chapter 2, you learned to add, subtract, multiply, and divide fractions. You may have noticed that sometimes the numerator of a fraction was less than the denominator while at other times the numerator was greater than or equal to the denominator. When the numerator is less than the denominator, the fraction is called a **proper fraction**. When the numerator of a fraction is greater than or equal to its denominator, it is called an **improper fraction** even though there is nothing "improper" or incorrect about such a fraction. Nonetheless, it is sometimes more meaningful to express an improper fraction as an equivalent mixed number. A **mixed number** is the sum of a whole number and a proper fraction which is indicated by writing the whole number and fraction without the plus sign. For example, $3 + \frac{5}{7} = 3\frac{5}{7}$ and is read "three and five-sevenths."

Example 1 a. $\frac{2}{3}$ is a proper fraction. b. $\frac{5}{3}$ is an improper fraction.

c. $\frac{5}{6}$ is a proper fraction. d. $10\frac{2}{3}$ is a mixed number.

e. $\frac{14}{14}$ is an improper fraction. f. $2\frac{1}{3}$ is a mixed number.

3.1 Equivalent Form for Mixed Numbers and Improper Fractions

The procedures for changing an improper fraction to a mixed number and a mixed number to an improper fraction are basic skills which every student should master. It is also important to recognize which form is best for a particular problem, since both are correct.

To change $5\frac{2}{3}$ to an improper fraction, think of $5\frac{2}{3}$ as $5 + \frac{2}{3}$. Since $5 = \frac{5}{1}$,

$$5 + \frac{2}{3} = \frac{5}{1} + \frac{2}{3}$$

$$= \frac{15}{3} + \frac{2}{3}$$

$$= \frac{17}{3}$$

Thus, $5\frac{2}{3} = \frac{17}{3}$.

Note that the denominator of the improper fraction is 3 and that the denominator of the fractional part of the mixed number is also 3. The numerator of the improper fraction is 17, which is obtained from $5\frac{2}{3}$ by adding 2 to $5 \cdot 3$. This example illustrates the rationale for the *rule for changing a mixed number to an improper fraction.*

RULE FOR CHANGING A MIXED NUMBER TO AN IMPROPER FRACTION	EXAMPLE
	Change $5\frac{2}{3}$ to an improper fraction.
	Solution:
1. Multiply the whole number by the denominator of the fraction.	$5 \cdot 3 = 15$
2. Add this product to the numerator of the fractional part.	$15 + 2 = 17$
3. This sum is the numerator of the improper fraction. The denominator of the improper fraction is the same as the denominator of the fractional part.	$\frac{17}{3}$
	Therefore, $5\frac{2}{3} = \frac{17}{3}$.

Example 2   Change $6\frac{1}{2}$ to an improper fraction.

$$6\frac{1}{2} = \frac{12 + 1}{2} = \frac{13}{2}$$

Thus, $6\frac{1}{2} = \frac{13}{2}$.

Example 3   Express $8\frac{7}{9}$ as an improper fraction.

$$8\frac{7}{9} = \frac{72 + 7}{9} = \frac{79}{9}$$

Thus, $8\frac{7}{9} = \frac{79}{9}$.

Example 4   Convert $14\frac{5}{13}$ to an equivalent improper fraction.

$$14\frac{5}{13} = \frac{182 + 5}{13}$$

$$\begin{array}{r} 14 \\ \times 13 \\ \hline 42 \\ 14 \\ \hline 182 \end{array}$$

$$= \frac{187}{13}$$

Thus, $14\frac{5}{13} = \frac{187}{13}$.

In these examples, an intermediate step was shown. However, you should be able to do most problems without showing any intermediate steps.

To change $\frac{30}{7}$ to a mixed number, recall that $\frac{30}{7}$ can be interpreted as $30 \div 7$.

$$7\overline{)30}\ \ ^4$$

$$\underline{28}\quad\text{(the largest multiple of 7 contained in 30)}$$
$$2$$

$$\frac{30}{7} = \frac{28}{7} + \frac{2}{7} = 4 + \frac{2}{7}\quad\text{Therefore, } \frac{30}{7} = 4\frac{2}{7}.$$

This procedure can be shortened according to the *rule for changing an improper fraction to a mixed number.*

RULE FOR CHANGING AN IMPROPER FRACTION TO A MIXED NUMBER	EXAMPLE
	Change $\frac{30}{7}$ to a mixed number.
	Solution:
1. Divide the numerator by the denominator. (This division should be done mentally if possible.)	$7\overline{)30}\ ^4$ $\underline{28}$ 2
2. The quotient will become the whole number part of the mixed number.	4 is the quotient.
3. The remainder will become the numerator of the fractional part of the mixed number.	2 is the remainder.
4. The denominator of the improper fraction will also be the denominator of the fractional part of the mixed number.	Therefore, $\frac{30}{7} = 4\frac{2}{7}.$

Example 5   Change $\frac{14}{3}$ to a mixed number.

$$3\overline{)14}\ ^4$$
$$\underline{12}$$
$$2$$

Thus, $\frac{14}{3} = 4\frac{2}{3}.$

Example 6   Express $\frac{522}{48}$ as a mixed number.

$$48\overline{)522}\ ^{10}\qquad \frac{522}{48} = 10\frac{42}{48} = 10\frac{7}{8}$$
$$\underline{48}$$
$$42$$

Thus, $\frac{522}{48} = 10\frac{7}{8}.$

Note that a mixed number is not in simplest form unless the fractional part is a proper fraction in lowest terms.

Example 7   Convert $\frac{64}{4}$ to an equivalent mixed number.

$$
\begin{array}{r}
16 \\
4\overline{\smash{)}64} \\
\underline{4} \\
24 \\
\underline{24}
\end{array}
$$

Thus, $\frac{64}{4} = 16$.

When working with mixed numbers and improper fractions, you can use either form of the number. For example, the numerals $\frac{11}{4}$ and $2\frac{3}{4}$ name the same number, and neither one is considered simpler than the other. In a particular problem, you should choose the most appropriate form. Generally if a problem is stated in terms of mixed numbers, then your answer should be expressed with mixed numbers. Always express the mixed number in simplest form.

3.1 Exercises

Change the mixed numbers in Exercises 1-20 to improper fractions. Use the rule stated in this section.

Change the improper fractions in Exercises 21-40 to mixed numbers in simplest form.

1. $2\frac{1}{2}$ 2. $3\frac{2}{3}$ 21. $\frac{14}{3}$ 22. $\frac{29}{3}$

3. $4\frac{3}{4}$ 4. $7\frac{3}{8}$ 23. $\frac{32}{4}$ 24. $\frac{37}{4}$

5. $5\frac{3}{7}$ 6. $1\frac{15}{16}$ 25. $\frac{51}{2}$ 26. $\frac{60}{9}$

7. $10\frac{3}{4}$ 8. $6\frac{7}{11}$ 27. $\frac{42}{8}$ 28. $\frac{11}{3}$

9. $7\frac{8}{9}$ 10. $8\frac{4}{5}$ 29. $\frac{8}{7}$ 30. $\frac{15}{11}$

11. $13\frac{2}{9}$ 12. $7\frac{1}{18}$ 31. $\frac{135}{17}$ 32. $\frac{304}{28}$

13. $100\frac{15}{16}$ 14. $564\frac{1}{2}$ 33. $\frac{252}{18}$ 34. $\frac{271}{23}$

15. $12\frac{7}{10}$ 16. $9\frac{23}{100}$ 35. $\frac{568}{124}$ 36. $\frac{16}{15}$

17. $249\frac{1}{3}$ 18. $57\frac{8}{11}$ 37. $\frac{15}{4}$ 38. $\frac{364}{12}$

19. $15\frac{2}{5}$ 20. $1000\frac{16}{17}$ 39. $\frac{100}{25}$ 40. $\frac{1530}{152}$

3.2 Addition and Subtraction of Mixed Numbers

When adding and subtracting mixed numbers, you may use the mixed number form or the equivalent improper fraction form. However, it is usually preferable to use the mixed number form. The procedure for adding is given in the *rule for adding mixed numbers.*

RULE FOR ADDING MIXED NUMBERS	EXAMPLE $\quad 4\frac{5}{9} + 2\frac{2}{3} = ?$
1. If the fractional parts do not have the same denominator, express the mixed numbers as equivalent mixed numbers whose fractional parts have the same denominator.	*Solution:* $$4\frac{5}{9} + 2\frac{2}{3} = 4\frac{5}{9} + 2\frac{6}{9}$$

2. Find the sum of
  a. the whole number parts
  b. the fractional parts
	$$= 6\frac{11}{9}$$

3. Simplify the result.	$$= 7\frac{2}{9}$$

Example 1 $11\frac{4}{9} + 3\frac{1}{9} = ?$

$$11\frac{4}{9} + 3\frac{1}{9} = 14\frac{5}{9}$$

When working problems involving addition of mixed numbers, you can write the problem vertically or horizontally.

Example 2 a. Find the sum of $2\frac{1}{3}$ and $5\frac{2}{3}$.

$$2\frac{1}{3}$$
$$+5\frac{2}{3}$$
$$\overline{7\frac{3}{3}} = 7 + 1 = 8$$

The sum is 8.

Why wasn't the answer left as $7\frac{3}{3}$?

b. $11\frac{8}{11} + \frac{2}{11} = ?$

$$11\frac{8}{11} + \frac{2}{11} = 11\frac{10}{11}$$

c. $11\frac{8}{9} + 15 = ?$

$$11\frac{8}{9} + 15 = 26\frac{8}{9}$$

Example 3

$$7\frac{2}{3} = 7\frac{4}{6}$$
$$+3\frac{5}{6} = 3\frac{5}{6}$$
$$10\frac{9}{6} = 10 + 1\frac{3}{6}$$
$$= 11\frac{3}{6} \quad \text{or} \quad 11\frac{1}{2}$$

Note that $10\frac{9}{6}$ is not in simplest form. The fractional part of the mixed number must be a proper fraction in lowest terms.

Example 4 $3\frac{1}{3} + 4\frac{3}{4}$

$$3\frac{1}{3} + 4\frac{3}{4} = 3\frac{4}{12} + 4\frac{9}{12}$$
$$= 7\frac{13}{12}$$
$$= 7 + 1\frac{1}{12}$$
$$= 8\frac{1}{12}$$

This problem may also be worked as follows:

$3\frac{1}{3} + 4\frac{3}{4}$

$$3\frac{1}{3} + 4\frac{3}{4} = \frac{10}{3} + \frac{19}{4}$$
$$= \frac{40}{12} + \frac{57}{12}$$
$$= \frac{97}{12} \quad \text{or} \quad 8\frac{1}{12}$$

Note: The answer is expressed as a mixed number since the problem is stated in terms of mixed numbers.

Example 5 $3\frac{1}{2} + 4\frac{2}{3} + 10\frac{3}{4} = ?$

$$3\frac{1}{2} + 4\frac{2}{3} + 10\frac{3}{4} = 3\frac{6}{12} + 4\frac{8}{12} + 10\frac{9}{12}$$
$$= 17\frac{23}{12}$$

$$= 17 + 1\frac{11}{12}$$

$$= 18\frac{11}{12}$$

Subtraction of mixed numbers is similar to addition. You subtract the fractional parts and you subtract the whole number parts. However, remember that the order in which you subtract is important.

Example 6   Subtract $1\frac{2}{7}$ from $4\frac{5}{7}$.

$$
\begin{array}{r}
4\frac{5}{7} \\
-1\frac{2}{7} \\
\hline
3\frac{3}{7}
\end{array}
$$

Therefore, $1\frac{2}{7}$ subtracted from $4\frac{5}{7}$ equals $3\frac{3}{7}$.

Example 7 $5\frac{4}{5} - 3\frac{1}{2} = ?$

$$5\frac{4}{5} - 3\frac{1}{2} = 5\frac{8}{10} - 3\frac{5}{10}$$

$$= 2\frac{3}{10}$$

Thus, $5\frac{4}{5} - 3\frac{1}{2} = 2\frac{3}{10}$.

In order to subtract mixed numbers, the fractional part of the minuend must be greater than or equal to the fractional part of the subtrahend. How then do we subtract $4\frac{5}{7}$ from $10\frac{2}{7}$? We use a process called borrowing. We **borrow** 1 from the 10, then add it to $\frac{2}{7}$.

$$10\frac{2}{7} = 9 + 1 + \frac{2}{7}$$

$$= 9 + \frac{7}{7} + \frac{2}{7}$$

$$= 9 + \frac{9}{7}$$

$$= 9\frac{9}{7}$$

Thus, we can rewrite the problem as $9\frac{9}{7} - 4\frac{5}{7}$.

Note that the object in borrowing is to increase the fractional part of the minuend so that the subtraction can be performed. Examples 8, 9, and 10 will familiarize you with this concept.

Example 8   Consider each of the following numbers. Borrow 1 from the whole number part and add it to the fractional part.

a. $4\dfrac{3}{8} = 3 + 1 + \dfrac{3}{8}$ b. $15\dfrac{4}{5} = 14 + 1 + \dfrac{4}{5}$

$\phantom{4\dfrac{3}{8}} = 3 + \dfrac{8}{8} + \dfrac{3}{8}$ $\phantom{15\dfrac{4}{5}} = 14 + \dfrac{5}{5} + \dfrac{4}{5}$

$\phantom{4\dfrac{3}{8}} = 3 + \dfrac{11}{8}$ $\phantom{15\dfrac{4}{5}} = 14 + \dfrac{9}{5}$

$\phantom{4\dfrac{3}{8}} = 3\dfrac{11}{8}$ $\phantom{15\dfrac{4}{5}} = 14\dfrac{9}{5}$

Example 9   Write each of the following numbers as a mixed number with the indicated denominator.

a. $7 = 6\dfrac{?}{3}$ b. $17 = 16\dfrac{?}{10}$

$\ 7 = 6 + 1$ $\ 17 = 16 + 1$

$ = 6 + \dfrac{3}{3}$ $ = 16 + \dfrac{10}{10}$

$ = 6\dfrac{3}{3}$ $ = 16\dfrac{10}{10}$

With practice you should be able to do this mentally.

Example 10   Find the missing numerators.

a. $13\dfrac{2}{5} = 12\dfrac{?}{5}$

Answer: 7

b. $122\dfrac{7}{18} = 121\dfrac{?}{18}$

Answer: 25

c. $85 = 84\dfrac{?}{12}$

Answer: 12

d. $42\dfrac{1}{43} = 41\dfrac{?}{43}$

Answer: 44

The procedure for subtracting mixed numbers is given in the *rule for subtracting mixed numbers.*

RULE FOR SUBTRACTING MIXED NUMBERS	EXAMPLE A $\quad 4\frac{2}{3} - 2\frac{5}{12} = ?$	EXAMPLE B $\quad 4\frac{1}{7} - 2\frac{4}{7} = ?$
1. If the fractional parts do not have the same denominator, express the mixed numbers as equivalent mixed numbers whose fractional parts have the same denominator.	*Solution:* $$4\frac{2}{3} - 2\frac{5}{12} = 4\frac{8}{12} - 2\frac{5}{12}$$	
2. If the fractional part of the subtrahend is less than the fractional part of the minuend: a. Subtract the whole number parts and the fractional parts. b. Simplify the result.	$$= 2\frac{3}{12}$$ $$= 2\frac{1}{4}$$ Therefore, $4\frac{2}{3} - 2\frac{5}{12} = 2\frac{1}{4}.$	*Solution:*
3. If the fractional part of the subtrahend is larger than the fractional part of the minuend: a. In the minuend borrow 1 from the whole number part and add it to the fractional part. b. Subtract the whole number parts and the fractional parts. c. Simplify the result.		$$4\frac{1}{7} = 3 + 1 + \frac{1}{7}$$ $$= 3 + \frac{7}{7} + \frac{1}{7} = 3\frac{8}{7}$$ $$4\frac{1}{7} = \quad 3\frac{8}{7}$$ $$\underline{-2\frac{4}{7} = -2\frac{4}{7}}$$ $$1\frac{4}{7}$$ Therefore, $4\frac{1}{7} - 2\frac{4}{7} = 1\frac{4}{7}$

Example 11 $10\frac{1}{3}$ less $3\frac{2}{3} = ?$

$$10\frac{1}{3} = 9 + 1 + \frac{1}{3} = 9 + \frac{3}{3} + \frac{1}{3} = \quad 9\frac{4}{3}$$
$$- 3\frac{2}{3} = \qquad\qquad\qquad\qquad\quad -3\frac{2}{3}$$
$$6\frac{2}{3}$$

Therefore, $10\frac{1}{3}$ less $3\frac{2}{3}$ equals $6\frac{2}{3}.$

Example 12   8 decreased by $2\frac{5}{8}$ = ?

$$8 - 2\frac{5}{8} = 7\frac{8}{8} - 2\frac{5}{8}$$

$$= 5\frac{3}{8}$$

Example 13 $6\frac{3}{4} - 2$ = ?

$$6\frac{3}{4}$$
$$\underline{-2}$$
$$4\frac{3}{4}$$

Example 14   If $2\frac{3}{4}$ is subtracted from $5\frac{4}{9}$, find the difference.

$$5\frac{4}{9} = \ 5\frac{16}{36} = \ 4\frac{52}{36}$$
$$\underline{-2\frac{3}{4} = -2\frac{27}{36} = -2\frac{27}{36}}$$
$$2\frac{25}{36}$$

This problem may also be worked as follows:

$$5\frac{4}{9} - 2\frac{3}{4} = \frac{49}{9} - \frac{11}{4}$$

$$= \frac{196}{36} - \frac{99}{36}$$

$$= \frac{97}{36} \ \text{ or } \ 2\frac{25}{36}$$

3.2 Exercises

Perform the additions in Exercises 1–12. Express your answers as mixed numbers.

1. $5\frac{7}{27} + 3\frac{2}{27}$

2. $2\frac{2}{5} + 2\frac{2}{5}$

3. $17\frac{8}{15} + 5\frac{2}{15}$

4. $4\frac{1}{2} + 7\frac{1}{2}$

5. $18\frac{5}{9} + 4\frac{4}{9}$

6. $5\frac{1}{8} + 70\frac{3}{8}$

7. $2\frac{3}{8} + 5\frac{7}{8} + 10\frac{1}{8}$

8. $15\frac{7}{10} + \frac{7}{10}$

9. $15 + 17\frac{2}{3}$

10. $3\frac{7}{20} + 10\frac{17}{20}$

11. $18\frac{5}{11} + \frac{10}{11}$

12. $50\frac{5}{12} + 17$

Perform the subtractions in Exercises 13–20. Express your answers as mixed numbers.

13. $7\frac{5}{7} - 2\frac{2}{7}$

14. $14\frac{3}{4} - 10\frac{1}{4}$

15. $72\frac{9}{50} - 50\frac{1}{50}$

16. $9\frac{7}{18} - 8\frac{1}{18}$

17. $8\frac{2}{3} - 8\frac{1}{3}$

18. $53\frac{4}{5} - 40\frac{2}{5}$

19. $15\frac{9}{13} - 10\frac{7}{13}$

20. $17\frac{8}{15} - 5\frac{2}{15}$

Find the missing numerators in Exercises 21-32.

21. $9\frac{8}{13} = 8 + 1 + \frac{8}{13}$

$\quad = 8 + \frac{13}{13} + \frac{8}{13}$

$\quad = 8\frac{?}{13}$

22. $15\frac{7}{10} = 14 + 1 + \frac{7}{10}$

$\quad = 14 + \frac{10}{10} + \frac{7}{10}$

$\quad = 14\frac{?}{10}$

23. $152\frac{1}{4} = 151 + 1 + \frac{1}{4}$

$\quad = 151 + \frac{4}{4} + \frac{1}{4}$

$\quad = 151\frac{?}{4}$

24. $5\frac{6}{7} = 5 + 1 + \frac{6}{7}$

$\quad = 5 + \frac{7}{7} + \frac{6}{7}$

$\quad = 5\frac{?}{7}$

25. $17\frac{5}{9} = 16 + 1 + \frac{5}{9}$

$\quad = 16\frac{?}{9}$

26. $2\frac{9}{20} = 1 + 1 + \frac{9}{20}$

$\quad = 1\frac{?}{20}$

27. $20\frac{7}{30} = 19 + 1 + \frac{7}{30}$

$\quad = 19\frac{?}{30}$

28. $49\frac{1}{49} = 48 + 1 + \frac{1}{49}$

$\quad = 48\frac{?}{49}$

29. $5\frac{7}{16} = 4\frac{?}{16}$

30. $7\frac{11}{12} = 6\frac{?}{12}$

31. $25\frac{47}{48} = 24\frac{?}{48}$

32. $9\frac{18}{19} = 8\frac{?}{19}$

Perform the subtractions in Exercises 33-44. Express your answers as mixed numbers.

33. $4\frac{3}{7} - 1\frac{5}{7}$

34. $14\frac{1}{3} - 11\frac{2}{3}$

35. $19\frac{5}{16} - 5\frac{7}{16}$

36. $7\frac{5}{12} - 2\frac{11}{12}$

37. $72\frac{1}{4} - 50\frac{3}{4}$

38. $33\frac{2}{9} - 13\frac{8}{9}$

39. $9\frac{11}{13} - 8\frac{12}{13}$

40. $12\frac{7}{10} - 11\frac{9}{10}$

41. $10\frac{4}{5} - 7$

42. $72\frac{1}{4} - 42$

43. $13 - 4\frac{3}{4}$

44. $27 - 10\frac{7}{8}$

Perform the indicated operations in Exercises 45-51. Express your answers as mixed numbers.

45. Find the sum of $6\frac{1}{7}$, $5\frac{4}{7}$, and 10.

46. Subtract $5\frac{8}{9}$ from $15\frac{4}{9}$.

47. Subtract $9\frac{4}{11}$ from $15\frac{5}{11}$.

48. $249\frac{11}{18} - 3\frac{7}{18} = ?$

49. Add $7\frac{8}{15}$ and $\frac{13}{15}$.

50. Add $53\frac{4}{9}$ to $7\frac{4}{9}$.

51. What is $6\frac{7}{9}$ added to $7\frac{5}{63}$?

52. A child weighed $6\frac{5}{8}$ pounds at birth. How much weight had she gained when she entered the first grade if her weight was then 35 pounds?

53. $6\frac{2}{3}$ less $4\frac{1}{4} = ?$

54. Find the sum of 15, $23\frac{1}{7}$, and $18\frac{5}{9}$.

55. The Chicago Bears needed 10 yards to make a first down. On the first play they gained $4\frac{1}{2}$ yards. On the second play they lost 2 yards. On the crucial third down, a pass netted $8\frac{1}{2}$ yards. Did they make a first down on that pass play?

56. A door measures 30 inches by 79 inches; however, a tolerance of $\frac{1}{32}$ inch is allowed on the width and on the length. What are the maximum possible dimensions for the door?

57. $249\frac{4}{9} - 125\frac{7}{8} = ?$

58. A chef is baking a large wedding cake which calls for $4\frac{3}{8}$ pounds of flour. He only has $2\frac{2}{3}$ pounds. How much more flour does he need?

59. If the reported rainfall for each week in February is $\frac{2}{100}$ inch, $\frac{1}{4}$ inch, $1\frac{7}{8}$ inches, and 2 inches, what was the total rainfall for the month?

60. What is the perimeter of a rectangular lot which measures $142\frac{3}{8}$ feet by $75\frac{1}{2}$ feet?

61. $3\frac{5}{8}$ plus $2\frac{5}{28}$ = ?

62. Decrease $122\frac{3}{5}$ by $99\frac{2}{3}$.

63. Over a period of ten weeks a dieter had the following weekly weight losses: $8\frac{1}{2}$, $3\frac{3}{4}$, $2\frac{1}{4}$, 4, $\frac{1}{2}$, $3\frac{1}{2}$, $\frac{1}{2}$, 2, 2, and $1\frac{3}{4}$. What was his weight at the end of the ten weeks if his initial weight was 246 pounds?

64. What is $6\frac{3}{5}$ less than $7\frac{1}{2}$?

65. The world record for longest fingernails is held by a man in India. Three fingers on his right hand have nails of length $11\frac{1}{2}$ inches, $10\frac{3}{4}$ inches, and $10\frac{1}{3}$ inches. What is the total length of these three fingernails?

66. What is $7\frac{8}{15}$ take away $\frac{9}{10}$?

67. $3\frac{3}{8}$ minus $\frac{5}{16}$ = ?

68. A $5\frac{3}{4}$-pound fish loses $1\frac{1}{2}$ pounds in cooking. How much will the cooked fish weigh?

69. Molding of the following lengths was required to complete a room: $6\frac{1}{2}$ inches, $42\frac{3}{4}$ inches, 8 inches, $132\frac{1}{16}$ inches, 144 inches, $72\frac{1}{2}$ inches, and $30\frac{5}{8}$ inches. How many inches would be left from a total of 44 feet of molding?

70. From the sum of $13\frac{2}{7}$ and $15\frac{5}{7}$ subtract $6\frac{3}{11}$.

3.3   Multiplication and Division of Mixed Numbers

When adding and subtracting mixed numbers, you may use the mixed number forms or the equivalent improper fractions. However, mixed numbers should usually be changed to improper fractions before multiplying or dividing.

Example 1   Multiply $4\frac{2}{3}$ by $5\frac{1}{2}$.

$$4\frac{2}{3} \cdot 5\frac{1}{2} = \frac{\overset{7}{\cancel{14}}}{3} \cdot \frac{11}{\underset{1}{\cancel{2}}}$$

$$= \frac{77}{3}$$

$$= 25\frac{2}{3}$$

$$\begin{array}{r} 25 \\ 3\overline{)77} \\ \underline{6} \\ 17 \\ \underline{15} \\ 2 \end{array}$$

Thus $4\frac{2}{3} \cdot 5\frac{1}{2} = 25\frac{2}{3}$. The answer is expressed as a mixed number since the problem was expressed in terms of mixed numbers.

The procedure in Example 1 is given in the *rule for multiplying mixed numbers.*

RULE FOR MULTIPLYING MIXED NUMBERS	*EXAMPLE* $2\frac{2}{3} \cdot 1\frac{1}{4} = ?$
	Solution:
1. Change the mixed numbers to equivalent improper fractions.	$2\frac{2}{3} = \frac{8}{3}$
	$1\frac{1}{4} = \frac{5}{4}$
2. Multiply using the rule for multiplying fractions given in Section 2.5.	$\frac{\overset{2}{\cancel{8}}}{3} \cdot \frac{5}{\underset{1}{\cancel{4}}} = \frac{10}{3}$ or $3\frac{1}{3}$
3. Express the answer as a mixed number.	Therefore, $2\frac{2}{3} \cdot 1\frac{1}{4} = 3\frac{1}{3}$.

Example 2 $6\frac{1}{2}$ multiplied by $\frac{4}{9} = ?$

$$6\frac{1}{2} \cdot \frac{4}{9} = \frac{13}{\underset{1}{\cancel{2}}} \cdot \frac{\overset{2}{\cancel{4}}}{9}$$

$$= \frac{26}{9} \text{ or } 2\frac{8}{9}$$

Thus, $6\frac{1}{2} \cdot \frac{4}{9} = \frac{26}{9}$ or $2\frac{8}{9}$.

Example 3   Find the product of $7\frac{3}{4}$ and 10.

$$7\frac{3}{4} \cdot 10 = \frac{31}{4} \cdot 10$$

$$= \frac{31}{\underset{2}{\cancel{4}}} \cdot \frac{\overset{5}{\cancel{10}}}{1}$$

$$= \frac{155}{2}$$

$$= 77\frac{1}{2}$$

Thus, the product of $7\frac{3}{4}$ and 10 is $77\frac{1}{2}$.

You can divide mixed numbers if you are able to divide fractions. The procedure is given in the *rule for dividing mixed numbers*.

RULE FOR DIVIDING MIXED NUMBERS	EXAMPLE $\quad 3\frac{2}{9} \div 2\frac{1}{3} = ?$
	Solution:
1. Change the mixed numbers to equivalent improper fractions.	$3\frac{2}{9} = \frac{29}{9}, \quad 2\frac{1}{3} = \frac{7}{3}$
2. Divide using the rule for dividing fractions given in Section 2.5.	$\frac{29}{9} \div \frac{7}{3} = \frac{29}{\overset{}{\underset{3}{9}}} \cdot \frac{\overset{1}{3}}{7}$
	$= \frac{29}{21} \quad \text{or} \quad 1\frac{8}{21}$
3. Express the answer as a mixed number.	Therefore, $3\frac{2}{9} \div 2\frac{1}{3} = 1\frac{8}{21}$.

Example 4 $\quad 2\frac{1}{2} \div 1\frac{3}{5} = ?$

$$2\frac{1}{2} \div 1\frac{3}{5} = \frac{5}{2} \div \frac{8}{5}$$

$$= \frac{5}{2} \times \frac{5}{8}$$

$$= \frac{25}{16}$$

$$= 1\frac{9}{16}$$

Thus $2\frac{1}{2} \div 1\frac{3}{5} = 1\frac{9}{16}$.

Example 5 $\quad$ Find the quotient when $5\frac{1}{2}$ is divided by $\frac{5}{6}$.

$$5\frac{1}{2} \div \frac{5}{6} = \frac{11}{2} \div \frac{5}{6}$$

$$= \frac{11}{\underset{1}{2}} \times \frac{\overset{3}{6}}{5}$$

$$= \frac{33}{5} \quad \text{or} \quad 6\frac{3}{5}$$

Therefore, the quotient obtained when $5\frac{1}{2}$ is divided by $\frac{5}{6}$ is $6\frac{3}{5}$.

Example 6   a. Divide 23 into $5\frac{3}{4}$.

$$5\frac{3}{4} \div 23 = \frac{23}{4} \div \frac{23}{1}$$

$$= \frac{\overset{1}{\cancel{23}}}{4} \cdot \frac{1}{\underset{1}{\cancel{23}}}$$

$$= \frac{1}{4}$$

b. $14 \div 3\frac{1}{7} = ?$

$$14 \div 3\frac{1}{7} = 14 \div \frac{22}{7}$$

$$= \frac{14}{1} \div \frac{22}{7}$$

$$= \frac{\overset{7}{\cancel{14}}}{1} \times \frac{7}{\underset{11}{\cancel{22}}}$$

$$= \frac{49}{11}$$

$$= 4\frac{5}{11}$$

3.3 Exercises

Multiply in Exercises 1–10. Express answers as mixed numbers.

1. $2\frac{1}{2} \cdot 1\frac{3}{4}$

2. $3\frac{4}{5} \cdot 2\frac{1}{2}$

3. $\left(2\frac{2}{3}\right)\left(5\frac{1}{2}\right)$

4. $5\frac{7}{8} \cdot 6\frac{2}{3}$

5. $1\frac{3}{7} \cdot 2\frac{4}{5}$

6. $5\frac{1}{2} \cdot 5\frac{1}{2}$

7. $(4)\left(2\frac{1}{3}\right)$

8. $10\frac{2}{3} \cdot \frac{3}{16}$

9. $\frac{7}{8} \cdot 1\frac{3}{4}$

10. $\left(3\frac{4}{5}\right)(25)$

Divide in Exercises 11–20. Express answers as mixed numbers.

11. $2\frac{2}{3} \div 5\frac{1}{2}$

12. $4\frac{5}{7} \div 4\frac{5}{7}$

13. $7\frac{1}{2} \div 1\frac{3}{4}$

14. $5\frac{2}{3} \div 2\frac{2}{3}$

15. $1\frac{3}{7} \div 2\frac{4}{5}$

16. $2\frac{1}{2} \div 3\frac{4}{5}$

17. $4\frac{3}{4} \div 3\frac{1}{6}$

18. $\frac{15}{16} \div 3\frac{1}{25}$

19. $4\frac{3}{8} \div 4\frac{3}{8}$

20. $3\frac{1}{2} \div 3\frac{1}{2}$

Perform the indicated operations in Exercises 21–32. Express answers as mixed numbers.

21. $4 \div 2\frac{2}{3}$

22. $36 \cdot 1\frac{1}{5}$

23. $\left(2\frac{1}{8}\right)\left(2\frac{1}{8}\right)$

24. $6 \div 3\frac{1}{2}$

25. $3\frac{5}{6} \div \frac{5}{12}$

26. $3\frac{5}{6} \div 3\frac{5}{6}$

27. $\left(25\frac{7}{9}\right)\left(10\frac{1}{2}\right)$

28. $4\frac{5}{9} \div 10\frac{1}{3}$

29. $\frac{5}{12} \div 3\frac{5}{6}$

30. $1\frac{1}{2} \cdot 2\frac{3}{8}$

31. $2\frac{2}{3} \div 4$

32. $2\frac{3}{4} \cdot 1\frac{1}{4}$

33. $\left(3\frac{3}{4}\right)\left(\frac{4}{9}\right) = ?$

34. $\left(2\frac{7}{24}\right)\left(2\frac{1}{5}\right) = ?$

35. $3\frac{3}{4} \div 5\frac{1}{4} = ?$

36. Multiply $3\frac{3}{4}$ by $5\frac{1}{4}$.

37. Divide 47 into $15\frac{2}{3}$.

38. $16 \div 2\frac{3}{4} = ?$

39. Divide 15 by $2\frac{5}{8}$.

40. $\frac{4}{15}$ times $5\frac{5}{8} = ?$

41. What is the product of $2\frac{1}{7}$ and $3\frac{1}{3}$?

42. $6\frac{1}{3} \div 7\frac{2}{3} = ?$

43. A board $7\frac{1}{2}$ feet long will be cut into 9 equal pieces. How long will each of the 9 pieces be?

44. Find the quotient if $3\frac{2}{13}$ is divided by 41.

45. $3\frac{5}{16} \div 1\frac{5}{12} = ?$

46. $2\frac{1}{2} \cdot 3\frac{3}{4} \cdot 9\frac{1}{2} = ?$

47. 16 times $2\frac{3}{4} = ?$

48. The planet Mars makes a complete revolution about the sun in $1\frac{22}{25}$ earth years. How many revolutions did Mars complete from 1900 through 1946? (*Hint:* Number of years = 47.)

49. Over a period of ten weeks, a dieter had the following weekly weight losses: $8\frac{1}{2}$, $3\frac{3}{4}$, $2\frac{1}{4}$, 4, $\frac{1}{2}$, $3\frac{1}{2}$, $\frac{1}{2}$, 2, 2, and $1\frac{3}{4}$. What was his average weekly loss over this period?

50. Divide $3\frac{3}{4}$ by $\frac{4}{9}$.

51. $\frac{4}{9} \div 3\frac{3}{4} = ?$

52. How many square inches of glass would be needed to cover a picture which measures 10 inches by $12\frac{1}{2}$ inches? What would be the cost, to the nearest cent, if glass sells for $5\frac{1}{2}¢$ per square inch?

53. How many square yards of carpeting would be needed to carpet a room which measures $11\frac{1}{2}$ feet by 12 feet? (1 square yard = 9 square feet)

54. A decorator is making a curtain which requires $7\frac{3}{8}$ yards of material. How much material will she need for 8 such curtains?

55. $\left(3\frac{3}{4}\right)(4)\left(2\frac{3}{5}\right) = ?$

56. If $19\frac{3}{11}$ is doubled, what is the result?

57. $2\frac{1}{4} \div 1\frac{3}{11} = ?$

58. $4\frac{5}{9} \cdot 3\frac{4}{9} = ?$

59. $6\frac{3}{7}$ divided by $1\frac{1}{9} = ?$

60. $\frac{15}{17} \cdot 3\frac{1}{11} = ?$

61. How many days would it take you to clear a 21-acre lot if you can clear $1\frac{3}{4}$ acres per day?

62. Multiply $1\frac{1}{3}$ by $2\frac{1}{2}$.

63. Divide $4\frac{3}{5}$ into 16.

64. What is $18\frac{1}{3}$ multiplied by $6\frac{5}{9}$?

65. $32\frac{1}{2} \cdot 40\frac{4}{5} = ?$

66. If the reported rainfall for each week in July was $\frac{2}{100}$ inch, $\frac{1}{4}$ inch, $1\frac{7}{8}$ inches, and 2 inches, what was the average weekly rainfall in July?

67. The cost of roofing a certain house in 1968 was $768. In 1973 the cost of reroofing was $1\frac{1}{2}$ times the 1968 cost. How much did it cost in 1973?

68. Find the product of $6\frac{1}{2}$, $4\frac{2}{3}$, and $5\frac{2}{9}$.

69. Add $2\frac{3}{4}$ to the product of $6\frac{1}{2}$ and $4\frac{2}{3}$.

70. Mr. Wood owns a plot of land which contains $3\frac{1}{2}$ acres. After he purchases an additional parcel of land, his total acreage will be $1\frac{1}{2}$ times his original acreage. If he divides the land equally among his 3 children, how many acres will each child receive?

Important Terms Used in this Chapter   improper fraction
mixed number
proper fraction

REVIEW EXERCISES

Express the following as improper fractions.

1. $7\frac{2}{3}$

2. $12\frac{1}{4}$

Express the following as mixed numbers.

3. $\frac{239}{41}$

4. $\frac{71}{8}$

5. Find the sum of $24\frac{1}{9}$ and $36\frac{5}{9}$.

6. A recipe for cranberry crunch which serves 20 people requires $\frac{3}{8}$ cup of milk. How much milk would be required if the recipe is increased to serve 50 people?

7. Divide $4\frac{1}{8}$ by $\frac{7}{8}$.

8. $2\frac{1}{2} \cdot 2\frac{2}{5} \cdot 4\frac{1}{4} = ?$

9. What is $6\frac{3}{7}$ subtracted from 24?

10. In 1964, the population of the U.S. was about 190 million people. If the per capita consumption of beef that year was $106\frac{3}{5}$ pounds, how many million pounds of beef were consumed in 1964 in the U.S.?

11. Find the product of $6\frac{3}{8}$ and $3\frac{1}{17}$.

12. Subtract 3 from the quotient $10\frac{2}{5} \div 2\frac{1}{3}$.

13. How many pieces of ribbon $2\frac{1}{2}$ inches long can be cut from a piece of ribbon of length $282\frac{1}{2}$ inches?

14. What is $10\frac{2}{3}$ divided by 6?

15. Multiply 3 times the sum of $10\frac{2}{5}$ and $2\frac{1}{3}$.

16. $7\frac{5}{9} + 12\frac{2}{3} - 10\frac{5}{6} = ?$

17. Subtract $3\frac{4}{9}$ from $7\frac{1}{9}$.

18. When Mac drives a nail, he sinks it $\frac{4}{15}$ of an inch with each blow. How many blows will be needed to sink the nail $1\frac{1}{3}$ inches?

19. A gardener decided to plant some vegetables. He needed to know the number of square feet in the garden in order to buy the correct amount of fertili-

zer. If the garden was 21 feet long and $18\frac{2}{3}$ feet wide, find the area in square feet.

20. In Exercise 19, if one bag of fertilizer covers 100 sq ft, how many bags should he buy?

21. Find the sum of $3\frac{1}{2}$, $2\frac{2}{5}$, and $4\frac{1}{4}$.

22. What is $10\frac{2}{3}$ diminished by 6?

23. A physical fitness buff jogs every day. During one week she jogged $1\frac{2}{3}$ miles on Monday, $1\frac{1}{4}$ miles on Tuesday, 4 miles on Wednesday, $2\frac{1}{2}$ miles on Thursday, and 1 mile on Friday. How far did she jog during the week?

24. $7\frac{3}{4} \div 6\frac{43}{60} = ?$

25. $4\frac{4}{15} + 5\frac{7}{10} + 3\frac{5}{6} = ?$

26. Multiply $3\frac{3}{4}$ by $2\frac{2}{5}$.

27. A farmer produced $175\frac{3}{4}$ acres of cotton, but he was able to harvest only 113 acres. How many acres of cotton does he have left in the field?

*28. Divide the sum $7\frac{1}{2} + 5\frac{3}{4}$ by the difference $7\frac{1}{2} - 5\frac{3}{4}$.

*29. Determine what fractional part 7 is of $10\frac{1}{2}$. $\left(Hint: \frac{7}{10\frac{1}{2}} = 7 \div 10\frac{1}{2}\right)$

*30. Divide the product $6 \cdot \frac{4}{5}$ by the quotient $7\frac{1}{2} \div 6\frac{1}{4}$.

*31. Find the sum of $7\frac{1}{2} \div 6\frac{1}{4}$ and $6 \cdot \frac{4}{5}$.

*32. $6\frac{3}{4} \div 10\frac{1}{2} \cdot 8\frac{2}{3} = ?$

*33. Multiply 2 times the sum of $7\frac{1}{2}$ and $7\frac{1}{2} - 5\frac{3}{4}$.

*34. $\frac{5\frac{3}{4}}{2\frac{1}{2}} = ?$ $\left(Hint: \frac{5\frac{3}{4}}{2\frac{1}{2}} = 5\frac{3}{4} \div 2\frac{1}{2}\right)$

*35. Determine what fractional part $2\frac{1}{3}$ is of $5\frac{5}{6}$.

*These exercises are optional.

4 PERCENT

PRETEST

1. Convert the following fractions to decimals (Objective 1).

   a. $\dfrac{29}{4}$ b. $\dfrac{3}{8}$ c. $\dfrac{47}{16}$

2. Convert the following decimals to fractions (Objective 2).

   a. 0.0625   b. 0.36   c. 2.05

3. Convert the following percents to decimals (Objective 3).

   a. 41.8%   b. 0.072%   c. $\dfrac{1}{4}\%$

4. Convert the following percents to fractional form (Objective 4).

   a. 408%   b. 86%   c. 0.6%

5. Calculate the following percentages (Objective 5).

   a. 45% of 96   b. 112% × 118   c. 0.05% × 118

6. Convert the following decimals to percents (Objective 6).

   a. 0.43   b. 0.015   c. 41.8

7. Convert the following fractions to percents (Objective 7).

   a. $\dfrac{3}{16}$ b. $\dfrac{5}{8}$ c. $\dfrac{13}{4}$

8. Solve each of the following (Objective 8).

   a. At the beginning of the 1976-77 basketball season, the University of Illinois had won seven out of nine games. What percent of the games did they win? (Round your answer to the nearest tenth of a percent.)

   b. Nancy's grade on a test was 86%. If there were 225 points possible on the test, how many points did she miss?

c. How much interest would be due on a loan of $3500 if the interest charged was 11.75% of the loan?

9. Work the following problems (Objective 9).

a. Write $\frac{1}{4}$ + 12.2 as a decimal.

b. Write $12.2 - \frac{1}{4}$ as a fraction.

c. Express $0.625 \times \frac{24}{25}$ as a fraction.

d. Write 17% of $\frac{3}{4}$ as a decimal.

It is common practice to use several different equivalent forms for fractional numbers. You have already looked at some of these forms—fractions, mixed numbers, and decimals. The different forms are used for different purposes. For example, if you were told that a room measured $\frac{41}{2}$ feet by $\frac{145}{4}$ feet, you would probably have to reinterpret this to mean that the room was $20\frac{1}{2}$ feet by $36\frac{1}{4}$ feet before the measurements were meaningful to you. However if you wanted to find the area of the room, it might be more convenient to have the measurements in fractional form for the multiplication. In addition to expressing mixed numbers as improper fractions and improper fractions as mixed numbers, it is often necessary to use other equivalent forms. In this chapter we will consider equivalent forms for fractions, decimals, and percents.

4.1 Expressing Fractions as Decimals

Although it is sometimes easier to work with fractions, it is often more convenient to use decimals. As a typist knows, decimals are much easier to type than fractions. For this reason decimals are frequently used in printed materials other than mathematics texts. Computers and calculators also readily accept decimals. Thus, banks and other businesses use numbers in decimal form more often than they use numbers in fractional form.

The purpose of this section is to convert fractions to equivalent decimals. Since the fraction bar can be interpreted as indicating division, you can transform a fraction to a decimal by dividing the numerator of the fraction by the denominator. This is specified in the *rule for converting a fraction to a decimal.*

RULE FOR CONVERTING A FRACTION TO A DECIMAL	*EXAMPLE* Convert $\frac{1}{2}$ to a decimal.
Divide the denominator into the numerator.	*Solution:* $$\begin{array}{r} 0.5 \\ 2\overline{)1.0} \\ \underline{1\,0} \end{array}$$ Thus, $\frac{1}{2} = 0.5$.

Example 1   a.   Convert $\frac{3}{4}$ to a decimal.

$$\begin{array}{r} .75 \\ 4\overline{\smash{)}3.00} \\ \underline{2\ 8} \\ 20 \\ \underline{20} \end{array}$$

Thus $\frac{3}{4} = 0.75$.

b.   Express $\frac{7}{8}$ as a decimal.

$$\begin{array}{r} .875 \\ 8\overline{\smash{)}7.000} \\ \underline{6\ 4} \\ 60 \\ \underline{56} \\ 40 \\ \underline{40} \end{array}$$

Thus $\frac{7}{8} = 0.875$.

Example 2   Convert $\frac{2}{3}$ to a decimal.

$$\begin{array}{r} .6666 \\ 3\overline{\smash{)}2.0000} \\ \underline{1\ 8} \\ 20 \\ \underline{18} \\ 20 \\ \underline{18} \\ 20 \\ \underline{18} \\ 2 \end{array}$$

This division is not exact since the 6 repeats forever. We indicate this by writing 0.6666... . Thus, $\frac{2}{3} = 0.6666...$.

In Example 1 we carried out the division until it was exact. In Example 2, the division will never be exact. In the number 0.6666... , the three dots following the last digit indicate that the 6s continue without end. The decimal 0.875 has a last or terminal digit and it is called a **terminating decimal**. Since 0.6666...

does not have a last or terminal digit and since the 6 repeats, this number is called a **nonterminating repeating decimal.**

Another notation that is frequently used to show that the 6s in the 0.6666... continue infinitely is to write $0.\overline{6}$ for this number. The bar over the 6 means that this digit repeats infinitely. Similarly, the number 0.07838383... can be denoted by $0.07\overline{83}$, where the bar extends over both the digit 8 and the digit 3. This means that this pair of digits continues to repeat. In this text, a repeating decimal will be indicated either by the bar or by writing the repeating digits three times followed by three dots.

Example 3   a.   2.4683 is a terminating decimal.

b.   7.864444... is a nonterminating repeating decimal.

c.   123 is a terminating decimal (consider 123 = 123.).

d.   487.5367777... is a nonterminating repeating decimal.

e.   4.80000... is a terminating decimal. Since the three dots indicate that only zeroes are continuing without end, 4.80000... is the same as 4.8 which terminates.

f.   0.07838383... is a nonterminating repeating decimal.

g.   81.364873209... is a nonterminating, nonrepeating decimal.

Example 4   Express each of the following fractions as decimals. If the division is not exact, express the number as a repeating decimal.

a. $\dfrac{5}{6} = 0.8\overline{3}$ Since

$$
\begin{array}{r}
0.8333 \\
6\overline{)5.0000} \\
\underline{4\ 8} \\
20 \\
\underline{18} \\
20 \\
\underline{18} \\
20 \\
\underline{18} \\
2
\end{array}
$$

b. $\dfrac{31}{250} = 0.124$ Since

$$
\begin{array}{r}
0.124 \\
250\overline{)31.000} \\
\underline{25\ 0} \\
6\ 00 \\
\underline{5\ 00} \\
1\ 000 \\
\underline{1\ 000}
\end{array}
$$

c. $\dfrac{142}{99} = 1.\overline{43}$　　　Since

$$
\begin{array}{r}
1.4343 \\
99\overline{)142.0000} \\
\underline{99} \\
43\;0 \\
\underline{39\;6} \\
3\;40 \\
\underline{2\;97} \\
430 \\
\underline{396} \\
340 \\
\underline{297} \\
43
\end{array}
$$

Instead of expressing the fraction as a repeating decimal when the division is not exact, you may round the decimal or express it as a mixed form. The numeral $0.66\frac{2}{3}$ is called a **mixed form** since it is a decimal which contains a fraction within the decimal part of the numeral. Mixed forms occur frequently in business applications.

Example 5　a.　Express $\dfrac{1}{3}$ as a decimal rounded to the nearest hundredth.

$$
\begin{array}{r}
.333 \\
3\overline{)1.000} \\
\underline{9} \\
10 \\
\underline{9} \\
10 \\
\underline{9} \\
1
\end{array}
$$
　Thus $\dfrac{1}{3} \doteq 0.33$

b.　Express $\dfrac{1}{3}$ as a mixed form containing two decimal places.

$$
\begin{array}{r}
0.33 \\
3\overline{)1.00} \\
\underline{9} \\
10 \\
\underline{9} \\
1
\end{array}
$$
　Therefore $\dfrac{1}{3} = 0.33\frac{1}{3}$.

Note that we carry the division two places and write the remainder over the divisor.

c. Express $\dfrac{3}{8}$ as a mixed form containing two decimal places.

$$\begin{array}{r} 0.37 \\ 8\overline{\smash)3.00} \\ \underline{2\,4} \\ 60 \\ \underline{56} \\ 4 \end{array}$$

Thus $\dfrac{3}{8} = 0.37\dfrac{4}{8}$

$= 0.37\dfrac{1}{2}$

If we had continued the division we would have obtained $\dfrac{3}{8} = 0.375$. So $0.37\dfrac{1}{2} = 0.375$.

Example 6 Convert each of the following to decimals. If the division is not exact express the result as a repeating decimal.

a. $\dfrac{1}{500}$

$$\begin{array}{r} 0.002 \\ 500\overline{\smash)1.000} \\ \underline{1\,000} \end{array}$$

So $\dfrac{1}{500} = 0.002$.

b. $1\dfrac{3}{4}$

$1\dfrac{3}{4} = \dfrac{7}{4}$ so $\begin{array}{r} 1.75 \\ 4\overline{\smash)7.00} \\ \underline{4} \\ 3\,0 \\ \underline{2\,8} \\ 20 \\ \underline{20} \end{array}$

Thus $1\dfrac{3}{4} = 1.75$.

c. $\dfrac{7}{12}$

$$\begin{array}{r} .5833 \\ 12\overline{\smash)7.0000} \\ \underline{6\,0} \\ 1\,00 \\ \underline{96} \\ 40 \\ \underline{36} \\ 40 \\ \underline{36} \\ 4 \end{array}$$

Thus $\dfrac{7}{12} = 0.58333...$ or $0.58\overline{3}$.

d. $\dfrac{104}{333}$

$$\begin{array}{r} 0.312312 \\ 333\overline{\smash)104.000000} \\ \underline{99\,9} \\ 4\,10 \\ \underline{3\,33} \\ 770 \\ \underline{666} \\ 1040 \\ \underline{999} \\ 410 \\ \underline{333} \\ 770 \\ \underline{666} \\ 104 \end{array}$$

Thus $\dfrac{104}{333} = 0.\overline{312}$.

In some applications involving both fractions and decimals, it is not practical to carry out the computations with the given numerals. First it may be necessary to convert the fractions to decimals or decimals to fractions before performing the indicated operations. In Example 7, the fractions are changed to decimals before the operations are performed.

Example 7   Perform the indicated operations, and express the answer as a decimal.

a. $\frac{1}{5} + 7.23$

Since the answer is to be written in decimal form, $\frac{1}{5}$ will be converted to a decimal before the addition is performed.

$$\frac{1}{5} = 0.2 \quad \text{since} \quad 5\overline{)\begin{array}{c} .2 \\ 1.0 \\ \underline{1\,0} \end{array}}$$

Thus $\frac{1}{5} + 7.23 = 0.2 + 7.23 = 7.43$.

b. $1.04 \cdot \frac{1}{16}$

First we express $\frac{1}{16}$ as the decimal 0.0625.

$$16\overline{)\begin{array}{c} .0625 \\ 1.0000 \\ \underline{96} \\ 40 \\ \underline{32} \\ 80 \\ \underline{80} \end{array}}$$

Then $1.04 \cdot \frac{1}{16} = 1.04 \cdot 0.0625 = 0.065$.

$$\begin{array}{r} 0.0625 \\ 1.04 \\ \hline 2500 \\ 625 \\ \hline .065000 \end{array}$$

4.1 Exercises

Convert each of the fractions in Exercises 1–10 to equivalent decimals.

1. $\frac{1}{2}$

2. $\frac{7}{4}$

3. $\frac{7}{5}$

4. $\frac{3}{8}$

5. $\frac{9}{10}$

6. $\frac{4}{5}$

7. $\frac{5}{8}$

8. $\frac{17}{10}$

9. $\frac{17}{16}$

10. $\frac{9}{16}$

Express the answer in each of Exercises 11–24 in exact decimal representation unless otherwise indicated.

11. Convert $\frac{3}{5}$ to an equivalent decimal.

12. Express $\frac{1}{1000}$ as a decimal.

13. Express $2\frac{3}{4}$ as a decimal.

14. Use a decimal to express the fractional part 14 is of 64.

15. Two-thirds of the people of the world go to bed hungry. Express this fraction as a decimal accurate to three decimal places.

16. Express $\dfrac{31}{10,000}$ as a decimal.

17. Use a decimal to express the fractional part 3 is of 4.

18. Express $\dfrac{2}{5}$ as a decimal.

19. Sixty-seven hundredths of the pollution of the waters of the world comes from motor and industrial oils. Express this number as a decimal.

20. Express $\dfrac{18}{6}$ as a decimal.

21. Convert $\dfrac{7}{8}$ to an equivalent decimal.

22. If 12 ounces of a 16-ounce solution are alcohol, express the fractional part of the solution which is alcohol as a decimal.

23. Express $3\dfrac{2}{8}$ as a decimal.

24. Express $\dfrac{28}{7}$ as a decimal.

25. Twenty of the twenty-five students in a class were females. Express the fractional part of the class which is female as a decimal.

26. On Saturday, May 31, 1975, the St. Louis baseball team had won 19 of 43 games. Express the fraction $\dfrac{19}{43}$ as a decimal accurate to the nearest thousandth. (If you wish, you may check your accuracy in the library by referring to the Sunday, June 1, 1975, edition of your paper.)

27. If a baseball player has 54 hits in 156 official at bats, his batting average is $\dfrac{54}{156}$. What is his batting average expressed as a decimal accurate to the nearest thousandth?

28. Express $\dfrac{26}{5}$ as a decimal.

29. Express $\dfrac{3}{400}$ as a decimal.

30. If you get 17 out of 20 problems correct on an assignment, express the fractional part you did correctly as a decimal.

31. Express $\dfrac{36}{8}$ as a decimal.

32. Express $\dfrac{7}{500}$ as a decimal.

Express each of Exercises 33–38 as a decimal.

33. $\dfrac{19}{1}$ 34. $\dfrac{19}{10}$

35. $\dfrac{19}{100}$ 36. $\dfrac{19}{1000}$

37. $\dfrac{19}{10,000}$ 38. $\dfrac{19}{100,000}$

After observing the correct answers for Exercises 33–38, convert Exercises 39–44 to decimals by inspection, without performing the division.

39. $\dfrac{31}{1}$ 40. $\dfrac{31}{10}$

41. $\dfrac{31}{100}$ 42. $\dfrac{31}{1000}$

43. $\dfrac{31}{1,000,000}$ 44. $\dfrac{31}{100,000,000}$

In Exercises 45–54, express the fractions as decimals and then perform the indicated operation. Express the answer as a decimal.

45. $\dfrac{1}{2}+0.415$ 46. $5.8+\dfrac{7}{4}$

47. $1.603-\dfrac{4}{5}$ 48. $\dfrac{26}{5}-4.37$

49. $\left(\dfrac{3}{8}\right)(11.02)$ 50. $11.02\div\dfrac{2}{5}$

51. $\dfrac{3}{5}\div1.6$ 52. $\dfrac{3}{2}+\dfrac{3}{4}+0.52$

53. $7.384\div\dfrac{26}{25}$ 54. $\left(0.0625\right)\left(\dfrac{7}{100}\right)$

Express each of the fractions in Exercises 55–65 as decimals.

55. $\dfrac{1}{3}$ 56. $\dfrac{7}{3}$

57. $\dfrac{2}{3}$ 58. $\dfrac{10}{11}$

59. $\dfrac{2}{9}$ 60. $\dfrac{6}{7}$

61. $\dfrac{5}{11}$ 62. $\dfrac{1}{33}$

63. $\dfrac{5}{7}$ 64. $\dfrac{17}{99}$

65. $\dfrac{5}{3}$

4.2 Expressing Terminating Decimals as Fractions

In this section we will express terminating decimals as fractions. You will see how to express repeating decimals as fractions in Chapter 12. If you recall how to *read* decimals, then expressing terminating decimals in fractional form is very easy. For example, 0.429 is read "four hundred twenty-nine thousandths." Writing this in fraction form, we obtain $\dfrac{429}{1000}$. The procedure is given in the *rule for converting a terminating decimal to a fraction.*

RULE FOR CONVERTING A TERMINATING DECIMAL TO A FRACTION	*EXAMPLE* Convert 0.5 to a fraction.

1. Form a fraction in which:
  a. the numerator is the whole number formed by the digits in the original decimal numeral, and
  b. the denominator is the multiple of ten which contains the same number of zeros as the number of digits to the right of the decimal point in the original decimal numeral.
 	
2. Reduce this fraction to lowest terms.
	Solution:
	
	The numerator is 5.
	
	There is one digit to the right of the decimal point so the denominator is 10.
	
	$0.5 = \dfrac{5}{10} = \dfrac{1}{2}$

Example 1   Express the following decimals as fractions.

a. 0.23

The numerator is 23.
There are two digits to the right of the decimal point so the denominator is 100.

Thus $0.23 = \dfrac{23}{100}$.

b. 0.0004

The numerator is 4.
There are four digits to the right of the decimal point so the denominator is 10000.

Thus $0.0004 = \dfrac{4}{10000} = \dfrac{1}{2500}$.

c. $1.3 = \dfrac{13}{10}$ d. $1.7 = \dfrac{17}{10}$ e. $1.73 = \dfrac{173}{100}$

f. $1.739 = \dfrac{1739}{1000}$ g. $246.6 = \dfrac{2466}{10} = \dfrac{1233}{5}$

h. $41 = \dfrac{41}{1}$

To express a whole number as a fraction, use a denominator of 1.

i. $1.7393 = \dfrac{17393}{10,000}$

In the previous section, we considered problems involving decimals and fractions where it was necessary to convert fractions to decimals before performing the operations. Sometimes it is necessary or easier to do the reverse—that is, convert decimals to fractions.

Example 2   In each of the following problems convert the decimals to fractions and then perform the indicated operation. Express your answer as a fraction.

a. $\dfrac{13}{20} - 0.25$

Since the answer is to be written in fractional form, 0.25 will be converted to a fraction before the subtraction is performed.

Since $0.25 = \dfrac{25}{100} = \dfrac{1}{4}$,

$$\frac{13}{20} - 0.25 = \frac{13}{20} - \frac{1}{4} = \frac{13}{20} - \frac{5}{20} = \frac{8}{20} = \frac{2}{5}.$$

Thus $\dfrac{13}{20} - 0.25 = \dfrac{2}{5}$.

b. $\dfrac{67}{64} \div 0.125$

First we express 0.125 as a fraction.

$$0.125 = \frac{125}{1000} = \frac{\cancel{5} \cdot \cancel{5} \cdot \cancel{5}}{2 \cdot 2 \cdot 2 \cdot \cancel{5} \cdot \cancel{5} \cdot \cancel{5}} = \frac{1}{8}$$

$$\frac{67}{64} \div 0.125 = \frac{67}{64} \div \frac{1}{8}$$

$$= \frac{67}{\underset{8}{\cancel{64}}} \times \frac{\overset{1}{\cancel{8}}}{1}$$

$$= \frac{67}{8}$$

Thus $\dfrac{67}{64} \div 0.125 = \dfrac{67}{8}$.

Sometimes you may encounter numerals which are of a mixed form containing a fraction within the decimal portion. The following example illustrates a procedure for expressing such numbers as fractions.

Example 3   Express $0.33\dfrac{1}{3}$ as a fraction.

We follow the same procedure as in Example 1.

The numerator is $33\dfrac{1}{3}$.

There are two digits to the right of the decimal point so the denominator is 100.

So $0.33\frac{1}{3} = \dfrac{33\frac{1}{3}}{100}$.

We use the fact that the fraction bar can be interpreted to mean division to simplify this fraction.

$$0.33\frac{1}{3} = \dfrac{33\frac{1}{3}}{100}$$

$$= 33\frac{1}{3} \div 100$$

$$= \frac{100}{3} \div \frac{100}{1}$$

$$= \frac{\overset{1}{\cancel{100}}}{3} \times \frac{1}{\underset{1}{\cancel{100}}}$$

$$= \frac{1}{3}$$

Thus $0.33\frac{1}{3} = \frac{1}{3}$.

4.2 Exercises

Express the numbers in Exercises 1–20 as fractions.

1. 0.2
2. 0.175
3. 2.5
4. 9.6
5. 1.45
6. 3.75
7. 0.37
8. 87
9. 9
10. 5.18
11. 125.2
12. 256.4
13. 0.875
14. 72.3
15. 15.15
16. 0.10
17. 53
18. 0.002
19. 10.01
20. 17

21. Probabilities and odds are often given as fractions. If the probability of a horse winning a race is 0.80, what fraction represents the probability that this horse will win?

22. The probability that all four children in a family will be boys is 0.0625. Express this probability as a fraction.

Express the numbers in Exercises 23–30 as fractions.

23. 400.04
24. 1.008
25. 759
26. 32
27. 5.0002
28. 0.708
29. 0.03125
30. $0.66\frac{2}{3}$

31. The probability of winning at a game in Las Vegas is 0.48. Express this probability of winning as a fraction.

32. If a multiple choice test has 5 possible answers, the probability of guessing the one correct choice is 0.2. Express this probability as a fraction.

In Exercises 33–45 convert the decimals to fractions before performing the indicated operation. Express your answers in fractional form.

33. $0.5 + \dfrac{3}{17}$
34. $1.5 - \dfrac{5}{8}$
35. $\dfrac{1}{3} - 0.25$

36. $\dfrac{5}{9} + 3.75$
37. $0.5625 - \dfrac{7}{16}$
38. $\left(\dfrac{1}{3}\right)(7.5)$

39. $\dfrac{64}{125} \div 0.625$
40. $\dfrac{1}{3} \div 7.5$
41. $\left(\dfrac{13}{20}\right)(0.52)$

42. $\left(0.66\dfrac{2}{3}\right)\left(\dfrac{15}{4}\right)$

43. Add the product $(1.8)\left(\dfrac{2}{3}\right)$ to $\dfrac{6}{7}$.

44. Divide the sum of 1.8 and $\dfrac{2}{3}$ by $0.33\dfrac{1}{3}$.

45. $\dfrac{3}{2} + 1.7 - \dfrac{2}{3} = ?$

4.3 The Meaning of Percent

The special form of writing fractional numbers which we shall consider in this section is percent. You should already be aware of the extremely wide usage of this particular notation from everyday experiences involving interest rates, cost of living increases, discount pricing, etc.

The word **percent** comes from the Latin *per centum* meaning "per hundred." Therefore, we can interpret the percent symbol (%) to mean hundredths, that is $\frac{1}{100}$ or equivalently 0.01. For example 69% means $69 \cdot \frac{1}{100} = \frac{69}{100}$ or (69)(0.01) = 0.69.

To change a percent to a fraction, use the *rule for expressing a percent as a fraction*.

RULE FOR EXPRESSING A PERCENT AS A FRACTION	EXAMPLE Express 96% as a fraction.
Replace the percent symbol by $\frac{1}{100}$, and multiply this times the number preceding the percent symbol. Reduce the fraction if it is not in lowest terms.	*Solution:* $96\% = 96 \cdot \frac{1}{100}$ Thus $96\% = \frac{96}{100}$ $= \frac{24}{25}$

Example 1   Convert each of the following percents to fractions.

a. 25%

$$25\% = 25 \cdot \frac{1}{100}$$

$$= \frac{25}{100}$$

$$= \frac{1}{4}$$

Thus $25\% = \frac{1}{4}$.

b. 90%

$$90\% = 90 \cdot \frac{1}{100}$$

$$= \frac{90}{100}$$

$$= \frac{9}{10}$$

Thus $90\% = \frac{9}{10}$.

c. 150%

$$150\% = 150 \cdot \frac{1}{100}$$

$$= \frac{150}{100}$$

$$= \frac{3}{2}$$

Thus $150\% = \frac{3}{2}$.

d. $\frac{1}{2}\%$

$$\frac{1}{2}\% = \frac{1}{2} \cdot \frac{1}{100}$$

$$= \frac{1}{200}$$

Thus $\frac{1}{2}\% = \frac{1}{200}$.

e. $33\frac{1}{3}\%$

$$33\frac{1}{3}\% = 33\frac{1}{3} \cdot \frac{1}{100}$$

$$= \frac{\overset{1}{\cancel{100}}}{3} \cdot \frac{1}{\underset{1}{\cancel{100}}}$$

$$= \frac{1}{3}$$

Thus $33\frac{1}{3}\% = \frac{1}{3}$.

To change a percent to a decimal, use the *rule for expressing a percent as a decimal.*

RULE FOR EXPRESSING A PERCENT AS A DECIMAL	EXAMPLE Express 96% as a decimal.
	Solution:
Replace the percent symbol by 0.01, and multiply this times the number preceding the percent symbol.	$96\% = 96 \cdot 0.01$
	Thus $96\% = 0.96$.

Example 2   Express each of the following percents as equivalent decimals.

a. 25%

$$25\% = (25)(0.01)$$
$$= 0.25$$

Thus $25\% = 0.25$.

b. 90%

$$90\% = 90(0.01)$$
$$= 0.90$$

Thus $90\% = 0.9$.

c. 150%

$$150\% = 150(0.01)$$
$$= 1.50$$

Thus $150\% = 1.5$.

d. 0.2%

$$0.2\% = 0.2(0.01)$$
$$= 0.002$$

Thus $0.2\% = 0.002$.

When changing a percent to a decimal, you multiply by 0.01. If you study Example 2, you will see that when you multiplied by 0.01, the decimal point was moved two places to the left. Thus the rule for changing a percent to a decimal is frequently stated as follows: *To change a percent to a decimal, reposition the decimal point two places to the left and omit the percent symbol.*

Example 3   a. Express 43% as a decimal.

$43\% = {}_\wedge 43.\% = 0.43$

b. Express 132% as a decimal.

$132\% = 1{}_\wedge 32.\% = 1.32$

c. Express 0.04% as a decimal.

$0.04\% = {}_\wedge 00.04\% = 0.0004$

Since you know how to convert decimals to fractions and fractions to decimals, you could also convert a percent to a decimal as follows:

$$\text{percent} \longrightarrow \text{fraction} \longrightarrow \text{decimal}$$

You could also change a percent to a fraction as follows:

$$\text{percent} \longrightarrow \text{decimal} \longrightarrow \text{fraction}$$

This is illustrated in Example 4.

Example 4   a. Express $\frac{1}{5}\%$ as a decimal.

$$\text{percent} \longrightarrow \qquad \text{fraction} \qquad \longrightarrow \text{decimal}$$

$$\frac{1}{5}\% \quad = \frac{1}{5} \cdot \frac{1}{100} = \frac{1}{500} = 0.002$$

This could also be worked as follows:

$$\frac{1}{5}\% = 0.2\% \quad \left(\frac{1}{5} = 0.2\right)$$
$$= 0.2(0.01)$$
$$= 0.002$$

b. Express 0.7% as a fraction.

$$\text{percent} \longrightarrow \qquad \text{decimal} \qquad \longrightarrow \text{fraction}$$

$$0.7\% \quad = 0.7(0.01) = 0.007 = \frac{7}{1000}$$

This could also be worked as follows:

$$0.7\% = \frac{7}{10}\% \quad \left(0.7 = \frac{7}{10}\right)$$
$$= \frac{7}{10} \cdot \frac{1}{100}$$
$$= \frac{7}{1000}$$

4.3 Exercises

Express each of the percents in Exercises 1-20 as fractions (or whole numbers).

1. 7%
2. 13%
3. 25%
4. 17%
5. 46%
6. 57%
7. 1%
8. 4%
9. $1\frac{1}{4}\%$
10. $\frac{1}{10}\%$
11. 102%
12. 237%
13. $\frac{8}{9}\%$
14. $2\frac{37}{100}\%$
15. 100%
16. 1700%
17. 513%
18. $\frac{1}{3}\%$

19. 2.37%
20. 0.1%

Express each of the percents in Exercises 21-40 as decimals (or whole numbers).

21. 7%
22. 13%
23. 25%
24. 17%
25. 46%
26. 5.7%
27. 1%
28. 4%
29. 0.7%
30. 1.5%
31. 102.73%
32. 237%
33. 0.004%
34. 0.502%
35. 100%
36. 1700%
37. 513%
38. 17.3%
39. $\frac{3}{5}\%$
40. $\frac{17}{8}\%$

4.4 An Application of Percent

In most problems involving percent, it is necessary to express the percent as a fraction or as a decimal before performing any other calculations. It is very easy to do this just by using the definition of percent. This is illustrated in Examples 1 and 2.

Example 1   a. 15% · 0.3 = ?

Since 0.3 is already a decimal, 15% is converted to a decimal before the multiplication is performed.

$$15\% \cdot 0.3 = 0.15 \cdot 0.3 = 0.045$$

b. (4.5)(10%) = ?

Since 10% = 0.10,

$$(4.5)(10\%) = (4.5)(0.10)$$
$$= 0.45$$

So (4.5)(10%) = 0.45.

c. (145%)(3.4) = ?

$$(145\%)(3.4) = (1.45)(3.4)$$
$$= 4.93$$

Thus (145%)(3.4) = 4.93.

Example 2   a. $\frac{2}{3} \cdot 5\frac{1}{7}\% = ?$

Since $\frac{2}{3}$ is a fraction, convert $5\frac{1}{7}\%$ to a fraction before multiplying.

$$5\frac{1}{7}\% = \frac{\overset{9}{\cancel{36}}}{7} \cdot \frac{1}{\cancel{100}} = \frac{9}{175}$$
$$\phantom{5\frac{1}{7}\% = \frac{36}{7} \cdot \frac{1}{100}}_{25}$$

Thus $\frac{2}{3} \cdot 5\frac{1}{7}\% = \frac{2}{\cancel{3}} \cdot \frac{\cancel{9}^{3}}{175}$

$$= \frac{6}{175}$$

b. $(15\%)\left(5\frac{1}{3}\right) = ?$

$$15\% = 15 \cdot \frac{1}{100} = \frac{15}{100} = \frac{3}{20}$$

Thus $(15\%)\left(5\frac{1}{3}\right) = \frac{\cancel{3}^{1}}{\cancel{20}_{5}} \cdot \frac{\cancel{16}^{4}}{\cancel{3}_{1}}$

$$= \frac{4}{5}$$

c. $(150\%)\left(54\frac{1}{2}\right) = ?$

$$(150\%)\left(54\frac{1}{2}\right) = \frac{3}{2} \cdot \frac{109}{2} \qquad\qquad \left(150\% = 150 \cdot \frac{1}{100} = \frac{150}{100} = \frac{3}{2}\right)$$

$$= \frac{327}{4}$$

Thus $(150\%)\left(54\frac{1}{2}\right) = \frac{327}{4} = 81\frac{3}{4}.$

A frequent application of percent involves taking a percent of a number. *To find such a percentage, you multiply the percent times the number (base).*

Example 3 a. Find 27% of $432.

We can change 27% to a decimal or fraction before multiplying. Since this problem involves dollars, it would make more sense to convert to a decimal.

Since 27% = 0.27,

$$27\% \cdot \$432 = (0.27)(\$432)$$
$$= \$116.64$$

b. Find 0.5% of 13.5.

Since 0.5% = 0.005,

$$0.5\% \cdot 13.5 = (0.005)(13.5)$$
$$= 0.0675$$

c. Find 175% of $20\frac{1}{2}$.

Since $175\% = 175 \cdot \frac{1}{100} = \frac{175}{100} = \frac{7}{4}$

$$175\% \left(20\frac{1}{2}\right) = \frac{7}{4} \cdot \frac{41}{2}$$

$$= \frac{287}{8} \text{ or } 35\frac{7}{8}$$

d. 16% of $33\frac{1}{3} = ?$

Since $16\% = 16 \cdot \frac{1}{100} = \frac{16}{100} = \frac{4}{25}$,

$$16\% \cdot 33\frac{1}{3} = \frac{4}{25} \cdot 33\frac{1}{3}$$

$$= \frac{4}{\overset{}{25}} \cdot \frac{\overset{4}{\cancel{100}}}{3}$$

$$= \frac{16}{3}$$

$$= 5\frac{1}{3}$$

The technique used in Example 3 is summarized below.

> **To find a percent of a number, multiply the percent times the number.**

Many word problems involving percents require finding a percent of a number. Some applications of this type are considered in Example 4.

Example 4 a. The sales tax in Atlanta, Georgia, is 4%. If you bought an item which cost $4.59, how much sales tax would you pay?

To determine this, you would need to find 4% of $4.59. Therefore, the tax is:

$$4\% \text{ of } \$4.59 = (4\%)(\$4.59)$$
$$= (0.04)(\$4.59)$$
$$= \$0.1836$$
$$\doteq \$0.18 \text{ or } 18\cancel{c}$$

Thus you would pay a sales tax of 18¢.

b. A concrete manufacturer produces precast concrete steps. Unfortunately, cracks and other defects cause him to reject about 17% of his steps as unsatisfactory. If he manufactures 3200 steps, how many can he expect to be defective?

17% of 3200 can be expected to be defective.

$$17\% \cdot 3200 = 0.17 \cdot 3200 = 544$$

Therefore he should expect about 544 defects.

c. A savings account pays $7\frac{1}{2}\%$ interest per year. How much interest is earned on $1000 in one year?

We need to find $7\frac{1}{2}\%$ of $1000. Since $7\frac{1}{2}\% = 7.5\% = 0.075$,

$$\left(7\frac{1}{2}\%\right)(\$1000) = 0.075(\$1000)$$

$$= \$75$$

Thus $75 interest is earned in one year.

d. In a class of 80 students, 20% made A's. How many students made A's? How many did not make A's?

20% of 80 is the number of students that made A's. Since 20% = 0.2,

$$20\% \cdot 80 = 0.2(80)$$

$$= 16$$

Thus 16 students made A's. Since there was a total of 80 students, $80 - 16 = 64$ did not make A's.

e. A woman who earns $15,850 annually has 34.3% of her salary withheld for taxes, insurance, and retirement. What is her yearly take-home pay?

To find her take-home pay, you must find the amount that is withheld, then subtract this from $15,850.

Since 34.3% = 0.343,

$$34.3\% \text{ of } \$15,850 = 0.343(\$15,850)$$

$$= \$5436.55$$

Thus her take-home pay is

$$\$15,850 - \$5436.55 = \$10,413.45$$

Alternate solution: Since 34.3% of her salary is withheld, she takes home 65.7% (100% − 34.3%) of her salary. Thus her take-home pay is

$$65.7\% \text{ of } \$15,850 = 0.657(\$15,850)$$

$$= \$10,413.45.$$

Other applications of percent are covered in Chapter 12.

4.4 Exercises

Perform the multiplications in Exercises 1–10. Express each answer as a decimal.

1. 7%(4)
2. 10%(15)
3. 100%(524.1)
4. 25% · 50
5. (0.1%)(72.5)
6. 300% · 48
7. (130%)(4.5)
8. (0.3%)(415)
9. 50%(74.5)
10. 100%(572.719)

Perform the multiplications in Exercises 11–20. Express each answer as a fraction.

11. $8\% \cdot \dfrac{25}{64}$
12. $2\frac{1}{2}\%\left(2\frac{1}{2}\right)$
13. $10\%\left(7\frac{3}{4}\right)$
14. $\dfrac{3}{10}\%(100)$
15. $6\frac{1}{2}\% \cdot \$400$
16. $100\%\left(50\frac{7}{18}\right)$
17. $\dfrac{25}{2}\% \cdot \dfrac{2}{3}$
18. $132\% \cdot 12\frac{1}{2}$
19. $100\%\left(4\frac{3}{7}\right)$
20. $\dfrac{1}{10}\%\left(\dfrac{100}{3}\right)$

Perform the multiplications in Exercises 21–34. Express each answer as a fraction or a decimal.

21. Find $12\frac{1}{2}\%$ of $760.
22. 43.5% of 102.56 = ?
23. 130% of $1\frac{2}{5}$ = ?
24. Calculate 18% of 72.
25. Find 11.25% of $1000.
26. 350% of 45 = ?
27. Calculate $37\frac{1}{2}\%$ of $7\frac{1}{9}$.
28. 82.3% of 7.15 = ?
29. $11\frac{3}{4}\%$ of 525 = ?
30. Find $66\frac{2}{3}\%$ of $7\frac{1}{2}$.
31. $\dfrac{8}{9}\%$ of $14\frac{3}{4}$ = ?
32. 450% of 14 = ?
33. Find $33\frac{1}{3}\%$ of 5602.
34. 5.5% of 17 = ?

35. The U.S. produces about 53 million tons of paper and paperboard every year. In 1969, only about 22% of the paper and paperboard produced was recycled. How many tons of paperboard and paper were recycled?

36. A bank pays 6% yearly interest on a savings account. How much interest is earned on $350 in one year?

37. Many mechanical products can be constructed so as to operate with less noise; however, this frequently imposes an extra cost on the consumer. If an air conditioner sells for $319, an improved model creating less noise might sell for 8% more. What would be the increase in cost to the consumer? What would the selling price be?

38. A solution contains 75% alcohol. How many ounces of alcohol are in 16 ounces of the solution?

39. What is 13% of 4.039?

40. $21\% \times \dfrac{3}{7}$ = ?

41. If a corporation paid $5\frac{3}{4}\%$ yearly interest on the par value of each share, how much would a shareholder earn if she had one share of par value $500?

42. If a quarterback completed $37\frac{1}{2}\%$ of his passes, how many of his 105 passes did he complete? (Round your answer to the nearest whole number.)

43. $1.45\% \times \dfrac{2}{3}$ = ?

44. What is 1.2% of 0.009?

45. Jerry West made 47.38% of the 17,234 field goals he attempted. How many field goals did he make? (Round your answer to the nearest whole number.)

46. In the 1968 presidential election, George C. Wallace received 13.5% of the 72,967,118 votes cast. How many votes did this third-party candidate receive? (Round your answer to the nearest whole number.)

47. In 1970 only 16.5% of Hawaii's population lived in rural areas. How many of the 769,913 residents lived in a rural area? (Round your answer to the nearest whole number.)

48. 38.6% of the 57 signers of the Declaration of Independence were lawyers. How many signers practiced law? (Round your answer to the nearest whole number.)

49. Find 3.2% of $\dfrac{3}{16}$.

50. Find 0.4% of the sum of 0.603 and 2.307.

51. Fifty-two percent of the 5130 students in a school are female. Find the number of females enrolled in the school. (Round your answer to the nearest whole number.)

52. If 18% of a person's salary is withheld each month for federal taxes, how much would a person whose salary

is $1280 per month receive after federal taxes are deducted?

53. A woman spends 10% of her monthly income on food, 19% on rent, and 7% on clothes. If she earns $976 per month, what is the total amount spent on these items?

54. Find the total cost (that is, cost plus tax) of a television set priced at $529 if the sales tax rate is 6%.

55. Harry can buy a car costing $5780 in one county which has a sales tax rate of 7% or he can buy an identical car for $5900 in another county which has a sales tax rate of 2%. Which is the better buy?

4.5   Expressing Decimals as Percents

Since we often find percents more meaningful than decimals or fractions, it is important that we have a method for expressing decimals and fractions as percents. The basis for this method is the meaning of the percent symbol. Recall that 1% = 0.01; thus, 100% = 1. Also recall that multiplying a number by one does not change its value. Both of these facts are used to convert decimals to percents in Example 1.

Example 1   Express each of the following decimals as percents.

a. $0.249 = (0.249)(1)$ multiply by one
$ = (0.249)(100\%)$ replace 1 with 100%
$ = 24.9\%$ multiply

b. $0.89 = (0.89)(1)$
$ = (0.89)(100\%)$
$ = 89\%$

c. $0.3 = (0.3)(1)$
$ = (0.3)(100\%)$
$ = 30\%$

d. $2.37 = (2.37)(1)$
$ = (2.37)(100\%)$
$ = 237\%$

e. $0.004 = (0.004)(1)$
$ = (0.004)(100\%)$
$ = 0.4\%$

f. $6 = (6)(1)$
$ = (6)(100\%)$
$ = 600\%$

If you study Example 1, you can see that when you change a decimal to a percent, you multiply by 100. This results in moving the decimal point two places to the right. This is stated in the *rule for expressing decimals as percents*.

RULE FOR EXPRESSING DECIMALS AS PERCENTS	EXAMPLE Express 0.031 as a percent.
	Solution:
Reposition the decimal point two places to the right and affix the percent symbol.	$0.031 = 0.03{\scriptstyle\wedge}1 = 3.1\%$

Example 2   Express the following decimals as percents.

    a. $0.134 = 0.13_\wedge 4 = 13.4\%$

    b. $0.25 = 0.25_\wedge = 25\%$

    c. $1.2 = 1.20_\wedge = 120\%$

    d. $23 = 23.00_\wedge = 2300\%$

    e. $0.0024 = 0.00_\wedge 24 = 0.24\%$

4.5 Exercises

Express each of the decimals in Exercises 1-30 as a percent.

1. 0.55	**2.** 0.22	**15.** 56.98	**16.** 1.23
3. 1.5	**4.** 0.001	**17.** 4.2	**18.** 0.075
5. 35	**6.** 3.1	**19.** 0.375	**20.** 4.25
7. 0.005	**8.** 4	**21.** 0.142	**22.** 1.5
9. 11	**10.** 0.0007	**23.** 0.0014	**24.** 0.125
11. 1	**12.** 145	**25.** 0.0007	**26.** 0.2
13. 1247	**14.** 0.0778	**27.** 0.375	**28.** 10
		29. $0.33\frac{1}{3}$	**30.** $0.66\frac{2}{3}$

4.6 Expressing Fractions as Percents

You have seen how to express a fraction as a decimal (how?) and a decimal as a percent (how?). We shall make use of these results in expressing fractions as percents. The procedure is given in the *rule for expressing fractions as percents.*

RULE FOR EXPRESSING FRACTIONS AS PERCENTS	*EXAMPLE* Express $\frac{1}{2}$ as a percent.
	Solution:
1. Express the fraction as a decimal.	$$2\overline{)1.0} \quad \frac{0.5}{}$$ $$\underline{1\ 0}$$
2. Express the decimal as a percent.	$0.5 = 0.50 = 50\%$
	So $\frac{1}{2} = 50\%$.

Example 1   Convert each of the following numbers to an equivalent percent.

a. $\dfrac{3}{4}$

fraction $\longrightarrow$ decimal $\longrightarrow$ percent

$$\dfrac{3}{4} \quad = \quad 0.75 \quad = \quad 75\%$$

$$\begin{array}{r} .75 \\ 4\overline{)3.00} \\ 2\,8 \\ \hline 20 \\ 20 \\ \hline \end{array}$$

b. $\dfrac{1}{500}$

fraction $\longrightarrow$ decimal $\longrightarrow$ percent

$$\dfrac{1}{500} \quad = \quad 0.002 \quad = \quad 0.2\%$$

$$\begin{array}{r} .002 \\ 500\overline{)1.000} \\ 1\,000 \\ \hline \end{array}$$

c. $\dfrac{7}{2}$

fraction $\longrightarrow$ decimal $\longrightarrow$ percent

$$\dfrac{7}{2} \quad = 3.5 = 3.50 \quad = \quad 350\%$$

$$\begin{array}{r} 3.5 \\ 2\overline{)7.0} \\ 6 \\ \hline 1\,0 \\ 1\,0 \\ \hline \end{array}$$

d. $2\dfrac{3}{4}$

fraction $\longrightarrow$ decimal $\longrightarrow$ percent

$$2\dfrac{3}{4} = \dfrac{11}{4} \quad = \quad 2.75 \quad = \quad 275\%$$

$$\begin{array}{r} 2.75 \\ 4\overline{)11.00} \\ 8 \\ \hline 3\,0 \\ 2\,8 \\ \hline 20 \\ 20 \\ \hline \end{array}$$

Example 2   Express $\dfrac{45}{13}$ to the nearest tenth of a percent.

$$\dfrac{45}{13} \doteq 3.462$$

$$= 346.2\%$$

$$\begin{array}{r} 3.4615 \\ 13\overline{)45.0000} \\ 39 \\ \hline 6\,0 \\ 5\,2 \\ \hline 80 \\ 78 \\ \hline 20 \\ 13 \\ \hline 70 \\ 65 \\ \hline 5 \end{array}$$

Thus $\dfrac{45}{13} \doteq 346.2\%$.

Note that in order to express $\frac{45}{13}$ to the nearest tenth of a percent, it is first necessary to express $\frac{45}{13}$ as a decimal to the nearest thousandth.

Example 3   Express $\frac{1}{3}$ as a percent.

The fraction $\frac{1}{3}$ does not have a terminating decimal representation, so to express it as a percent we will use a repeating decimal or a mixed form.

$$\frac{1}{3} = 0.33\frac{1}{3} \text{ or } 0.333...$$

$$= 33\frac{1}{3}\% \text{ or } 33.333...\%$$

$$\begin{array}{r} .33 \\ 3\overline{)1.00} \\ \underline{9} \\ 10 \\ \underline{9} \\ 1 \end{array}$$

So $\frac{1}{3} = 33\frac{1}{3}\%$ or $33.333...\%$

Some applications of percent involve determining what percent one number is of another number. The first step in working such problems is to determine what fractional part one number is of another number. We did this in Chapter 2. Consider how this is used in the following example.

Example 4   a. Seven is what percent of 10?

First, determine what fractional part 7 is of 10. Then express this as a decimal, then as a percent.

$$\frac{7}{10} = 0.70 = 70\%$$

Thus 7 is 70% of 10.

b. What percent of 40 is 15?

$$\frac{15}{40} = \frac{3}{8} = 0.375 = 37.5\%$$

c. If a $100 suit is marked down $20, what is the percent of the reduction?

We need to determine what percent 20 is of 100.

$$\frac{20}{100} = \frac{1}{5} = 0.20 = 20\%$$

Thus the reduction is 20%.

4.6   Exercises

Express the numbers in Exercises 1–20 as percents.

1. $\dfrac{1}{2}$ 2. $\dfrac{3}{4}$ 3. $\dfrac{5}{8}$

4. $\dfrac{3}{2}$ 5. $4\dfrac{1}{2}$ 6. $1\dfrac{19}{50}$

7. $\dfrac{1}{3}$ 8. $\dfrac{3}{5}$ 9. $\dfrac{9}{100}$

10. $5\dfrac{3}{4}$ 11. 1 12. 10

13. $\dfrac{6}{25}$ 14. $\dfrac{3}{8}$ 15. $\dfrac{2}{3}$

16. $\dfrac{3}{16}$ 17. $\dfrac{13}{50}$ 18. $\dfrac{17}{25}$

19. $\dfrac{9}{16}$ 20. $\dfrac{7}{8}$

21. Express $\dfrac{1}{4}$ as a percent.

22. Express $\dfrac{1}{1000}$ as a percent.

23. If $\dfrac{1}{8}$ of a manufacturer's products have at least some defect, what percent of defectives are being produced?

24. Express $2\dfrac{3}{4}$ as a percent.

25. Express $\dfrac{18}{3}$ as a percent.

26. Thirty-two is what percent of 100?

27. Express $1\dfrac{4}{5}$ as a percent.

28. A store advertised $\dfrac{1}{5}$ off all items in stock. What was the percent of discount?

29. Two-sevenths of the class failed the last test. Find to the nearest tenth of a percent the percent of the class which passed.

30. Express $\dfrac{17}{5}$ as a percent.

31. Fourteen is what percent of fifty?

32. Express $\dfrac{3}{20}$ as a percent.

33. Express $\dfrac{2}{13}$ as a percent accurate to the nearest hundredth of a percent.

34. A bank collected $18 interest on a $200 loan. Thus the interest rate was $\dfrac{18}{200}$ for the year the money was loaned. What was the interest rate as a percent?

35. During a 50-day quarter, a student was absent 8 times. What percent of the time was he absent?

36. Express $\dfrac{3}{8}$ as a percent.

37. A student answered 17 out of 20 questions correctly on a test. Thus, the student answered $\dfrac{17}{20}$ of the questions correctly. Express this grade as a percent.

38. Express $\dfrac{1}{16}$ as a percent.

39. Express $3\dfrac{1}{5}$ as a percent.

40. A suit originally priced at $200 was marked down $60. What percent was the suit marked down?

41. Express $\dfrac{2}{3}$ as a percent.

42. A contractor finished his job in $\dfrac{7}{8}$ of the allotted time. What percent of the allotted contract period did he use?

43. Express $\dfrac{3}{2}$ as a percent.

44. Express $\dfrac{15}{4}$ as a percent.

45. If 20 of the 30 students in a class are male, what percent of the class is male?

Important Terms   base   percentage
Used in this Chapter   nonterminating repeating decimal   terminating decimal
                           percent

REVIEW EXERCISES

Complete the following table for Exercises 1-30. The first ten problems give some common fraction–decimal–percent equivalents which are frequently used. You may find it helpful to memorize these equivalents.

	Fraction	Decimal	Percent
1.	$\frac{1}{2}$		
2.			$33\frac{1}{3}\%$
3.			25%
4.	$\frac{2}{3}$		
5.		0.4	
6.			20%
7.		1	
8.	$\frac{4}{5}$		
9.		0.75	
10.		0.6	
11.		0.0025	
12.			3.125%
13.			1%
14.		0.34	
15.			98%
16.		4	
17.			0.85%
18.	$\frac{17}{25}$		
19.		6.25	
20.			392%
21.	$\frac{65}{4}$		
22.		0.0065	
23.	$\frac{7}{1000}$		
24.*	$\frac{5}{11}$		

	Fraction	Decimal	Percent
25.			5%
26.		0.125	
27.	$\frac{5}{16}$		
28.			0.5%
29.	$\frac{5}{8}$		
30.		0.375	

31. A man who earns $12,500 annually pays 15% of his salary in taxes. How much tax does he pay?

32. A student received a grade of 80% on a test. There were 150 points possible on the test. How many points did he receive on the test?

33. Testing showed that 2.5% of the TV parts produced by a certain machine were defective. For every 1000 parts, how many were defective?

34. The authors of an English text collected a twelve percent royalty on each copy sold. The first year the publisher sold three thousand copies at $6.95 per copy. If one of the authors received one-half of the royalty payment, how much did this author receive?

35. A son inherited twenty-five percent of his father's four-fifths interest in a filling station. What fractional portion of the station does the son own?

36. A real estate broker collected a 7% fee on a house which sold for $37,500. How much did the broker collect?

37. The current inflation rate is 12.3% per year. If your salary is $15,000 this year, what would it have to be next year in order to keep up with the inflation rate?

38. A new technique for oil well production can improve well production by 3%. How many extra barrels can a company expect to produce if its current production is 3,400,000 barrels?

39. Sixty-two percent of the graduates of Monroe High School attend college. If the school has had 1500 graduates in the last ten years, how many have *not* gone to college?

40. Beef prices have increased 32% within the past four months. What is the present cost of a pound of beef which cost $1.98 per pound four months ago?

41. Seventeen is what percent of eighty-five?

*Round the decimal to the nearest thousandth and the percent to the nearest tenth.

42. Calculate $\frac{1}{2}$% of 1000.

43. 6.75% of $2500 = ?

44. Express the sum $5\frac{5}{6} + 1.625$ as a mixed number.

45. Express the quotient $36.27 \div \frac{9}{20}$ as a decimal.

46. A student correctly answered 65% of the questions on a test. What fractional part of the questions did he answer correctly?

*47. The total capacity of the world's largest aquarium is 2,450,000 gallons. Only 450,000 gallons are in display tanks while the remaining 2,000,000 gallons are in reserve. What percent of the total capacity is in the display tanks? (Round your answer to the nearest percent.)

*48. What is thirty-two percent of the sum of three-fourths and four hundred seventeen thousandths?

*49. Find 0.2% of the sum 4.5 and 0.12.

*50. Two and one-third is what percent of $5\frac{5}{6}$?

*These exercises are optional.

5

OPERATIONS WITH SIGNED NUMBERS

OBJECTIVES

Upon completion of this chapter you should be able to:

1. Represent the graphs of rational numbers on the number line.	5.1
2. Write the opposite of a number, and calculate the absolute value of a number.	5.1
3. Recognize that division by zero is undefined.	5.6
4. Evaluate squares and square roots.	5.2
5. Add two or more signed numbers.	5.3
6. Subtract signed numbers.	5.4
7. Evaluate expressions in simplified notation.	5.5
8. Translate expressions to simplified notation.	5.5
9. Multiply two or more signed numbers.	5.6
10. Divide signed numbers.	5.6
11. Evaluate expressions which contain grouping symbols.	5.7
12. Evaluate expressions in which the order of performing operations must be determined.	5.7
13. Work word problems involving signed numbers.	5.5

PRETEST

1. Draw a number line, and represent the graphs of the following numbers (Objective 1).

    a. -5
    b. $3\frac{1}{2}$
    c. $-2\frac{3}{4}$
    d. -1.9

2. For each of the following numbers, first write the opposite of the number and then write the absolute value of the number (Objective 2).

    a. -17
    b. 0
    c. 17
    d. $1\frac{3}{4}$
    e. -403
    f. $-\frac{2}{9}$

3. Answer the following true (T) or false (F) (Objective 3).

    a. $0 \div 3 = 0$
    b. $0 \div 3 = 3$
    c. $3 \div 0 = 0$
    d. $3 \div 0 = 1$
    e. $3 \div 0$ is undefined.

4. Evaluate the following (Objective 4).

    a. 7^2
    b. $\sqrt{81}$
    c. $\sqrt{42}$ (round to the nearest tenth)
    d. $\sqrt{33}$ (round to the nearest tenth)
    e. $(20)^2$
    f. $\sqrt{576}$

5. Perform the following additions (Objective 5).

    a. $(-10) + (+8)$

b. (+7.8) + (+18.05)

c. (+4.03) + (−3.04)

d. (−28) + (−391)

e. (−14.9) + (−8.4) + (26.2)

f. $\left(\dfrac{-4}{7}\right) + \left(\dfrac{3}{7}\right)$

6. Perform the following subtractions (Objective 6).

    a. (+108.3) − (+97.6)   **b.** (−108.3) − (+97.6)

    c. (−108.3) − (−97.6)   **d.** (+97.6) − (+108.3)

    e. (+97.6) − (−108.3)   **f.** $\left(\dfrac{-17}{19}\right) - \left(\dfrac{-18}{19}\right)$

7. Evaluate each of the following expressions (Objective 7).

    a. 23 − 35   **b.** −23 − 35

    c. −35 −23   **d.** −35 + 23

    e. 14 − 18 + 13 − 29

8. Rewrite each of the following expressions so that only one sign appears between the two numbers (Objective 8).

    a. −11 + (−8)   **b.** +11 − (+8)

    c. −11 − (−8)   **d.** −11 + (+8)

9. Perform the following multiplications (Objective 9).

    a. (−3) · (−17.08)   **b.** $\left(\dfrac{+7}{8}\right) \cdot \left(\dfrac{+16}{49}\right)$

    c. (+4.3) · (−9.2)   **d.** (−0.008) · (+0.054)

    e. (−1)(−2)(−3)(−4)(−5)

10. Perform the following divisions (Objective 10).

    a. (−427) ÷ (7)   **b.** (+0.02415) ÷ (−1.05)

    c. (−3.731) ÷ (−0.041)   **d.** $\left(\dfrac{-2}{3}\right) \div \left(\dfrac{+14}{15}\right)$

    e. $\left(+\dfrac{15}{36}\right) \div \left(+\dfrac{25}{66}\right)$

11. Evaluate the following expressions (Objective 11).

    a. −12 −(11 − 8)

    b. 15 − [40 + (−11)]

    c. (4 − 5) − (−7 + 8 − 13)

    d. 6 − 8 − (7 − 6 − [3 − 9]) − 2

12. Evaluate the following expressions (Objective 12).

    a. 2 − 6 × 5

    b. 2 × 7 − 4

    c. 5 × (−8) − 11 × (−6)

    d. −15 ÷ 5 + 4 − 3 × 2

    e. 11 + 2(7 − 18) − 5(3 − 15)

    f. 1 − 3[4 − 6{8 − 4(7 − 9)} − 5]

13. Solve the following (Objective 13).

A company sold ten machines at a loss of five dollars per machine. The company also sold five thousand fifty items at a profit of one dollar and ten cents per item. What was the company's profit?

Your work in Chapters 1, 2, and 3 should enable you to add, subtract, multiply, and divide positive numbers. This chapter will give the rules for performing these operations with signed numbers. We will use the term **signed numbers** to mean the set of rational numbers. This set includes all the numbers you worked with in Chapters 1-4 and all the opposites of these numbers.

The first task of this chapter will be to present some terminology and notation that will be needed later. Then the rules and procedures used for calculating with positive numbers will be extended to signed numbers. These new rules will be consistent with all rules and procedures given for positive numbers. That is, these rules will not contradict or change any rules or procedures which you have already learned.

5.1 The Number Line

Visual representations are frequently used in mathematics to clarify abstract concepts. One of the simplest of these is the **number line**. It is customary to draw the number line as indicated below.

    1. A horizontal line is drawn.

2. A point on the line is chosen to correspond with the number 0. This point is called the **origin**.

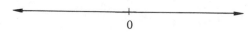

3. A convenient unit of measure is chosen; then beginning at the origin, equally-spaced points are marked off along the line to the right and to the left.

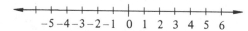

Even though only a few of the points on the number line have been labeled, every point on the number line is associated with a number. The numbers to the right of 0 on the number line are called **positive numbers,** and the numbers to the left of 0 are called **negative numbers.** The symbol "−," called the negative sign, is used to indicate that a number is negative. The symbol "−4" is read "negative four." The symbol "+," called the plus sign, is used to indicate that a number is positive. If a number is written without a sign, it is understood to be positive. For example, $6 = +6$. The number 0 is neither positive nor negative.

The number associated with a point on the number line is called the **coordinate** of the point, and the point is called the **graph** of the number. The graph of a number is usually indicated by a solid dot. The graph of 2 is shown in the figure below.

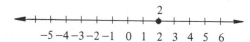

Example 1   On a number line, represent the graphs of each of the following numbers: 2, 5, −3, −1, $7\frac{1}{2}$.

Example 2   Represent the graphs of the following numbers: −20, 0, 20, 100.

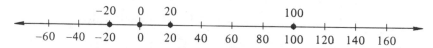

Note: The value of 20 for the scale for labeling the number line was used because of the magnitude of the numbers involved.

Example 3   Represent the graphs of 300, 302, and 305.

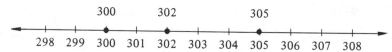

Note: It may be necessary to omit the origin when large numbers are involved.

Example 4   Represent the graphs of the natural numbers: 1, 2, 3, 4, 5, etc.

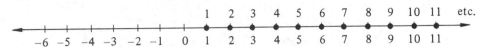

Since it is not possible to show the graph of all the numbers, "etc." is used.

Example 5   Represent the graphs of the following numbers: . . ., −3,−2,−1,0,1,2,3,4, . . .

Note: These numbers are called **integers**.

The points on the number line associated with 2 and −2 are the same distance from 0 but are on opposite sides of 0. These numbers are called opposites of each other. In general, if two points on the number line are the same distance from 0, the numbers associated with these points are called **opposites**. Since the points are the same distance from the origin, we say that the numbers have the same magnitude or **absolute value**. Thus, 2 and −2 both have absolute value 2. A pair of vertical lines is used to denote the absolute value of a number. That is, the absolute value of −2 would be denoted as |−2|. Thus, |2| = |−2| = 2.

Example 6   a. $-\frac{1}{3}$ and $\frac{1}{3}$ are opposites. The absolute value of both $\frac{1}{3}$ and $-\frac{1}{3}$ is $\frac{1}{3}$.

That is, $\left|\frac{1}{3}\right| = \left|-\frac{1}{3}\right| = \frac{1}{3}$.

b. 100 is the opposite of −100. The absolute value of each is 100.
Thus, |100| = |−100| = 100.

c. −1.5 is the opposite of 1.5, and the absolute value of each number is 1.5.
|−1.5| = |1.5| = 1.5

d. $\left|-11\frac{7}{8}\right| = 11\frac{7}{8}$ e. |5| = 5

f. |0| = 0   g. |−7| = 7

The symbol "−5" is usually read "negative five." However, since −5 is the opposite of 5, "−5" can be read "the opposite of five." This particular interpretation is useful in certain situations. For example, what is −(−10) equal to? Since −(−10) is the opposite of −10, it is equal to 10.

Example 7   a. The opposite of +4 is −4, or −(+4) = −4.

b. −(−4) = 4 since the opposite of −4 is 4.

c. −(−32) = 32 and −(+32) = −32.

The number line offers an excellent visual representation of the concepts of **greater than** and **less than**. These order relations will be examined in more detail in Chapter 13. For now we will merely introduce these concepts and their relationship to the number line. As you know, 3 is greater than 2. On the number line, this is illustrated by the fact that the graph of 3 lies to the right of the graph of 2.

Similarly, 2 is less than 3 and the graph of 2 lies to the left of the graph of 3. In fact, if the graphs of any two numbers are represented on the number line, the number to the right is greater than the number to the left. Of course, this means that the number to the left is always less than the number to the right.

Example 8   Fill in the blank with " is greater than " or " is less than."

a. 4 _____ 5.71

4 __is less than__ 5.71.

b. −4 _____ −9

Locate both numbers on the number line.

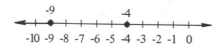

Since the graph of −4 lies to the right of the graph of −9,
−4 __is greater than__ −9.

c. −9 _____ −4

Since the graph of −9 lies to the left of the graph of −4,
−9 __is less than__ −4.

d. 7.82 _____ 7.11

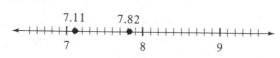

7.82 __is greater than__ 7.11.

e. 3.45 _____ 3.4   (3.4 = 3.40)

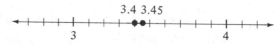

3.45 __is greater than__ 3.4.

f. −3.4 _____ −3.2

−3.4 __is less than__ −3.2.

g. 59 _____ $58\frac{1}{2}$

59 <u>is greater than</u> $58\frac{1}{2}$.

h. $3\frac{7}{8}$ _____ $3\frac{5}{8}$

$3\frac{7}{8}$ <u>is greater than</u> $3\frac{5}{8}$.

i. $-29\frac{4}{5}$ _____ $-29\frac{1}{5}$

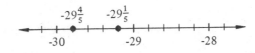

$-29\frac{4}{5}$ <u>is less than</u> $-29\frac{1}{5}$.

5.1   Exercises

1. Represent the graphs of the following numbers on a number line: $-4, 0, 3.5, \frac{1}{2}$.

2. Represent the graphs of 500, 600, and 750.

3. Represent the graphs of the following numbers: 0, 1, 2, 3, 4, etc.

4. Represent the graphs of the opposites of the following numbers: 1, 2, 3, 4, 5, etc.

5. Represent the graph of $5\frac{5}{6}$. *Hint:* Divide the portion of the number line between 5 and 6 into six equal parts.

6. Represent the graphs of the following even numbers: 2, 4, 5, 6, etc.

7. Represent the graphs of the following odd numbers: 1, 3, 5, 7, 9, etc.

8. Represent the graph of $4\frac{3}{8}$.

9. List the numbers whose graphs are represented below.

10. Find the coordinates of the points indicated on the number line below.

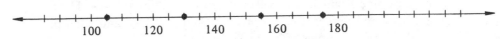

11. Give the coordinates of the points indicated on the number line below.

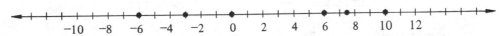

Give the opposite and absolute value of each of the numbers in Exercises 12–19.

12. 10

13. $-4\frac{1}{2}$

14. 0

15. -2

16. 3.59

17. 8

18. $-5\frac{1}{3}$

19. 0.07

20. What is the set of numbers whose graph is shown below?

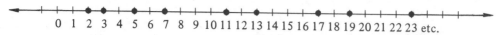

0 1 2 3 4 5 6 7 8 9 10 11 12 13 14 15 16 17 18 19 20 21 22 23 etc.

Use the negative sign to write each expression in Exercises 21-24 in *symbols*.

21. The opposite of $\frac{2}{3}$

22. The opposite of -10.7

23. The opposite of 113,405

24. The opposite of -0.0049

Answer each of Exercises 25-31 with 36 or -36.

25. What is the opposite of 36?

26. What is the opposite of -36?

27. $-(-36) = ?$

28. $-(+36) = ?$

29. What is the opposite of a positive thirty-six?

30. What is the opposite of a negative thirty-six?

31. What is the absolute value of thirty-six?

Evaluate the expressions in Exercises 32-40.

32. $|-42|$

33. $|35|$

34. $|-17.8|$

35. $-|14|$

36. $|15|$

37. $\left|-5\frac{1}{2}\right|$

38. $\left|\frac{13}{4}\right|$

39. $|2.7|$

40. $-|-32|$

Fill each of the blanks in Exercises 41-55 with either the words "is greater than" or the words "is less than." Draw graphs of the numbers to check your answer.

41. 5 _____ 4.2

42. -4 _____ -5

43. -6 _____ -3

44. -1.2 _____ -1.1

45. 2.03 _____ 2.30

46. -3 _____ 0

47. -17 _____ -20

48. $4\frac{8}{9}$ _____ $4\frac{2}{9}$

49. 7.81 _____ 7.8

50. -3.245 _____ -3.24

51. 0 _____ $-2\frac{1}{2}$

52. -50 _____ -21

53. $-\frac{14}{39}$ _____ 1

54. $-\frac{14}{15}$ _____ $-\frac{7}{15}$

55. $-15\frac{1}{3}$ _____ $-15\frac{2}{3}$

5.2 Squares and Square Roots

Every point on the number line is associated with a number, and the numbers which correspond to the points are called **real numbers**. The real numbers we have considered up to this point are named rational numbers. A **rational number** is any number which can be written as the quotient (ratio) of two integers where the denominator is not 0. Thus the rational numbers include not only the real numbers which are already expressed as fractions, but also those numbers which can be rewritten in fractional form. The decimals, mixed numbers, and percents considered in Chapters 1-4, and the opposites of these numbers, are examples of rational numbers.

Even though every rational number corresponds to a point on the number line, the points associated with the rational numbers do not "fill up" the line. The real numbers associated with the points which remain are called **irrational numbers**. Each of these collections of numbers—the rational numbers and the irrational numbers—contains an infinite number of elements. The following chart shows the relationship among the real numbers, the rational numbers, and the irrational numbers.

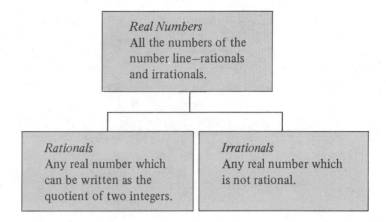

Before considering some specific examples of irrational numbers, we will look at some notation and terminology which will be useful.

When a number is multiplied by itself, it is said to be **squared.** A small numeral 2 placed above and to the right of a number indicates that the number is to be squared. For example, $5^2 = 5 \cdot 5 = 25$.

Example 1   Evaluate each of the following.

a. $9^2 = 9 \cdot 9 = 81$

b. $(-4)^2 = (-4)(-4) = 16$

c. $(2.4)^2 = (2.4)(2.4) = 5.76$

The product obtained when a natural number is squared is called a **perfect square.** Thus 25 is a perfect square since $5^2 = 25$.

Example 2   a. 81 is a perfect square since $9^2 = 81$.

b. 100 is a perfect square since $(10)^2 = 100$.

c. 3 is not a perfect square because there is no natural number whose square is 3.

It would be helpful for you to be familiar with the following perfect squares:

$$
\begin{array}{ll}
1 = 1^2 & 49 = 7^2 \\
4 = 2^2 & 64 = 8^2 \\
9 = 3^2 & 81 = 9^2 \\
16 = 4^2 & 100 = (10)^2 \\
25 = 5^2 & 121 = (11)^2 \\
36 = 6^2 & 144 = (12)^2
\end{array}
$$

Irrational numbers are *not* rational, so they *cannot* be written in fractional form. One of the most familiar examples of an irrational number is $\sqrt{2}$, read "the square root of two." The symbol $\sqrt{}$ is called the **radical symbol** and the number under the radical is the **radicand**. To find the **square root** of a number, you must find the number whose square is equal to the given number. For example, since $6^2 = 36$, $\sqrt{36} = 6$. Also, $\sqrt{64} = 8$ since $8^2 = 64$. Thus $\sqrt{2}$ is a number whose square is 2. Is it possible to find a rational number whose square is 2? Consider the squares of the following numbers:

$$(1.4)^2 = 1.96$$
$$(1.41)^2 = 1.9881$$
$$(1.414)^2 = 1.999396$$
$$(1.4142)^2 = 1.99996164$$
$$(1.41421)^2 = 1.9999899241$$
$$(1.414214)^2 = 2.000001237796$$

Note that the square of each of the numbers is fairly close to 2 but none is exactly equal to 2. In fact, there is no rational number whose square is 2. Thus $\sqrt{2}$ is an irrational number. However, 1.4, 1.41, 1.414, 1.4142, 1.14121, and 1.414214 are rational approximations of $\sqrt{2}$.* (There is a formal mathematical proof that $\sqrt{2}$ is irrational which is often presented in more advanced courses.)

Another irrational number with which you may be familiar is π, read "pi." This irrational number occurs in the formula for the circumference of a circle, $C = \pi d$, and in the formula for the area of a circle, $A = \pi r^2$. The rational approximations most frequently used for π are 3.14 and $\frac{22}{7}$. Some other examples of irrational numbers are:

$$\sqrt{3}, \sqrt{24}, \sqrt{\frac{2}{3}}, \sqrt{1.6}, \sqrt{5}, \sqrt{6}, \sqrt{7}$$

Additional examples will be given in Chapter 12.

Most of the examples of irrational numbers which have been given are square roots. However, while many square roots are irrational numbers, many are also rational. For example, we have already seen that $\sqrt{36} = 6$ and that $\sqrt{64} = 8$. In general, how do you evaluate a square root? Several techniques are available. These include the use of tables, the use of calculators, and the use of various algorithms which have been developed. In this section we will consider an algorithm (procedure) which can be used to determine the square root of a number if the number is a perfect square, or to find a rational approximation of the square root if the number is not a perfect square. First, we will consider square roots which are rational. You should be able to evaluate the square roots in Example 3 from your knowledge of perfect squares.

---

*A more accurate approximation was calculated by Dr. Dutka at Columbia University. In a two-year research project, he used a computer to calculate the square root of 2 accurate to one million decimal places. It begins 1.41421356237309504880168872.... . You can multiply this number by itself to check his accuracy!

Example 3   Evaluate each of the following.

a. $\sqrt{64} = 8$, since $8^2 = 64$.

b. $\sqrt{100} = 10$, since $(10)^2 = 100$.

c. $\sqrt{1} = 1$ d. $\sqrt{25} = 5$

e. $\sqrt{16} = 4$ f. $\sqrt{121} = 11$

If you are evaluating square roots of larger numbers, you may need to use a procedure involving trial and error.

Example 4   Evaluate each of the following.

a. $\sqrt{361}$

If you do not know what number squared is 361, guess.

Try 10. $(10)^2 = 100$ so 10 is too small.
Try 20. $(20)^2 = 400$ so 20 is too large.
Try 19. $(19)^2 = 361$

Therefore, $\sqrt{361} = 19$.

b. $\sqrt{2209}$

$(20)^2 = 400$, so 20 is too small.
$(30)^2 = 900$, so 30 is too small.
$(40)^2 = 1600$, so 40 is too small.
$(50)^2 = 2500$, so 50 is too large.

We can see that $\sqrt{2209}$ must be between 40 and 50. Thus we check the numbers between 40 and 50.

$(41)^2 = 1681$ $(45)^2 = 2025$
$(42)^2 = 1764$ $(46)^2 = 2116$
$(43)^2 = 1849$ $(47)^2 = 2209$
$(44)^2 = 1936$

Thus $\sqrt{2209} = 47$.

The square root of a natural number which is not a perfect square is irrational. A procedure similar to the one above can be used to approximate these square roots. The procedure is given in the *rule for approximating square roots of natural numbers.*

RULE FOR APPROXIMATING SQUARE ROOTS OF NATURAL NUMBERS	*EXAMPLE* Evaluate $\sqrt{11}$. Round the answer to the nearest hundredth.

RULE FOR APPROXIMATING SQUARE ROOTS OF NATURAL NUMBERS

1. Determine the two consecutive natural numbers between which the square root lies.
2. The smaller of the natural numbers is the whole number part of the square root.
3. Guess the digit in the tenths position in order to obtain an approximation of the square root.
4. Square this approximation.
5. a. If the square of this approximation is less than the radicand, square the number which is one-tenth larger and continue increasing by one-tenth until the square is larger than the radicand. If you are rounding to tenths, the number whose square is closest to the radicand is the desired approximation. If you are not rounding to tenths, the tenths digit identified in the *next to last squaring* is the tenths digit in the approximation.

   b. If the square of the approximation is greater than the radicand, square the number which is one-tenth smaller and continue decreasing by one-tenth until the square is less than the radicand. If you are rounding to tenths, the number whose square is closest to the radicand is the desired approximation. If you are not rounding to tenths, the tenths digit in the *last squaring* is the tenths digit in the approximation.
6. Repeat steps 3, 4, and 5 replacing tenths with hundredths, then tenths with thousandths, and so forth until the desired accuracy is obtained.

EXAMPLE   Evaluate $\sqrt{11}$. Round the answer to the nearest hundredth.
Solution:

$$\sqrt{9} = 3 \text{ and } \sqrt{16} = 4$$

Thus, $\sqrt{11}$ lies between 3 and 4.

$$\sqrt{11} \doteq 3.\underline{?}\,\underline{?}$$

If you guess 0.1, the approximation is 3.1.

$$(3.1)^2 = 9.61$$

9.61 is less than 11.

$$(3.2)^2 = 10.24$$
$$(3.3)^2 = 10.89$$
$$(3.4)^2 = 11.56$$

Thus, $\sqrt{11} \doteq 3.3\,\underline{?}$

If you guess 0.5 the approximation is 3.5.

$$(3.5)^2 = 12.25$$

12.25 is greater than 11.

$$(3.4)^2 = 11.56$$
$$(3.3)^2 = 10.89$$

Thus, $\sqrt{11} \doteq 3.3\,\underline{?}$

To obtain the next digit in the square root, guess the digit in the hundredths position. If you guess 0.01, the approximation is 3.31.

$$(3.31)^2 = 10.9561$$
$$(3.32)^2 = 11.0224$$
$$11 - 10.9561 = 0.0439$$
$$11.0224 - 11 = 0.0224$$

Therefore, 11.0224 is closest to 11 so $\sqrt{11} \doteq 3.32$.

Example 5   Evaluate $\sqrt{39}$ to one decimal place.

$\sqrt{39}$ is between 6 and 7 since $6^2 = 36$ and $7^2 = 49$.

$$6.1^2 = 37.21 \quad \text{(less than 39)}$$
$$6.2^2 = 38.44 \quad \text{(less than 39)}$$
$$6.3^2 = 39.69 \quad \text{(more than 39)}$$

Since $39 - 38.44 = 0.56$ and $39.69 - 39 = 0.69$, 38.44 is closest to 39.

Thus $\sqrt{39} \doteq 6.2$.

Example 6   Evaluate $\sqrt{20}$ to the nearest tenth.

$\sqrt{20}$ is between 4 and 5 since $4^2 = 16$ and $5^2 = 25$.

$$(4.2)^2 = 17.64$$
$$(4.3)^2 = 18.49$$
$$(4.4)^2 = 19.36 \qquad 20 - 19.36 = 0.64$$
$$(4.5)^2 = 20.25 \qquad 20.25 - 20 = 0.25$$

Thus $\sqrt{20} \doteq 4.5$.

Example 7   Evaluate $\sqrt{46,656}$.

$\sqrt{46,656}$ is between 200 and 300 since $(200)^2 = 40,000$ and $(300)^2 = 90,000$.

$$(210)^2 = 44,100$$
$$(220)^2 = 48,400$$

Therefore, $\sqrt{46,656}$ is between 210 and 220.

$$(211)^2 = 44,521 \qquad (214)^2 = 45,796$$
$$(212)^2 = 44,944 \qquad (215)^2 = 46,225$$
$$(213)^2 = 45,369 \qquad (216)^2 = 46,656$$

Thus, $\sqrt{46,656} = 216$.

Square roots can also be determined by using tables. A table of squares and square roots is located in Appendix C. The excerpt below is from this table. Unless the number in the column headed "n" is a perfect square, the square root in the table is an approximation.

n	n^2	$\sqrt{n}$
35	1,225	5.916
36	1,296	6.000
37	1,369	6.083
38	1,444	6.164
39	1,521	6.245
40	1,600	6.325
41	1,681	6.403
42	1,764	6.481
43	1,849	6.557
44	1,936	6.633
45	2,025	6.708

Example 8 Use the table to approximate each of the following square roots to the nearest tenth.

    a. $\sqrt{38}$

        Locate 38 in the column headed *n*. Go across this row until you reach the column headed $\sqrt{n}$. In this column, you should find 6.164. Since you are asked to find the square root to the nearest tenth, round this approximation to tenths.

        Thus $\sqrt{38} \doteq 6.2$.

    b. $\sqrt{44} \doteq 6.6$

    c. $\sqrt{36} = 6$

5.2 Exercises

Evaluate each of the squares in Exercises 1–10 without using a table.

1. 8^2	**2.** $(12)^2$
3. $(21)^2$	**4.** $(3.5)^2$
5. 4^2	**6.** $(0.02)^2$
7. $(6.51)^2$	**8.** $(300)^2$
9. $(40)^2$	**10.** $(0.008)^2$

Evaluate each of the square roots in Exercises 11–20.

11. $\sqrt{49}$	**12.** $\sqrt{81}$
13. $\sqrt{400}$	**14.** $\sqrt{9}$
15. $\sqrt{225}$	**16.** $\sqrt{900}$
17. $\sqrt{144}$	**18.** $\sqrt{289}$
19. $\sqrt{196}$	**20.** $\sqrt{169}$

Approximate the square roots in Exercises 21–30 to the nearest tenth using the table in Appendix C.

21. $\sqrt{28}$	**22.** $\sqrt{44}$
23. $\sqrt{10}$	**24.** $\sqrt{50}$
25. $\sqrt{56}$	**26.** $\sqrt{72}$
27. $\sqrt{89}$	**28.** $\sqrt{52}$
29. $\sqrt{24}$	**30.** $\sqrt{84}$

Approximate the square roots in Exercises 31–40 to the nearest tenth without using the table.

31. $\sqrt{7}$	**32.** $\sqrt{18}$
33. $\sqrt{21}$	**34.** $\sqrt{529}$
35. $\sqrt{75}$	**36.** $\sqrt{89}$
37. $\sqrt{15}$	**38.** $\sqrt{12}$
39. $\sqrt{2304}$	**40.** $\sqrt{78}$

5.3 Addition of Signed Numbers

Our development of the number line lead us to a discussion of both rational and irrational numbers. However, we will limit our treatment of operations in this chapter to rational numbers. First, we will consider addition of positive and negative numbers.

We have observed that each negative number is the opposite of a positive number. Although it is natural to associate some numbers with physical objects, this is not the case with negative numbers. For example, the shepherd may see seven lambs on a hillside, but his sanity would probably be questioned if he said, "Look at my negative five lambs." Nonetheless, the shepherd might find the concept of a negative five appropriate if a wolf were to ravage his flock. A negative five would mean, in this case, a loss of five sheep—the opposite of a gain of five sheep. Similarly, "−12" yards could be used to indicate a loss of 12 yards by a football team. The fact that the absolute value of both 12 and −12 is 12 indicates that the yardage is the same whether the team gains or loses. The positive number represents a gain while the negative number indicates a loss.

You can probably perform many simple computations without the aid of formal rules. Before we examine the rules which work for all signed numbers, try to determine the net result of the following business transactions.

1. A gain of $20 followed by a second gain of $30.

2. A loss of $20 followed by a second loss of $30.

3. A gain of $20 followed by a loss of $30.

4. A loss of $20 followed by a gain of $30.

The number line gives us a basis for illustrating these four problems as follows:

1. A gain of $20 and a second gain of $30.

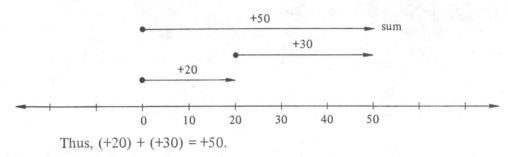

Thus, $(+20) + (+30) = +50$.

2. A loss of $20 and a second loss of $30.

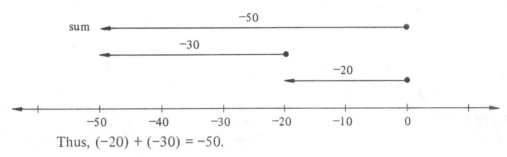

Thus, $(-20) + (-30) = -50$.

3. A gain of $20 and a loss of $30.

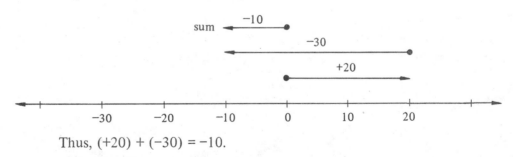

Thus, $(+20) + (-30) = -10$.

4. A loss of $20 and a gain of $30.

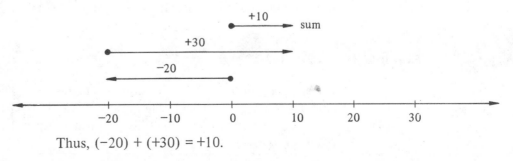

Thus, $(-20) + (+30) = +10$.

These graphs illustrate every possible sum of signed numbers: a positive plus a positive, a negative plus a positive, a positive plus a negative and a negative plus a negative. However, it would be inconvenient to use the number line each time we need to add signed numbers. A careful analysis of the graphs suggests the *rule for adding signed numbers.*

RULE FOR ADDING SIGNED NUMBERS	EXAMPLES
Like Signs: Find the sum of the absolute values of the numbers, and prefix the common sign.	$(+20) + (+30) = +50$ $(-20) + (-30) = -50$
Unlike Signs: Find the difference in the absolute values of the numbers and prefix the sign of the number having the larger absolute value.	$(+20) + (-30) = -10$ $(-20) + (+30) = +10$

Example 1 $(+20) + (-30) = ?$

The signs are unlike so we use the second part of the rule. Since the absolute value of $+20$ is 20 and the absolute value of -30 is 30, the difference in the absolute values is 10.

Also, since the negative number has the larger absolute value, the sum will be negative. Thus, $(+20) + (-30) = -10$.

Example 2 $\left(-2\frac{3}{4}\right) + \left(-5\frac{5}{8}\right) = ?$

The signs are like so we add the absolute values of $-2\frac{3}{4}$ and $-5\frac{5}{8}$, which are $2\frac{3}{4}$ and $5\frac{5}{8}$ respectively.

Now,

$$2\frac{3}{4} + 5\frac{5}{8} = 2\frac{6}{8} + 5\frac{5}{8}$$

$$= 7\frac{11}{8}$$

$$= 8\frac{3}{8}$$

Since the like signs were negative, $\left(-2\frac{3}{4}\right) + \left(-5\frac{5}{8}\right) = -8\frac{3}{8}$.

Example 3 $(+6.2) + (+10.31) = ?$

The signs are both "+" so the sum will be positive. The absolute values of the numbers are 6.2 and 10.31. The sum of 6.2 and 10.31 is 16.51. Thus, $(+6.2) + (+10.31) = +16.51$. Note that this problem is equivalent to the problem 6.2 + 10.31.

5.3   Exercises

Add in Exercises 1–70.

1. $(+7) + (+3)$ **+ 10**
2. $(-15) + (-4)$
3. $(-43) + (-17)$ **– 60**
4. $(+8) + (+80)$
5. $(+15) + (+13)$ **+ 28**
6. $(+50) + (+42)$
7. $(+104) + (+720)$ **+ 824**
8. $(-17) + (+17)$
9. $(-450) + (-75)$
10. $(-4) + (-68)$
11. $(+7) + (-3)$
12. $(-15) + (+4)$
13. $(-43) + (+17)$
14. $(+8) + (-80)$
15. $(+17) + (-17)$
16. $(-104) + (+720)$
17. $(-450) + (+75)$
18. $(+50) + (-42)$
19. $(+4) + (-68)$
20. $(+50) + (-50)$
21. $(-6) + (+2)$
22. $(-10) + (-15)$
23. $(+54) + (-21)$
24. $(+10) + (+15)$
25. $(-69) + (-13)$
26. $(+10) + (-15)$
27. $(+31) + (+7)$
28. $(-10) + (+15)$
29. $(-14) + (+10)$
30. $(+35) + (+79)$
31. $(-21) + (-14)$
32. $(-92) + (+10)$
33. $(-75) + (-18)$
34. $(+5) + (+801)$
35. $(-82) + (+17)$
36. $(-14) + (+73)$
37. $(+41) + (+18)$
38. $(+14) + (-73)$
39. $(+69) + (-43)$
40. $(-19) + (-14)$
✗ 41. $(+0.004) + (-0.02)$
42. $(-0.52) + (-2.2)$

43. $\left(+3\frac{1}{3}\right) + \left(+6\frac{2}{3}\right)$
44. $\left(+8\frac{2}{3}\right) + \left(-11\frac{1}{2}\right)$
45. $(+6.4) + (-2.5)$
46. $(-6.4) + (+6.4)$
47. $\left(-5\frac{2}{7}\right) + (+3)$
48. $(+0.02) + (-0.004)$
49. $(+3) + (+2)$
50. $\left(+6\frac{7}{8}\right) + \left(+3\frac{1}{8}\right)$
51. $(-6) + (+4)$
52. $\left(-\frac{9}{10}\right) + \left(-\frac{3}{10}\right)$
✗ 53. $(-6.4) + (-.35)$
54. $(+15) + (-6)$
55. $\left(-\frac{7}{9}\right) + \left(-\frac{2}{3}\right)$
56. $\left(+\frac{5}{11}\right) + \left(+\frac{6}{11}\right)$
57. $(+2.5) + (+7.1)$
58. $(-61) + (-16)$
59. $\left(-\frac{5}{11}\right) + \left(-\frac{7}{11}\right)$
60. $(-50) + (+40)$
✗ 61. $\left(+7\frac{1}{5}\right) + \left(-9\frac{2}{3}\right)$
62. $(+17) + (+1.7)$
63. $(+9) + (-4)$
64. $\left(-\frac{5}{8}\right) + \left(-\frac{3}{4}\right)$
65. $(-3.45) + (+3.45)$
66. $\left(-4\frac{2}{3}\right) + (+1)$
67. $\left(+\frac{12}{19}\right) + \left(+\frac{7}{19}\right)$
68. $(+3.5) + (-4)$
✗ 69. $(+2.3) + \left(-\frac{1}{2}\right)$
70. $\left(-1\frac{3}{4}\right) + (-0.75)$

5.4   Subtraction of Signed Numbers

Before a general rule is stated for subtracting signed numbers, write a signed number which indicates the result of each of the following changes in temperature. A positive number should be used to denote a rise in temperature and a negative number should be used to denote a drop in temperature.

1. A change in temperature from +20 degrees to +10 degrees.
2. A change in temperature from −20 degrees to −10 degrees.
3. A change in temperature from +20 degrees to −10 degrees.
4. A change in temperature from −20 degrees to +10 degrees.

The following figures illustrate these four problems.

1. A change from +20 to +10.   2. A change from −20 to −10.

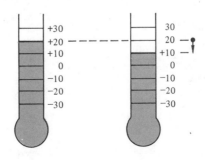

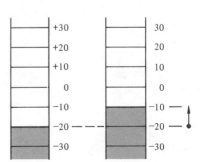

3. A change from +20 to −10.   4. A change from −20 to +10.

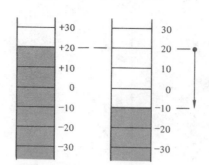 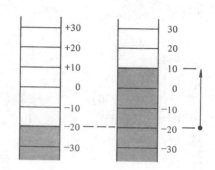

Each of these problems can be interpreted as a subtraction problem in which the original temperature is the subtrahend and the final temperature is the minuend. That is, the change when there is a drop from +20 degrees to +10 degrees could be written as (+10) − (+20) = −10. Similarly, the change in temperature when there is a rise from −20 degrees to −10 degrees is (−10) − (−20) = +10.

This interpretation can be used to find the answer to any signed number subtraction problem. For example, (+15) − (−20) could be interpreted as the change in temperature when there is a rise from −20 degrees to +15 degrees. Thus, (+15) − (−20) = +35. Since this model is cumbersome, we commonly use the procedure which is given in the *rule for subtracting signed numbers.*

RULE FOR SUBTRACTING SIGNED NUMBERS	EXAMPLE (−6) − (−2) = ?

1. Rewrite the problem as an addition problem by
  a. changing the operation to addition
  b. replacing the subtrahend with its opposite (the minuend is not changed)
2. Perform the addition by using the rule for adding signed numbers.
	Solution:
	
	(−6) + (−2)
	+2
	
	(−6) + (+2) = −4

Example 1   a. (+10) − (+20) = (+10) + (−20) = −10

b. (−10) − (−20) = (−10) + (+20) = +10

c. (−10) − (+20) = (−10) + (−20) = −30

d. (+10) − (−20) = (+10) + (+20) = +30

Note that each subtraction problem is changed to an equivalent addition problem in order to determine the answer.

5.4 Exercises

Rewrite each of Exercises 1–40 as an addition problem, and *then* evaluate this expression.

1. (−6) − (+14)

3. (−4) − (−10)

5. (+64) − (−14)

2. (+12) − (−4)

4. (−25) − (+13)

6. (−42) − (−42)

7. (+8) − (−10)

9. (−25) − (+13)

11. (+12) − (+13)

13. (+64) − (−64)

15. (+81) − (+54)

8. (+42) − (+42)

10. (−8) − (−10)

12. (−8) − (+10)

14. (−64) − (−64)

16. (−18) − (−45)

17. $(+24) - (+21)$ 18. $(-7) - (+10)$ 29. $(+25) - (+71)$ 30. $(-61) - (-16)$

19. $(+10) - (-10)$ 20. $(+34) - (-5)$ 31. $(-55) - (-77)$ 32. $(-50) - (+40)$

21. $(+7) - (+2)$ 22. $(+48) - (+25)$ 33. $(+35) - (-27)$ 34. $(+17) - (+17)$

23. $(-6) - (+4)$ 24. $(-90) - (-30)$ 35. $(+9) - (-4)$ 36. $(-40) - (-40)$

25. $(-64) - (-35)$ 26. $(+15) - (-6)$ 37. $(-345) - (+345)$ 38. $(-14) - (+1)$

27. $(-63) - (-6)$ 28. $(+55) - (+66)$ 39. $(-12) - (-12)$ 40. $(+35) - (-4)$

5.5 Expressions in Simplified Notation

Expressions involving addition and subtraction of signed numbers are said to be written in **simplified notation** if only a single sign appears between numbers in the expression. The expression $-6 - 4$ is written in simplified notation while $-6 + (-4)$ is not since two signs appear between the 6 and the 4.

Signed number problems are usually written in simplified notation in textbooks and in frequently-used formulas since there are fewer symbols to type and to read with this notation. To work effectively with signed number problems, you need to be able to interpret this notation. In the expression $-6 - 4$, the negative sign can be interpreted to mean subtraction. The 4 is then understood to be positive. That is,

$$-6 - 4 = -6 - (+4)$$
$$= -6 + (-4)$$
$$= -10$$

However, the same result is obtained if you interpret the simplified notation as follows: *take each sign with the number it precedes, and consider the operation to be addition.* That is, $-6 - 4$ would be interpreted as $-6 + (-4)$ which is equal to -10. With this interpretation, it is not necessary to rewrite the problem.

$$-6 - 4 = -10 \qquad (\textit{Think: } -6 \text{ plus } -4.)$$

RULE FOR EVALUATING EXPRESSIONS IN SIMPLIFIED NOTATION	*EXAMPLE* $-3 - 7 = ?$
1. Associate each sign in the expression with the number it precedes. 2. Add the numbers obtained in step 1 following the rule for addition of signed numbers.	Consider the numbers as -3 and -7. $(-3) + (-7) = -10$ Thus, $-3 - 7 = -10$.

Example 1 $6 - 10 = ?$

$$6 + (-10) = -4 \qquad (\textit{Think: } 6 \text{ plus } -10.)$$

Thus, $6 - 10 = -4$

Note: Recall that 6 is understood to be positive since no sign precedes it.

Example 2 $7 - 3 = ?$

$$7 + (-3) = 4 \qquad (\textit{Think: } 7 \text{ plus } -3.)$$

Example 3   a. $-6 + 2 = -4$ (*Think:* -6 plus $+2$.)

b. $-6 - 2 = -8$ (*Think:* -6 plus -2.)

c. $6 - 2 = 4$ (*Think:* $+6$ plus -2.)

d. $6 + 2 = 8$ (*Think:* $+6$ plus $+2$.)

Expressions which are written in simplified notation can frequently be evaluated more easily than expressions containing two signs between numbers. The *rule for changing to simplified notation* can be used to rewrite problems in which there are two signs between numbers.

RULE FOR CHANGING TO SIMPLIFIED NOTATION	EXAMPLES
Like Signs: If two adjacent signs are alike, they may be replaced by a plus symbol. *Unlike Signs:* If two adjacent signs are unlike they may be replaced by a minus symbol.	$6 + (+4) = 6 + 4$ $6 - (-4) = 6 + 4$ $6 + (-4) = 6 - 4$ $6 - (+4) = 6 - 4$

Example 4   Write each of the following expressions in simplified notation and then evaluate the expression.

a. $11 + (+19)$

$$11 + (+19) = 11 + 19$$
$$= 30$$

Thus $11 + (+19) = 30$.

b. $11 - (-19)$

$$11 - (-19) = 11 + 19$$
$$= 30$$

Thus $11 - (-19) = 30$.

c. $11 + (-19)$

$$11 + (-19) = 11 - 19$$
$$= -8$$

Thus $11 + (-19) = -8$.

d. $11 - (+19)$

$$11 - (+19) = 11 - 19$$
$$= -8$$

Thus $11 - (+19) = -8$.

Example 5   Evaluate.

a. $-10 - (-3)$

$$-10 - (-3) = -10 + 3$$
$$= -7$$

Thus $-10 - (-3) = -7$.

b. $10.3 + (-0.7)$

$$10.3 + (-0.7) = 10.3 - 0.7$$
$$= 9.6$$

Thus $10.3 + (-0.7) = 9.6$.

c. $-6\frac{2}{3} - \left(-\frac{1}{3}\right)$

$$-6\frac{2}{3} - \left(-\frac{1}{3}\right) = -6\frac{2}{3} + \frac{1}{3}$$

$$= -6\frac{1}{3}$$

Thus $-6\frac{2}{3} - \left(-\frac{1}{3}\right) = -6\frac{1}{3}$.

In addition problems, you can evaluate from double notation or you can convert to simplified notation before evaluating. A double notation subtraction problem *must be* either converted to an equivalent addition problem or rewritten in simplified notation before evaluating. However, it is important that you learn to evaluate expressions written in simplified notation without rewriting.

5.5 Exercises

Evaluate the expressions in Exercises 1–40.

1. $7 - 9$
2. $-18 + 9$
3. $-34 - 16$
4. $-5 - 10$
5. $-50 + 32$
6. $52 - 75$
7. $16 - 12$
8. $15 - 15$
9. $-7 - 18$
10. $-17 + 80$
11. $(-72) - (+32)$
12. $(-15) - (+6)$
13. $72 + (-32)$
14. $13 - (-13)$
15. $-25 - (-45)$
16. $-7 + (-3)$
17. $5 + (-6)$
18. $(+7) + (-10)$
19. $72 - (-32)$
20. $(-7) - (-3)$
21. $18 - 40$
22. $(-15) - (-17)$
23. $85 + 100$
24. $14 - (+14)$
25. $22 - (-7)$
26. $(+32) - (-42)$
27. $(-8) + (-8)$
28. $75 - 60$
29. $(-8) - (-8)$
30. $-43 - 42$
31. $-29 - 29$
32. $(-15) + (-17)$
33. $-15 - 33$
34. $43 - (-7)$
35. $23 - 35$
36. $-15 + 17$
37. $-72 + 80$
38. $15 - 17$
39. $(-18) + 5$
40. $42 + (+73)$

Evaluate each of the expressions in Exercises 41–70. Translate each of the word problems into an expression involving signed numbers, and express the answer as a signed number.

41. $\frac{5}{6} + \left(-\frac{7}{9}\right) = ?$

42. $\frac{2}{3} - (-4) = ?$

43. What is 7 minus 3 equal to?

44. Subtract −4 from −12.

45. Decrease 13 by −13.

46. A thermometer registers −32 degrees. The temperature rises 30 degrees. What does the thermometer then register?

47. Death Valley, the lowest spot in the continental U.S., is 282 feet below sea level. Mt. Whitney has an altitude of 14,495 feet above sea level. How high above Death Valley is the top of Mt. Whitney?

48. Find the result if −5 is increased by 7.

49. $-8.7 - (+2.53) = ?$

50. Mrs. C. Smith has a credit of $17.45 at Rich's. What was the state of her account after she charged a dress which cost $25.75?

51. $\left(-7\frac{2}{9}\right) + \left(-3\frac{7}{9}\right) = ?$

52. What is −100 plus 0.43?

53. A 190-pound man lost 10 pounds one week, lost 4 pounds the second week, and gained 2 pounds the third week. What was his weight at the end of the three weeks?

54. What is 20 more than −11?

55. A football team lost 7 yards on a first down, gained 10 yards on the second down, and gained $2\frac{1}{2}$ yards on the third down. What was the net yardage on the three plays?

56. What is $-\frac{1}{2}$ minus $\frac{3}{2}$ equal to?

57. Subtract 8 from -8.

58. $(+72) - (-72) = ?$

59. $7.9 - 1.3 = ?$

60. From -25, subtract 8.

61. What is 0.02 more than 0.46?

62. $\left(+24\frac{1}{9}\right) + \left(+22\frac{5}{9}\right) = ?$

63. Bill had expenses of $5.50 on Monday and $22.32 on Tuesday. On Wednesday he earned $25.00. Did his earnings on Wednesday cover his expenses from Monday and Tuesday? Express his financial condition as a signed number.

64. Increase $-16\frac{1}{2}$ by $3\frac{1}{2}$.

65. Find the sum of $-3\frac{1}{3}$ and $5\frac{2}{3}$.

66. $-16\frac{1}{2} - 3\frac{1}{2} = ?$

67. Subtract $2\frac{5}{8}$ from -8.

68. Add $\frac{4}{5}$ to $-\frac{3}{5}$.

69. What is the sum of $+5.7$ and -4.5?

70. Subtract $-\frac{3}{5}$ from $\frac{4}{5}$.

5.6 Multiplication and Division of Signed Numbers

We need to consider four cases for multiplication. They are:

1. A positive number times a positive number.
2. A positive number times a negative number.
3. A negative number times a positive number.
4. A negative number times a negative number.

The following discussion will give some basis for the *rule for multiplying two signed numbers.*

Case 1: A positive number times a positive number.

You are already familiar with this case from your work with the positive rationals in Chapters 1, 2, and 3. A positive number times a positive number is a positive number. For example, $3 \times 4 = 12$.

Case 2: A positive number times a negative number.

Recall that 3×4 can be interpreted in terms of addition, that is, $3 \times 4 = 4 + 4 + 4 = 12$. Similarly, $3 \times (-4) = (-4) + (-4) + (-4) = -12$. This illustrates that a positive number times a negative number is a negative number.

Case 3: A negative number times a positive number.

Recall that two numbers can be multiplied in any order, that is, 4×3 is the same as 3×4. It is also true that $(-4) \times 3 = 3 \times (-4)$. In Case 2 we showed that $3 \times (-4) = -12$; thus, $(-4) \times 3 = -12$. We have illustrated that the product of a negative number and a positive number is negative; that is, Case 3 and Case 2 are the same.

Case 4: A negative number times a negative number.

To determine the product of two negative numbers, consider the pattern illustrated by the following products.

$$(3) \times (-3) = -9$$
$$(2) \times (-3) = -6$$
$$(1) \times (-3) = -3$$
$$(0) \times (-3) = 0$$
$$(-1) \times (-3) = ?$$

The answer to $(-1) \times (-3)$ is $+3$. This suggests that the product of two negative numbers is positive.

RULES FOR MULTIPLYING TWO SIGNED NUMBERS	EXAMPLES
Like Signs: Find the product of the absolute value of the two numbers, and prefix a plus sign.	$(+3) \times (+4) = +12$ $(-3) \times (-4) = +12$
Unlike Signs: Find the product of the absolute values of the two numbers, and prefix a negative sign.	$(-4) \times (+3) = -12$ $(+4) \times (-3) = -12$

Example 1   Evaluate the following.

a. $(+5) \times (+7)$

Since the signs are like, the first part of the rule states that the product will be positive. The product of the absolute values is 35. Thus $(+5) \times (+7) = 35$.

b. $(-5) \times (-7)$

Since the signs are like, the product will be positive. The product of the absolute values, 5×7, is 35. Thus $(-5) \times (-7) = +35$.

c. $(-5) \times (+7)$

Since the signs are unlike, the second part of the rule states that the product will be negative. The product of the absolute values is 35. Thus $(-5) \times (+7) = -35$.

d. $(+5) \times (-7)$

Since the signs are unlike, the product will be negative. Thus $(+5) \times (-7) = -35$.

Observe that the rule for multiplication is given for two factors. If the product of more than two factors is to be found, the rule must be applied to only two factors at a time. However, note that an odd number of negative factors will yield a negative product while an even number of negative factors will yield a positive product.

Example 2   Evaluate.

    a. $(-2)(3)(-4) = (-6)(-4)$
                                 $= +24$

    b. $(-2)(-3)(-4) = (+6)(-4)$
                                 $= -24$

    c. $(-1)(-2)(1)(-3)(-1)(-1) = -6$

       First, since there are five negative factors, the answer will be negative. Then, multiply the absolute values (1, 2, 1, 3, 1, 1) to obtain the answer.

    d. $(-4)(17)(0)(-1) = 0$

       You must multiply, but the answer can be easily determined without multiplying all the factors since 0 times any number is equal to 0.

The *rule for dividing signed numbers* is very similar to the rule for multiplication.

RULE FOR DIVIDING SIGNED NUMBERS	EXAMPLES
Like Signs: Find the quotient of the absolute values of the two numbers, and prefix a plus sign.	$(+6) \div (+3) = +2$ $(-6) \div (-3) = +2$
Unlike Signs: Find the quotient of the absolute values of the two numbers, and prefix a negative sign.	$(+6) \div (-3) = -2$ $(-6) \div (+3) = -2$

Example 3   Evaluate.

    a. $(+21) \div (+7)$

       The signs are like so the answer is positive.
       Thus $(+21) \div (+7) = +3$.

    b. $(-21) \div (-7)$

       The signs are like so the answer is positive. Since the absolute values are 21 and 7, $(-21) \div (-7) = +3$.

    c. $(-21) \div (+7)$

       The signs are unlike so the answer will be negative.
       Thus $(-21) \div (+7) = -3$.

    d. $(+21) \div (-7)$

       Since the signs are unlike, $(+21) \div (-7) = -3$.

Example 4 $24 \div (-0.6) = ?$

Since the signs are unlike, the answer will be negative.

$$\begin{array}{r} 4\,0. \\ .6_{\wedge}\overline{)24.0_{\wedge}} \end{array} \quad \text{Thus, } 24 \div (-0.6) = -40.$$

Example 5 $-\dfrac{2}{3} \div \left(-\dfrac{4}{3}\right) = ?$

$$\left(-\frac{2}{3}\right) \div \left(-\frac{4}{3}\right) = +\left(\frac{2}{3} \div \frac{4}{3}\right)$$

Since the signs are like, the quotient will be positive. The absolute values are $\dfrac{2}{3}$ and $\dfrac{4}{3}$.

$$= \frac{2}{3} \div \frac{4}{3}$$

$$= \frac{\overset{1}{\cancel{2}}}{\underset{1}{\cancel{3}}} \times \frac{\overset{1}{\cancel{3}}}{\underset{2}{\cancel{4}}}$$

$$= \frac{1}{2}$$

Thus $-\dfrac{2}{3} \div \left(-\dfrac{4}{3}\right) = +\dfrac{1}{2}$.

There are two special division problems with which you should be familiar. They are the type $3 \div 0$ $\left(\text{or } \dfrac{3}{0}\right)$ and $0 \div 3$ $\left(\text{or } \dfrac{0}{3}\right)$. Consider $3 \div 0$, that is, $0\overline{)3}$. The quotient must be some number that we can multiply by 0 to obtain 3. That is, $? \cdot 0 = 3$. There is no such number, so we say that $\dfrac{3}{0}$ is **undefined**. In general, division by zero is undefined. This important fact is stated in the box below.

> **Division by zero is undefined.**

Now consider $0 \div 3$, that is, $3\overline{)0}$. The quotient is some number that we can multiply by 3 to obtain 0. Since $0 \cdot 3 = 0$, $\dfrac{0}{3} = 0$. Thus, $\dfrac{3}{0}$ is undefined and $\dfrac{0}{3} = 0$.

Example 6   Evaluate each of the following.

    a. $\dfrac{4}{0}$ is undefined.

    b. $\dfrac{0}{-5} = 0$

    c. $\dfrac{-10}{0}$ = undefined

    d. $\dfrac{0}{7} = 0$

5.6   Exercises

1. $3 \cdot 5$

2. $(-7) \cdot 2$

3. $(-3)(-3)$

4. $100(-7)$

5. $71 \cdot 5$

6. $(-9)(-1000)$

7. $-15(1000)$

8. $(4)(20)$

9. $10(-150)$

10. $(-8)(-8)$

11. $10 \div 5$

12. $\dfrac{9}{3}$

13. $(-10) \div (+2)$

14. $64 \div (0)$

15. $(-12) \div (-6)$

16. $36 \div (-9)$

17. $\dfrac{-42}{7}$

18. $100 \div 25$

19. $15 \div (-15)$

20. $(-71) \div (-71)$

21. $-14 \cdot 5$

22. $(-2)(-2)$

23. $51 \div (-3)$

24. $(-11)(0)(6)$

25. $(-75) \div (0)$

26. $-70 \cdot 70$

27. $(-1)(1)(-1)(-1)$

28. $(-8) \div (-4)$

29. $\dfrac{57}{-19}$

30. $8 \div (-4)$

31. $(-3)(-5)(2)$

32. $51 \div (-17)$

33. $(0) \div (7)$

34. $0 \div (-7)$

35. $(-4)(3)(0)$

36. $\dfrac{14}{-14}$

37. $(-13)(-7)$

38. $(-1)(-1)(-1)(-1)$

39. $15 \div (-5)$

40. $11 \cdot 11$

41. Find the product of -152.72 and 100.

42. $(-3)\left(-\dfrac{1}{9}\right)$

43. Divide 12.5 by -0.5.

44. Divide -5 into 55.

45. $\left(\dfrac{2}{3}\right)\left(-\dfrac{3}{4}\right)$

46. -10 times $-7 = ?$

47. $(-3.9) \div (-0.3)$

48. $-25 \div 2.5$

49. Multiply 3.2 by -50.

50. Find the product of 7, $-\dfrac{1}{7}$, and 9.

51. $(-36) \div 1.2$

52. $\left(-\dfrac{2}{3}\right)\left(\dfrac{7}{5}\right)\left(-\dfrac{3}{4}\right)$

53. $\left(-\dfrac{1}{3}\right)\left(\dfrac{6}{5}\right)(-21)$

54. Divide $-2\dfrac{1}{22}$ by $1\dfrac{2}{11}$.

55. $(-12.5) \div 1.25$

56. $\left(-\dfrac{2}{3}\right) \cdot \left(-\dfrac{3}{4}\right)$

57. -15 times $\dfrac{5}{3} = ?$

58. $(-2)(-3)(-4)5$

59. $(5)(5)(-2)(-1)$

60. $\left(-\dfrac{2}{3}\right) \div \left(-\dfrac{3}{4}\right)$

61. $(15)(-20)(17)(0)(-1)$

62. Multiply -7 by 52.

63. Divide $3\dfrac{2}{3}$ into $-\dfrac{5}{9}$.   ANSWER 0

64. Divide -14 by $3\dfrac{1}{2}$.

65. $\dfrac{5}{6} \div \left(-\dfrac{7}{12}\right)$

66. $-36 \div (-1.8)$

67. $\left(-\dfrac{1}{3}\right)(-9)(-9)$

68. $\dfrac{-18}{-0.6}$

69. $9 \div 18$

70. $(-9) \div 36$

5.7   Order of Operations and Symbols of Grouping

Consider the problem $2 + 6 \times 3$. Does this mean $(2 + 6) \times 3$ or does it mean $2 + (6 \times 3)$? Mathematicians have agreed that $2 + 6 \times 3$ is understood to mean $2 + (6 \times 3)$. The procedure for solving such problems is given in the *order of performing operations* rule.

ORDER OF PERFORMING OPERATIONS WHEN THE ORDER IS NOT INDICATED BY GROUPING SYMBOLS	EXAMPLE $\quad 10 \div 2 \times 4 - 3 \times 1$
1. Perform all multiplications and divisions in the order in which they appear from left to right. 2. After performing all multiplications and divisions, perform all additions and subtractions as they appear from left to right.	*Solution:* $\begin{aligned} 10 \div 2 \times 4 - 3 \times 1 &= 5 \times 4 - 3 \times 1 \\ &= 20 - 3 \times 1 \\ &= 20 - 3 \\ &= 17 \end{aligned}$

Example 1   Give the operation which should be performed first in each of the following.

   a. $4 + 6 - 2 + 8$ Addition
   b. $6 - 6 \times 2 \div 8$ Multiplication
   c. $4 + 6 \div 2 \times 8$ Division
   d. $12 \div 2 \times 3 + 7$ Division

Example 2   Evaluate $2 \times 8 + 6 \div 2$.   operation

$$
\begin{aligned}
2 \times 8 + 6 \div 2 &= 16 + 6 \div 2 \qquad \times \\
&= 16 + 3 \qquad\qquad \div \\
&= 19 \qquad\qquad\quad +
\end{aligned}
$$

In Example 2, 2×8 was replaced with the single quantity 16, the quantity $6 \div 2$ was replaced with 3, and then the resulting expression $16 + 3$ was evaluated. In replacing 2×8 with 16 and $6 \div 2$ with 3 we have used an important principle of mathematics known as *the substitution principle.*

Substitution Principle: A quantity may be substituted for its equal in a mathematical expression without changing the value of the expression.

Example 3   Evaluate each of the following.   operation

   a.
$$
\begin{aligned}
8 \div 2 \times 3 + 1 &= 4 \times 3 + 1 \qquad \div \\
&= 12 + 1 \qquad\quad \times \\
&= 13 \qquad\qquad +
\end{aligned}
$$

   b.
$$
\begin{aligned}
2 + 6 \times 3 &= 2 + 18 \qquad \times \\
&= 20 \qquad\quad +
\end{aligned}
$$

   c.
$$
\begin{aligned}
8 - 3 \times 2 &= 8 - 6 \qquad \times \\
&= 2 \qquad\quad +
\end{aligned}
$$

   d.
$$
\begin{aligned}
4 + 6 \div 2 \times 8 &= 4 + 3 \times 8 \qquad \div \\
&= 4 + 24 \qquad\quad \times \\
&= 28 \qquad\qquad +
\end{aligned}
$$

   e.
$$
\begin{aligned}
4 + 6 - 2 + 8 &= 10 - 2 + 8 \qquad + \\
&= 8 + 8 \qquad\qquad - \\
&= 16 \qquad\qquad\quad +
\end{aligned}
$$

Example 4 $-8 + 4 - 2 = ?$

$$-8 + 4 - 2 = -4 - 2$$
$$= -6$$

In Example 4 we have used the order of operations rule and performed the addition in order from left to right. Another way of calculating the sum in Example 4 is as follows:

$$-8 + 4 - 2 = -8 - 2 + 4$$
$$= -10 + 4$$
$$= -6$$

This particular solution is based on two important properties of real numbers called the **commutative property** and the **associative property**. *Together, these properties allow us to *add* numbers in any order that is convenient.

> *The Generalized Associative and Commutative Property.* When finding the sum of several numbers, you may add the numbers in any order that is convenient.

Example 5 $-10 + 4 - 8 = ?$

Here, we are adding -10, $+4$, and -8. Any of the following solutions would be correct.

Solution 1: $-10 + 4 - 8 = -6 - 8$
$$= -14$$

Solution 2: $-10 + 4 - 8 = -10 - 4$
$$= -14$$

Solution 3: $-10 + 4 - 8 = -10 - 8 + 4$
$$= -18 + 4$$
$$= -14$$

Example 6   Evaluate $23 - 6 - 2 + 7$.

$$23 - 6 - 2 + 7 = 23 + 7 - 6 - 2$$
$$= 30 - 8$$
$$= 22$$

---

*See Appendix D for a statement of each of these properties.

Example 7   Evaluate $69 - 4 - 7 - 6$.

$$69 - 4 - 7 - 6 = 69 - 4 - 6 - 7$$
$$= 69 - 10 - 7$$
$$= 69 - 17$$
$$= 52$$

In many formulas and problems involving applications of signed numbers, symbols of grouping are frequently used to indicate the order of operation for some but not necessarily all of the calculations. Three symbols of grouping in common use are **parentheses** ( ), **brackets** [ ] , and **braces** {}. The fraction bar, —, is also used. These symbols may have several uses. We have already used them in expressions such as $(-3)(+4)$ to denote multiplication and in expressions such as $+3 - (-4)$ to separate signs in a problem. These symbols may also be used to enclose mathematical expressions which are to be treated as one quantity. Consider the following example.

Example 8 $2 + (3 \times 4) = 2 + 12$
$$= 14$$

Example 9 $(2 + 3) \times 4 = 5 \times 4$
$$= 20$$

Note: When the expression within a grouping symbol is replaced with a single number, the grouping symbol is usually omitted, except where it is needed to indicate multiplication or to separate two adjacent signs.

Example 10 $$\frac{(2 + 3) \times 4}{10} = \frac{5 \times 4}{10}$$
$$= \frac{20}{10}$$
$$= 2$$

Notice that in this example the fraction bar is considered a symbol of grouping.

Whenever more than one symbol of grouping appears in a problem, you should start from the inside grouping symbols and work toward the outside grouping symbols. In other words, the substitution principle should be used to rewrite the problem by replacing the expression within the innermost grouping symbols with a single number. This process should be continued until the value of the entire expression is obtained.

Example 11   Evaluate $6 - \{4 + 3 - [5 - 10] + 4\}$

$$
\begin{aligned}
6 - \{4 + 3 - [5 - 10] + 4\} &= 6 - \{4 + 3 - [-5] + 4\} \\
&= 6 - \{4 + 3 + 5 + 4\} \\
&= 6 - \{+16\} \\
&= 6 - 16 \\
&= -10
\end{aligned}
$$

Therefore, $6 - \{4 + 3 - [5 - 10] + 4\} = -10$.

Note: In the second line of the solution, $5 - 10$ has been replaced by -5. However, the brackets have been retained to separate the two adjacent negative symbols.

Example 12   Evaluate $7 + (-8 + 2) - [6 + \{3 - 9\} + 8]$

$$
\begin{aligned}
&7 + (-8 + 2) - [6 + \{3 - 9\} + 8] \\
&= 7 + (-6) - [6 + \{-6\} + 8] \\
&= 7 - 6 - [6 - 6 + 8] \\
&= 7 - 6 - 8 \\
&= -7
\end{aligned}
$$

Therefore, $7 + (-8 + 2) - [6 + \{3 - 9\} + 8] = -7$.

If a number is written next to a symbol of grouping with no operation symbol between the number and the grouping symbol, multiplication is implied.

Example 13 $2\{6 - [4 + 3]\}$

$$
\begin{aligned}
2\{6 - [4 + 3]\} &= 2\{6 - 7\} \\
&= 2\{-1\} \\
&= -2
\end{aligned}
$$

In many problems, you must use both the procedure for removing symbols of grouping and the rule on order of operations This procedure is summarized as follows:

Start with the expression within the innermost grouping symbol. Perform the indicated operations within this expression as specified in the order of operations rule. Replace this expression with the number obtained. Then repeat the above procedure until the original expression has been evaluated.

Example 14   Evaluate $3 + 4(6 - 9)$.

$$
\begin{aligned}
3 + 4(6 - 9) &= 3 + 4(-3) \\
&= 3 - 12 \\
&= -9
\end{aligned}
$$

Example 15 $-2(7-4)-5(2-12)=?$

$$-2(7-4)-5(2-12)=-2(3)-5(-10)$$
$$=-6+50$$
$$=44$$

Example 16   Evaluate $3-5\{2+3(6+4)+1\}$

$$3-5\{2+3(6+4)+1\}$$
$$=\ 3-5\{2+3(10)+1\}$$
$$=\ 3-5\{2+30+1\}$$
$$=\ 3-5\{33\}$$
$$=\ 3-165$$
$$=-162$$

5.7 Exercises

Evaluate each of the expressions in Exercises 1-10.

1. $16-12+28$ **2.** $11+17+4$

3. $-21-18-39$ **4.** $13-2+4-14$

5. $-3+8-7-10+4$ **6.** $-2-3-6$

7. $17+13-2-15$

8. $15-3-4-7+8+2$

9. $-10-12+20$ **10.** $-1-2-3+4$

Name the operation which should be performed first in each of Exercises 11-20.

11. $-3+3(-42)$ **12.** $(-8+7)(-3)$

13. $16\div4\times3+1$ **14.** $6+\{7+(4-8)\}$

15. $\dfrac{(-4+7)(-2)}{6}$ **16.** $\dfrac{5}{16}+\dfrac{1}{2}\times\dfrac{3}{8}$

17. $5\{4-2(4+6)\}$ **18.** $7-5(-15)$

19. $\dfrac{1}{5}+\left\{\dfrac{2}{3}-\left(\dfrac{5}{8}-\dfrac{2}{10}\right)\right\}$ **20.** $8\div32+4(-3)$

Evaluate each of the expressions in Exercises 21-100.

21. $6+2(4)$ **22.** $-3+3(-42)$

23. $7-5(-15)$ **24.** $-7-4(-13)$

25. $-3(10)+17$ **26.** $4(-3)-(-7)(-10)$

27. $(-8)(-3)+(-3)(7)$ **28.** $5(-15)-13$

29. $6+3(-2)-6(-4)$ **30.** $7-3(2)+4(-5)$

31. $-36+6\div3$ **32.** $100\div10-4\div7$

33. $-21\div7-4$ **34.** $-4+30\div(-3)$

35. $12\div(-12)-36\div12$ **36.** $-51\div17-10$

37. $4-17\div(-1)+10$

38. $-60\div15+18\div(-6)$ **39.** $-24-4\div4$

40. $-72-10\div(-2)$ **41.** $16\div4\times3+1$

42. $3\times(-6)\div2+1$ **43.** $2\times8\div(-4)+4$

44. $7-16\div4\times(-3)$ **45.** $(-12)\div3+1-8(2)$

46. $\dfrac{(-4+7)(-2)}{6}$

47. $8+4\div2+8\times2-6\div3$ **48.** $8\div32+4(-3)$

49. $\dfrac{(-4\times7)-2}{10}$ **50.** $-4\div2-7-3(-2)$

51. $-(20+9)+(8+2)$ **52.** $6+[7+(-5+1)]$

53. $4-[5+(2-4)]$

54. $2\{(5-3)+(1-3)\}$ **55.** $5+\{3-(2+8)\}$

56. $2+(3-8)-(8-2)$ **57.** $8-(4+2)+1$

58. $48-[16-(-8)]$ **59.** $2+[7-(1+1)]$

60. $-(15-3)-(19-26)$ **61.** $-4-4(4+4)$

62. $8+2(7-9)$ **63.** $(-8+7)(-3)$

64. $(-12+88)100+(-12+88)(-10)$

65. $2+3(6+7)+3$

66. $-4\{2-6[3-7]\}+7$

67. $(-12+88)(100-10)$ **68.** $25-8(-6+4)$

69. $(2-5)+[7+(3-1)]$

70. $22-2(222-22)-22$ **71.** $-10+10(-10-10)$

72. $-4[8-(-9)]\div(-2)$

73. $10-5\{4-2(4+6)\}$

74. $10\div5\{4+2[4-6]\}$ **75.** $6(3-3)+30$

76. $(-4)(-2)+8\div(-2)$

77. $-4(10-12)+8(15-20)$ **78.** $-3(-1+8)-10$

79. $2-[1+2(3-4)]$ **80.** $6-\{7-(4-8)\}$

81. $6\dfrac{5}{9}+2\dfrac{1}{9}-3\dfrac{2}{9}$

82. $18.1 - 17.3 - 2.3 - 18.1$ **83.** $\dfrac{6}{5} - \dfrac{1}{5}\left(\dfrac{7}{5} + \dfrac{3}{5}\right)$

84. $\dfrac{5}{12} - \dfrac{1}{3}\left(-\dfrac{5}{6} + \dfrac{1}{12}\right)$

85. $(-4.01)\,[0.7 + 100(-0.105 - 0.002)]$

86. $2\dfrac{1}{6} - 5\dfrac{1}{6} - 4\dfrac{1}{6}$ **87.** $-0.3 - 0.04 - 1.4$

88. $\dfrac{5}{16} + \dfrac{1}{2} \times \dfrac{3}{8}$ **89.** $\left(\dfrac{4}{5} - \dfrac{3}{8}\right) - \dfrac{1}{2}\left(\dfrac{2}{5} - \dfrac{1}{2}\right)$

90. $8.8 - (-6.6 - 2.2)$

91. $(-15)\,[-0.9 + (-0.08 + 0.98)]$

92. $2 + 5\,[3.3 + (1.8 - 2.9)]$

93. $6.1 + (-2.03)\,[0.012 - 0.023]$

94. $\dfrac{1}{5} + \left\{\dfrac{2}{3} - \left(\dfrac{5}{8} - \dfrac{2}{10}\right)\right\}$ **95.** $\dfrac{7}{6} + 3 - \dfrac{1}{6}$

96. $\dfrac{8}{3} - 4\left(\dfrac{5}{12} - \dfrac{1}{6}\right)$

97. $\dfrac{1}{2} - \left(\dfrac{1}{3} + \dfrac{2}{3}\right)\left(\dfrac{4}{7} + \dfrac{1}{14}\right)$

98. $0.72 - 2(-0.04 + 0.4)$

99. $\dfrac{1}{2} - \left\{\dfrac{1}{3} + \dfrac{2}{3}\left(\dfrac{4}{7} + \dfrac{1}{14}\right)\right\}$

100. $-\dfrac{3}{4}\left[\dfrac{7}{8} - \left(-\dfrac{5}{6}\right)\right]$

Important Terms Used in this Chapter		
	absolute value	origin
	associative property	parentheses
	braces	perfect square
	brackets	positive number
	commutative property	radical
	coordinate	radicand
	graph	rational number
	greater than	real number
	integers	real number line
	irrational number	signed number
	like signs	simplified notation
	less than	squared
	negative number	square root
	number line	substitution principle
	opposite	symbols of grouping
	order of operation	unlike signs

Important Symbols Used in this Chapter

( ) parentheses

[ ] brackets

{ } braces

$\sqrt{}$ indicates the square root

| | indicates the absolute value of a number

REVIEW EXERCISES

1. Indicate the graphs of the following numbers on the number line.

$0, \dfrac{3}{2}, -11, 3.4, 7$

2. Give the coordinates of the points indicated on the number line below.

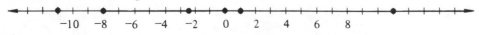

3. Find the opposite of 251.

4. Find the opposite of −29.

5. What is the opposite of −2.365?

Fill each of the blanks in Exercises 6-10 with either the words *is greater than* or the words *is less than*.

6. 4.031 _____ 4.032

7. −2.07 _____ −2.08

8. $\frac{11}{17}$ _____ $\frac{10}{17}$

9. $-\frac{11}{23}$ _____ $-\frac{23}{47}$

10. 5 _____ −17

(*Hint:* First rewrite each fraction as a decimal accurate to three decimal places.)

In Exercises 11-22, evaluate the given expression.

11. $\left|-2\frac{3}{4}\right|$

12. $|\,34\,|$

13. $|\,5.56\,|$

14. $|\,0\,|$

15. $\frac{6}{0}$

16. $\frac{0}{4+5}$

17. $(15)^2$

18. $\sqrt{169}$

19. Round $\sqrt{42}$ to the nearest tenth.

20. $(102)^2$

21. $(0.02)^2$

22. Round $\sqrt{115}$ to the nearest tenth.

23. Add +4.5 to +0.61.

24. $16 - 32 = ?$

25. $(+1.2) - \left(+\frac{2}{5}\right) = ?$

26. $-3 + 5 - 6 - 2 = ?$

27. $-12(-4) = ?$

28. Divide −16 by +32.

29. $(-3) \times (-4) + 2 = ?$

30. Subtract $-1\frac{4}{5}$ from $-2\frac{1}{5}$.

31. The highest point in the world is Mount Everest in Nepal-Tibet with an altitude of 29,028 ft. The Dead Sea in Israel is 1290 ft below sea level. What is the difference in altitude between these two points?

32. $\left(-\frac{5}{2}\right) + \left(+2\frac{1}{2}\right) = ?$

33. Multiply −10 times itself.

34. $(-2.3) \div (-100) = ?$

35. Find the sum of +10 and −32.

36. $-12 + 4 = ?$

37. Dawn's bank balance was $132.56. She wrote checks for the following amounts: $25, $15.14, and $97.52. How much money should Dawn deposit so that her account will not be overdrawn?

38. $4\{6 - (-4) + (-8 + 3)\} = ?$

39. Find the product of $+2\frac{1}{22}$ and $-\frac{11}{5}$.

40. Translate $(-10) + (-10)$ to simplified notation, and evaluate.

41. $-5 - 12 = ?$

42. Decrease −0.36 by +0.36.

43. $(+10) - (-10) = ?$

44. $18 \div 3 \times 6 = ?$

45. Multiply −8 by 6.

46. $(+1.51) \times (+10) = ?$

47. The highest temperature ever recorded in the shade was 136.4 degrees Fahrenheit. This temperature was recorded in Azizia, Libya, on Sept. 13, 1922. The record low temperature of 89.9 degrees below zero was recorded at Oimekon, Siberia, on February 6, 1933. Find the difference between these two extreme temperatures.

48. $2\frac{1}{2} - 2\frac{2}{5} - 4\frac{1}{4} = ?$

49. Subtract −17 from −6.

50. $12 - 7 = ?$

51. $7 - 3(6 - \{4 + 2\} - 3) = ?$

52. $-6(17) = ?$

53. Increase −12 by 4.

54. $-6 - \{-4 - [-6 - (2 - 4)]\} = ?$

55. Divide −16 by 4.

56. $[-4 + 10(-0.324 - 0.076)] \div 4 = ?$

57. The average closing price for Marathon stock during one week was $17. The daily variations from the $17 were +$3, +$5, −$2, −$4, −$2. Find the sum of the variations.

58. $\frac{3}{7} + \frac{6}{7} - \frac{2}{7} = ?$

59. $(124)(-123)(0)(-1)(2) = ?$

60. $4 - \{6 + 2[6 - (8 + 2)] - 4\} = ?$

61. $(-4 - 2)(3 + 2) = ?$

62. $(-4)(-2)(3)(-1)(-1)(-1) = ?$

63. $(-3 + 2) \div 10 = ?$

64. $0 \div 7 = ?$

65. $\frac{-35}{-5} = ?$

66. $6 + 2(8 - 4) = ?$

67. $4\left\{\frac{4}{5} - \left(-\frac{1}{5}\right) + \frac{1}{3}\left(\frac{5}{6} - \frac{1}{12}\right)\right\} = ?$

68. Translate $12 - (-3)$ to simplified notation and evaluate.

69. Multiply $\frac{-2}{3}$ times the sum of $\frac{1}{2}$ and $\frac{4}{3}$.

70. $-7 + 0 = ?$

71. $6 + 2(-8) = ?$

In Exercises 72-74 arrange the numbers in order from smallest to largest.

*72. 4.12, 4.1, 4.07, 4.1205

*73. $1.5, \pi, \sqrt{3}, \sqrt{2}$

*74. $-3, -4\frac{1}{8}, -4\frac{7}{8}, -4\frac{3}{4}, -4\frac{5}{8}, -4\frac{3}{8}, -5$

*75. The variations from the average daily temperature for a week were −7, +2, −3, +9, 0, +13, and −6. What was the average variation?

*These exercises are optional.

OPERATIONS WITH POLYNOMIALS

2

TERMINOLOGY ASSOCIATED WITH POLYNOMIALS

OBJECTIVES

Upon completion of this chapter you should be able to:

1. Discriminate between constants and variables. — 6.1
2. Evaluate an algebraic expression when given values for all variables in the expression. — 6.1, 6.2
3. Translate a verbal expression to an algebraic expression. — 6.1
4. Translate an algebraic expression to a verbal expression. — 6.1
5. Correctly interpret and use exponents in algebraic expressions. — 6.2
6. Express numbers greater than one in scientific notation. — 6.3
7. Use the terminology associated with polynomials. — 6.4

PRETEST

1. a. Which of the following are constants (Objective 1)?

$ab, \quad xy, \quad 21, \quad r, \quad 2.41, \quad m, \quad \pi, \quad \sqrt{3}$

b. Which of the following are variables?

$a, \quad x, \quad 21, \quad r, \quad 2.41, \quad m, \quad \pi, \quad \sqrt{3}$

c. In the term $3abc$, identify the constant factor(s) and the variable factor(s).

2. Evaluate each of the following expressions if $a = 3$ and $b = -2$ and $c = 0$ (Objective 2).

a. abc **b.** $ab + 4c$ **c.** $-a^2$

d. $(a + b)^2$ **e.** $\dfrac{3ab}{c}$ **f.** $-b^3$

g. $2a^2 - 4a + 6$ **h.** $-4 + 3b - 4b^2$ **i.** $2a^3 - 6a^2$

3. Translate each of the following verbal expressions into symbolic form (Objective 3).

a. six times the quantity x plus five
b. one-third the sum of nine and r
c. three times x plus eight divided by x
d. the cube of the sum of a and b

4. Translate each of the following algebraic expressions into verbal statements (Objective 4).

a. $4m + 3$ **b.** $4(m + 3)$

c. $\dfrac{a + b}{2}$ **d.** $a + \dfrac{b}{2}$

5. Write each of the following in expanded form (Objective 5).

a. $3a^2$ **b.** $(3a)^2$ **c.** x^3y

d. $(x + y)^3$ **e.** $-2ab^3$ **f.** $3a^2(bc)^4$

Write each of the following in exponential form.

g. $-2a \cdot a \cdot b \cdot b \cdot b$ **h.** $3a \cdot 3a \cdot 3a$

i. $(a + b)(a + b)$ **j.** $x \cdot x \cdot x \cdot x \cdot y \cdot y \cdot y$

k. $(-m)(-m)(-m)(-m)$ **l.** $3 \cdot a \cdot b \cdot 3 \cdot a$

Evaluate each of the following.

m. 2^3 **n.** -3^4 **o.** $2^2 \cdot 3^3$

p. $(-5)^2$ **q.** -5^2 **r.** $(-1)^{2465}$

6. Express each of the following numbers in scientific notation (Objective 6).

   a. 2400   **b.** 256.2
   c. 29.234   **d.** 34500
   e. 500,000   **f.** 32,000

7. Identify the given polynomial as a monomial, binomial, trinomial, or none of these (Objective 7).

   a. $x^2 - 3x$ **b.** $2abc$

   c. $-6a + 3b + c$ **d.** $a^3 + a^2b + ab^2 + b^3$
   e. $8m^5$ **f.** $a^2b - 3ab$

Consider the polynomial $3a^2 + ab - b^4$

   g. List each term.
   h. Give the numerical coefficient of each term.
   i. Give the degree of each term and the degree of the polynomial.

In the arithmetic review in Unit 1, we considered only expressions involving fundamental operations with specific numbers. In this introduction to algebra we will consider expressions involving letters of the alphabet. They enable us to consider general properties of numbers and to state formulas. This makes algebra a powerful tool exhibiting both generality and conciseness. Since this tool is more complex than arithmetic, you will probably not need to use algebra as often as you use arithmetic. Nonetheless, algebraic skills are frequently used and are essential in many other courses, especially in the sciences.

6.1 Variables

The symbols which we have used to represent real numbers are called **constants** since they have a fixed or constant value. For example, $2, -3, \frac{1}{2}, 1013, \pi$, and 0 are constants. A symbol which is used to represent any number in a set that contains more than one number is called a **variable**. Suppose we wish to represent the sum of any real number and the number 3. This can be done very easily if we let a variable, say x, represent any real number. The desired sum could then be indicated by the expression $x + 3$. If x is equal to 4, the value of the expression can be found by replacing x with 4; that is, the sum would be $4 + 3$ or 7. If x is equal to -2.5, the value of the expression is $-2.5 + 3$ or 0.5. Consider the following examples.

Example 1   Find the value of the expression $4 \cdot x$ if $x = 2; x = \frac{1}{2}; x = -1.33$.

            If $x = 2, \quad 4 \cdot x = 4 \cdot 2 = 8$

            If $x = \frac{1}{2}, \quad 4 \cdot x = 4 \cdot \frac{1}{2} = 2$

            If $x = -1.33, \quad 4 \cdot x = 4 \cdot (-1.33) = -5.32$

In Example 1, we call $2, \frac{1}{2}$, and -1.33 **values** of the variable. The expression has been **evaluated** for $x = 2$, $x = \frac{1}{2}$, and $x = -1.33$, respectively.

Example 2   Evaluate the expression $-2 \cdot t + 7$ for $t = 3$ and $t = -3$.

If $t = 3$,
$$-2 \cdot t + 7 = -2 \cdot (3) + 7$$
$$= -6 + 7$$
$$= 1$$

If $t = -3$,
$$-2 \cdot t + 7 = -2 \cdot (-3) + 7$$
$$= 6 + 7$$
$$= 13$$

Fewer errors will be made if the substitutions for the variables are made using parentheses before the operations involved are performed.

Example 3   Evaluate $2 \cdot a + b$ for $a = 4$ and $b = -3$.
$$2 \cdot a + b = 2 \cdot (4) + (-3)$$
$$= 8 + (-3)$$
$$= 5$$

In the first paragraph of this section we wrote $x + 3$ to indicate the sum of any real number and 3. The addition in the expression cannot be performed unless we replace x with one of its values. We say that in the expression the addition is indicated and that $x + 3$ is the **indicated sum**. Similarly, $5 \cdot y$ is the **indicated product** of 5 and y. Some other common ways of indicating the product are $5y$, $5(y)$, and $(5)(y)$. In all four of these methods of indicating products, 5 and y are factors of the product.

Note that the expression $5y$ means $5 \cdot y$. In general, if a constant and a variable, or two or more variables, are written with no sign between them, then this expression is understood to mean multiplication. For example, xy means $x \cdot y$, 32t means 32 times t. However, a symbol must be used between constants to indicate multiplication since 32 means thirty-two, not 3 times 2, and $3\frac{1}{2}$ means $3 + \frac{1}{2}$, not 3 times $\frac{1}{2}$. Although the symbol $\times$ is often used to indicate multiplication of real numbers, this symbol is generally not used to indicate the product of variables because it may not be clear whether the $\times$ represents a variable or indicates the operation of multiplication. Also, it is customary for the product of a constant and a variable to be written so that the constant precedes the variable, that is, the product of y and 5 would be written 5y, not y5.

Example 4   If t represents a number, write an expression containing t for each of the following.

a. 2 times t   2t

b. t divided by 2 $\frac{t}{2}$ or $t \div 2$

c. the sum of t and 4 $t + 4$
d. two t plus 4 $2t + 4$
e. twice the sum of t and 4 $2(t + 4)$
f. two t minus 4 $2t - 4$
g. twice the quantity t minus 4 $2(t - 4)$

Example 5   Give a verbal translation of each of the following algebraic expressions.

a. $x + y$ The sum of x and y.
b. $3x - 10$ Three x minus ten.
c. $3(x - 10)$ Three times the quantity x minus 10.

Example 6   Consider the expression 4x. Evaluate this expression for $x = 1$, $x = 2$, $x = 3$, and $x = 4$.

$$\text{If } x = 1, \quad 4x = 4(1) = 4$$
$$\text{If } x = 2, \quad 4x = 4(2) = 8$$
$$\text{If } x = 3, \quad 4x = 4(3) = 12$$
$$\text{If } x = 4, \quad 4x = 4(4) = 16$$

The set of numbers which may replace the variable is called the **replacement set** of the variable. In this text, if a replacement set is not specified, it is understood to be the set of real numbers.

Example 7   The area of a rectangle is $l \cdot w$ where l represents the length of the rectangle and w represents the width. What is the area in square meters of a rectangle with length 4 meters and width 10.4 meters?

The area is equal to $l \cdot w$, which is 4 meters $\cdot$ 10.4 meters or 41.6 square meters.

6.1 Exercises

1. Which of the following are constants?
$$-3, \quad m, \quad \sqrt{3}, \quad y, \quad 25$$

2. Which of the following are variables?
$$.25, \quad n, \quad \pi, \quad y, \quad \frac{-10}{30}$$

Evaluate $5x + 6$ for the values of x given in Exercises 3-7.

3. $x = 0$ 4. $x = \frac{5}{2}$

5. $x = -7$ 6. $x = 0.3$

7. $x = \frac{-9}{4}$

Evaluate 2,384,562x for the values of x given in Exercises 8-12.

8. $x = 0$ 9. $x = -2$

10. $x = -1$ 11. $x = -0.001$

12. $x = 10,000$

Evaluate $\frac{a + 2}{b}$ for the values of a and b given in Exercises 13-16.

13. $a = -1$ and $b = 1$ 14. $a = 16$ and $b = 9$

15. $a = 8$ and $b = 0$ 16. $a = 0.4$ and $b = 0.12$

Evaluate the expressions in Exercises 17-22 for $a = -2$, $b = 3$, and $c = \frac{3}{4}$.

17. $a - b + 4c$ 18. $a - (b + 4c)$

19. $6b + ac$ 20. $c + 2(a + b)$

21. $\frac{b + 4c}{a + 2}$ 22. $(a + b)(b + c)$

23. If the sum of a number w and 3 is doubled, which of the following is the correct form: $2w + 3$ or $2(w + 3)$?

24. If 5 is added to twice a certain number and the sum is divided by 3, which is the correct form: $\frac{2n + 5}{3}$ or $\frac{2n}{3} + 5$?

25. If the sum of x and seven is subtracted from eighteen, which is the correct form: $(x + 7) - 18$ or $18 - x + 7$ or $18 - (x + 7)$?

26. If the sum of x and y is multiplied times itself, which of the following is correct: $x \cdot x + y \cdot y$ or $(x + y)(x + y)$ or $2(x + y)$?

Write an expression containing a variable to represent each of the verbal expressions in Exercises 27–46.

27. The product of 3 and x

28. Two u added to 4

29. The quantity a times n

30. Subtract 5 from t.

31. Six more than m

32. The product of c and d

33. Divide 6 by n.

34. Nine times g

35. Subtract t from 5.

36. Divide n by 6.

37. The sum of twice x and y

38. Twice the sum of x and y

39. Subtract nine from three times y.

40. Divide 6 into m.

41. Divide m into 6.

42. Multiply m times the quantity a − b.

43. Twice the sum of x and three

44. Subtract six from two times m.

45. Divide the quantity t plus 2 by four.

46. One-half the sum of x and three

Translate each of the algebraic expressions in Exercises 47-56 into a verbal expression.

47. $6 + t$

48. $6t$

49. $\frac{6}{t}$

50. $78 + 5t$

51. $5(b - 3)$

52. $\frac{b - 3}{5}$

53. $5b - 3$

54. $7 - 3x$

55. $\frac{y - 2}{3}$

56. $\frac{7 + b}{5}$

57. Let *w* represent the width of a rectangle. If the length is 5 feet more than the width, write an expression containing *w* which would represent the length.

58. If the width of a rectangle is twice its length, *l*, write an expression containing *l* which will represent the width.

59. One side of a triangle is x inches long and a second is y inches long. The length of the third side is 7 inches. Write a phrase for the number of inches in the perimeter of the triangle. (The perimeter is the sum of the lengths of the sides.)

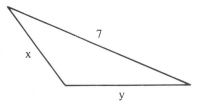

60. Of the three sides of a triangle, the second side is 3 inches longer than the first side. If *s* represents the first side, write an expression containing *s* to represent the second side.

61. The width of a rectangle is 4 feet shorter than its length. If *l* represents the length, write an expression containing *l* to represent the width.

62. Write an expression for the number of meters in the perimeter of the following figure. (The dimensions shown are in meters.)

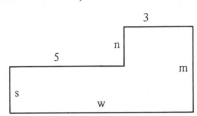

63. If x = 6, evaluate
$x(x - 1)(x - 2)(x - 3)(x - 4)(x - 5)(x - 6)$.

64. If x = 5, evaluate
$x(1 - x)(2 - x)(3 - x)(4 - x)(5 - x)$.

65. If x = 50, evaluate $x - 49$.

6.2   Exponents

Much of the notation in mathematics has been developed for convenience. Exponential notation, which we shall now consider, is an example of a notation which was introduced for convenience. Note how much more concise it is to write 5^{10}, than $5 \cdot 5 \cdot 5 \cdot 5 \cdot 5 \cdot 5 \cdot 5 \cdot 5 \cdot 5 \cdot 5$. In the expression 5^{10}, the 10 indicates that 5 is to be used as a factor ten times. The 10 is called the **exponent** and the 5 is called the **base**. The symbol 5^{10}, read "five to the tenth power," is called a **power** and is said to be in **exponential form**, while $5 \cdot 5 \cdot 5 \cdot 5 \cdot 5 \cdot 5 \cdot 5 \cdot 5 \cdot 5 \cdot 5$ is referred to as the **expanded form**. If a number or variable does not have an exponent indicated, the exponent is understood to be 1, that is, x is x^1 and 4 means 4^1.

Example 1   What does each of the following mean?

    a. 4^3 means $4 \cdot 4 \cdot 4$ and is read "four to the third power" or "four **cubed**."

    b. x^3 means $x \cdot x \cdot x$. How is this read?

    c. $\left(\dfrac{2}{3}\right)^2$ means $\dfrac{2}{3} \cdot \dfrac{2}{3}$. This is read "two thirds to the second power" or "two-thirds **squared**."

    d. $(xy)^4$ means $(xy)(xy)(xy)(xy)$. This is read "xy raised to the fourth power" or "the fourth power of xy."

    e. a^6 means $a \cdot a \cdot a \cdot a \cdot a \cdot a$, and this is read "a to the sixth power."

Example 2   Write each of the following in expanded form.

    a. $10^5 = 10 \cdot 10 \cdot 10 \cdot 10 \cdot 10$

    b. $-2t^2 = -2 \cdot t \cdot t$

    c. $(x + y)^4 = (x + y) \cdot (x + y) \cdot (x + y) \cdot (x + y)$

    d. $x^2 y = x \cdot x \cdot y$

    e. $xz^4 y^3 = x \cdot z \cdot z \cdot z \cdot z \cdot y \cdot y \cdot y$

    f. $(2x)^3 = (2x) \cdot (2x) \cdot (2x)$

    g. $2x^3 = 2 \cdot x \cdot x \cdot x$

    h. $-3^2 = -3 \cdot 3$

    i. $(-3)^2 = (-3) \cdot (-3)$

Note: In Example h, the base is 3, not −3. However, in Example i parentheses are used to show that the base is −3.

It is important in working with exponents that you learn to identify the base correctly. In the expression ab^3, the base of the exponent 3 is b, *not* ab. Only the single constant or variable immediately to the left of the exponent is the base. In the expression -3^2, the base is 3, *not* −3. Thus the expanded form is $-(3)(3)$ which equals −9. If −3 is to be squared, it is written $(-3)^2$. The expanded form is $(-3)(-3)$ or 9.

Example 3   Identify the base of the exponent 4 in each of the following expressions.

    a. x^4 base x

    b. $5x^4$ base x

    c. $(5x)^4$ base 5x

    d. $-y^4$ base y

    e. xy^4 base y

    f. $-(xy)^4$ base xy

    g. -2^4 base 2

    h. $(-2)^4$ base −2

Example 4 Write each of the following in exponential form.

    a. $x \cdot x \cdot y \cdot y \cdot y = x^2 y^3$

    b. $10 \cdot 10 \cdot 10 \cdot 10 = 10^4$

    c. $(a + 4) \cdot (a + 4) = (a + 4)^2$

    d. $(6p) \cdot (6p) \cdot (6p) \cdot (6p) \cdot (6p) = (6p)^5$

    e. $6 \cdot p \cdot p \cdot p \cdot p \cdot p = 6p^5$

    f. $(-2)(-2)(-2)(-2)(-2)(-2) = (-2)^6$

    g. $\dfrac{1}{3 \cdot 3 \cdot 3 \cdot 3} = \dfrac{1}{3^4}$

Example 5 Evaluate each of the following.

    a. $2^3 = 2 \cdot 2 \cdot 2 = 8$

    b. $(-3)^2 = (-3) \cdot (-3) = 9$

    c. $-3^2 = -3 \cdot 3 = -9$

    d. $1^{50} = 1$

    e. $\left(\dfrac{3}{4}\right)^3 = \dfrac{3}{4} \cdot \dfrac{3}{4} \cdot \dfrac{3}{4} = \dfrac{27}{64}$

    f. $(3.1)^2 = (3.1)(3.1) = 9.61$

    g. $4(3)^2 = 4(3)(3) = 36$

    h. $3(4)^2 = 3(4)(4) = 48$

Example 6 a. $10^1 = 10$

    b. $10^2 = 10 \cdot 10 = 100$

    c. $10^3 = 10 \cdot 10 \cdot 10 = 1000$

    d. $10^4 = 10 \cdot 10 \cdot 10 \cdot 10 = 10,000$

Example 7 a. $3 \times 10^2 = 3 \cdot 10 \cdot 10 = 3 \cdot 100 = 300$

    b. $3.12 \times 10^2 = 3.12(10)(10) = 3.12 \cdot 100 = 312$

    c. $3.12 \times 10^3 = 3.12(1000) = 3,120$

    d. $3.12 \times 10^7 = 3.12(10,000,000) = 31,200,000$

Example 8 Evaluate each of the following expressions for $x = -1$ and $y = 4$.

    a. $(x + y)^2$

$$(x + y)^2 = [(-1) + (4)]^2$$
$$= [3]^2$$
$$= 9$$

    b. $x^2 + y^2$

$$x^2 + y^2 = (-1)^2 + (4)^2$$
$$= 1 + 16$$
$$= 17$$

c. $-x^2$

$$-x^2 = -(-1)^2$$
$$= -(1)$$
$$= -1$$

d. $x + 2y$

$$x + 2y = (-1) + 2(4)$$
$$= -1 + 8$$
$$= 7$$

e. $(-x)^2$

$$(-x)^2 = [-(-1)]^2 = [1]^2 = 1$$

6.2 Exercises

Write the expressions in Exercises 1–20 in expanded form.

1. 5^3

2. $\left(\dfrac{2}{3}\right)^2$

3. x^4

4. 10^5

5. $(-2)^4$

6. y^3

7. $\left(\dfrac{3}{7}\right)^3$

8. 7^2

9. 10^3

10. $(-x)^3$

11. y^9

12. $(-h)^5$

13. $5t^3$

14. $-h^5$

15. $(5t)^3$

16. $m^4 + s^4$

17. $(m+s)^4$

18. $x^{10}y^7$

19. a^2b^3

20. $(xy)^4$

Write the expressions in Exercises 21–40 in exponential form.

21. $x \cdot x \cdot x \cdot x \cdot x$

22. $(-y)(-y)(-y)$

23. $x \cdot t \cdot t \cdot t \cdot z$

24. $-y \cdot y \cdot y$

25. $\dfrac{1}{4 \cdot 4 \cdot 4 \cdot 4 \cdot 4 \cdot 4}$

26. $a \cdot a \cdot a + b \cdot b \cdot b$

27. $(a+b)(a+b)(a+b)$

28. $(-5)(-5)(-5)(-5)(-5)(-5)$

29. $-5 \cdot 5 \cdot 5 \cdot 5 \cdot 5 \cdot 5$

30. $a \cdot a \cdot a \cdot b \cdot b$

31. $3 \cdot a \cdot a$

32. $(-6)(-6)(-6)$

33. $a \cdot a \cdot b \cdot b \cdot b \cdot c$

34. $4 \cdot 4 \cdot 4 \cdot 4 \cdot 4$

35. $6 \cdot 6 \cdot 6 \cdot 6$

36. $6 \cdot b \cdot b$

37. $(3a)(3a)$

38. $a \cdot b \cdot b \cdot b \cdot c \cdot c$

39. $(-4)(-4)$

40. $(6b)(6b)$

Evaluate each of the expressions in Exercises 41–60.

41. 3^2

42. 5^2

43. 2^3

44. 4^3

45. 6^2

46. 7^2

47. -6^2

48. -7^2

49. $(-6)^2$

50. $(-7)^2$

51. $-2 \cdot 2 \cdot 2 \cdot 2 \cdot 2 \cdot 2 \cdot 2 \cdot 2 \cdot 2 \cdot 2$

52. $(-2)^5$

53. $\left(\dfrac{2}{3}\right)^4$

54. $(0.07)^2$

55. $(-0.01)^3$

56. $(-1)^5$

57. $(-1)^7$

58. $\left(\dfrac{-4}{3}\right)^2$

59. $\left(2\dfrac{1}{3}\right)^2$

60. $(-1)^8$

Evaluate each of the expressions in Exercises 61–67 for $x = 2$ and $y = -3$.

61. $x^2 + y^2$

62. $x^3 - y^3$

63. $(x+y)^2$

64. $(x-y)^3$

65. $4x^2$

66. $-2y^3$

67. $(-2y)^3$

Write each of the verbal expressions in Exercises 68–75 as an algebraic expression, and then evaluate this expression.

68. Five squared

69. Negative four raised to the third power

70. Two to the sixth power

71. Six to the second power

72. Seven cubed

73. Eight squared

74. The opposite of (minus) the quantity seven to the fourth power.

75. The cube of the sum of 2 and 3

Evaluate each of the expressions in Exercises 76–80.

76. 4×10^2

77. 3.1×10^3

78. 34.5×10^1

79. 2×10^4

80. 3.5×10^5

6.3 Scientific Notation

One application of exponential notation is scientific notation which involves representing a number using powers of ten. This notation is used extensively in the sciences because it is very compact and simplifies computations with very large and very small numbers. Although many rational numbers can be expressed in scientific notation, we will limit our treatment to numbers greater than 1. A number is written in **scientific notation** when it is expressed as the product of a number between 1 and 10 and a power of ten. For example, 3100 is written in scientific notation as 3.1×10^3. We write 801 as 8.01×10^2 and 423,000,000 as 4.23×10^8. A number can be expressed in this form by following the *rule for expressing a number greater than one in scientific notation.*

RULE FOR EXPRESSING A NUMBER GREATER THAN ONE IN SCIENTIFIC NOTATION	*EXAMPLE* Write 3456.2 in scientific notation.
	Solution:
1. Reposition the decimal point to the right of the first digit in the number.	$3\,456.2$
2. Count the number of decimal places from the original decimal point to the repositioned decimal point.	3 places
3. Use the number of places found in step 2 as the exponent for 10.	10^3
4. Multiply the number found in step 1 by the power of 10 from step 3.	3.4562×10^3
	Thus, $3456.2 = 3.4562 \times 10^3$

Example   Express each of the following in scientific notation.

a. $389 = 3\,89. = 3.89 \times 10^2$

b. $312.5 = 3\,12.5 = 3.125 \times 10^2$

c. $1234 = 1\,234. = 1.234 \times 10^3$

d. $56789.2 = 5\,6789.2 = 5.67892 \times 10^4$

6.3 Exercises

Express each of the numbers in Exercises 1–20 in scientific notation.

1. 2560	**2.** 54.24	**11.** 38.564	**12.** 1235.4
3. 3896.1	**4.** 567800	**13.** 56780	**14.** 324.56
5. 127.56	**6.** 12.756	**15.** 2900	**16.** 81.456
7. 25.3	**8.** 425000	**17.** 34000	**18.** 300000
9. 176.89	**10.** 59000	**19.** 38.571	**20.** 56000

6.4 Polynomials

You must know the notation and terminology associated with mathematics in order to read and understand it. In this section we will introduce some of the basic terminology which is used in algebra. You should become familiar with the terms which are in dark type.

You may recall, from working with signed numbers in Chapter 5, that $-4 - 8 + 3 - 10$ is interpreted as $(-4) + (-8) + (+3) + (-10)$. It is convenient to make the same interpretation with algebraic expressions. For example, $2t + 4y - 6z$ is the same as $(+2t) + (+4y) + (-6z)$. Thus, if a single sign, a "+" or a "−," appears between two algebraic expressions, the sign denotes the sign of the term and the operation is understood to be addition. The parts of the expression which are added are called **terms**. Thus, $2t$, $4y$, and $-6z$ are the terms in the expression $2t + 4y - 6z$. In the term $2t$, the factor 2 is called the **numerical coefficient** of the term. The numerical coefficient of the term $4y$ is 4 and of the term $-6z$ is -6. If the numerical coefficient is not indicated, it is understood to be 1. For example, the numerical coefficient of t^2 is 1.

Example 1   Give the numerical coefficient of each term in the following expressions.

a. $2x^3yz$ has coefficient 2.

b. $a^2 + 5$ has coefficients 1 and 5.

c. $5y^7 + 89y^5 + 14y - 69y^9 - 90y^2$ has coefficients 5, 89, 14, −69, −90.

d. $\dfrac{2y^2}{3} + 4y$ has coefficients $\dfrac{2}{3}$ and 4.

Note: In the first term, 2 is the coefficient of the *numerator,* but the coefficient of the *term* is $\dfrac{2}{3}$, since $\dfrac{2}{3}y^2 = \dfrac{2y^2}{3}$.

In this unit we are interested in a special kind of algebraic expression called a polynomial. A **polynomial** is an expression which is formed by adding, subtracting, and multiplying constants and variables. This implies that only natural numbers may appear as exponents of variables and that no variable may appear in a denominator.

Example 2   The following algebraic expressions are polynomials.

a. $2y$

b. $\dfrac{4u^4}{3} - \dfrac{2u}{4}$

c. $3y^2 + 3y - 4$

d. 0

e. $\dfrac{1}{4}t - \dfrac{7}{4}$

f. $2x^2yz + 5y$

Example 3   The following algebraic expressions are *not* polynomials.

a. $\dfrac{2}{y}$ is not a polynomial because a variable appears in the denominator of a fraction.

b. $t^{1/2} - 3$ is not a polynomial because the exponent is not a natural number.

Polynomials containing one, two, and three terms are called respectively **monomials, binomials,** and **trinomials.**

Example 4   Classify each of the following polynomials as a monomial, binomial, or trinomial.

a. 2y is a monomial.

b. $\dfrac{4u^4}{3} - 2u$ is a binomial. The terms are $\dfrac{4u^4}{3}$ and $-2u$.

c. 0 is a monomial.

d. $3y^2 + 3y - 4$ is a trinomial. $3y^2$, $3y$, and -4 are the terms.

e. $\dfrac{1}{4}t - \dfrac{7}{4}$ is a binomial. The terms are $\dfrac{1}{4}t$ and $-\dfrac{7}{4}$.

f. $2x^2yz + 5y$ is a binomial. $2x^2yz$ and $5y$ are the terms.

The **degree of a monomial** in one variable is the exponent of the variable in the monomial. The monomial $2x^5$ has degree 5 while 2y has degree 1. A nonzero constant is understood to have degree equal to 0. *No degree is assigned to the special monomial 0.* If a monomial contains more than one variable, such as $3x^2yz^4$, the degree is given by the sum of the exponents of the variables. In the monomial $3x^2yz^4$ the exponents of the variables are 2, 1, and 4 so the degree of the monomial is $2 + 1 + 4$ or 7.

Example 5   Give the degree of each of the following monomials.

a. 2y has degree 1 since the exponent of y is 1.

b. $5u^4$ has degree 4 since the exponent of u is 4.

c. $-\dfrac{7}{4}$ has degree 0 since it is a constant.

d. $2x^2yz$ has degree 4 since the exponents of the variables are 2, 1, and 1.

e. 0 has no degree assigned to it.

f. xyz has degree 3 since the exponents of the variables are 1, 1, and 1.

The **degree of a polynomial** is the degree of the term of the highest degree. For example, the polynomial $3y^2 + 2y - 4$ has three terms: $3y^2$ which has degree 2, 2y which has degree 1, and -4 which has degree 0. Thus, the degree of the polynomial is 2. Students often confuse the procedure for finding the degree of a polynomial of more than one term with the procedure for finding the degree of a monomial. Compare these procedures, and notice that in finding the degree of a polynomial, you do *not* find the sum of the degrees of its terms.

Example 6   Give the degree of each of the following polynomials.

   a. $x^4 - 7x^3y^2 + 8x^2y^3z - 3xyz^2$

x^4 has degree 4.
$-7x^3y^2$ has degree 5 since the exponents of the variables are 3 and 2.
$8x^2y^3z$ has degree 6 since the exponents of the variables are 2, 3, and 1.
$-3xyz^2$ has degree 4 since the exponents of the variables are 1, 1, and 2.

Since the highest degree of any term is 6, the degree of the polynomial is 6.

   b. $x^3y + 2x^2y^2 + 4xy^3$

x^3y has degree 4 since the exponents of the variables are 3 and 1.
$2x^2y^2$ has degree 4.
$4xy^3$ also has degree 4.

Since the highest degree of any term is 4, the degree of the polynomial is 4.

   c. 4 is a constant and thus has degree 0.

   d. $-4x^{10}y^{15}z$ has degree 26 since this is a monomial and the exponents of the variables are 10, 15, and 1.

   e. $5y^7 + 89y^5 + 14y - 69y^9 - 90y^2$ has degree 9.

In algebra it is frequently necessary to evaluate polynomials for specific values of the variable. This is illustrated in the next example.

Example 7   Evaluate each of the following polynomial expressions for $x = -1$.

   a. $\begin{aligned} x^2 + 1 &= (-1)^2 + 1 \\ &= 1 + 1 \\ &= 2 \end{aligned}$

   b. $\begin{aligned} 4 + x^3 &= 4 + (-1)^3 \\ &= 4 + (-1) \\ &= 3 \end{aligned}$

   c. $\begin{aligned} 2x^2 - 3x - 7 &= 2(-1)^2 - 3(-1) - 7 \\ &= 2(1) - 3(-1) - 7 \\ &= 2 + 3 - 7 \\ &= -2 \end{aligned}$

   d. $\begin{aligned} -x^2 &= -(-1)^2 \\ &= -(1) \\ &= -1 \end{aligned}$

   e. $\begin{aligned} [-x]^2 &= [-(-1)]^2 \\ &= [1]^2 \\ &= 1 \end{aligned}$

6.4 Exercises

1. Give the degree of each of the following monomials.
   $3x^2$, $2x^3$, and $-5x^4$

2. What is the degree of the polynomial
   $3x^2 + 2x^3 - 5x^4$?

3. What is the degree of the polynomial $x^5 - x^4 + x^3 - x$?

4. Give the degree of each of the following monomials.
   x^5, $-x^4$, x^3, and $-x$

5. What is the degree of x^3y^2z?

6. What is the degree of the polynomial $x^3 + y^2 + z$?

7. Give the degree of each term of the polynomial $x^3y + x^2y^2 + y^3$.

8. Give the degree of the polynomial $x^3y + x^2y^2 + y^3$.

9. What is the degree of $2x^4 - y^3 + z^2$?

10. What is the degree of $2x^4y^3z^2$?

The following list of polynomials is referred to in Exercises 11–31.

   a. $3x^2y + 4xy^2$ **b.** $\frac{12}{13}t^5sr^{20}$

   c. $\frac{12}{13}t^5 + s - r^{20}$ **d.** $x - y + z$

   e. $4 - 3x + 7x^2$ **f.** 2

   g. $25t^4n^3 + 3t^9$

   h. $5x^{25} + 3x^{20} + 2x^5 + x^4 + 6x^3 + 2$

Answer the following questions about the polynomials given in a–h.

11. Which of the polynomials are monomials?

12. Which of the polynomials are binomials?

13. Which of the polynomials are trinomials?

14. Which of the polynomials are constants?

15. Which polynomial has the highest degree?

16. What is the coefficient of the term of highest degree in g?

17. List the terms of the polynomial given in e.

18. Which polynomial has the largest number of terms?

19. What is the sign of the second term of the polynomial given in d?

20. Which polynomial contains the term with the largest coefficient?

21. Give the degree of the polynomial in b.

22. Give the constant term of the polynomial in e.

23. Give the third term in h.

24. Which polynomial has the lowest degree?

25. What is the coefficient of the first term in c?

26. List the terms of the polynomial in c.

27. Which polynomial contains the term with the smallest coefficient?

28. What is the sign of the second term of the polynomial given in e?

29. Which polynomials have the smallest number of terms?

30. What is the second term in c?

31. Give the degree of the polynomial in h.

Evaluate each of the polynomial expressions in Exercises 32–45 for $x = -1$.

32. $5x - 1$ 33. $-x + 4$

34. $2x^2 - x + 3$ 35. $x^2 - x - 7$

36. $3 - x - x^3$ 37. $3x^4 - 12$

38. x^{101}

39. $5x^6 - x^5 + 10x^3 - 45x^2 - 100$

40. x^{200} 41. $-x^4 + (-x)^4$

42. $-x^6 + 3(-x)^6$ 43. x^{75}

44. $5x^4 - 3x^3 + 7x - 11$ 45. $10x^{80} - 5x^{49}$

Evaluate $yx^3 + 4$ if x and y have the values given in Exercises 46–50.

46. $x = 0$ and $y = 1,000,256$ 47. $x = -1$ and $y = 4$

48. $x = \frac{1}{2}$ and $y = 4$ 49. $x = 43$ and $y = 0$

50. $x = 2$ and $y = 0.3$

Important Terms Used in this Chapter	base	monomial
	binomial	numerical coefficient
	constants	polynomial
	cubed	power
	degree of a monomial	replacement set
	degree of a polynomial	scientific notation
	evaluate	squared
	expanded form	term
	exponent	trinomial
	exponential form	value of a variable
	indicated product	variable
	indicated sum	

REVIEW EXERCISES

In Exercises 1–10, identify the given polynomial as a monomial, binomial, trinomial, or none of these. List the terms of each polynomial; list the numerical coefficients of these terms; and give the degree of each polynomial.

1. $x^2 - 7x + 6$

2. $4t^5 - 2t^3 + \frac{1}{3}t^2 - 7$

3. $3x^5 + 4x^3 - 75x$

4. $-0.3u^3 - u$

5. $3a^6 + \frac{3}{4}$

6. $-16b$

7. $-16 + b$

8. $x^4 + 4x^3y + 6x^2y^2 + 4xy^3 + y^4$

9. -2

10. $8n^2$

Write the expressions in Exercises 11–14 in expanded form.

11. $15t^3rs^2$

12. $(2x + 3)^4$

13. $(x - 3y)^3$

14. $7x^5y^2z$

Write the expressions in Exercises 15–18 in exponential form.

15. $5 \cdot m \cdot n \cdot n \cdot n \cdot n \cdot n$

16. $(us)(us)(us)(us)(us)$

17. $-x \cdot x \cdot x \cdot y \cdot y$

18. $(-x)(-x)(-x)(-x)$

Evaluate each of the expressions in Exercises 19–24 for $x = 2$.

19. $5x - 1$

20. $2x^2 - x + 3$

21. $3 - x - x^3$

22. $3x^4 - 12$

23. $-x^4$

24. $(-x)^4$

Evaluate $mt - 2m + 3t^2$ for the indicated values in Exercises 25–28.

25. $m = 0$ and $t = -1$

26. $m = \frac{1}{2}$ and $t = 0$

27. $m = -4$ and $t = -5$

28. $m = \frac{3}{4}$ and $t = \frac{1}{3}$

Write an algebraic expression to represent each of the verbal expressions in Exercises 29–35.

29. Two x plus eight

30. The product of x and y

31. m subtracted from n

32. One-half the sum of x and one

33. The quantity three x minus two

34. m divided by n

35. x cubed minus three times y squared

Translate each of the algebraic expressions in Exercises 36–42 into verbal expressions.

36. $b - 5$

37. $\frac{b - 5}{4}$

38. $m + 3$

39. $x^2 - 5y^2$

40. x^{17}

41. $3m$

42. $(x + 2)^3$

Evaluate $x^{1,234,999}$ for the values of x in Exercises 43–45.

43. $x = 1$

44. $x = -1$

45. $x = 0$

46. The total revenue received from a motorcycle stunt was two million five hundred sixty-seven thousand two hundred dollars. The revenue came from TV rights and gate receipts. If d represents the amount in dollars that each person paid to see the event, and 37,400 people bought tickets, write an expression containing d to represent the gate receipts.

47. All votes cast in an election were cast for candidate A or for candidate B. If x represents the number of votes candidate A received and y represents the number of votes candidate B received, write an expression to stand for the total number of votes cast.

48. Write an algebraic expression for the perimeter (the sum of lengths of sides) of the following figure.

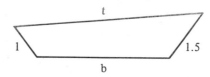

49. Evaluate $-x^5$ for $x = -2$, $x = 0$, and $x = 2$.

50. Which of the following are constants?
$-17, \ \frac{1,000,007}{43}, \ x, \ k, \ \pi, \ y, \ \sqrt{2}$

51. Which of the following are variables?
$m, \ n, \ \frac{6}{3}, \ \pi$

Express each of the numbers in Exercises 52–60 in scientific notation.

52. 256.3

53. 5000

54. 3600000

55. 546

56. 589.61

57. 5200000

58. 60000

59. 68

60. 7291

7

ADDITION AND SUBTRACTION OF POLYNOMIALS

OBJECTIVES   Upon completion of this chapter you should be able to:

1. Distinguish like terms from unlike terms, and collect the like terms.   7.1
2. Add polynomials and write the result in standard form.   7.2
3. Subtract one polynomial from another and write the result in standard form.   7.3
4. Simplify expressions which contain symbols of grouping.   7.4

PRETEST

1. In each of the following, indicate whether the terms are like or unlike (Objective 1).
   a. $3ab, -5ab$ **b.** $6a^2b, 6ab$
   c. $a^3b^2c, -2a^3b^2c$ **d.** $7x^5y^2, 7x^2y^5$
   e. $2x^2y, xy^2$

2. Add the following polynomials, and write the result in standard form (Objective 2).
   a. $3x^2 + 6x - 5$ and $6 - 7x + x^2$
   b. $(x^3 + 9x^2 - 4) + (5x + 6)$
   c. $6 - 5x + 2x^2, 3x^2 - 5,$ and $-6x + 9$
   d. $\quad x^3 + 2x^2 - 5x + 4$ **e.** $\quad h^2 + hk + k^2$
   $\quad\underline{-2x^3 + 6x^2 + 10x + 5}$ $\quad\underline{6h^2 - 5hk + 7k^2}$

3. Perform the following subtractions, and write the result in standard form (Objective 3).

   a. Subtract $3x^2 + 6x - 5$ from $6x - 9$.
   b. $(h^2 + 6hk - 7k^2) - (3h^2 + 6hk - 10k^2)$
   c. $(x^3 + 6x^2 - 3x + 9) - (9x^2 + 3x^3 - 5x + 5)$
   d. $\quad h^2 + hk + k^2$
   $\quad\underline{-(6h^2 - 5hk + 7k^2)}$

   e. $\quad x^3 + 2x^2 - 5x + 4$
   $\quad\underline{-(-2x^3 - 6x^2 + 10x + 5)}$

4. Simplify each of the following, and write the result in standard form (Objective 4).
   a. $5x - (3x - 4)$ **b.** $-(6x - y) + (-4x + y)$
   c. $(3x - y) - \{6 + (3x + y) - 5y\} + 9$
   d. $5x - \{3x - (2x + y)\}$
   e. $-(2x + 3) - \{5x - (3 + 8x)\}$

In previous chapters we have studied the addition and subtraction of numbers. Since polynomials are expressions which represent real numbers, many of the techniques, procedures, and properties developed earlier can be applied to the addition and subtraction of polynomials.

7.1   Like Terms

Before we add or subtract polynomials, it will be necessary to consider the concept of like or similar terms. The terms 3xy and 2xy are like terms while the terms 3xy and x^2y are *not* like terms. **Like terms** are terms which contain identical variable factors. Note that no restrictions are placed on the numerical factors. In our examples, 3xy and 2xy are like terms since the variable factors of 3xy are x and y, and the variable factors of 2xy are also x and y. The requirement that the variable factors be identical requires that the exponents of the like variable factors be the same. This can be easily seen by writing each term in expanded form. Since $3xy = 3 \cdot x \cdot y$ and $x^2y = x \cdot x \cdot y$, the variable factors of 3xy are x and y while the variable factors of x^2y are x, x, and y. From this we note that x^2y has an extra x factor. Thus, these two terms do not have exactly the same variable factors and are unlike terms. We also consider all constant terms to be like terms. Thus, 17 and −0.13 are like terms.

Example 1   For each set of terms, list the terms, if any, which are like terms.

a. 3x, −2y, z, 2x

> 3x and 2x are like terms.

b. 10ab, 5bc, 7abc, −4ab, cd

> 10ab and −4ab are similar.

c. $4n^2$, 4n, 4

> None of these terms has the same variable factors so they are unlike terms.

d. $\frac{1}{2}x^2y^2$, $\frac{1}{2}xy^2$, $\frac{1}{2}y^2x$

> $\frac{1}{2}xy^2$ and $\frac{1}{2}y^2x$ are similar.

e. 4x, 13z, −3y, −8, 8y, 7x, −6z, −8x, −5z, 2y, 10

> 4x, 7x, and −8x are like terms; 13z, −6z, and −5z are like terms; −3y, 8y, and 2y are like terms; and −8 and 10 are like terms.

The sum of two real numbers, such as 8 + 15, can be written as a third real number. However, as we pointed out in Chapter 6, there is no shorter form for the indicated sum x + 3 unless we evaluate the expression for a specific value of x. However, when the algebraic terms to be added are like terms the sum can be simplified.

Consider 5x + 2x. Recall that $3 \cdot 4$ can be interpreted to mean 4 + 4 + 4. Similarly, $5x = x + x + x + x + x$ and $2x = x + x$. Thus,

$$5x + 2x = \underbrace{\overbrace{x + x + x + x + x}^{\text{five x's}} + \overbrace{x + x}^{\text{two x's}}}_{\text{seven x's}} = 7x$$

Note that 7 is the sum of the coefficients. Now consider the difference 5x − 2x. If you have five x's and take away two x's, you have three x's left. Thus, 5x − 2x = 3x. Note that 3 is the sum of the coefficients 5 and −2. These examples suggest the *rule for adding like terms.*

RULE FOR ADDING LIKE TERMS	EXAMPLE 2x + 5x = ? *Solution:*
1. Add the numerical coefficients of the terms to obtain the coefficient of the result.	2 + 5 = 7 (do this mentally)
2. Identify and retain the common variable factor(s).	x (do this mentally)
3. Form the product of the coefficient (obtained in step 1) and the common variable factor(s).	7x Thus, 2x + 5x = (2 + 5)x = 7x

Example 2   If possible express each of the following as a single term.

   a. 2a + 4a

$$2a + 4a = (2 + 4)a = 6a$$

   b. 3t − t

$$3t - t = 3t - 1t = (3 - 1)t = 2t$$

   c. $\frac{1}{2}x^2y + 6x^2y$

$$\frac{1}{2}x^2y + 6x^2y = \left(\frac{1}{2} + 6\right)x^2y = \left(\frac{1}{2} + \frac{12}{2}\right)x^2y = \frac{13}{2}x^2y$$

   d. $4n^2 - 4n$

$4n^2 - 4n$
This expression cannot be simplified since the terms are unlike.

   e. −13z − 5z

$$-13z - 5z = (-13 - 5)z = -18z$$

   f. 4n + 6n − 3n

$$4n + 6n - 3n = (4 + 6 - 3)n = 7n$$

   g. $3t^3 + t^2$

$3t^3 + t^2$ This expression cannot be simplified.

Note: When an expression can be simplified by adding like terms, the addition should be done mentally whenever possible.

The instruction *simplify* occurs frequently in mathematics. The precise meaning of this instruction depends upon the problem. When operations are indicated, the results should be calculated where possible. The following examples illustrate some of the ways in which the instruction **simplify** is used.

Example 3   Simplify each of the following expressions.

a. $\dfrac{16}{32} = \dfrac{1}{2}$

Fractions are not in simplest form until reduced to lowest terms.

b. $\dfrac{7}{8} - \dfrac{5}{8} = \dfrac{2}{8} = \dfrac{1}{4}$

To simplify, you must subtract, and then reduce the fraction.

c. $2x + 3x = 5x$

The operation is addition so like terms should be added.

7.1 Exercises

For each set of terms in Exercises 1–10, list the terms, if any, which are like terms.

1. $2a, 3b, 3a$

2. $\dfrac{1}{2}x, 3x^2, \dfrac{2}{3}x^2, 0.17x$

3. x^2, y, y^2, x

4. $-3w^2, 4w, -6w, 11w^2$

5. $3ab, -4ba$

6. w^3, z, w^2, z^2

7. $4x, 3y, -7x, 11z, 9z, \dfrac{1}{2}y, -0.7x$

8. $7xy, 8yx$

9. $w^2y, 2wy^2, 3wy$

10. $5x^2y^2z, 7x^2zy^2, 9xy^2z, 11xy^2z^2$

Perform the indicated operations in Exercises 11–30.

11. $3a + 2a$

12. $4a - a$

13. $6a - 3a$

14. $2x^2 - 3x^2$

15. $6x^2 + 3x^2$

16. $5a + 4a$

17. $-6m + 2m$

18. $5m - 2m$

19. $7a + a$

20. $-4a + a$

21. $3x^2 - 3x^2$

22. $7n + 10n$

23. $11y^2 - y^2$

24. $3ab + 5ab$

25. $-7n + 10n$

26. $7a^2bc - 9a^2bc$

27. $a + a$

28. $2m - 2m$

29. $3x^2 + 4x^2$

30. $-3m^2n + m^2n$

Simplify Exercises 31–50 where possible by performing the indicated operations.

31. Subtract $8ab$ from $6ab$.

32. Add $2a$, $-3a$, and $4a$.

33. $\dfrac{1}{2}x - \dfrac{13}{4}x$

34. $-3x^2y + 9x^2y - 5x^2y$

35. $3x^2y - 9x^2y + 5x^2y$

36. $x^3 + 3x^2 + 3$

37. $0.25t + t - 3.33t$

38. $3n + 3n + 3n + 3n + 3n + 3n$

39. $8a + 8$

40. $3ab - abc$

41. $5n + 5n + 5n + 5n + 5n + 5n + 5n$

42. $x^3y^2 + x^2y^3$

43. $17.3 - 14.01$

44. $5x - 4$

45. $5xyz - 2xy$

46. $\dfrac{5}{3} - \dfrac{7}{4}$

47. $\dfrac{1}{2}x - \dfrac{2}{3}x + \dfrac{3}{5}x$

48. $5y - 7y + 0.3y$

49. $2n - 3n - 2n + 3n$

50. $4x^3 - 3x^4$

7.2 Addition of Polynomials

Suppose we wish to find the sum of $4x^3 + 3$ and $2x^3 + 7$. Since x represents a real number, we may use properties of real numbers in order to find the sum. We know from the generalized associative and commutative property that numbers can be added in any order that is convenient. Thus, $(4x^3 + 3) + (2x^3 + 7)$ is equal to $4x^3 + 2x^3 + 3 + 7$. Note that the terms have been rearranged so that the like terms are next to each other. From the previous section we know that $4x^3 + 2x^3 = 6x^3$. Since $3 + 7 = 10$, the sum of $4x^3 + 3$ and $2x^3 + 7$ is $6x^3 + 10$.

The process of reordering the terms so that like terms are next to each other is sometimes called **collecting like terms.** This procedure is summarized in the *rule for adding polynomials.*

RULE FOR ADDING POLYNOMIALS	EXAMPLE
	$(4x^3 + 3x) + (2x^3 + 7x + 9) = ?$ *Solution:*
1. Reorder the terms so that like terms are next to each other. (This step may be done mentally.)	$4x^3 + 2x^3 + 3x + 7x + 9$
2. Combine like terms.	$6x^3 + 10x + 9$ Thus, $(4x^3 + 3x) + (2x^3 + 7x + 9) =$ $6x^3 + 10x + 9$.

Example 1   Find the sum of $3x - 2y$ and $4y + 2x$.

$$3x - 2y + 4y + 2x = 3x + 2x - 2y + 4y$$
$$= 5x + 2y$$

Example 2 $4m^2 + 4m$ plus $4m + 4$

$$4m^2 + 4m + 4m + 4 = 4m^2 + 8m + 4$$

Example 3 $(4x + 13z - 3y) + (8y + 7x - 6z) + (-8x - 5z + 2y)$

$$(4x + 13z - 3y) + (8y + 7x - 6z) + (-8x - 5z + 2y)$$
$$= 4x + 7x - 8x - 3y + 8y + 2y + 13z - 6z - 5z$$
$$= 3x + 7y + 2z$$

Example 4 $(3t^3 + 4t^4 - t + 4) + (-2t^4 + 3t - 7)$

$$(3t^3 + 4t^4 - t + 4) + (-2t^4 + 3t - 7)$$
$$= 3t^3 + 4t^4 - 2t^4 - t + 3t + 4 - 7$$
$$= 3t^3 + 2t^4 + 2t - 3$$
$$= 2t^4 + 3t^3 + 2t - 3$$

In Example 4, the terms of $3t^3 + 2t^4 + 2t - 3$ were rearranged so that the degrees of the terms are decreasing. It is often convenient to arrange the terms of polynomials so that the degrees are decreasing in a particular variable. A polynomial in one variable is said to be in **standard form** if the terms are arranged in **decreasing powers** of the variable. A polynomial involving two variables is in standard form if (1) the variables are written in alphabetical order in each term and (2) the terms are arranged in decreasing powers of the first variable. This concept can be extended to polynomials involving three or more variables. In any polynomial written in standard form, the constant term is written last.

Example 5   Express in standard form.

a. $5 + a^4 + a^2 + a$

If the terms are arranged in decreasing powers of a, we get

$a^4 + a^2 + a + 5$

b. $b^2a + a^3b^2 + a^2b$

If each term is alphabetized and the polynomial is arranged in decreasing powers of a, we get

$a^3b^2 + a^2b + ab^2$

In Examples 1–4, the problems were written in a horizontal format, but a vertical format may also be used to add polynomials. The rearrangement and reordering of terms is accomplished by rewriting the polynomials in standard form and placing them in rows, one under the other, so that like terms are in the same column. For example, the polynomials $4x^3 + 3$ and $2x^3 + 7$ can be added as follows:

$$\begin{array}{r} 4x^3 + 3 \\ 2x^3 + 7 \\ \hline 6x^3 + 10 \end{array}$$

The sums in Examples 1–4 are now worked in vertical format in Examples 6–9.

Example 6 $(3x - 2y) + (2x + 4y) = ?$

$$\begin{array}{r} 3x - 2y \\ 2x + 4y \\ \hline 5x + 2y \end{array}$$

Example 7 $(4m^2 + 4m) + (4m + 4) = ?$

$$\begin{array}{r} 4m^2 + 4m \\ 4m + 4 \\ \hline 4m^2 + 8m + 4 \end{array}$$

Note: Since only like terms are placed in each column, $4m^2$ and 4 must be in separate columns.

Example 8 $(4x + 13z - 3y) + (8y + 7x - 6z) + (-8x - 5z + 2y) = ?$

$$\begin{array}{r} 4x - 3y + 13z \\ 7x + 8y - 6z \\ -8x + 2y - 5z \\ \hline 3x + 7y + 2z \end{array}$$

Example 9 $(3t^3 + 4t^4 - t + 4) + (-2t^4 + 3t - 7) = ?$

$$\begin{array}{r} 4t^4 + 3t^3 - t + 4 \\ -2t^4 + 3t - 7 \\ \hline 2t^4 + 3t^3 + 2t - 3 \end{array}$$

Addition problems may be worked in either horizontal format or vertical format. The vertical format will be needed in some multiplication problems. However, in general, problems should be worked using the format in which they are written.

7.2 Exercises

Collect like terms in each of the polynomials in Exercises 1–10. Express the results in standard form.

1. $3m + 2n + 2m$
2. $2x^2 - x + 3x^2$
3. $-6x^2 + x^2 - 5x^2$
4. $3x - 4y - 7x$
5. $y^3 - 2y^2 + 3y^2$
6. $m^2 - 4m^2 + m^3$
7. $3a + 6b - 5a + 2b$
8. $x^2 - x - 5x^2 + 2x^2$
9. $-6x^2 - 2x^2 - 3x^2 + x$
10. $m^3 - n^2 + m^2 - mn$

Perform the additions in Exercises 11–40 and express the results in standard form.

11. $5a + 3$
    $\underline{2a + 2}$
12. $6m + 5$
    $\underline{2m + 1}$
13. $3a - 6$
    $\underline{2a - 2}$
14. $-4m + 5$
    $\underline{-2m + 3}$
15. $12a - 5$
    $\underline{-3a + 3}$
16. $16 - \ n$
    $\underline{2 + 3n}$
17. $\ \ \ x - 4$
    $\underline{-6x - 5}$
18. $-3m - 10$
    $\underline{-2m + \ \ 2}$
19. $\ \ \ 6x + 2$
    $\underline{-3x - 4}$
20. $-2x - 3$
    $\underline{-5x - 1}$

21. $(2m + 3) + (m + 5)$
22. $(2x - 5y) + (-3x - 2y)$
23. Add $(21rs - 14r)$ and $(-21rs + 14r)$.
24. $(10ab + 5bc - 6cd) + (-2bc + 5ab - 3cd)$

25. $(5x - 3y + 10) + (4x + 2y - 9)$
26. Find the sum of
    $(-2x^6 + 2x^4 - 4x^3 + 2x^2 - 1)$ and
    $(4x^6 + 3x^5 - 2x^4 + 5x^3 - 2x^2 + x + 1)$.
27. $(2.2x^3 + 0.2) + (1.2x^2 - 4.6)$
28. $\left(\dfrac{2}{3}t - 8\right) + \left(-\dfrac{10}{3}t + \dfrac{7}{2}\right)$
29. Add $(t^3 - 4t)$ and $(5t^3 + 4t^2)$.
30. $(y^2 + 4) + (3n + 9) + (-3y^2 - 4n - 9)$
31. $\ \ \ 10x^3 + 5x^2y - 9xy^2 + y^3$
    $\underline{+ (4x^3 - 6x^2y - 3xy^2 - y^3)}$
32. $\ \ \ -4wy - \ \ 3wz$
    $\underline{+ (\ \ \ \ \ \ - 24wz + 3yz)}$
33. $\ \ \ \ \ \ z^5 \ \ \ \ \ \ - 9z^3 + \ 3z^2 \ \ \ \ \ \ - 11$
    $\underline{+ (6z^5 - 3z^4 \ \ \ \ \ \ + 11z^2 - 4z \ \ \ \ \ \ )}$
34. $\ \ \ \ \ 4a + 3b - 7$
    $\underline{+ (-4a - 3b + 7)}$
35. $\left(\dfrac{1}{2}x + \dfrac{2}{3}y\right) + \left(\dfrac{3}{4}x - \dfrac{1}{3}y\right)$
36. $(8.4x - 1.3y) + (-7.06x + 4.57)$
37. $\left(\dfrac{1}{4}x - \dfrac{1}{5}y + \dfrac{1}{7}\right) + \left(\dfrac{3}{4}x + \dfrac{1}{5}y + \dfrac{3}{7}\right)$
38. Find the sum of $(5r + 8 - 7s)$ and $(13s - 4r + 11)$.
39. $(a^2 + 3) + (a^2 - 4a) + (a^2 - 9)$
40. $(w^4 - 17w^2) + (w^3 - 3w^2 - w + 2)$

7.3 Subtraction of Polynomials

Before considering subtraction of polynomials, we will review subtraction of signed numbers which was introduced in Chapter 5. Recall that the problem $2 - (-3)$ is worked as follows:

$$2 - (-3) = 2 + (+3)$$
$$= 5$$

In other words, to subtract, we form the opposite of the subtrahend and change the operation to addition. Likewise, to find $(4x^2 + 3x + 2) - (2x^2 - x + 1)$, we add the opposite of $(2x^2 - x + 1)$ to $4x^2 + 3x + 2$. The opposite of $(2x^2 - x + 1)$

is $-2x^2 + x - 1$. In general the **opposite of a polynomial** is the polynomial formed by replacing each term with its opposite. So

$$(4x^2 + 3x + 2) - (2x^2 - x + 1)$$
$$= (4x^2 + 3x + 2) + (-2x^2 + x - 1)$$
$$= 4x^2 - 2x^2 + 3x + x + 2 - 1$$
$$= 2x^2 + 4x + 1$$

RULE FOR SUBTRACTING POLYNOMIALS	*EXAMPLE*
	$\underbrace{(4x^2 + 3x + 2)}_{\text{minuend}} - \underbrace{(2x^2 - x + 1)}_{\text{subtrahend}} = ?$
	Solution:
1. Form the opposite of the subtrahend.	$-(2x^2 - x + 1) = -2x^2 + x - 1$
2. Add the opposite of the subtrahend to the minuend following the rule for adding polynomials.	$(4x^2 + 3x + 2) + (-2x^2 + x - 1)$ $\qquad\qquad = 2x^2 + 4x + 1$

Example 1   Simplify $(4x^2 + 3x - 2) - (3x^2 - 7x + 4)$.

$$(4x^2 + 3x - 2) - (3x^2 - 7x + 4) = (4x^2 + 3x - 2) + (-3x^2 + 7x - 4)$$
$$= 4x^2 - 3x^2 + 3x + 7x - 2 - 4$$
$$= x^2 + 10x - 6$$

Example 2   Subtract $4x^2 - 3x - 3$ from $2 - 3x + 6x^2$.

$$(2 - 3x + 6x^2) - (4x^2 - 3x - 3) = (2 - 3x + 6x^2) + (-4x^2 + 3x + 3)$$
$$= 2 + 3 - 3x + 3x + 6x^2 - 4x^2$$
$$= 5 + 0x + 2x^2$$
$$= 5 + 2x^2$$
$$= 2x^2 + 5$$

Example 3   Simplify $(3x^2 - 7x + 4) - (4x^2 + 3x - 2)$.

$$(3x^2 - 7x + 4) - (4x^2 + 3x - 2) = (3x^2 - 7x + 4) + (-4x^2 - 3x + 2)$$
$$= 3x^2 - 4x^2 - 7x - 3x + 4 + 2$$
$$= -x^2 - 10x + 6$$

As with addition, a vertical format may be used in subtraction. Write the minuend in the first row and the subtrahend in the second row so that like terms are in the same column.

$$\begin{array}{r} 4x^2 + 3x + 2 \\ -(2x^2 - x + 1) \\ \hline \end{array} \quad \text{is the same as} \quad \begin{array}{r} 4x^2 + 3x + 2 \\ +(-2x^2 + x - 1) \\ \hline 2x^2 + 4x + 1 \end{array}$$

To keep from rewriting the problem, we may change the sign of every term in the subtrahend and add the like terms using the new signs. To avoid errors it is suggested that you write the new signs in circles above the original signs. For example, to subtract $2x^2 - x + 1$ from $4x^2 + 3x + 2$ we consider the work as

$$
\begin{array}{r}
4x^2 + 3x + 2 \\
\ominus \quad \oplus \quad \ominus \\
2x^2 - \ x + 1 \\
\hline
2x^2 + 4x + 1
\end{array}
$$

Avoid writing over the signs in the original problem. This makes it very difficult to check your work, and often leads to errors.

Examples 1–3 are reworked in Examples 4–6 using the vertical format.

Example 4   Subtract $3x^2 - 7x + 4$ from $4x^2 + 3x - 2$.

$$
\begin{array}{r}
4x^2 + \ 3x - 2 \\
\ominus \quad \oplus \quad \ominus \\
3x^2 - \ 7x + 4 \\
\hline
x^2 + 10x - 6
\end{array}
$$

Example 5   Subtract $4x^2 - 3x - 3$ from $2 - 3x + 6x^2$.

$$
\begin{array}{r}
2 - 3x + 6x^2 \\
\oplus \quad \oplus \quad \ominus \\
-3 - 3x + 4x^2 \\
\hline
5 \quad\ + 2x^2 = 2x^2 + 5
\end{array}
$$

Example 6   Subtract $4x^2 + 3x - 2$ from $3x^2 - 7x + 4$.

$$
\begin{array}{r}
3x^2 - \ 7x + 4 \\
\ominus \quad \ominus \quad \oplus \\
4x^2 + \ 3x - 2 \\
\hline
-x^2 - 10x + 6
\end{array}
$$

Subtraction problems may be worked in either horizontal or vertical format. The vertical format will be used in long division. In general, the vertical format should not be used unless the problem is written vertically.

7.3   Exercises

Subtract in Exercises 1–30. Express the results in standard form.

1. $\quad 5a + 3$
$\quad\ \underline{2a + 2}$

2. $\quad 6m + 5$
$\quad\ \underline{2m + 1}$

3. $\quad 3a - 6$
$\quad\ \underline{2a - 2}$

4. $\quad -4m + 5$
$\quad\ \underline{-2m + 3}$

5. $\quad 12a - 5$
$\quad\ \underline{-3a + 3}$

6. $\quad 16 - \ n$
$\quad\ \underline{2 + 3n}$

7. $\quad x - 4$
$\quad\ \underline{-6x - 5}$

8. $\quad -3m - 10$
$\quad\ \underline{-2m + \ 2}$

9. $\quad 6x + 2$
$\quad\ \underline{-3x - 4}$

10. $\quad -2x - 3$
$\quad\ \ \underline{-5x - 1}$

11. $(2a + 6) - (a + 2)$

12. $(5a + 3) - (6a + 5)$

13. $(-6a + 2) - (3a - 1)$

14. $(2m - 5) - (3a - 1)$

15. $(6m - 10) - (2m + 3)$

16. $(-2a - 6) - (-3a + 1)$

17. $(-3m + 2) - (-3m + 2)$

18. $(3m + 10) - (5m + 12)$

19. $(8a - 6) - (a + 9)$ 20. $(x - 4) - (x - 4)$

21. $(4y + 5) - (2y + 4)$ 22. $(11a - 9b) - (6a + 7b)$

23. Subtract $(-7g + 2h - 9)$ from $(-4g + 3h - 5)$.

24. $(7x^2 + 3x - y + 2) - (2x^2 + 3x - y)$

25. $(-8x^2 + 14x^2 - 3) - (-8x^2 + 12x^2 - 2x)$

26. $(-4wy - 3wz) - (24wz - 3yz)$

27. $(5r + 8 - 7s) - (13s - 7r - 80)$

28. Subtract $\left(\dfrac{1}{5}x + \dfrac{2}{3}\right)$ from $\left(\dfrac{2}{7}x - \dfrac{1}{2}\right)$.

29. $(12.06x^2 - 5) - (9.3x^2 - x - 4.73)$

30. $(4.8w^3 - 7) - (-2.07w^3 + 3w^2 - 8.1)$

In each of Exercises 31–40, subtract the lower expression from the upper expression.

31. $\begin{array}{l} 2M + 3 \\ \underline{M + 5} \end{array}$ 32. $\begin{array}{l} 21NS - 14N \\ \underline{-21NS + 14N} \end{array}$

33. $\begin{array}{l} 10ab + 5bc - 6cd \\ \underline{5ab - 2bc - 3cd} \end{array}$ 34. $\begin{array}{l} 4t^3 + 4t^2 + 4t \\ \underline{5t^3 + 4t^2 - 4t + 4} \end{array}$

35. $\begin{array}{l} 2.2x^3 + 0.2 \\ \underline{1.2x^3 - 4.6} \end{array}$

36. $\begin{array}{l} 11xy - 9xz + 4yz - 3 \\ \underline{-9xy - 9xz - 9yz - 9} \end{array}$

37. $\begin{array}{l} -2w^5 - 14w^4 + 3w^2 - 8 \\ \underline{7w^5 -4w^3 - 3w^2 - 2w} \end{array}$

38. $\begin{array}{l} 4.3s^2 - 2.7s + 4.9 \\ \underline{8.1s^2 + 3.4s - 0.008} \end{array}$

39. $\begin{array}{l} -a^4 + 3a^3 - 7a^2 - 17 \\ \underline{7a^4 - 8a^3 -4a + 13} \end{array}$

40. $\begin{array}{l} 1.27346x^2 - 0.0008x + 4.103 \\ \underline{1.27346x^2 - 0.0008x + 4.103} \end{array}$

7.4   Removing Symbols of Grouping

Simplifying an expression that involves only addition or subtraction of polynomials can be thought of in terms of the *rule for removing symbols of grouping* and then combining like terms. There are two cases of this rule that must be considered.

RULE FOR REMOVING SYMBOLS OF GROUPING	EXAMPLES
1. To remove parentheses preceded by a plus sign or no sign, rewrite the terms inside the parentheses omitting the parentheses and the preceding plus sign if one appears.	no sign: $(-3a + 4b) = -3a + 4b$ plus sign: $+(-3a + 4b) = -3a + 4b$
2. To remove parentheses preceded by a minus sign, write the opposite of each term which appears inside the parentheses and omit both the parentheses and the preceding minus sign.	minus sign: $-(-3a + 4b) = +3a - 4b$
Note: This rule is stated only for parentheses, but it also applies for other symbols of grouping such as brackets or braces.	

Example 1   Remove the parentheses from each of the following polynomials.

a. $+(x + y) = +x + y$ b. $-(x + y) = -x - y$

c. $-(-x + y) = +x - y$ d. $+(-x - y) = -x - y$

e. $(2x + 3y - z) = 2x + 3y - z$

Example 2   Remove the symbols of grouping from each of the following polynomials.

a. $-[5x - 3y + 7] = -5x + 3y - 7$

b. $-\{-7a + 8b\} = 7a - 8b$

c. $[-11w - 8.1z] = -11w - 8.1z$

d. $+\left\{\dfrac{2}{3}k^2 - \dfrac{1}{6}k + 7\right\} = \dfrac{2}{3}k^2 - \dfrac{1}{6}k + 7$

Consider the following examples which illustrate how this rule is applied to simplify sums and differences of polynomials.

Example 3   Simplify $6x - (2x + 2)$.

$$6x - (2x + 2) = 6x - 2x - 2$$
$$= 4x - 2$$

Note: $-(2x + 2)$ was replaced with $-2x - 2$.

Example 4   Simplify $-(2x + y) - (-5x - y)$.

$$-(2x + y) - (-5x - y) = -2x - y + 5x + y$$
$$= -2x + 5x - y + y$$
$$= 3x + 0$$
$$= 3x$$

Note: $-(2x + y) = -2x - y$ and $-(-5x - y) = +5x + y$

Example 5   Simplify $-(2x + y) + (-5x - y)$.

$$-(2x + y) + (-5x - y) = -2x - y - 5x - y$$
$$= -7x - 2y$$

Frequently parentheses, braces, and brackets appear in a single expression. The same techniques applied in Chapter 5 for removing symbols of grouping are appropriate here. *Remove the symbols of grouping one at a time, starting with the innermost.* This is illustrated by the following examples.

Example 6   Simplify $-\{6 + (3x + y) - 5y\}$.

$$-\{6 + (3x + y) - 5y\} = -\{6 + 3x + y - 5y\}$$
$$= -6 - 3x - y + 5y$$
$$= -3x + 4y - 6$$

Example 7   Simplify $2a - [3c + 2b - (a + c) + (5a - 4b)]$.

$$2a - [3c + 2b - (a + c) + (5a - 4b)]$$
$$= 2a - [3c + 2b - a - c + 5a - 4b]$$
$$= 2a - 3c - 2b + a + c - 5a + 4b$$
$$= 2a + a - 5a - 3c + c - 2b + 4b$$
$$= -2a + 2b - 2c$$

Alternate Solution:

$$2a - [3c + 2b - (a + c) + (5a - 4b)]$$
$$= 2a - [3c + 2b - a - c + 5a - 4b]$$
$$= 2a - [2c + 4a - 2b]$$
$$= 2a - 2c - 4a + 2b$$
$$= -2a + 2b - 2c$$

Example 8   Evaluate $2a - \{3 + 6b\}$ for $a = 2$ and $b = -5$.

$$2a - \{3 + 6b\} = 2(2) - \{3 + 6(-5)\}$$
$$= 4 - \{3 - 30\}$$
$$= 4 - \{-27\}$$
$$= 4 + 27$$
$$= 31$$

7.4   Exercises

Remove the symbols of grouping from each of the polynomials in Exercises 1–10.

1. $+(-3a + 8b)$
2. $-[-4z - 9]$
3. $-\{11k - 1.03m\}$
4. $-\{-11k - 1.03m\}$
5. $+[3y^3 - 4y^2 - y + 7]$
6. $+(8h^4 - 7h^3 - 9h^2 + 4)$
7. $-\{-w^4 - 3w^2 + 2\}$
8. $-\left[\frac{2}{3}x - 4\right]$
9. $-(-3g + 2)$
10. $+\{-0.0049z^2 + 3z - 2\}$

Remove the symbols of grouping, and combine like terms in Exercises 11–30.

11. $(3a + 2b) - (a + b)$
12. $(4x - 2) - x$
13. $(2a + b) + (a - 3b)$
14. $(m - n) - (2m + n)$
15. $(5a - 6) - (3a + 2)$
16. $-(3a + 2) - (a + 6)$
17. $-(6a - 1) + (a + 3)$
18. $(6a + 2) + (a - 1)$
19. $-(3a - 1) - (-2a + 3)$
20. $(4m - 12) + (-m - 3)$
21. $(6m + 1) - (m + 3)$
22. $-(2a + 3) + (-a + 6)$
23. $(-2y - 4) - (-y + 5) + (3y - 7)$
24. $(3t^4 + 6t^3) - (2t^2 + 3t - 4) + (t^4 + t^3 - 3)$
25. $-(3a + 2b) + (a - b) - (-a + b)$
26. $[-7x + 3] - 4$
27. $(4w - 3z) - (w - z)$
28. $(x^2 + xy - y^2) - (-x^2 - xy - 2y^2) - (2x^2 + 2xy + 4y^2)$
29. $(-3x - y) - (4x - y) + (-x + y)$
30. $19 - (17x - y) - (-9x + y - 3)$

Simplify the expressions in Exercises 31–40.

31. $(7ab + 3bc - 9cd) - (-8ab - 4bc - 7cd)$
32. $(t^3 - 4t) - (5t^3 + 4t^2) + (4t^3 + 4t^2 + 4t)$
33. $(y^2 + 4) - (3n + 9) - (-3y^2 - 4n - 9)$
34. $2t - [4r + s + (r - t)]$
35. $p - \{r + (p - 4) - (r - 5)\}$
36. $\{3x - (4y + 5x)\} + 15y$
37. $4 + \{7x + [5x - (5 - 2x) + 7x] - 8\}$
38. $(4x^2 + 3x) + \{(3x^2 - 7x) - (4x + 3)\}$
39. $[y + (x - z)] - 2x + \{x - (y - z)\}$
40. $3a - \{-4b + c - [a + b + (a - b + c)]\}$

Evaluate each of the expressions in Exercises 41–50 for $a = 0$, $b = -2$, and $c = 4$.

41. $\dfrac{a^2 + b^2 + c^2}{5c}$
42. $\dfrac{4ab - c}{a}$
43. $5a^2b + 4a$
44. $2[a - (b + c)]$
45. $-3(a - b) + 5a$
46. $1{,}000{,}485abc$
47. $\dfrac{47ab - 3c}{c - 4}$
48. $-2 + 3(b - c)$
49. $a - 3b(a - c)$
50. $a^3 - b^2 - 7c$

Important Terms Used in this Chapter

collecting like terms
decreasing powers
like terms

opposite of a polynomial
simplify
standard form

REVIEW EXERCISES

Write the polynomials in Exercises 1–10 in standard form.

1. $a^2 + 3a^3 - 6$

2. $m^2 - m + m^4$

3. $b^2a + a^2b$

4. $7 - a^2 + 3a$

5. $m^5 + m^4 - m + m^2$

6. $y^2 - y^3 + 3$

7. $5a + 6b - a^2$

8. $c - b + a$

9. $m - n^2 + m^3$

10. $h^2 - 6 - h$

Answer true (T) or false (F) in Exercises 11–16.

11. 5x and 5y are like terms since they both have a co-efficient of 5.

12. $\frac{1}{2}$ and $\frac{3}{7}$ are like terms since they are both constants.

13. 3xyw and 17xyz are unlike terms.

14. $13x^3y^2$ and $72x^2y^3$ are like terms.

15. 1.5ab and $\frac{2}{3}$ba are similar terms even though the last term is not written in standard form.

16. None of the terms 4x, 3y, $-7z$, $2x^2$, $-9xy$, or 5z are similar.

Perform the additions in Exercises 17–20.

17. $2x^2 - 4x - 2$
    $5x^2 + 4x + 4$

18. $6a \quad\quad + 10c$
    $-5a - 5b - \ 7c$

19. $\ 4x^2 + 5x - 9$
    $-6x^2 + 2x - 4$

20. $4a - 2b + \ c$
    $\ a - 5b + 3c$

In each of Exercises 21–24, subtract the lower expression from the upper expression.

21. $\ \ 5a^2bc - 3ab^2c$
    $-2a^2bc - 7ab^2c$

22. $\ \ -9xy^2z + y^3$
    $-10xy^2z + y^3$

23. $0.83s^3 + 0.009s^2 - 0.1s$
    $0.83s^3 + 0.009s^2 - 0.1s$

24. $\ \ 9.073x^4 - 3x^2 \quad\quad + 1.07$
    $11.073x^4 \quad\quad\ \ - 2x + 7.07$

Perform the indicated operations in Exercises 25–44 and simplify the results.

25. $a^3 + a^3$

26. $3a - 2a$

27. $3x^5 - x^5$

28. $7y^2 - 7y^2$

29. $x^2 + 3x^3$

30. $(3a + 2b - 6) + (3b + a - 4)$

31. $(7x^2 + 3y) - (3x^2 - 4y)$

32. $a^2 - 6a + 4a - 24$

33. $6a + [a - (2 + 6a)]$

34. $2x - \{y + (x - 3)\} - 4$

35. $3ab + 4ac - 5bc - (7ab + 3bc - 5ac)$

36. $(2x^2y - xy^2 + 3y^3) - (-2y^3 - x^2y + xy^2)$

37. $(a^2 - 6ab + 10) + (2a^2 - 7)$

38. $(a^2 - 6ab + 10) - (2a^2 - 7)$

39. $-(2a^2 - 7) - (a^2 - 6ab + 10)$

40. $4a - [9 - (2a - 3)]$

41. $-(3a - 6x - 7) + (6a + 8x) - (-3x + 2a + 9)$

42. $5x^2 - 3x^3 + 4x - 9 + 7x^3 + 6$

43. $[x - (3 + 4x)] + 7x - \{3 + x\}$

44. $\frac{1}{2}x^2 - \frac{3}{4}x^3 + 2x - 6 + \frac{7}{4}x^3 + \frac{1}{3}$

45. Add $3ab - 5bc + 6cd$ and $-3cd + 3bc - 2ab$ and $cd - ab - 2bc$.

46. Subtract $3x^2 - 5x + 4$ from $7x^2 + 5x$.

47. Find the sum of $x^2 + 6x + 7$ and $x^2 - 3x + 2$.

48. $7x^3 - 16$ minus $10x^3 - 5x + 7$

49. Add $10x^3 + 6x^2 - 5x + 3$ and $4x^3 - 3 + 4x - 5x^2$.

50. Collect like terms: $2x^2 - 16x + 14 - 2x^2 - 6x - 9 + 7y$.

Simplify in Exercises 51–60.

*51. $-[6 - (a + 2 - b) + 3a]$

*52. $3n - \{6 + 3n - [7 - (2 + m)] + 6m\}$

*53. $12m + 6 - \{6 - [6 - (-6m)]\}$

*54. $a^2 + b - [3c + 2a^2 - (3b + a^2 - 2c) + b]$

*55. $a - \{b + c - [d - (e - f)]\}$

*56. $12x^a - 16x^a$

*57. $2m^{3a} + 6m^{3a}$

*58. $2a^m + 3a^m + 6a^m$

*59. $12m^{2n} - 8n^{3m} + 3m^{2n} + n^m$

*60. $3x^a - 5y^a - 2y^a + 7x^a$

*These exercises are optional.

8

MULTIPLICATION OF POLYNOMIALS

OBJECTIVES

Upon completion of this chapter you should be able to:

1. Multiply powers with the same base.	8.1
2. Raise a power to a power.	8.1
3. Raise a product to a power.	8.1
4. Multiply monomials.	8.2
5. Multiply a polynomial by a monomial.	8.3
6. Multiply two or more polynomials.	8.4
7. Simplify expressions involving multiplication, addition, and subtraction of polynomials.	8.4

PRETEST

1. Simplify the following (Objective 1).
   a. $a^3 \cdot a^6$ **b.** $a^3 \cdot a^2$
   c. $a^3 \cdot a^4$ **d.** $a \cdot a^2 \cdot a^3$
   e. $b \cdot b^3 \cdot b^4$

2. Simplify the following (Objective 2).
   a. $(x^3)^2$ **b.** $-(a^3)^4$
   c. $(-2)^3(a^3)^2$ **d.** $(a^4)^3(b^2)^4(c^3)^6$
   e. $3a(b^2)^3(-c)^2$

3. Simplify the following (Objective 3).
   a. $(ab)^3$ **b.** $(-3ab)^2$
   c. $-(3ab)^2$ **d.** $(abc)^4$
   e. $(a^2b^3c)^4$

4. Simplify the following (Objective 4).
   a. $(2a)(3ab)$ **b.** $(2a^2b)(3ab^2)(5a)$

   c. $(5a)(-2ab)$ **d.** $(3a)^2(2a^2b)$
   e. $5a^2b(-3a)^3$

5. Simplify the following (Objective 5).
   a. $3(a + b)$ **b.** $-4(a - b)$
   c. $3a^2b(a^2 - 4ab)$ **d.** $(-3a^2)^2(2a^2b - 3ab^3)$
   e. $-(4a)^3(3a^2 - 6ab^3)$

6. Simplify the following (Objective 6).
   a. $(x - 3)(x + 4)$ **b.** $(2x - 1)(x + 6)$
   c. $(3a + 4)^2$ **d.** $(x - 3)(x + 2)(x - 5)$
   e. $(2x + 3)(5x - 1)^2$

7. Simplify the following (Objective 7).
   a. $(2a)(a - b) + 4(3a + 2)$ **b.** $4[x - 2(x + y)] - 5x$
   c. $(2a - 3)^2 - [4a^2 + 9]$ **d.** $2c - 3[a + 2(c - 3)]$
   e. $(a + 5)^2 - 2(a^2 + 25)$

In the arithmetic review, you attained skill in multiplying numbers. We will now reexamine multiplication and extend it to expressions containing variables. Many of the techniques and properties introduced earlier will be applicable to the multiplication of polynomials. These include the rules for multiplication of signed numbers, the commutative and associative laws, and the procedure for removing symbols of grouping. It will also be helpful to recall some relationships mentioned in Chapter 6. They are

1. The meaning of x^3 is $x \cdot x \cdot x$.
2. The coefficient of the term x^3 is 1.
3. The coefficient of the term $-x^2$ is -1.
4. The exponent of x in the term 4x is 1.

8.1 Laws of Exponents Related to Multiplication

Exponents are used extensively in mathematics. Knowledge of how to use them will be necessary in much of your later work. It will be useful for you to learn how to use several generalizations before we proceed. Recall that 3^2 means $3 \cdot 3$ and that 3^7 means $3 \cdot 3 \cdot 3 \cdot 3 \cdot 3 \cdot 3 \cdot 3$. Now consider $3^2 \cdot 3^7$:

$$3^2 \cdot 3^7 = \underbrace{\overbrace{(3 \cdot 3)}^{2 \text{ factors}} \cdot \overbrace{(3 \cdot 3 \cdot 3 \cdot 3 \cdot 3 \cdot 3 \cdot 3)}^{7 \text{ factors}}}_{9 \text{ factors}} = 3^9$$

The number of times 3 is used as a factor in the product can be found by adding the exponents. This example suggests the *rule for multiplying powers with the same base.*

RULE FOR MULTIPLYING POWERS WITH THE SAME BASE	EXAMPLE $y^5 \cdot y^3 = ?$
	Solution:
1. Identify the common base.	y
2. Add the exponents.	$5 + 3 = 8$
3. Form the power which has the base identified in step 1 and has as its exponent the sum obtained in step 2.	y^8

As mentioned earlier, one of the most useful applications of algebra is to replace lengthy verbal statements with concise algebraic statements. This application can be illustrated by expressing rules in symbolic form. Such expressions are called **formulas** and are used extensively in mathematics. Compare the *formula for multiplying powers with the same base* with the above rule.

FORMULA FOR MULTIPLYING POWERS WITH THE SAME BASE	EXAMPLE $y^5 \cdot y^3 = ?$
	Solution:
$x^a \cdot x^b = x^{a+b}$	$y^5 \cdot y^3 = y^{5+3} = y^8$

The algebraic *formula* gives you exactly the same information as the *rule* if you know how to interpret it. In the formula, x represents the common base and a and b represent the exponents. The formula indicates that the product

$x^a \cdot x^b$ is the same as x^{a+b}. In other words, in order to multiply two powers with the same base, you retain the common base and add the exponents to obtain the exponent of the product. Thus, $y^5 \cdot y^3 = y^{5+3} = y^8$.

Since the algebraic form for writing rules is more concise and often easier to remember, rules will be stated as formulas whenever possible throughout the remainder of this text.

Example 1   In each of the following identify the common base and the exponents, and then use the *formula for multiplying powers with the same base* to simplify each expression.

a. $z^3 z^{10}$

z is the common base and 3 and 10 are the exponents. Thus,

$$z^3 \cdot z^{10} = z^{3+10} = z^{13}$$

b. $5^2 \cdot 5^4 \cdot 5^6$

5 is the common base and 2, 4 and 6 are exponents. Thus,

$$5^2 \cdot 5^4 \cdot 5^6 = 5^{2+4+6} = 5^{12}$$

Note that the rule for multiplying powers is stated for the product of two powers, but it can be extended to any number of powers.

Example 2   Simplify each of the following expressions.

a. $(-t)^4(-t)^3 = (-t)^{4+3} = (-t)^7 = -t^7$

b. $m^3 \cdot m^3 = m^{3+3} = m^6$

c. $t^2 \cdot n^3 = t^2 n^3$

The factors do not have a common base, so $t^2 \cdot n^3$ cannot be simplified.

The expression $(b^3)^4$ can be written without parentheses by considering the fact that the 4 indicates that b^3 is to be used as a factor 4 times, that is,

$$(b^3)^4 = b^3 \cdot b^3 \cdot b^3 \cdot b^3 = b^{12}$$

Similarly, $(a^7)^2 = a^7 \cdot a^7 = a^{14}$. In both cases, the exponent in the resulting power can be found by multiplying the exponents in the problem.

As with the product of powers, these examples suggest a rule for raising a power to a power. Note that this rule is stated in formula form. Can you translate it into a verbal statement?

FORMULA FOR RAISING A POWER TO A POWER	EXAMPLE $(y^3)^4 = ?$
$(x^a)^b = x^{ab}$	Solution: $(y^3)^4 = y^{3 \cdot 4} = y^{12}$

Example 3   In each of the following, identify the base and the exponents, and then use the *formula for raising a power to a power* to simplify each expression.

     a. $(n^2)^3$

        n is the base and 2 and 3 are exponents. Thus,

$$(n^2)^3 = n^{2 \cdot 3} = n^6$$

     b. $(y^5)^5$

        y is the base and 5 and 5 are exponents. Thus,

$$(y^5)^5 = y^{5 \cdot 5} = y^{25}$$

Example 4   Simplify each of the following expressions:

     a. $(t^{10})^7 = t^{10 \cdot 7} = t^{70}$

     b. $(s^3)^2 = s^{3 \cdot 2} = s^6$

     c. $-(s^3)^2 = -(s^{3 \cdot 2}) = -s^6$

        Note that in this problem the minus indicates the *opposite* of the result when s^3 is raised to the second power.

Consider $(xy)^2$. This means $(xy) \cdot (xy)$. Now, since x and y represent real numbers, the properties for real numbers apply here. A generalized associative and commutative property for addition was stated in Chapter 5. We have a similar property for multiplication, that is, *in multiplication, the factors in a product may be arranged in any convenient order.* For example, the product $2 \cdot 4 \cdot 5 \cdot 10$ may be found in each of the following ways:

$$2 \cdot 4 \cdot 5 \cdot 10 = 8 \cdot 50 \quad \text{or} \quad 4 \cdot 5 \cdot 2 \cdot 10 = 20 \cdot 20$$
$$= 400 \qquad\qquad\qquad\qquad = 400$$

Consider this example also:

$$(xy)^2 = (xy) \cdot (xy)$$
$$= x \cdot x \cdot y \cdot y$$
$$= x^2 y^2$$

It conveniently leads us to the *formula for raising a product to a power.*

FORMULA FOR RAISING A PRODUCT TO A POWER	EXAMPLE $(uv)^2 = ?$
$(xy)^a = x^a y^a$	*Solution:* $(uv)^2 = u^2 v^2$

Example 5   Identify the factors in the product and the exponent in each of the following. Then simplify each expression using the formula for raising a product to a power.

     a. $(3m)^2$

        3 and m are the factors of the product, and 2 is the exponent. Thus,

$$(3m)^2 = 3^2 \cdot m^2 = 9m^2$$

b. $(-3t)^4$

-3 and t are the factors of the product, and 4 is the exponent. Thus,

$$(-3t)^4 = (-3)^4 \cdot t^4 = 81t^4$$

Example 6   Simplify each of the following expressions.

a. $(abc)^{15} = a^{15}b^{15}c^{15}$

Note that the rule is stated for two factors, but it can be extended to any number of factors.

b. $-(rs)^4 = -(r^4s^4) = -r^4s^4$

It is often necessary to use more than one of these laws of exponents in order to simplify an expression. This is illustrated in Examples 7 and 8.

Example 7   Identify the rule or rules needed and simplify:

a. $(x^2)^4x^5 = x^8x^5$ (raising a power to a power)

$\qquad\quad\; = x^{13}$ (multiplying powers with the same base)

b. $(x^2y)^3 = (x^2)^3y^3$ (raising a product to a power)

$\qquad\quad\; = x^6y^3$ (raising a power to a power)

Example 8   Simplify each of the following.

a. $(-x^2)^{13} = (-1 \cdot x^2)^{13}$

$\qquad\qquad\;\; = (-1)^{13} \cdot (x^2)^{13}$

$\qquad\qquad\;\; = -1 \cdot x^{26}$

$\qquad\qquad\;\; = -x^{26}$

b. $(-2xy^4)^4y^7 = (-2)^4(x^4)(y^4)^4y^7$

$\qquad\qquad\qquad\; = 16x^4y^{16}y^7$

$\qquad\qquad\qquad\; = 16x^4y^{23}$

8.1 Exercises

Use the formula $x^a \cdot x^b = x^{a+b}$ to simplify each of Exercises 1–20.

1. $x^5 \cdot x^2$
2. $a^3 \cdot a^2$
3. $m^5 \cdot m$
4. $n^{10} \cdot n^2$
5. $a^5 \cdot a^3$
6. $a^5 \cdot a^4$
7. $y^{10} \cdot y^3$
8. $y \cdot y$
9. $x \cdot x^3$
10. $m^4 \cdot m^3$
11. $m^3 \cdot m^2$
12. $a^3 \cdot a^6$
13. $a \cdot a$
14. $y \cdot y^2$
15. $x^{10} \cdot x^5$
16. $y^2 \cdot z^3$
17. $a^6 \cdot a$
18. $c \cdot c$
19. $m^2 \cdot n^5$
20. $b \cdot b^4$

Use the formula $(x^a)^b = x^{ab}$ to simplify each of Exercises 21–40.

21. $(x^5)^2$
22. $(a^3)^2$
23. $(m^5)^3$
24. $(n^{10})^2$
25. $(a^3)^5$
26. $(a^5)^4$
27. $(y^{10})^3$
28. $(y^3)^3$
29. $(x)^3$
30. $(m^3)^4$
31. $(a^3)^6$
32. $(a^2)^5$
33. $(y)^2$
34. $(x^{10})^5$
35. $(b^3)^2$
36. $(m)^5$
37. $(b^5)^3$
38. $(y^4)^3$
39. $(y)^6$
40. $(n^{12})^2$

Use the formula $(xy)^a = x^ay^a$ to simplify each of Exercises 41–55.

41. $(ab)^3$
42. $(ab)^2$
43. $(2n)^5$
44. $(3y)^2$
45. $(mny)^6$
46. $(abc)^7$
47. $(-2yz)^4$
48. $(-3bcd)^2$
49. $(wxyz)^3$
50. $\left(\dfrac{2}{5}a\right)^2$
51. $\left(\dfrac{1}{2}ab\right)^5$
52. $(4b)^2$

53. $\left(\frac{2}{3}ab\right)^3$ **54.** $\left(-\frac{1}{2}ab\right)^3$ **55.** $(-2ab)^5$

81. $(a^5)^{10}$ **82.** a^2b^2 **83.** $(ms)^{10}$

84. $t \cdot t$ **85.** $2^2 \cdot 3^2$ **86.** $(b^3)^{15}$

Use $(xy)^a = x^a y^a$ and $(x^a)^b = x^{ab}$ to simplify each of Exercises 56–70.

87. Multiply a^3 and a^{12}. **88.** $(t^{10})^3$

89. $(m^4 m^3)^7$ **90.** $(2t^3)^2$

56. $(a^2b)^3$ **57.** $(m^2n^3)^4$ **58.** $(mn^2)^3$

91. $(-2x^2y^4)^5$ **92.** $(4a^5y^2z)^4$

59. $(a^3b)^5$ **60.** $(4a^3b)^2$ **61.** $(-3a^6b^2)^3$

93. $(x+y)^6(x+y)^7$ **94.** $t \cdot t^2 \cdot t^3$

62. $\left(\frac{2}{3}a^2\right)^3$ **63.** $(4ab^2)^2$ **64.** $(2a^2b^3c)^4$

95. $-(x^2)^5$ **96.** $(3a)^2$

97. $(-3ab)^2$ **98.** $(-m^3)^2$

65. $\left(\frac{1}{2}mn^2\right)^5$ **66.** $(m^3n^2)^5$ **67.** $\left(\frac{3}{5}m^2\right)^2$

99. $4^2 \cdot 2^3$ **100.** $(-2rs)^3$

101. Write a verbal statement of the formula for raising a power to a power.

68. $\left(\frac{3}{4}mn\right)^3$ **69.** $(3a^2b^3)^3$ **70.** $(6a^2b^6)^2$

102. $-(y^3)^2$

103. Find the cube of 2a.

If possible, simplify each of Exercises 71–100. State the name of the rule that is used in working Exercises 71–75.

104. $(a+b)^2(a+b)^5$ **105.** $(-n^5)^3$

106. $a^3 \cdot b^2$ **107.** $(a+2)^4(a+2)^3$

71. $n^3 \cdot n^5$ **72.** $(m^3)^5$ **73.** $(mn)^3$

108. Write a verbal statement of the formula for raising a product to a power.

74. $(hk)^4$ **75.** $(h^3)^2$ **76.** h^3k^2

77. $2^3 \cdot 2$ **78.** $(2^2)^4$ **79.** $(ab)^2$

109. $(xy)^4(xy)^{16}$ **110.** $(2a)^2(ab)^3$

80. Find the product of a^5 and a^{10}.

8.2 Multiplying Monomials

Now we are ready to multiply monomials. We will use the laws of exponents stated in Section 8.1 and the generalized commutative and associative laws for multiplication. Consider the product of the monomial $5x^2y$ and the monomial $-4x^4y^6z^3$.

$$(5x^2y) \cdot (-4x^4y^6z^3) = 5(-4)x^2x^4yy^6z^3$$
$$= -20x^6y^7z^3$$

The procedure to be followed is indicated in the *rule for multiplying monomials.*

RULE FOR MULTIPLYING MONOMIALS	*EXAMPLE* $(5x^2y)(-4x^4y^6z^3) = ?$

1. Indicate
  a. the product of the constant factors.
  b. the product of the powers with the same base.
  c. the factors not used in a or b.
2. Indicate the product of the results from step 1.
3. Simplify the product from step 2.
	Solution:
	
	$5(-4)$
	$x^2 \cdot x^4, y \cdot y^6$
	z^3
	
	$5(-4) \cdot x^2 \cdot x^4 \cdot y \cdot y^6 \cdot z^3$
	$-20x^6y^7z^3$
	
	Thus, $(5x^2y)(-4x^4y^6z^3) = -20x^6y^7z^3$.

Consider the following examples which illustrate how the product of monomials is found. Step 1 of the rule is done mentally. The reordering of the factors is written here to clarify the procedure, but this can and should also be done mentally in most problems.

Example 1
$$(3a)(5a^2) = 3 \cdot 5 \cdot a \cdot a^2$$
$$= 15a^3$$

Example 2
$$(2m^2)(6m^3) = 2 \cdot 6 \cdot m^2 \cdot m^3$$
$$= 12m^5$$

Example 3
$$(5ab)(4a^2) = 5 \cdot 4 \cdot a \cdot a^2 \cdot b$$
$$= 20a^3b$$

Example 4
$$(-2m^2n)(mn^3) = -2m^2 \cdot m \cdot n \cdot n^3$$
$$= -2m^3n^4$$

Example 5
$$2v^3(3u^2vt^3)(5t^7) = 2 \cdot 3 \cdot 5 \cdot v^3 \cdot v \cdot u^2t^3t^7$$
$$= 30v^4u^2t^{10}$$

Example 6
$$(x^2y)^2(4xy^4) = (x^2)^2y^2 \cdot 4xy^4$$
$$= x^4y^2 \cdot 4xy^4$$
$$= 4 \cdot x^4 \cdot x \cdot y^2 \cdot y^4$$
$$= 4x^5y^6$$

Example 7
$$(-3x^2y^3)(2y^3z)(-xz^2) = -3 \cdot 2 \cdot -1 \cdot x^2xy^3y^3zz^2$$
$$= 6x^3y^6z^3$$

8.2 Exercises

Simplify the results in Exercises 1–60.

1. $(+x)(+y)$
2. $(+x)(-y)$
3. $(-x)(+y)$
4. $(-x)(-y)$
5. $(-x)(-y)(-2)$
6. $(x)(-y)(-2)$
7. $(3x)(5x)$
8. $(2x^2)(5x)$
9. $(5m^3)(3m^2)$
10. $(3x)(4y)$
11. $\left(\frac{1}{2}a^2\right)(8a)$
12. $(5a^2)(2a^3)$
13. $(5a)(6b)$
14. $(3m^3)(4m^4)$
15. $(3a^2)(2a)(2a^3)$
16. $(a^3)(2a^5)(5a^2)$
17. $(m^6)(3m^2)(m^5)$
18. $(x^4)(4x^2)(x^8)$
19. $(4x)(cy)$
20. $4a(2z^2)$
21. $-3u(u^4v)$
22. $x^2y(xy^2)$
23. $(-4a)(-3b)$
24. $5ab\left(-\frac{1}{5}ab\right)$
25. $6a\left(-\frac{1}{2}a^2b\right)$
26. $-5a(-6a)$
27. $4m(m^2n)$
28. $5a^2\left(-\frac{1}{10}a^3\right)$
29. $-2m^2(-3m^3)$
30. $m^4n(mn^4)$
31. $-(-x^2y)(-xy^3)$
32. $-(4x)(3y)$

33. Multiply $-3a^2$ times a^2b.
34. $2x^2(-3y^2)$
35. $-8c^3d^2(d^2)$
36. $5r^2s(-5rs^2)$
37. $-9(2x^2y)(-3xy^2)$
38. Find the product of $4a^2$ and $-2ab$.
39. $-9^2(5a^2b^4)(7a^3b)$
40. $(m^4p)(-3p)\left(\frac{1}{3}m^2\right)$
41. $-4xy(5x^3y)$
42. $-4xy(-2y)$
43. $-4xy(9)$
44. $(1.2c^2d)(3.42f^5)$
45. $-4a(a^3b)$
46. $(-2a^2)(-6ab)$
47. $(5x)(-2xy)$
48. Find the opposite of the product obtained when $-2a$ is multiplied by the square of $-3a$.
49. $(-2x^2y)^2\left(\frac{1}{4}xy\right)^2(-2y^2)^3$
50. $(xy)^2xy^2$
51. $(3a)^2 \cdot ab^2$
52. $(-3a)^2(-ab)$
53. Cube (ab), and multiply the result by ab^3.
54. $5a^2(2a)^3$
55. $-(3a)^2 \cdot ab$
56. $(4x)^2(3y)$
57. $(5x^3y)^2$
58. $(-3a)^2(-2a)^3$
59. $-(2ab)^4$
60. $(xy)^4(xy)^{16}$

8.3 Multiplying a Polynomial by a Monomial

The property of real numbers which enables us to multiply a polynomial by a monomial is called the **distributive property**. You may not recognize this property by name even though you have used it in arithmetic computations. It gives us the procedure we need for calculating the product of a polynomial and a monomial. To find the product of the monomial $4x^2$ and the polynomial $3x^3 + 5x^2 + 6$, you multiply each term of the polynomial by the monomial and then add the resulting products. For example,

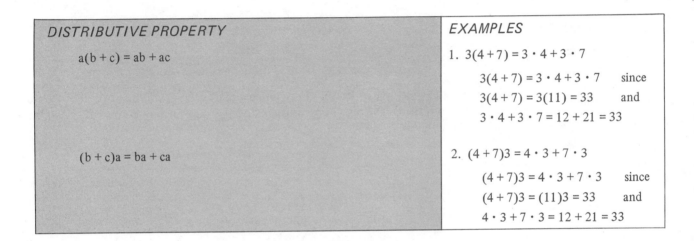

$$4x^2(3x^3 + 5x^2 + 6) = 4x^2(3x^3) + 4x^2(5x^2) + 4x^2(6)$$
$$= 12x^5 + 20x^4 + 24x^2$$

DISTRIBUTIVE PROPERTY	*EXAMPLES*

$a(b + c) = ab + ac$	1. $3(4 + 7) = 3 \cdot 4 + 3 \cdot 7$
	
	     $3(4 + 7) = 3 \cdot 4 + 3 \cdot 7$ since
	     $3(4 + 7) = 3(11) = 33$ and
	     $3 \cdot 4 + 3 \cdot 7 = 12 + 21 = 33$

$(b + c)a = ba + ca$	2. $(4 + 7)3 = 4 \cdot 3 + 7 \cdot 3$
	
	     $(4 + 7)3 = 4 \cdot 3 + 7 \cdot 3$ since
	     $(4 + 7)3 = (11)3 = 33$ and
	     $4 \cdot 3 + 7 \cdot 3 = 12 + 21 = 33$

Here is the *rule for multiplying a polynomial by a monomial.*

RULE FOR MULTIPLYING A POLYNOMIAL BY A MONOMIAL	*EXAMPLE* $3a(a^2 + b - c) = ?$ *Solution:*
1. Indicate the products obtained when each term of the polynomial is multiplied by the monomial.	$3a(a^2)$ (Do this mentally.) $3a(b)$ $3a(-c)$
2. Indicate the sum of the products from step 1.	$3a(a^2) + 3a(b) + 3a(-c)$
3. Perform the indicated multiplications.	$3a^3 + 3ab - 3ac$
	Thus, $3a(a^2 + b - c) = 3a^3 + 3ab - 3ac.$

The *rule for multiplying a polynomial by a monomial* is used in the following examples to find the product of a monomial and a polynomial. Step 1 is done mentally. With practice, intermediate steps can also be done mentally. They are shown in the examples for the purpose of illustration.

Example 1

$$6x(3x + 4) = 6x(3x) + 6x(4)$$
$$= 18x^2 + 24x$$

Example 2

$$6x(3x^2 + 5x + 4) = 6x(3x^2) + 6x(5x) + 6x(4)$$
$$= 18x^3 + 30x^2 + 24x$$

Example 3

$$6xy(3x^2 - 5x + 4) = 6xy(3x^2) + 6xy(-5x) + 6xy(4)$$
$$= 18x^3y - 30x^2y + 24xy$$

Example 4

$$(3x^2 - 5x + 4)6xy = (3x^2)6xy + (-5x)6xy + (4)6xy$$
$$= 18x^3y - 30x^2y + 24xy$$

Example 5

$$-6xy(-3x^2 + 5x - 4) = -6xy(-3x^2) + (-6xy)(5x) + (-6xy)(-4)$$
$$= 18x^3y - 30x^2y + 24xy$$

Examples 6–8 show that it is often necessary to combine several rules in order to simplify algebraic expressions. See if you can name the rule used in each step.

Example 6 $(x^2 + 3x)(2x)^2 = ?$

$$(x^2 + 3x)(2x)^2 = (x^2 + 3x)(4x^2)$$
$$= (x^2)(4x^2) + (3x)(4x^2)$$
$$= 4x^4 + 12x^3$$

Example 7   Simplify $2a(a + b) - b(a + b)$.

$$2a(a + b) - b(a + b) = 2a^2 + 2ab - ab - b^2$$
$$= 2a^2 + ab - b^2$$

Example 8   Simplify $3(x + y) - \{4x - 2(-x + 2y)\}$.

$$3(x + y) - \{4x - 2(-x + 2y)\} = 3x + 3y - \{4x + 2x - 4y\}$$
$$= 3x + 3y - 4x - 2x + 4y$$
$$= -3x + 7y$$

8.3 Exercises

Simplify the expressions in Exercises 1-60.

1. $3(a + b)$
2. $-4(a + b)$
3. $-3(a - b)$
4. $4(a - b)$
5. $3(a - 2b)$
6. $-3(2a - b)$
7. $a(a + b)$
8. $a^2(a - b)$

9. $(a - 6)b$
10. $(m + 2n)m$
11. $3a(x - y)$
12. $2m(m - 2n)$
13. $2(x + y)$
14. $3(a + 2)$
15. $a(2x - y)$
16. $10(x - 2y)$
17. $(a - 5)4$
18. $2(x - y)$

19. $4(a - 6)$	20. $x(a - 3b)$	40. $abc(a - b + c)$	41. $a^3(2ab - c)$
21. $12(x + 3y)$	22. $(x - 6)5$	42. $(x^2 + 2y - 3)(-xy)$	43. $-ab(2a^2 + 3b)$
23. $3(a^2 + 2a - 5)$	24. $5(4a^2 - 3a + 9)$	44. $(3a^2 + 2b - c)(abc)$	45. $(ab)(a - 3b)$

25. What is the product of $2x^2$ and $(x + y + z)$?

46. $3(a + b) - 6(a - b)$

26. $(4 - 6t - 2t^2)(-4)$

47. $(a - b)(-2) + (2a + b)(3)$

27. $4np^2(2np - 4np^2 + 3p^3)$

48. $3x(x^3y)(4x - 6y)$

28. $12c^2d(cd + 3)$

49. $(-2ab)(a + b)$

29. $-4x^3y^2(x^2 + y^2 - 3y + 4x)$

50. $7 + 3[-t - 2(t + 5)]$

30. Multiply $(4m^2 + 3m + 10)$ by $-5m$.

51. $(5x^7 - 3x^2 - 0.23)(x^2)$

31. $\frac{1}{2}z(6z + 4)$	32. $-\frac{2}{3}z(9 - 3z)$	52. $(-ab)(3a + b)$	53. $2(x + 5) - 3(x - 4)$
		54. $a^3(54a^2 - 81)$	55. $x^2(4x^2 - 16)$
33. $(a^2 + 2ab + b^2)(-5a^5b^4)$	34. $2a^3b^2\{-3a^2 - b^3\}$	56. $2a^2(ab) + 4a^3b$	57. $2a^2(ab)(4a^3 + b)$

35. Subtract 56 from 5t, and multiply the result by $-4t$.

58. $3a(-4a^2b)(2a^3 + b^3)$

59. $5x^2y^3(3x^3 - 5y^2)$

36. $b^4(3ab - b^2)$ 37. $abc(a^2 - b + c^2)$

60. $m^3n^5(6mn^2 + 5m^2n)$

38. $(a^2 + 2ab + b^2)(ab)$ 39. $-3x^2(x^2 + 2x - 3)$

8.4   Multiplying a Polynomial by a Polynomial

We will now apply the distributive property to multiply polynomials by polynomials. If the second polynomial is treated as a single expression, the second form of the distributive property can be applied as illustrated in the following example.

Example 1 $(x + 2)(4x^2 + 3x + 6) = ?$

$$(x + 2)(4x^2 + 3x + 6) = x(4x^2 + 3x + 6) + 2(4x^2 + 3x + 6)$$
$$= x(4x^2) + x(3x) + x(6) + 2(4x^2) + 2(3x)$$
$$+ 2(6)$$
$$= 4x^3 + 3x^2 + 6x + 8x^2 + 6x + 12$$
$$= 4x^3 + 11x^2 + 12x + 12$$

Consider the rule to be used when multiplying polynomials.

RULE FOR MULTIPLYING A POLYNOMIAL BY A POLYNOMIAL	*EXAMPLE* $(2a + 3)(3a - 6) = ?$ *Solution:*
1. Indicate the product of each term of the first polynomial and the second polynomial.	$2a(3a - 6)$ $3(3a - 6)$
2. Indicate the sum of these products.	$2a(3a - 6) + 3(3a - 6)$
3. Perform the multiplications.	$6a^2 - 12a + 9a - 18$
4. Combine like terms.	$= 6a^2 - 3a - 18$
	Thus, $(2a + 3)(3a - 6) = 6a^2 - 3a - 18$.

Example 2   a. $(2x + 3y)^2 = ?$

$$(2x + 3y)(2x + 3y) = 2x(2x + 3y) + 3y(2x + 3y)$$
$$= (2x)(2x) + (2x)(3y) + (3y)(2x) + (3y)(3y)$$
$$= 4x^2 + 6xy + 6xy + 9y^2$$
$$= 4x^2 + 12xy + 9y^2$$

b. $(2x - 3y)(2x + 3y) = ?$

$$(2x - 3y)(2x + 3y) = 2x(2x + 3y) + (-3y)(2x + 3y)$$
$$= 2x(2x) + 2x(3y) + (-3y)(2x) + (-3y)(3y)$$
$$= 4x^2 + 6xy + (-6xy) + (-9y^2)$$
$$= 4x^2 + 6xy - 6xy - 9y^2$$
$$= 4x^2 - 9y^2$$

After you have learned the procedure, the detailed solution shown in Example 2 is not necessary. Steps in this procedure may be omitted whenever they can be done mentally.

Example 3 $(x + y - z)(x - y + z) = ?$

$$(x + y - z)(x - y + z) = x(x - y + z) + y(x - y + z) + (-z)(x - y + z)$$
$$= x^2 - xy + xz + yx - y^2 + yz - zx + zy - z^2$$
$$= x^2 - y^2 + 2yz - z^2$$

Many times the actual computation of multiplying polynomials is unwieldy. A device which may prove useful is a vertical format. This format is illustrated in the following example. (Note that the multiplication is done from left to right.)

Example 4 $(x + 7)(4x^2 + 3x + 6) = ?$

$$
\begin{array}{r}
4x^2 + 3x + 6 \\
x + 7 \\
\hline
4x^3 + 3x^2 + 6x \qquad \leftarrow x(4x^2 + 3x + 6) \\
+ 28x^2 + 21x + 42 \leftarrow 7(4x^2 + 3x + 6) \\
\hline
4x^3 + 31x^2 + 27x + 42
\end{array}
$$

In the second row of work, note that the terms are arranged under the terms in the first row so that similar terms are in the same columns. This arrangement makes it easier to add similar terms and to express the answer in standard form.

The rule for multiplying a polynomial by a polynomial can be summarized by the following statement:

To multiply two polynomials, multiply each term in one polynomial by each term in the other polynomial.

Example 5 $(2x - 3y)(2x + 3y) = ?$

$$
\begin{array}{r}
2x \; - \; 3y \\
2x \; + \; 3y \\
\hline
4x^2 - 6xy \\
+ \, 6xy - 9y^2 \\
\hline
4x^2 \qquad\quad - 9y^2
\end{array}
$$

Example 6 $(x + y - z)(x - y + z) = ?$

$$
\begin{array}{l}
x \; + \; y - \; z \\
x \; - \; y + \; z \\
\hline
x^2 + xy - xz \\
\quad\; - xy \qquad\quad - y^2 + \; yz \\
\quad\qquad\quad + xz \qquad\quad + \; yz - z^2 \\
\hline
x^2 \qquad\qquad\qquad - y^2 + 2yz - z^2
\end{array}
$$

Example 7 $3(a + b)(a - b) + (a + b)^2 = ?$

$$
\begin{aligned}
3(a + b)(a - b) + (a + b)^2 &= 3(a^2 - b^2) + a^2 + 2ab + b^2 \\
&= 3a^2 - 3b^2 + a^2 + 2ab + b^2 \\
&= 4a^2 + 2ab - 2b^2
\end{aligned}
$$

8.4   Exercises

Find the products in Exercises 1–44. Express the results in standard form.

1. $(x + 2)(x + 3)$
2. $(x + 2)(x - 3)$
3. $(x - 2)(x + 3)$
4. Multiply $(x - 2)$ times $(x - 3)$.
5. $(a + b)(a - b)$
6. $(2t - 3)(t - 4)$
7. $(x + 2)^2$ *Hint:* $(x + 2)^2 = (x + 2)(x + 2)$.
8. $(x + 2)(x^2 - 3x - 4)$
9. Find the product of $(t - 4)$ and $(2t - 3)$.
10. $(2x + 1)(x - 5)$
11. $(a + 7)(3a - 1)$
12. $x - 3$ multiplied by $x + 5$
13. $(3a + 2)(a - 5)$
14. $(x - 8)(2x - 1)$
15. $(2x - 3)(x - 2)$
16. $(3a + 1)(a + 2)$
17. The product of $x + 4$ and $2x - 1$
18. $(x - y)(2x - y)$
19. $(2a + 5)(a + 6)$
20. $(2a^2 + 3b^2)^2$
21. $(2x + 3y)(2x - 3y)$
22. $(x^2 - y^2)(x^2 + y^2)$
23. Square $(x - y)$.
24. $(x^2 + x - 3)(x - 1)$
25. $(-2x - 3y)(2x + 3y)$
26. $(7y - 2)(y - 3)$
27. $(x^2 + y^2)(x^4 - 2x^2y^2 + y^4)$

28. $(a + 2b + c)(2a - b + 2c)$
29. $(t - 1)(t^4 + t^3 + t^2 + t + 1)$
30. $(2x - 3y + 4)^2$
31. $3(x + 2)(x - 3)$
32. Multiply -3 times the product of $(3a + b)$ and $(a - 1)$.
33. $(2x - 4)^2$
34. $(x - 2)(x^3 - 2x^2 + 3x - 4)$
35. $(x^2 + x - 1)(x - 5)$
36. $(3b)^2(b^2 - b + 1)$
37. $(3a + 5)(2a - 6)$
38. $(3b^2 - 2)(5b^2 - 1)$
39. $\left(\dfrac{1}{2}a^2 + 1\right)(2a^2 - 6)$
40. $(0.5a + 3)(0.2a - 5)$
41. $3m(m - n)^2$
42. $-2a(a + b)^2$
43. $3ab(a + 2b)(2a - 3b)$
44. $2xy(3x + y)(2x - 3y)$

Perform the indicated operations in Exercises 45–48. Simplify the results.

45. $(3a)(a - b)^2 + 3(6a + 3)$
46. $3(x - 2)(x + 4) - 10x$
47. $(3x + 4)[8x - (6x + 8)]$
48. $(3a - b)^2 - [9a^2 + b^2]$

Evaluate each of Exercises 49–60 for $a = 1$, $b = -2$, $c = 0$.

49. $(a - 3)(a + 4)$ **50.** $(2a + b)(3a - b)$ **55.** $(5a^2b)(ab^2)$ **56.** $(a + 5)^2 + a^2 - 25$

51. $2c(2a - 3)$ **52.** $3b(2a + b - c)$ **57.** $5a^7b^5$ **58.** $2a^2 + 10a$

53. $3b^2(a - 2c)$ **54.** $5a^3(b^2 + 5c)$ **59.** $2c - (6 + b)$ **60.** $-2a(a + 3c)$

Important Terms   distributive property
Used in this Chapter   formula

REVIEW EXERCISES

Use the laws of exponents to simplify each of Exercises 1–10. Indicate the rule or rules needed for each problem.

1. $a^2 \cdot a^5$ **2.** $(2xy)^4$

3. $(a^2)^3$ **4.** $(5x^3y)^2$

5. $x^{17}x^2$ **6.** $(x^6)^3$

7. x^5x^6 **8.** $(x^2)^3x^4$

9. $(-2a)^3$ **10.** $(a^2b^3c^4)^2$

In Exercises 11–20, identify the type of problem as (1) the product of monomials, (2) the product of a monomial and a polynomial, or (3) the product of polynomials. Then find the product and express the result in standard form.

11. $(x + 2y)^2$ **12.** $(a - 6)(a + 4)$

13. $(4y)^2(y - 5)$ **14.** $-3ab(4a^3b - 4b + 10)$

15. $-4^2(4n^2t^4)(8n^3t)$ **16.** $(3x - 4)(3x^2 + 2x + 4)$

17. $(2a + 3)(a - 1)$ **18.** $(2a + 1)(a^2 - 3a + 5)$

19. $(3x)(4y)$ **20.** $(2a^2 + 3a - 6)9a^2$

21. Multiply $3ax^2y$ by $4a^2y^3$.

22. Find the product of $2a$, $-5a^2b^4$ and $7a^3b$.

23. Cube $-2a$, and multiply the result by a^3b.

24. Raise $5x^2$ to the second power and yz^2 to the fourth power, and then find the product of these results.

25. If the trinomial $5x^3y - 2z + 9$ is multiplied by $-4xy$, what is the product?

26. Square the binomial $x - 3$.

27. Cube the binomial $x - 3$.

28. If the product of $3a^2$ and $-a^2$ is added to $10a^2$, what is the result?

Perform the indicated operations in Exercises 29–45 and simplify the results.

29. $-\{5a + a^2 - 6\} + 4a$ **30.** $6a + 3[a - (2 + 6a)]$

31. $-3\{y + 2(x - 3)\}$ **32.** $(a + 5)^2 + a^2 - 25$

33. $(2a^2 - 3b)3a - 2a(a^2 + 3b)$

34. $(5a^2 - 6b)2b - 5b(a^2 + 3b)$

35. $(a + b)^2 - (a - b)^2$

*36. $(4n^2t^3)^3(5n^4t^2)^2$

*37. $(a - b)(2a + b)(a + b)$

*38. $(x - 1)^4$

*39. $-3\{6 - 2a[6 - 3(a + b) + b] - 4b\}$

*40. $6[a - (3 + b) - a]^2$

*41. $x^n(x^n - 4)$

*42. $(x^n + 1)(x^n - 5)$

*43. $3x^{2n}(x^{3n} - 3x^{2n} + 6x^n - 12)$

*44. $(3x^n + 1)(2x^n - 7)$

*45. $(x^n + 1)(x^n - 1)$

---

*These exercises are optional.

9 DIVISION OF POLYNOMIALS

OBJECTIVES

Upon completion of this chapter you should be able to:

1. Divide powers with the same base.	9.1
2. Raise a fraction to a power.	9.1
3. Simplify the quotient of two monomials.	9.2
4. Divide a polynomial by a monomial.	9.3
5. Divide a polynomial by a polynomial.	9.4

PRETEST

1. Simplify the following (Objective 1).

   a. $\dfrac{a^4}{a}$

   b. $a^3 \div a^2$

   c. $\dfrac{a^2}{a^6}$

   d. $a^3 \div a^6$

   e. $a^8 \div a^4$

2. Simplify the following (Objective 2).

   a. $\left(\dfrac{a}{b}\right)^3$

   b. $\left(\dfrac{2a}{b^2}\right)^4$

   c. $\left(\dfrac{-3a^2}{b^3}\right)^3$

   d. $-\left(\dfrac{3a}{b}\right)^2$

3. Simplify the following (Objective 3).

   a. $3a^3bc \div abc$

   b. $(5a^3b^2c^4) \div (-a^3bc)$

   c. $\dfrac{24a^4b^6cd^2}{-8ab^3c^4e}$

   d. $\dfrac{-49y^{10}z^2}{-343yz^6}$

   e. $\dfrac{(-2x^3y^2)^4}{-(3xy^5)^3}$

4. Simplify the following (Objective 4).

   a. $\dfrac{4a + 6b}{2}$

   b. $\dfrac{b^3 + b^2 - b}{b}$

   c. $(5ab^2 + ab) \div (-b)$

   d. $(18a^3 - 6a^4) \div (-3a^2)$

   e. $\dfrac{77x^3 - 55x^2 - 22x}{11x}$

5. Simplifying the following (Objective 5).

   a. $(3a^2 - 5a - 2) \div (a - 2)$

   b. $(27a^3 - 1) \div (3a - 1)$

   c. $\dfrac{20a^2 - 21a - 10}{4a - 5}$

   d. $\dfrac{3x^4 - 6x + 5}{x - 1}$

Before discussing division of polynomials, let us recall some of the relationships and techniques related to division that were presented previously. Since algebra is an extension of arithmetic, it is much easier to learn the procedures in algebra if the arithmetic has been mastered. The following arithmetic facts are especially important in studying division:

1. The fraction bar may be used to indicate division.

$$8 \div 4 = \frac{8}{4} \text{ or } 2 \qquad 12 \div 16 = \frac{12}{16} \text{ or } \frac{3}{4} \text{ or } 0.75$$

2. Division problems may be rewritten as multiplication problems.

$$8 \div 4 = \frac{\overset{2}{\cancel{8}}}{1} \cdot \frac{1}{\cancel{4}} = 2 \qquad 12 \div 16 = \frac{\overset{3}{\cancel{12}}}{1} \cdot \frac{1}{\cancel{16}} = \frac{3}{4}$$

3. A nonzero number divided by itself yields a quotient of 1.

$$4 \div 4 = \frac{4}{4} = 1$$

4. Division by 0 is undefined. If an expression contains variables in the denominator, we will assume that the values are such that the denominator is not zero.

5. Zero divided by any nonzero number yields a quotient of 0.

Since the fraction bar may be used to indicate division, the quotient of two polynomials may be written in fractional form. For example, $(x^2 + 3x + 4) \div (x + 3)$ may be written as $\dfrac{x^2 + 3x + 4}{x + 3}$ and $27x^3y^2 \div 3xy$ may be written as $\dfrac{27x^3y^2}{3xy}$. Thus, it is possible to view such problems in two ways—as finding the quotient in a division problem or as simplifying a fraction. We shall consider both of these interpretations in this chapter.

9.1   Laws of Exponents Related to Division

Just as multiplication of polynomials required a special consideration of laws of exponents, we must also establish rules for exponents when dealing with quotients.

Example 1 $a^5 \div a^3 = ?$

Since $a^5 = a \cdot a \cdot a \cdot a \cdot a$, and $a^3 = a \cdot a \cdot a$,

$$a^5 \div a^3 = \frac{a^5}{a^3} \text{ or } \frac{\overset{1}{\cancel{a}} \cdot \overset{1}{\cancel{a}} \cdot \overset{1}{\cancel{a}} \cdot a \cdot a}{\underset{1}{\cancel{a}} \cdot \underset{1}{\cancel{a}} \cdot \underset{1}{\cancel{a}}} = \frac{a^2}{1} = a^2$$

Thus, $a^5 \div a^3 = a^2$.

Example 2 $a^4 \div a^7 = \dfrac{a^4}{a^7}$

$$= \frac{\overset{1}{\cancel{a}} \cdot \overset{1}{\cancel{a}} \cdot \overset{1}{\cancel{a}} \cdot \overset{1}{\cancel{a}}}{\underset{1}{\cancel{a}} \cdot \underset{1}{\cancel{a}} \cdot \underset{1}{\cancel{a}} \cdot \underset{1}{\cancel{a}} \cdot a \cdot a \cdot a} = \frac{1}{a^3}$$

Example 3 $a^3 \div a^3 = \dfrac{a^3}{a^3}$

$$= \frac{\overset{1}{\cancel{a}} \cdot \overset{1}{\cancel{a}} \cdot \overset{1}{\cancel{a}}}{\underset{1}{\cancel{a}} \cdot \underset{1}{\cancel{a}} \cdot \underset{1}{\cancel{a}}} = \frac{1}{1} = 1$$

Examples 1–3 illustrate the rules for dividing powers with the same base. These rules can be stated as formulas.

FORMULAS FOR DIVIDING POWERS WITH THE SAME BASE	EXAMPLES
1. If a is greater than b, $\dfrac{x^a}{x^b} = x^{a-b}$	$\dfrac{m^5}{m^2} = m^{5-2} = m^3$
2. If b is greater than a, $\dfrac{x^a}{x^b} = \dfrac{1}{x^{b-a}}$	$\dfrac{m^2}{m^5} = \dfrac{1}{m^{5-2}} = \dfrac{1}{m^3}$
3. If a = b, $\dfrac{x^a}{x^b} = 1$	$\dfrac{m^3}{m^3} = 1$

Example 4   If possible, simplify each of the following using the formulas for dividing powers with the same base.

a. $\dfrac{n^7}{n^{10}} = \dfrac{1}{n^{10-7}} = \dfrac{1}{n^3}$

The rule may be applied here since the base in the numerator is the same as the base in the denominator. The second formula is used since the exponent in the denominator is greater than the exponent in the numerator.

b. $\dfrac{n^{10}}{n^7} = n^{10-7} = n^3$

Since the bases are the same and the exponent in the numerator is greater than the exponent in the denominator, the first formula is used.

c. $\dfrac{n^{10}}{m^7}$

The rule cannot be applied here since the base in the numerator is not the same as the base in the denominator.

d. $\dfrac{t^7}{t^7} = 1$

The third formula is used since the bases are the same and the exponents are equal.

The following example suggests another law of exponents.

Example 5

$$\left(\frac{x}{y}\right)^4 = \left(\frac{x}{y}\right)\left(\frac{x}{y}\right)\left(\frac{x}{y}\right)\left(\frac{x}{y}\right)$$

$$= \frac{x \cdot x \cdot x \cdot x}{y \cdot y \cdot y \cdot y}$$

$$= \frac{x^4}{y^4}$$

FORMULA FOR RAISING A FRACTION TO A POWER	EXAMPLE
$\left(\dfrac{x}{y}\right)^a = \dfrac{x^a}{y^a}$	$\left(\dfrac{m}{n}\right)^3 = \dfrac{m^3}{n^3}$

Example 6

$$\left(\frac{4}{m}\right)^2 = \frac{4^2}{m^2} = \frac{16}{m^2}$$

Example 7

$$\left(\frac{m^6}{m^2}\right)^5 = (m^4)^5 = m^{20}$$

Alternate Solution:

$$\left(\frac{m^6}{m^2}\right)^5 = \frac{m^{30}}{m^{10}} = m^{20}$$

The procedure used in the alternate solution in Example 7 should be avoided even though it yields the correct result. When like bases appear in both the numerator and denominator of a fraction, the fraction should be simplified before it is raised to a power. This is the procedure used in the first solution of Example 7. When the alternate procedure is used the computation may become unnecessarily difficult.

9.1 Exercises

Use the formulas for dividing powers with the same base to evaluate Exercises 1–21.

1. $\dfrac{m^6}{m^3}$

2. $\dfrac{m^3}{m^6}$

3. $\dfrac{m^3}{m^3}$

4. $\dfrac{a^6}{a^2}$

5. $\dfrac{a^2}{a^6}$

6. $\dfrac{a^9}{a^4}$

7. $\dfrac{a^8}{a^2}$

8. $\dfrac{a^2}{a^8}$

9. $\dfrac{a^4}{a^9}$

10. $\dfrac{-x^6}{x^4}$

11. $\dfrac{3^5}{3^4}$

12. $\dfrac{a^{10}}{a^2}$

13. $-\dfrac{a^{10}}{a^5}$

14. $\dfrac{-x^4}{-x^6}$

15. $\dfrac{a^6}{a^6}$ 16. $\dfrac{4^3}{4^3}$

17. $\dfrac{a^2}{a^{10}}$ 18. $\dfrac{-5^2}{5^3}$

19. $\dfrac{-x^3}{-x^6}$ 20. $\dfrac{a^6}{a^{12}}$

21. $\dfrac{5^3}{-5^6}$

Use the formula for raising a fraction to a power to evaluate Exercises 22-42.

22. $\left(\dfrac{a}{b}\right)^2$ 23. $\left(\dfrac{3}{b}\right)^3$

24. $\left(\dfrac{-2}{5}\right)^2$ 25. $\left(\dfrac{-a}{4}\right)^3$

26. $\left(\dfrac{a}{2}\right)^5$ 27. $\left(\dfrac{2}{3}\right)^2$

28. $\left(\dfrac{1}{3}\right)^3$ 29. $\left(\dfrac{-a}{3}\right)^2$

30. $\left(\dfrac{-2}{3}\right)^2$ 31. $\left(\dfrac{x}{y}\right)^4$

32. $\left(\dfrac{-x}{y}\right)^2$ 33. $\left(\dfrac{-3}{x}\right)^3$

34. $\left(\dfrac{-5}{x}\right)^2$ 35. $\left(\dfrac{x}{3}\right)^4$

36. $\left(\dfrac{1}{2}\right)^5$ 37. $\left(\dfrac{-x}{4}\right)^2$

38. $\left(\dfrac{2}{7}\right)^2$ 39. $\left(\dfrac{3}{5}\right)^3$

40. $\left(-\dfrac{5}{6}\right)^2$ 41. $\left(\dfrac{1}{6}\right)^3$

42. $\left(-\dfrac{3}{5}\right)^2$

If possible, use the laws of exponents to express each of Exercises 43-65 in simplest form.

43. $\dfrac{a}{a^{10}}$ 44. Divide a^2 by a^6.

45. $\left(\dfrac{a}{3}\right)^4$ 46. $\dfrac{a^6}{a}$

47. $\dfrac{a^7}{a^7}$ 48. $\dfrac{c^{10}}{b^2}$

49. $\left(\dfrac{a}{-4b}\right)^2$ 50. $a^6 \div a^6$

51. Divide a into a^{10}.   52. $\dfrac{x^{15}}{x^{12}}$

53. $x^7 \div x^9$ 54. $\left(\dfrac{3}{x}\right)^5$

55. x^{10} divided by x^3 56. $\left(\dfrac{a}{b}\right)^{50}$

57. $x^5 \div x^5$ 58. $x^{10} \div x^{15}$

59. $y \div y^5$ 60. Divide b into b.

61. $\dfrac{c^2}{x^2}$ 62. $\left(\dfrac{-2x}{y}\right)^3$

63. $\left(\dfrac{m^6}{m^3}\right)^3$ 64. $\left(\dfrac{b^2}{b^6}\right)^4$

65. $\left(\dfrac{a^4}{a^3}\right)^3$

9.2   Simplifying the Quotient of Two Monomials

We will now use the laws of exponents to simplify quotients of monomials. The following example uses the exponent laws and the definition of multiplication of fractions.

Example 1   Express $\dfrac{12a^5}{3a^2}$ in simplest form.

$$\dfrac{12a^5}{3a^2} = \dfrac{12}{3} \cdot \dfrac{a^5}{a^2}$$
$$= 4 \cdot a^3$$
$$= 4a^3$$

The solution in Example 1 illustrates the method given in the *rule for simplifying the quotient of two monomials.*

RULE FOR SIMPLIFYING THE QUOTIENT OF TWO MONOMIALS

1. Indicate

   a. the quotient of the constant factors,

   b. the quotient of the powers with the same base,

   c. the factors which appear only in the numerator or only in the denominator.

2. Indicate the product of these results from step 1.

3. Simplify the indicated quotients.

4. Express the product as a single fraction.
   (All steps can be done mentally.)

EXAMPLE $\dfrac{24a^4b^3c^2}{18a^5b^3d} = ?$

Solution:

$$\dfrac{24}{18}$$

$$\dfrac{a^4}{a^5}, \dfrac{b^3}{b^3}$$

$$\dfrac{c^2}{1}, \dfrac{1}{d}$$

$$\dfrac{24}{18} \cdot \dfrac{a^4}{a^5} \cdot \dfrac{b^3}{b^3} \cdot \dfrac{c^2}{1} \cdot \dfrac{1}{d}$$

$$\dfrac{4}{3} \cdot \dfrac{1}{a} \cdot 1 \cdot \dfrac{c^2}{1} \cdot \dfrac{1}{d}$$

$$\dfrac{4c^2}{3ad}$$

Example 2

$$\dfrac{10x^2y^3}{2x^5} = \dfrac{10}{2} \cdot \dfrac{x^2}{x^5} \cdot \dfrac{y^3}{1}$$

$$= \dfrac{5}{1} \cdot \dfrac{1}{x^3} \cdot \dfrac{y^3}{1}$$

$$= \dfrac{5y^3}{x^3}$$

Example 3

$$\dfrac{x^2y}{3yz^3} = \dfrac{1}{3} \cdot \dfrac{x^2}{1} \cdot \dfrac{y}{y} \cdot \dfrac{1}{z^3}$$

$$= \dfrac{1}{3} \cdot \dfrac{x^2}{1} \cdot 1 \cdot \dfrac{1}{z^3}$$

$$= \dfrac{x^2}{3z^3}$$

Whenever possible the entire problem should be done mentally and only the answer written down. This is illustrated in Example 4 where no work is shown.

Example 4   Simplify $\dfrac{45x^3y^2z^3}{9xy^4}$.

$$\dfrac{45x^3y^2z^3}{9xy^4} = \dfrac{5x^2z^3}{y^2}$$

Example 5 illustrates another method which is frequently used to simplify quotients of monomials. This method uses the principle of reducing fractions by dividing out common factors.

Example 5   Simplify $\dfrac{45x^3y^2z^3}{36xy^4z^3w^2}$.

$$\frac{\overset{5}{\cancel{45}}\overset{x^2}{\cancel{x^3}}\overset{1}{\cancel{y^2}}\overset{1}{\cancel{z^3}}}{\underset{4}{\cancel{36}}\underset{y^2}{\cancel{x}}\underset{1}{\cancel{y^4}\cancel{z^3}}w^2} = \frac{5x^2}{4y^2w^2}$$

The common factors in this example are 9, x, y^2, and z^3.

Example 6
$$\frac{(3a^2b)^3}{(4a^3b)^2} = \frac{3^3(a^2)^3b^3}{4^2(a^3)^2b^2}$$
$$= \frac{27a^6b^3}{16a^6b^2}$$
$$= \frac{27}{16} \cdot \frac{a^6}{a^6} \cdot \frac{b^3}{b^2} = \frac{27}{16} \cdot 1 \cdot b = \frac{27b}{16}$$

In Example 6, note that the exponents outside the parentheses are different. Thus, the numerator must be raised to the third power and the denominator must be raised to the second power before the division is performed.

Example 7
$$\left(\frac{m^2}{n^3}\right)^4 = \frac{(m^2)^4}{(n^3)^4} = \frac{m^8}{n^{12}}$$

In Example 7, the formula for raising a fraction to a power was used first, and then the formula for raising a power to a power was used. Care should be taken, however, to simplify the fraction inside the parentheses first when possible, as was emphasized in the previous section. This is illustrated again in the following example.

Example 8
$$\left(\frac{6m^6}{3m^2}\right)^5 = (2m^4)^5 = 32m^{20}$$

As you will see, the following alternate solution is much more difficult, and should be avoided.
$$\left(\frac{6m^6}{3m^2}\right)^5 = \frac{6^5(m^6)^5}{3^5(m^2)^5} = \frac{7776m^{30}}{243m^{10}} = 32m^{20}$$

9.2 Exercises

Simplify each of the expressions in Exercises 1–40.

1. $6a^2 \div 2a$

2. $\dfrac{-4x^5}{8x^2}$

3. $36a^6$ divided by $12a^5$.

4. $(xyz) \div (xyz)$

5. $x^5y^3 \div 4x^3y$

6. $(12a^5) \div (-3a^3)$

7. $\dfrac{12a^3}{4a^6}$

8. $\dfrac{-3x^3}{6x^5}$

9. $\dfrac{10a^2}{25a^6}$

10. $\dfrac{36a^5}{48a}$

11. $\dfrac{24m^{18}}{-12m^9}$

12. $3a^2 \div 3a^2$ '

13. $\dfrac{3x^2y}{-3x^2y}$

14. $\dfrac{15a^3b^4}{10a^5b}$

15. $(3a^2b) \div (-6a)$

16. $\dfrac{3a^5b^4}{ab^4}$

17. $\dfrac{x^2}{x^6y^2}$

18. $\dfrac{-5ab^3}{5ab^3}$

19. $\dfrac{10a^4b^{10}}{5a^4b^5}$

20. $\dfrac{20a^6}{15a^7}$

21. $\dfrac{24a^4b^6c}{-16ab^2c^4}$

22. $\dfrac{10xyz^3}{100x^3y^3z^3}$

23. $-343x^2y^5 \div (-49y^{10}z^2)$

24. Cube $(-ab)$ and divide the result into $15a^2b^4$.

25. $-49y^{10}z^2 \div (-343x^2y^5)$

26. $\dfrac{a^2b^6c^3}{a^8b^8c^8}$

27. $-75a^6b \div (-25ab)$

28. $-4x^3y \div x^5y^3$

29. Find the quotient when $100x^4y^7z^7$ is divided by 100.

30. $\left(\dfrac{-5a^{10}bc}{-5a^4b^4d}\right)^7$

31. $\dfrac{(-2mx^2)^4}{32m^8x^2}$

32. Divide $48x^5y$ by $24x^2y$, and raise the quotient to the fifth power.

33. $\dfrac{3a^2b^5c}{6ab^7c^5}$

34. $\dfrac{14a^3b^7}{12a^6b}$

35. $\dfrac{-5ab^3c}{ab^2c}$

36. $\dfrac{100a^3bc^4}{10a^3bc}$

37. $\dfrac{(3a^5b^7c)^3}{-(4ab^4)^2}$

38. $\dfrac{(-2x^5y)^4}{-(3xy^5)^3}$

39. $\left(\dfrac{6a^2b^3}{2ab^5}\right)^4$

40. $\left(\dfrac{10x^2y^3z}{20x^6y^6z}\right)^5$

9.3   Dividing a Polynomial by a Monomial

In Chapter 8 we used the distributive property to multiply a monomial by a polynomial. Since division can be expressed in terms of multiplication, the same property can be used to divide a polynomial by a monomial. This is illustrated in the following example.

Example 1 $(6x^4 + 4x^2 + 2x) \div (2x)$

$$(6x^4 + 4x^2 + 2x) \div (2x) = (6x^4 + 4x^2 + 2x) \cdot \dfrac{1}{2x}$$

$$= 6x^4\left(\dfrac{1}{2x}\right) + 4x^2\left(\dfrac{1}{2x}\right) + 2x\left(\dfrac{1}{2x}\right)$$

$$= \dfrac{6x^4}{2x} + \dfrac{4x^2}{2x} + \dfrac{2x}{2x}$$

$$= 3x^3 + 2x + 1$$

An alternate solution which uses the definition of addition of fractions would be:

$$\dfrac{6x^4 + 4x^2 + 2x}{2x} = \dfrac{6x^4}{2x} + \dfrac{4x^2}{2x} + \dfrac{2x}{2x} = 3x^3 + 2x + 1$$

The second solution in Example 1 provides the technique given in the *rule for dividing a polynomial by a monomial.*

RULE FOR DIVIDING A POLYNOMIAL BY A MONOMIAL	*EXAMPLE* $(3a^2 + 6ab^5) \div (3ab)$
1. Indicate the sum of the quotients obtained when each term of the polynomial is divided by the monomial. 2. Simplify each quotient using the rule for dividing monomials.	*Solution:* $\dfrac{3a^2b}{3ab} + \dfrac{6ab^5}{3ab}$ (This can be done mentally.) $= a + 2b^4$ Thus, $(3a^2b + 6ab^5) \div (3ab) = a + 2b^4$.

Example 2 $(8a^8 + 6a^6) \div 2a^2 = ?$

$$\frac{8a^8 + 6a^6}{2a^2} = \frac{8a^8}{2a^2} + \frac{6a^6}{2a^2} = 4a^6 + 3a^4$$

Example 3 $(5x^4 - 2x^2) \div x^3 = ?$

$$\frac{5x^4 - 2x^2}{x^3} = \frac{5x^4}{x^3} + \frac{-2x^2}{x^3} = 5x - \frac{2}{x}$$

Notice that the monomial may not divide each term of the polynomial exactly. However, there are other techniques which may be used to simplify the expression when this occurs. One such technique involving factoring will be discussed in Chapter 16.

9.3 Exercises

Perform the indicated operations in Exercises 1–40.

1. $\dfrac{2a + 6b}{2}$

2. $\dfrac{4x^2 - 2x}{2}$

3. $\dfrac{5a + 10b}{5}$

4. $\dfrac{10a - 12b}{2}$

5. $\dfrac{ax + bx}{x}$

6. $\dfrac{ab - ac}{a}$

7. $\dfrac{3a - 12b + 6c}{3}$

8. $\dfrac{2ac - 3ab + 6a}{a}$

9. $\dfrac{3x - 3y}{-3}$

10. $\dfrac{2x^2 - 3x}{-x}$

11. $\dfrac{x^3 - 2x^2 + 3x}{x}$

12. $\dfrac{a - a^2 + a^3}{a}$

13. $\dfrac{3ab + b}{b}$

14. $\dfrac{abc - 3bc}{bc}$

15. $\dfrac{mn - m^2}{-m}$

16. $\dfrac{x^2y - y^2}{-y}$

17. $\dfrac{5x^3 + 10x^2}{5x^2}$

18. $\dfrac{mn^3 - m^2n^2}{mn^2}$

19. $\dfrac{-a^2x + 2ax}{-ax}$

20. $\dfrac{-a^2bc - ab^2c}{abc}$

21. $\dfrac{10m^2n^2 - m^2n}{m^2n}$

22. $\dfrac{5ab - 10b^2}{5b}$

23. $(5b + bx) \div (-b)$

24. $\dfrac{-35x^2 + 14xy}{-7x}$

25. $(18a^3 - 6a^4)$ divided by $(-3a)$

26. $\dfrac{-15a^2 - 15b^2}{25a^2}$

27. $(a^2x^3y - 5ax^3y^2) \div (ax^2y)$

28. $\dfrac{2xy - 3xz}{-x}$

29. $\dfrac{x^4y^3 - x^2y^2 + xy}{-xy}$

30. Divide $(8x - 6y - 2)$ by 2.

31. $\dfrac{a^3 + a^2 + a}{a}$

32. Find the quotient if $56x^4 - 28$ is divided by -28.

33. $(9a - 3b + 6c) \div 3$

34. $\dfrac{-x^4y^3 + x^2y^2 - xy}{-xy}$

35. $(a^2b^3c + ab^2c^3) \div (ab^2c)$

36. $(36x^5y - 12xy^5) \div (4xy)$

37. Divide $-25x$ into $(5x^3 - 25x^4y^3)$.

38. $\dfrac{x^4 + x^3 + x^2}{-x}$

39. $\dfrac{39x^3yz^2 - 65xy^2z^3 - 78x^2y^2z^2}{13xyz^2}$

40. $\dfrac{15a^2b^2 - 10a^2b^3 + 25a^3b^2}{-5a^2b^2}$

9.4 Dividing a Polynomial by a Polynomial

When a polynomial of more than one term is divided by a second polynomial of more than one term, a procedure called the *long division algorithm* may be used to find the quotient. (In some problems factoring, which is discussed in Chapter 16, may be used.) This procedure follows the pattern of long division in arithmetic if the polynomials are in standard form.

PROCEDURE FOR DIVIDING A POLYNOMIAL BY A POLYNOMIAL	EXAMPLE

PROCEDURE FOR DIVIDING A POLYNOMIAL BY A POLYNOMIAL

EXAMPLE

$(6x^3 - 28x + 3x^2 + 15) \div (2x - 3)$
Solution:

1. Arrange the polynomials in standard form.

$(6x^3 + 3x^2 - 28x + 15) \div (2x - 3)$
dividend   divisor

2. Set up the divisor and the dividend for long division as in arithmetic.

$2x - 3\overline{)6x^3 + 3x^2 - 28x + 15}$

3. Divide the first term of the divisor into the first term of the dividend.

$6x^3 \div 2x = 3x^2$

4. The result of this division is the first term in the quotient.

$$\begin{array}{r} 3x^2 \\ 2x - 3\overline{)6x^3 + 3x^2 - 28x + 15} \end{array}$$

5. Multiply the first term of the quotient by *every* term in the divisor.

$3x^2(2x - 3) = 6x^3 - 9x^2$

6. Write this product under the dividend, and subtract as in arithmetic, bringing down the next term. (Align like terms to facilitate the subtraction.)

$$\begin{array}{r} 3x^2 \\ 2x - 3\overline{)6x^3 + 3x^2 - 28x + 15} \\ \ominus \quad \oplus \\ \underline{6x^3 - 9x^2} \\ 12x^2 - 28x \end{array}$$

7. Consider the difference from the previous step as a new dividend and repeat steps 3–6 until the remainder is 0 or the first term of the remainder is of lower degree than the first term of the divisor.

$$\begin{array}{r} 3x^2 + 6x - 5 \\ 2x - 3\overline{)6x^3 + 3x^2 - 28x + 15} \\ \ominus \quad \oplus \\ \underline{6x^3 - 9x^2} \\ 12x^2 - 28x \\ \ominus \quad \oplus \\ \underline{12x^2 - 18x} \\ -10x + 15 \\ \oplus \quad \ominus \\ \underline{-10x + 15} \\ 0 \end{array}$$

Example 1 $(3a^2 - 5a - 2) \div (a - 2)$

$$\begin{array}{r} 3a + 1 \\ a - 2\overline{)3a^2 - 5a - 2} \\ \ominus \quad \oplus \\ \underline{3a^2 - 6a} \\ a - 2 \\ \ominus \oplus \\ \underline{a - 2} \end{array}$$

Example 2 $(2a^2 - 18 - 9a) \div (a - 6)$

$$\begin{array}{r} 2a + 3 \\ a - 6\overline{)2a^2 - 9a - 18} \\ \ominus \quad \oplus \\ \underline{2a^2 - 12a} \\ + 3a - 18 \\ \ominus \qquad \oplus \\ \underline{+ 3a - 18} \end{array}$$

Notice that the dividend is written in standard form before the division is begun.

Example 3 $(2a^3 - 6a - 10) \div (a^2 + 2a + 1)$

$$
\begin{array}{r}
2a \quad - 4 \\
a^2 + 2a + 1\overline{)2a^3 \qquad - 6a - 10} \\
\ominus \quad \ominus \qquad \ominus \\
\underline{2a^3 + 4a^2 + 2a} \\
-4a^2 - 8a - 10 \\
\oplus \quad \oplus \quad \oplus \\
\underline{-4a^2 - 8a - 4} \\
- 6
\end{array}
$$

Notice that space is left for the second-degree term so that like terms will be properly aligned.

The quotient is $2a - 4$ with a remainder of -6. This answer may be written as

$$2a - 4 + \frac{-6}{a^2 + 2a + 1}$$

Example 4 $(a^4 - 2b^4) \div (a + b)$

$$
\begin{array}{r}
a^3 - a^2b + ab^2 - b^3 \\
a + b\overline{)a^4 \qquad\qquad - 2b^4} \\
\ominus \quad \ominus \\
\underline{a^4 + a^3b} \\
- a^3b \\
\oplus \quad \oplus \\
\underline{-a^3b - a^2b^2} \\
a^2b^2 \\
\ominus \quad\quad \ominus \\
\underline{a^2b^2 + ab^3} \\
- ab^3 - 2b^4 \\
\oplus \quad \oplus \\
\underline{-ab^3 - b^4} \\
- b^4
\end{array}
$$

Notice that a space is left in the dividend since no terms containing a^3, a^2, or a occur in it.

Thus, $(a^4 - 2b^4) \div (a + b) = a^3 - a^2b + ab^2 - b^3 + \dfrac{-b^4}{a + b}$.

9.4 Exercises

Fill in the boxes in Exercises 1–4.

1.
$$
\begin{array}{r}
a \quad + 6 \\
a - 3\overline{)a^2 + 3a - 5} \\
\ominus \quad \oplus \\
\underline{\boxed{} - 3a} \\
\boxed{} - 5 \\
\ominus \quad\quad \oplus \\
\underline{6a - 18}
\end{array}
$$

2.
$$
\begin{array}{r}
\boxed{} + \;\;3y \\
5x - 4y\overline{)35x^2 - 13xy - 2y^2} \\
\underline{35x^2 \boxed{}} \\
\boxed{} - 2y^2 \\
\underline{15xy - 12y^2}
\end{array}
$$

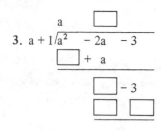

3. $a + 1\overline{)a^2 \quad -2a \quad -3}$

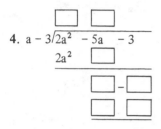

4. $a - 3\overline{)2a^2 \quad -5a \quad -3}$

Find the quotients in Exercises 5-14.

5. $a + 6\overline{)a^2 + 4a - 12}$ 6. $x - 5\overline{)x^2 - 7x + 10}$

7. $x + 2\overline{)3x^2 + 5x - 2}$ 8. $x + 3\overline{)5x^2 + 13x - 6}$

9. $m + 4\overline{)2m^2 + 11m + 12}$ 10. $m - 4\overline{)3m^2 - 11m - 4}$

11. $m - 1\overline{)2m^2 - 5m + 3}$ 12. $m + 5\overline{)5m^2 + 28m + 15}$

13. $(3a^2 - 5a - 2) \div (a - 2)$

14. $(2x^2 + 7x - 15) \div (x + 5)$

Use the long division algorithm to find the quotients in Exercises 15-35.

15. $(2b^2 - b - 6) \div (2b + 3)$ 16. $\dfrac{10b^2 - 11b + 3}{2b - 1}$

17. $(x^2 + 4x - 3) \div (x + 5)$

18. Divide $(8x^3 - 1)$ by $(2x - 1)$.

19. $(2a^2 - 20 - 3a) \div (a - 4)$

20. $(x^3 + 6x + 4) \div (x + 2)$

21. $\dfrac{a^5 + a^4 - 2a^3 + a^2 - 3a + 1}{a - 1}$

22. $\dfrac{2x^5 + 4x^4 - x^3 - 5x + 12}{x^2 + 2x - 1}$

23. $2a^3 - 11a^2 + 17a - 15$ divided by $2a - 5$

24. $(x^2 + xy - 2y^2) \div (x - y)$ 25. $\dfrac{a^3 - b^3}{a - b}$

26. $(6a^2 - 11a - 10) \div (3a + 2)$

27. Divide $(a - 3b)$ into $(a^2 - 6ab + 9b^2)$.

28. $(6a^2 - a - 15) \div (2a + 3)$

29. $\dfrac{2x^5 - 3x^4 + 5x^2 - 4x - 4}{2x^2 - x - 1}$

30. $(6a^2 - 8a - 30) \div (3a + 5)$ 31. $\dfrac{p^3 - 125}{p - 5}$

32. $(9a^2 - 16) \div (3a + 4)$ 33. $\dfrac{a^3 + 6a^2 - 2a - 12}{a^2 - 2}$

34. Find the quotient if $(3a + 2b)$ is divided into $(3a^3 + 8a^2b + 7ab^2 + 5b^3)$.

35. $\dfrac{a^5 - 1}{a - 1}$

REVIEW EXERCISES

Simplify each of Exercises 1-35.

1. $x^5 \div x^3$ 2. $\dfrac{x^3}{x^5}$

3. $2a^4 \div 48a^8$ 4. $\dfrac{24a^4}{3a}$

5. Divide $24x^2y^3z$ by $6x^2yt$.   6. $\dfrac{2x^2 - 5x - 3}{x - 3}$

7. $\dfrac{4r^2s^2 - 4rs^2 + 8rs}{-2rs}$ 8. $\dfrac{4x^2 - 4}{x - 1}$

9. $\dfrac{-10x^3y^4}{xy^2}$ 10. $\dfrac{3x^2 + xy - x}{x}$

11. Divide $x^2 - 5xy - 6y^2$ by $x - 6y$.

12. Raise $\dfrac{x^3y^2}{z^4}$ to the fifth power.

13. $\dfrac{8x^5y^7}{4x^3y^5}$ 14. $\dfrac{5a^4b^4 - 20a^3b^3}{4a^2b^3}$

15. $(12t^3 - 24t^2 + 72t) \div 4t$ 16. $\left(\dfrac{3a}{b^2}\right)^2$

17. $\dfrac{-28x^2 - 14xy}{-7x}$ 18. $\dfrac{a^3 + 6a + 6}{a + 1}$

19. Find the quotient if $2a^3 - 10a^2 + 14a - 6$ is divided by $a^2 - 2a + 1$.

20. $\dfrac{(3x^5y^7)^3}{(4x^3y^5)^2}$ 21. $\dfrac{32x^6y^4}{16x^5y^4}$

22. $\dfrac{24a^3b^2c - 24a^3bc}{6abc}$ 23. $\dfrac{-15a^2b}{(3ab^2)^3}$

24. Divide $-28x$ into $-28x^2 - 14xy$.

25. $\dfrac{xyz^3 - xy^3z - x^3yz}{xyz}$

*26. Divide $w^5 - 32$ by $w - 2$.   *27. $\left(\dfrac{6a^2b}{3a^3b^4}\right)^5$

*28. $\left(\dfrac{7a^3b^6c}{49a^6b^6c^5}\right)^3$

*29. Find the quotient if $m^6 - 1$ is divided by $m - 1$.

*30. $\dfrac{(6a^2b)^2}{(3a^3b^4)^3}$

*31. $\dfrac{x^{4n}}{x^n}$

*32. $10x^{3a} \div 2x^a$

*33. $\dfrac{m^a}{m^{2a}}$

*34. $\dfrac{x^{2n} - 3x^n}{x^n}$

*35. $x^n + 1\overline{\smash{\big)}x^{2n} + 2x^n + 1}$

*These exercises are optional.

EQUATIONS AND INEQUALITIES IN ONE VARIABLE

FIRST-DEGREE EQUATIONS IN ONE VARIABLE

10

PRETEST

1. Answer each of the following true (T) or false (F) (Objective 1).
   a. $2a + 3$ is an equation.
   b. An identity is an equation that is always true.
   c. -3 is a solution for $x - 2 = -5$.
   d. A root of an equation is said to satisfy the equation.
   e. $x + 1 = x$ is a conditional equation.
   f. $2x + 5x = 10x$ is an equation.

2. Answer each of the following with contradiction, identity, or conditional equation (Objective 2).
   a. If the only solution of $x + 4 = 6$ is 2, then the equation $x + 4 = 6$ is a (an) _____.
   b. If the equation $x + 4 = x$ has no solution, then the equation $x + 4 = x$ is a (an) _____.
   c. If the equation $2x - x + 4 = x + 4$ is always true, then this equation is a (an) _____.
   d. If the only root of the equation $5x = 4x$ is 0, then the equation $5x = 4x$ is a (an) _____.
   e. The equation $6 + 4 = 10$ is a (an) _____.
   f. The equation $8 + 2 = 11$ is a (an) _____.

3. Each of the following equations has a root of either -6, -2, 0, or 6. Determine which, if any, of the following equations are equivalent (Objective 3).
   a. $x + 4 = 10$ b. $x - 8 = -10$
   c. $3x = 18$ d. $2x + 24 = 3x + 18$
   e. $10x + 20 = 0$ f. $\frac{x}{2} = -3$

4. Identify the equations below which are linear equations in one variable (Objective 4).
   a. $2x - 3 = x + 7$ b. $x = y + 4$
   c. $p = 0$ d. $2v^2 = 50$
   e. $3(a - 2) = 7(a - 5)$ f. $p = a + b + c$

5. Solve each of the following equations and check the solutions (Objective 5).

   a. $6w = -12$

   b. $\dfrac{x}{6} = -12$

   c. $-\dfrac{2}{3}y = 30$

   d. $\dfrac{-z}{3} = 12$

   e. $1.4t = 2.8$

   f. $2s = 5$

6. Solve each of the following equations and check the solutions (Objective 6).

   a. $3x - 4 = 16$

   b. $-2y - 5 = 15$

   c. $\dfrac{2}{3}z + 4 = -6$

   d. $1.2s + 0.5 = -0.7$

   e. $3.04r + 0.59 = 0.59$

   f. $\dfrac{v}{4} - 10 = 2$

7. Solve each of the following equations and check the solutions (Objective 7).

   a. $6x = 0.4x - 14$

   b. $-2n = 4n - 12$

   c. $4h = 2h + 1$

   d. $6k = 6k + 1$

8. Solve each of the following equations and check the solutions (Objective 8).

   a. $2v + 5 = 4v + 10$

   b. $-6m - 2 = -4m - 10$

   c. $5s + 3 = 2s - 10$

   d. $-1.2n - 7 = 0.8n + 1.3$

9. Solve each of the following equations and check the solutions (Objective 9).

   a. $2t + 3(t - 4) = 2t - 5$

   b. $-0.6r + 0.52 + 2.4r = 0.1 - 0.3r$

   c. $2(x + 4) - 3 = 2x + 5$

   d. $3s + 2s + 10 - 8 = 7s - 4 + 10$

10. Solve each of the following equations and check the solutions (Objective 10).

   a. $-\dfrac{x}{3} + 4 = \dfrac{2x}{3} - 5$

   b. $2z + \dfrac{2}{3} = \dfrac{z}{6} + \dfrac{35}{3}$

   c. $-\dfrac{2}{3}(y + 2) = y + 4$

   d. $\dfrac{w}{2} + \dfrac{w}{3} + \dfrac{w}{4} = -13$

11. Solve each of the following equations for x (Objective 11).

   a. Three more than four times x is thirteen.

   b. Seven x plus 18 equals 74.

   c. Two x minus 3 is the same as 5x.

   d. Five equals 10x plus 3.

12. Solve each of the following equations for the variable indicated (Objective 12).

   a. $A = lw$ for w

   b. $I = prt$ for t

   c. $x = 2y - 5$ for y

   d. $\dfrac{a + b}{c} = 5$ for b

In our daily lives, we use a variety of sentences to describe our emotions, to give orders or directions, and to ask questions. In mathematics, we are less concerned with emotions and more concerned with making judgments based upon facts or assumptions. Therefore, the literature of mathematics is dominated by statements, that is, sentences which are either true or false. The interrogative sentence "Where were you last night when the lights went out?" has no truth value. The statement "A triangle has three sides" has a truth value which is true. The statement "Four divided by two is equal to three" has a truth value which is false. Thus, we note that in mathematics statements do not have to be true, but they must be either true or false.

10.1 Terminology Associated with Equations

A particularly useful type of mathematical statement is the **equation,** which is a statement that two quantities are equal. You have already used equations numerous times in the preceding chapters. For example, when you found the sum of 2.13 and 14.06 you wrote $2.13 + 14.06 = 16.19$. In this case, the equation is used to give two different names for the same quantity.

One of the primary uses of equations is problem solving. Equations can be used to state frequently-used formulas such as $A = lw$ for the area of a rectangle, $C = 2\pi r$ for the circumference of a circle, and $V = \pi r^2 h$ for the volume of a cylinder. Of course, many other equations play an important role in our lives. The sales tax on your purchases is read from a table prepared through use of equations; your payroll deductions are probably calculated on a computer which has stored certain equations; and the orbit of a weather satellite overhead was prescribed before launch by using many equations.

The material in this chapter is the culmination of your earlier efforts, and it would be difficult to overestimate its importance. If you have mastered the previous material, you should find working with equations relatively easy.

Example 1   a. $2 + 3 = 5$

The equation "Two plus three equals five" is a true statement.

b. $7 + 1 = 9 - 5$

The equation "Seven plus one equals nine minus five" is a false statement.

c. $3x = 6$

This equation is true if the variable x is replaced by two but is false if x is replaced by any other value such as four.

These examples do more than show you how to read an equation. They give you examples of three different types of equations. Example **a** is an **identity**, an equation that is always true. The equation $2x = 2x$ is also an identity since it is always true. In other words, it is true for every value of the variable. Example **b** is a **contradiction**, an equation that is always false. The equation $x = x + 1$ is also a contradiction. It is always false since any replacement for x results in a false statement. Example **c** is a **conditional equation**, an equation that involves a variable and is true for only certain values of that variable.

Very little work will be done here with contradictions. Most of our work will be determining values for variables that will make a conditional equation true. These values are said to **satisfy** the equation and are called **roots** or **solutions** of the equation.

Example 2   a. $3x = 6$

Two is a root or solution of this conditional equation because 2 makes the equation true when it is substituted for x.

b. $2x + 3x = 5x$

If any real number is substituted for x, the result is a true statement. Since every value of x satisfies the equation, it is an identity.

c. $y^2 = 9$

The roots of this equation are 3 and −3.

d. $z = z + 1$

No value of z will satisfy this equation. Since the equation is always false it is a contradiction.

e. $w = 2w$

The only root of this equation is 0.

f. $3x + 5(x - 7) = 3(x - 5)$

Four is the only value for x which makes this equation true.

Identities are frequently used in mathematics to state general properties, as illustrated in Example 3.

Example 3   **a.** $a + b = b + a$

This identity is a statement of the *commutative property of addition*. For example, $2 + 3 = 3 + 2$.

b. $(a + b) + c = a + (b + c)$

This identity is a statement of the *associative property of addition*. For example, $(2 + 3) + 4 = 2 + (3 + 4)$.

Example 4   Determine, by substituting for x, whether -1, 0, or 1 is a root of $3x - 5 = 7(x - 1) + 2$.

$$\text{Substituting } -1: \quad 3(-1) - 5 = 7(-1 - 1) + 2$$
$$-3 - 5 = 7(-2) + 2$$
$$-8 = -14 + 2$$
$$-8 = -12 \text{ is false}$$

Therefore, -1 is not a root.

$$\text{Substituting } 0: \quad 3(0) - 5 = 7(0 - 1) + 2$$
$$0 - 5 = 7(-1) + 2$$
$$-5 = -7 + 2$$
$$-5 = -5 \text{ is true}$$

Therefore, 0 is a root.

$$\text{Substituting } 1: \quad 3(1) - 5 = 7(1 - 1) + 2$$
$$3 - 5 = 7(0) + 2$$
$$-2 = 0 + 2$$
$$-2 = 2 \text{ is false}$$

Therefore, 1 is not a root.

We can conclude that 0 is the only one of the values tried which is a root of $3x - 5 = 7(x - 1) + 2$.

One of the most important skills to develop is the ability to translate verbal statements into algebraic statements. Some of the key words and phrases that represent "=" are "is equal to," "equals," "is the same as," and "is."

Example 5   Translate each of the following statements into an algebraic equation.

a. Three x equals seven.

$$3x = 7$$

b. Two more than five x is the same as eleven.

$$5x + 2 = 11$$

c. three times x equals y minus four.

$$3x = y - 4$$

d. Three less than w is nine.

$$w - 3 = 9$$

Example 6   Write a verbal statement for each of the following algebraic statements.

a. $5s = t + 7$ Five s is equal to t plus 7.
b. $y - 4 = x + 2$ y minus four equals x plus 2.
c. $2m + 4 = 11$ Two m plus four is eleven.
d. $a + b = b + a$ a plus b is the same as b plus a.

10.1 Exercises

Translate each statement in Exercises 1–10 into an algebraic equation.

1. Twice x equals eighteen.
2. Two more than x is eighteen.
3. One half of h is equal to the sum of h and 9.
4. Two thirds of x equals y plus three.
5. Five minus x is the same as y divided by four.
6. One times n equals n.
7. Three more than twice x is twelve.
8. x plus the opposite of x is equal to 0.
9. 0 equals x times 0.
10. If b plus c is multiplied by a, the result equals ba plus ca.

Write a verbal statement for each of the algebraic statements in Exercises 11–15.

11. $3y = 11$ 12. $z + 5 = 8$
13. $x - 7 = 5x$ 14. $x + y = y + x$
15. $a - 3 = \dfrac{b}{3}$

Identify each of the equations in Exercises 16–20 as an identity or a contradiction.

16. $5 = 7 - 2$ 17. $6 = 19 - 11$
18. $6 + 7 = 7 + 6$ 19. $(5 + 3) - 2 = 5 + (3 - 2)$
20. $5 = -5$

Fill in the blanks in Exercises 21–30 with one of the following: identity, contradiction, or conditional equation.

21. If the only root of $5x = 7 - 2$ is 1, then this equation is a (an) _____.
22. If the only root of $m = 2m$ is 0, then this equation is a (an) _____.
23. If the equation $7k = 6(k + 1) + k - 6$ is always true, then this equation is a (an) _____.
24. If the only roots of $y^2 = 9$ are 3 and -3, then this equation is a (an) _____.
25. If the equation $w = w + 1$ has no solution, then this equation is a (an) _____.
26. If the solution of $6x = -19 + 11$ is $-\dfrac{4}{3}$, then this equation is a (an) _____.
27. If the solution of $3m = 4m$ is 0, then this equation is a (an) _____.
28. If the equation $5(x + 3) = 4(x + 4) - (1 - x)$ is always true, then this equation is a (an) _____.
29. If the equation $2t + 1 = 2t + 1$ is always true, then this equation is a (an) _____.
30. If the equation $7k = 6(k + 1) + k$ is always false, then this equation is a (an) _____.

Answer Exercises 31–40 as either true (T) or false (F).

31. Seventeen is a root of $k = k$.
32. Zero satisfies $u = u + 1$.

33. Three satisfies $y^2 = y$.

34. Zero is a root of $3m = 4m$.

35. A solution of $0.4w = 1.6$ is 4.

36. Negative four is a solution of $x^2 = 16$.

37. The expression $3(x + 2) + 7$ is an equation.

38. An identity has no solution.

39. A contradiction has no solution.

40. One-seventh satisfies $49w + 2 = 9$.

Substitute each number listed in Exercises 41–50 in the equation to the right to determine which, if any, of these values is a solution of the given equation.

41. 4, 8, 16 $2w = 8$

42. 4, 8, 16 $\dfrac{w}{2} = 8$

43. 15, 12, 9 $\dfrac{1}{3}z = 4$

44. −4, 0, 4 $y^2 = 16$

45. −4, 0, 4 $8p = 5p$

46. −4, 0, 4 $5x = 5x + 17.3$

47. 7, 1, 5 $2(w - 1) + 3(w + 2) = 5w + 4$

48. −6, 18, 6 $\dfrac{z}{2} + \dfrac{z - 9}{3} = z - 6$

49. $-\dfrac{1}{2}, 0, \dfrac{1}{2}$ $x^2 = \dfrac{1}{4}$

50. −2, 0, 7, 10 $5(x + 1) = 3(x + 6) + 1$

10.2 Equivalent Equations

Try to guess the roots of the following equations:

a. $x = 7$

b. $2x = 14$

c. $5x + 5 = 3x + 19$

d. $5(x + 1) = 3(x + 6) + 1$

It is rather obvious that the root for equation **a** is **7**, and you probably guessed that equation **b** also has 7 as a root. It is unlikely that you guessed that 7 is also the root for equations **c** and **d**. However, since solutions are frequently needed for equations such as **c** and **d**, we will need a technique for finding the roots of these equations. The process of finding these roots is called **solving the equation**. If you can determine the solution by *looking* at the equation, then you have **solved** the equation **by inspection**.

In order to develop a technique for solving equations, first notice that the above equations are all **equivalent**, that is, they have the same solution, 7. The technique for solving more complicated equations such as equation **d** is systematically to produce equivalent equations, such as equations **c**, **b**, and finally **a** until the solution is obvious. The solution is obvious when the variable is isolated on one side of the equals sign and a constant is isolated on the other side.

Example   Each of the following equations has as its solution one or more of the values −2, 0, 2. Evaluate the equations for *each* of these values, and determine if any of the equations are equivalent.

a. $3x = 6$ **b.** $x^2 = 4$

c. $-4x = 8$ **d.** $-3x = +6$

e. $-4x = -8$ **f.** $4x = 3x$

Sample Solution: **a.** $3x = 6$ $3(-2) = -6$ which is not equal to 6

$3(0) \ \ = \ \ 0$ which is not equal to 6

$3(2) \ \ = \ \ 6.$ Therefore, 2 is a root of $3x = 6$

The roots of other equations can be determined in the same way. The solutions of these equations are:

a. 2   **b.** 2, −2   **c.** −2   **d.** −2   **e.** 2   **f.** 0

Since equations **a** and **e** have the same solution, they are equivalent. Likewise equations **c** and **d** are equivalent.

10.2 Exercises

Substitute each number listed in Exercises 1–20 in both of the equations on the right to determine whether the equations are equivalent.

1. −2, 2	$3x = 6$	$x + 10 = 8$	**10.** 28, 38	$\dfrac{x + 5}{3} = 11$	$x + 5 = 33$
2. 0, 1	$x + x = x$	$x \cdot x = x$	**11.** $\dfrac{1}{2}, \dfrac{1}{3}$	$6x = 3$	$6x + 5 = 8$
3. 1, 3	$4x = 12$	$x - 2 = 5$	**12.** 0, 10	$3x = 0$	$5x = 2x$
4. −11, 11	$-3x = 33$	$4x = 44$	**13.** $-\dfrac{1}{2}, \dfrac{1}{2}$	$2x + 1 = 0$	$2x = 1$
5. 3, 5	$5x = 15$	$x - 11 = 14$	**14.** −2, 3	$x^2 - x - 6 = 0$	$x^2 = x + 6$
6. 0, 7	$8x = 0$	$8x + 3 = 3$	**15.** −2, 5	$x^2 - 3x - 10 = 0$	$3x = -6$
7. −1, −4	$5x + 20 = 0$	$5x + 30 = 25$	**16.** −1, 4	$x^2 - 3x - 4 = 0$	$x + 7 = 6$
8. −3, −1	$3(x + 2) = 4 + x$	$3x + 6 = x + 4$	**17.** −3, 3, 4	$x^2 - 7x + 12 = 0$	$x^2 = 9$
			18. −2, 2	$3(x + 1) = 4x + 1$	$7x = -14$
9. $-\dfrac{1}{2}, \dfrac{1}{2}$	$2(x + 4) = 7$	$2x + 8 = 7$	**19.** −2, 2	$x^2 = 4$	$|x| = 2$
			20. −3, 3	$x^2 = 9$	$|x| = 3$

10.3 Solving Equations of the Form ax = b

The subject of this chapter is the solution of first-degree equations. Later in this text, and in other books, you may solve higher degree equations or equations which contain more than one variable. First-degree equations with only one variable are the simplest conditional equations and thus are the first which we shall solve. A first-degree equation in one variable is called a **linear equation** in one variable.

Example 1 Identify the equations below which are linear equations in one variable.

    a. $5w = 30$ This is a linear equation in one variable since w has an exponent of 1.

    b. $x^2 = 16$ This is not a linear equation since it is not first degree. The exponent on x is 2.

    c. $3V - 5 = 18 - 4V$ This is a linear equation in one variable. The only variable is V and the degree is 1.

    d. $13x - 5 = 4y + 7$ This equation contains two variables, not one. The variables are x and y.

Every linear equation in one variable can be written in the form $ax = b$ where a and b represent constants and $a \neq 0$. The letter x represents a variable.

When we say an equation is in this form, we mean there is one variable term in the left member and one constant in the right member. $3x = 12$, $5y = 0$, $3x = -6$, and $\frac{1}{3}w = 2$ are all in this form. The procedure for solving these equations, as well as other equations, is to isolate the variable on one side of the equation and some constant on the other side. For linear equations already in this form ($ax = b$), isolating the variable on one side of the equation is a simple matter. The principle which we shall use to produce an equivalent equation with the variable isolated is given below.

MULTIPLICATION–DIVISION PRINCIPLE OF EQUALITY

If both members of an equation are multiplied or divided by the same nonzero number, the result is an equivalent equation.

RULE FOR SOLVING EQUATIONS OF THE FORM $ax = b$, $a \neq 0$	*EXAMPLE* Solve $\frac{3}{4}x = 3$.
	Solution:
Using Multiplication: Form the reciprocal of the coefficient of x and multiply both sides of the equation by this reciprocal.	The reciprocal of $\frac{3}{4}$ is $\frac{4}{3}$.
	$$\frac{4}{3}\left(\frac{3}{4}x\right) = \frac{4}{3}\left(\frac{3}{1}\right)$$
	$x = 4$
	EXAMPLE Solve $2x = 6$.
	Solution:
Using Division: Divide both sides of the equation by the coefficient of x.	The coefficient of x is 2.
	$$\frac{2x}{2} = \frac{6}{2}$$
	$x = 3$

Although these two techniques are equivalent, we generally use multiplication if the coefficient of the variable is a fraction and use division otherwise.

Example 2   Solve $12x = 48$.

Solution: $12x = 48$

$$\frac{1}{12}(12x) = \frac{1}{12}(48)$$

$$x = 4$$

Alternate Solution:

$$12x = 48$$

$$\frac{12x}{12} = \frac{48}{12}$$

$$x = 4$$

Example 3   Solve 10r = 25

Solution: 10r = 25

$$\frac{10r}{10} = \frac{25}{10}$$

$$r = \frac{5}{2}$$

Example 4   Solve 0.023y = −4.6 for y.

Solution: 0.023y = −4.6   *Check:*   0.023y = −4.6

$$\frac{0.023y}{0.023} = \frac{-4.6}{0.023}$$ 0.023(**−200**) = −4.6

                                                                                          −4.6 = −4.6

y = −200

Note: You may check your work by substituting the value of the variable in the original equation. If the calculations have been performed correctly, the resulting equation should be true. If it is false, you have made an error in solving or in checking the equation.

Example 5   Solve $4\frac{2}{3}$h = 14 and check the solution.

Solution: $4\frac{2}{3}$h = 14   *Check:* $4\frac{2}{3}$h = 14

$$\frac{14h}{3} = 14$$ $$4\frac{2}{3}(3) = 14$$

$$\left(\frac{\cancel{3}}{\cancel{14}}\right)\left(\frac{\cancel{14}h}{\cancel{3}}\right) = \left(\frac{3}{\cancel{14}}\right)\left(\frac{\cancel{14}}{1}\right)$$ $$\frac{14}{\cancel{3}}\left(\frac{\cancel{3}}{1}\right) = 14$$

h = 3   14 = 14

In Examples 6 and 7, the variable term is in the right member, but the solution still involves dividing both members by the coefficient of the variable (Example 6) or multiplying both members by the reciprocal of the coefficient of the variable (Example 7).

Example 6   Solve 4 = 2w.

Solution: 4 = 2w

$$\frac{4}{2} = \frac{2w}{2}$$

2 = w

Thus, w = 2.

Example 7   Solve and check: $\frac{1}{2} = -\frac{3}{4}v$.

Solution: $\frac{1}{2} = -\frac{3}{4}v$ *Check:* $\frac{1}{2} = -\frac{3}{4}v$

$$\left(-\frac{\overset{2}{\cancel{4}}}{3}\right)\left(\frac{1}{\underset{1}{\cancel{2}}}\right) = \left(-\frac{\overset{1}{\cancel{4}}}{3}\right)\left(-\frac{\overset{1}{\cancel{3}}}{\underset{1}{\cancel{4}}}v\right)$$ $$\frac{1}{2} = -\frac{\overset{1}{\cancel{3}}}{\cancel{4}}\left(-\frac{\overset{1}{\cancel{2}}}{3}\right)$$

$$-\frac{2}{3} = v$$ $$\frac{1}{2} = \frac{1}{2}$$

Thus, $v = -\frac{2}{3}$.

Example 8   Solve $-x = 4$ and check the solution.

Solution: $-x = 4$ *Check:* $-x = 4$

$(-x)(-1) = 4(-1)$ $-(-4) = 4$

$x = -4$ $4 = 4$

10.3   Exercises

In Exercises 1–10, identify the equations which are linear equations in one variable.

1. $11x = 143$
2. $x^3 = 8$
3. $2r - 5t = 18$
4. $4z + 7 = 3z - 9$
5. $x^2 = 3x - 5$
6. $2y - 5 = 3(y + 2)$
7. $2(w + 7) = 5(w - 3)$
8. $4a - 3b = 5c$
9. $5a - 8 + 2a = 5(a - 3)$
10. $6c - 7(c - 4) = 3c + 1$

Solve the equations in Exercises 11–20 by dividing both members of the equation by the coefficient of the variable, then check your solutions.

11. $7x = 42$
12. $3x = 18$
13. $45y = 5$
14. $-121w = -11$
15. $-2.4v = -0.24$
16. $-12t = 8$
17. $6a = 0$
18. $-b = 9$
19. $-15n = 10$
20. $50m = 100$

Solve the equations in Exercises 21–30 by multiplying both sides of the equation by the reciprocal of the coefficient of the variable, then check your solutions.

21. $\frac{2w}{7} = 14$
22. $\frac{x}{5} = 15$
23. $\frac{-5h}{3} = 25$
24. $\frac{2n}{3} = 6$
25. $\frac{x}{13} = 54$
26. $-\frac{7}{2}t = -14$

27. $4\frac{3}{7}v = 2\frac{20}{21}$
28. $\frac{-x}{10} = 0$
29. $-\frac{4}{3}w = -36$
30. $\frac{b}{13} = \frac{-4}{39}$

Solve the equations in Exercises 31–50 by using the multiplication-division principle of equality.

31. $-m = 18$
32. $\frac{n}{2} = 18$
33. $\frac{2}{3}x = 8$
34. $\frac{3}{2}y = 8$
35. $0.1s = 2$
36. $0.2a = 1$
37. $-5t = 20$
38. $20u = -5$
39. $\frac{v}{20} = 5$
40. $\frac{w}{5} = 20$
41. $2\frac{1}{3}z = 14$
42. $3\frac{1}{2}a = 21$
43. $0.12b = 13.2$
44. $1.2c = 13.2$
45. $12.0m = 13.2$
46. $120x = 13.2$
47. $\frac{n}{17} = \frac{50}{34}$
48. $\frac{p}{11} = \frac{2}{33}$
49. $5\frac{1}{7}x = 0$
50. $3.17s = 1.585$

Solve the equations in Exercises 51–70. Translate the word problems into equations and then solve the equations.

51. $\frac{x}{3} = 7$ **52.** $7b = \frac{21}{4}$

53. Negative six is equal to two-thirds x.

54. $3d = 8$ **55.** $\frac{3f}{2} = -4$

56. 2.22 times x equals 222.

57. The result of multiplying a number x by five is one-fifth.

58. $-\frac{2h}{3} = -4$ **59.** $-2k = \frac{3}{4}$

60. $8y = 3$

61. Three t divided by seven equals three.

62. $\frac{1}{7}w = \frac{1}{105}$

63. Three x is equal to fifty-three and one-tenth.

64. $\frac{3}{2}t = \frac{3}{2}$ **65.** $-\frac{1}{4}x = 9$

66. Negative two-fifths is equal to negative four-tenths x.

67. $-4t = 28$ **68.** $360 = 1.8u$

69. The result of halving a number x is 562.

70. The opposite of z equals negative 11.

10.4 Solving Equations of the Form ax + b = c or ax = bx + c

If a linear equation is not already in the form ax = b, it should be converted to an equivalent equation of this form. The main idea is to isolate all the terms containing the variable on one side of the equation and all constant terms on the other side of the equation. This is accomplished using the principle of equality given below.

ADDITION-SUBTRACTION PRINCIPLE OF EQUALITY

If the same number is added to or subtracted from both members of an equation, the result is an equivalent equation.

Our first illustration of the addition principle will be solving equations which are of the form *ax + b = c,* where a, b, and c represent constants and a ≠ 0. Again, x represents a variable. In this form there is a variable term and a constant term in one member, but only a constant term in the other member. The equations 3x − 4 = 10, 5y + 2 = 6, and 3m − 1 = 0 are all in this form. *To solve such equations, first isolate the constants on one side of the equation by adding the opposite of b (the constant in the member with the variable term) to both members of the equation.*

Example 1 Solve 4x + 3 = 11, and check the solution.

$$\begin{aligned} \textit{Solution:} \qquad 4x + 3 &= 11 \\ 4x + 3 + (-3) &= 11 + (-3) \\ 4x + 0 &= 8 \\ 4x &= 8 \\ \frac{4x}{4} &= \frac{8}{4} \\ x &= 2 \end{aligned}$$

$$\begin{aligned} \textit{Check:} \quad 4x + 3 &= 11 \\ 4(2) + 3 &= 11 \\ 8 + 3 &= 11 \\ 11 &= 11 \end{aligned}$$

Alternate Solution: This is an alternate format which may be used when the addition–subtraction principle is applied.

$$4x + 3 = 11$$
$$\underline{-3 = -3}$$
$$4x = 8$$

$$\frac{4x}{4} = \frac{8}{4}$$

$$x = 2$$

Example 2   Solve $4 = 3v - 6$

Solution: $4 = 3v - 6$

$$4 + \mathbf{6} = 3v - 6 + \mathbf{6}$$

$$10 = 3v$$

$$\frac{10}{3} = \frac{3v}{3}$$

$$\frac{10}{3} = v$$

$$v = \frac{10}{3}$$

The addition principle can also be used to solve equations of the form $ax = bx + c$, where there are variable terms on both sides and a single constant on the right. *To solve such equations, first isolate the variable on one side of the equation by adding the opposite of bx (the variable term on the right) to both members of the equation.*

Example 3   Solve $5x = 21 - 2x$ and check the solution.

Solution: $5x = 21 - 2x$ *Check:* $5x = 21 - 2x$

$5x + \mathbf{2x} = 21 - 2x + \mathbf{2x}$ $5(3) = 21 - 2(3)$

$7x = 21$ $15 = 21 - 6$

$\frac{1}{7}(7x) = \frac{1}{7}(21)$ $15 = 15$

$\phantom{\frac{1}{7}(7x) =}x = 3$

Example 4   Solve $2y = -4y - 4$ and check the solution.

Solution: $2y = -4y - 4$ *Check:* $2y = -4y - 4$

$\underline{+4y = +4y}$ $2\left(-\frac{2}{3}\right) = -4\left(-\frac{2}{3}\right) - 4$

$6y = -4$ $\frac{-4}{3} = \frac{8}{3} - 4$

$\frac{6y}{6} = \frac{-4}{6}$ $\frac{-4}{3} = \frac{8}{3} - \frac{12}{3}$

$\phantom{\frac{6y}{6} =}y = -\frac{2}{3}$ $\frac{-4}{3} = \frac{-4}{3}$

Example 5   Solve $0.37w - 0.013 = 0.098$, and check the solution.

Solution:

$$0.37w - 0.013 = 0.098$$
$$0.37w - 0.013 + (\mathbf{0.013}) = 0.098 + (\mathbf{0.013})$$
$$0.37w = 0.111$$
$$\frac{0.37w}{\mathbf{0.37}} = \frac{0.111}{\mathbf{0.37}}$$
$$w = 0.3$$

Check:

$$0.37w - 0.013 = 0.098$$
$$0.37(\mathbf{0.3}) - 0.013 = 0.098$$
$$0.111 - 0.013 = 0.098$$
$$0.098 = 0.098$$

Example 6   Solve $-3x = 5 - 2x$ and check the solution.

Solution: $-3x = 5 - 2x$ *Check:* $-3x = 5 - 2x$

$-3x + \mathbf{2x} = 5 - 2x + \mathbf{2x}$ $-3(\mathbf{-5}) = 5 - 2(\mathbf{-5})$

$-x = 5$ $15 = 15 + 10$

$-x(\mathbf{-1}) = 5(\mathbf{-1})$ $15 = 15$

$x = -5$

Example 7   Solve $\frac{3}{4}z + 1 = 10$, and check the solution.

Solution: $\frac{3}{4}z + 1 = 10$ *Check:* $\frac{3}{4}z + 1 = 10$

$\frac{3}{4}z + 1 = 10$ $\frac{3}{4}\left(\frac{12}{1}\right) + 1 = 10$

$\underline{-1 = -1}$ $9 + 1 = 10$

$\frac{3}{4}z = 9$ $10 = 10$

$\frac{4}{3}\left(\frac{3}{4}z\right) = \frac{4}{3}\left(\frac{9}{1}\right)$

$z = 12$

10.4   Exercises

Solve the equations in Exercises 1–20 and check your solutions.

1. $x - 11 = 13$

2. $a + 11 = 13$

3. $-b + 13 = 11$

4. $x - 13 = 11$

5. $2y - 11 = 13$

6. $-2c + 11 = 13$

7. $2v + 13 = 11$

8. $2r - 13 = 11$

9. $7t + 8 = 15$

10. $8w + 7 = 15$

11. $\frac{1}{2}n + 4 = 6$

12. $\frac{1}{4}v + 2 = 6$

13. $\frac{1}{6}z - 2 = 4$

14. $\frac{1}{4}m - 2 = 6$

15. $\frac{1}{4}a + 6 = 2$

16. $2b + 6 = 3$

17. $2v - 6 = 3v$

18. $-3x - 6 = -2x$

19. $-3y + 6 = -2y$

20. $5z + 20 = 3z$

Solve the equations in Exercises 21–60. Translate the word problems into equations and then solve.

21. $x - 7 = 2$

22. $x + 15 = -25$

23. $-8 - m = -2$

24. The sum of x and negative ten equals two-thirds.

25. $\frac{1}{2} = a - \frac{1}{2}$

26. $3b = b + 15$

27. The difference of three x minus four is four.

28. $8c = 3c + 10$

29. 0.5h equals the sum of h and 9.

30. $-d = d - 7$

31. $-f = f + 7$

32. Two x is equal to four x plus 2.

33. $x = -x - 7$

34. $5x + 11 = 6$

35. $x = -x + 7$

36. If six is added to twice x, the result is sixteen.

37. $3t = 4t$

38. $-2x - 7 = 11$

39. 2.5w is the same as the sum of 4.8 plus w.

40. $17x = 15 + 2x$

41. $1.23x - 0.668 = 8.065$

42. $5y = 3y + 12$

43. $29x = 4x - 0.50$

44. $83x = -57x + 14$

45. Two x is equal to negative four x minus 6.

46. $6.78 + 6.78x = 13.56$

47. $\frac{5}{7}x = 36 - x$

48. $11t + 21 = 32$

49. $\frac{2}{3}x - 1 = 4$

50. $-17x = -2x + 45$

51. If negative two is subtracted from three x, the result is five x.

52. Negative x plus ten equals zero.

53. $-5u = -3u + 12$

54. $y + 1.8 = 3.4$

55. $\frac{x}{5} - \frac{4}{3} = -\frac{1}{2}$

56. $\frac{3}{5}x + 1 = 22$

57. $5x = 7x + 3$

58. $36.2 - 5x = 36.2$

59. $3x + \frac{3}{2} = 6$

60. 2.3m is equal to the sum of 3.3m and 7.

10.5   Solving Equations of the Form ax + b = cx + d

Frequently, linear equations will have variables and constants in both members of the equation. Again, the addition principle will allow us to isolate variables in one member and the constants in the other member. *In equations of the form ax + b = cx + d, this can be done by adding the opposite of b (the constant on the left) to both members of the equation, and by adding the opposite of cx (the variable term on the right) to both members of the equation.* For example, to solve $9x + 7 = 5x - 13$, add the opposite of 7 to both members and add the opposite of 5x to both members.

Example 1   Solve $9x + 7 = 5x - 13$.

Solution:
$$9x + 7 = 5x - 13$$
$$9x + 7 + (-7) = 5x - 13 + (-7)$$
$$9x = 5x - 20$$
$$9x + (-5x) = 5x - 20 + (-5x)$$
$$4x = -20$$
$$\frac{4x}{4} = \frac{-20}{4}$$
$$x = -5$$

Alternate Solution: With practice much of this work can be done mentally.

$$9x + 7 = 5x - 13$$

$9x = 5x - 20$ −7 added to both sides

$4x = -20$ −5x added to both sides

$x = -5$ both sides divided by 4

Example 2   Solve $23v - 14 = 18v + 56$ and check the solution.

Solution:
$$23v - 14 = 18v + 56$$
$$23v - 14 + \mathbf{14} = 18v + 56 + \mathbf{14}$$
$$23v = 18v + 70$$
$$23v - \mathbf{18v} = 18v + 70 - \mathbf{18v}$$
$$5v = 70$$
$$\frac{1}{5}(5v) = \frac{1}{5}(70)$$
$$v = 14$$

Check:
$$23v - 14 = 18v + 56$$
$$23(\mathbf{14}) - 14 = 18(\mathbf{14}) + 56$$
$$322 - 14 = 252 + 56$$
$$308 = 308$$

Example 3   Solve $12 - 4z = 3z + 15$

Solution:
$$12 - 4z = 3z + 15$$
$$12 - 4z + \mathbf{4z} = 3z + 15 + \mathbf{4z}$$
$$12 = 7z + 15$$
$$12 - \mathbf{15} = 7z + 15 - \mathbf{15}$$
$$-3 = 7z$$
$$\frac{-3}{7} = z$$
$$z = -\frac{3}{7}$$

Example 4   Solve $0.35x - 1.7 = 1.85x + 4$ and check the solution.

Solution:
$$0.35x - 1.7 = 1.85x + 4$$
$$0.35x - 1.7 - \mathbf{0.35x} = 1.85x + 4 - \mathbf{0.35x}$$
$$-1.7 = 1.5x + 4$$
$$-1.7 - \mathbf{4} = 1.5x + 4 - \mathbf{4}$$
$$-5.7 = 1.5x$$
$$\frac{-5.7}{1.5} = \frac{1.5x}{1.5}$$
$$-3.8 = x$$
$$x = -3.8$$

Check:
$$0.35x - 1.7 = 1.85x + 4$$
$$0.35(\mathbf{-3.8}) - 1.7 = 1.85(\mathbf{-3.8}) + 4$$
$$-1.33 - 1.7 = -7.03 + 4$$
$$-3.03 = -3.03$$

Example 5   Solve $16z - 13 = -13 - 17z$ and check the solution.

Solution:
$$16z - 13 = -13 - 17z$$
$$16z - 13 + 13 = -13 - 17z + 13$$
$$16z = -17z$$
$$16z + 17z = -17z + 17z$$
$$33z = 0$$
$$\frac{33z}{33} = \frac{0}{33}$$
$$z = 0$$

Check:
$$16z - 13 = -13 - 17z$$
$$16(0) - 13 = -13 - 17(0)$$
$$-13 = -13$$

Up to this point we have only solved conditional equations. The special cases of contradictions and identities will be considered now.

Example 6   Solve $4x + 2 = 4x + 1$.

$$4x + 2 = 4x + 1$$
$$4x + 2 - 2 = 4x + 1 - 2$$
$$4x = 4x - 1$$
$$4x - 4x = 4x - 1 - 4x$$
$$0 = -1 \quad \text{No solution.}$$

Note that all variables were eliminated in the process of solving the equation in Example 6. Since the last equation is false, it is a contradiction. This implies that the original equation is also a contradiction since it is equivalent to $0 = -1$. Thus, the given equation has no solution.

Example 7   Solve $4w + 1 = 4w + 1$.

$$4w + 1 = 4w + 1$$
$$4w + 1 - 1 = 4w + 1 - 1$$
$$4w = 4w$$
$$4w - 4w = 4w - 4w$$
$$0 = 0 \quad \text{Every real number is a solution.}$$

Note that all variables were eliminated in the process of solving the equation in Example 7. The last equation is easily recognized as an identity. Since this equation is equivalent to the original equation, the original equation is also an identity. This means that every real number is a root of the equation.

10.5   Exercises

Solve each of the equations in Exercises 1–20 and then check your solutions.

1. $5b + 3 = 2b + 2$
2. $7x - 8 = 6x - 10$
3. $3z + 2 = 5z + 2$
4. $4m - 3 = 6m + 3$
5. $11w + 4 = 13w - 4$
6. $7c - 2 = 3c + 6$
7. $10t + 11 = 11t + 10$
8. $4z - 3 = 3z - 4$
9. $4v + 3 = 3v + 4$
10. $16y + 1 = 4y + 25$
11. $11a - 5 = 5a - 2$
12. $0.2z - 5 = -0.8z + 7$
13. $0.7k + 3 = 6 - 0.3k$
14. $19w - 1 = 8w + 120$
15. $15x - 3 = 8x + 60$
16. $\frac{1}{2}x - 8 = 6 - \frac{1}{2}x$
17. $5x + 7 = 7 + 5x$
18. $3y - 6 = -6 + 3y$
19. $11n + 1 = 11n + 2$
20. $5x - 4 = 5x + 4$

Solve each of the equations in Exercises 21–50. Translate the word problems into equations and then solve the equations.

21. $5 - x = x + 5$
22. $3x + 5 = 7 + x$
23. Negative two x plus two is equal to three minus x.
24. $2x + 6 = x + 7$
25. $-2y + 6 = 11 - y$
26. $4w + 13 = 5w + 16$
27. The sum of negative one and negative four x equals three x plus twenty.
28. $5x + 4 = 4 - 5x$
29. Negative three x increased by six is equal to negative three x decreased by six.
30. $-22y + 4 = -2y - 6$
31. $-17w + 5 = -9w + 5$
32. x minus ten equals x plus negative ten.
33. $0.3x - 0.17 = 0.3 + 0.1x$
34. $-2a + 5 = 4a + 6$
35. If three x is increased by one-half, the result is equal to x decreased by one-half.
36. $12t - 12 = -t + 14$
37. $5y - 4 = 3y + 8$
38. $x - 5 = 5 - x$
39. $5 - x = -x + 5$
40. Fifteen more than negative six x equals forty x minus seventy-seven.
41. $5 - x = -x - 5$
42. $47y + 11 = 35y - 13$
43. $-5 + x = -x + 5$
44. $-17w - 5 = -6w - 5$
45. One and three thousand four hundred seventy-eight ten-thousandths decreased by 10x is the same as 12x decreased by eight thousand five hundred twenty-two ten-thousandths.
46. Four and seven-tenths x minus five and two-tenths equals eight plus five x.
47. $-5 - x = -x + 5$
48. $5x + 21 = 13 - 3x$
49. $-5 - x = -x - 5$
50. One-half x decreased by 5 is equal to three-halves x plus 2.

10.6   Solving More General Forms of First-Degree Equations

In Chapter 8 you learned to simplify expressions such as $5(2x - 3) - (5 - 4x)$. If you encounter a linear equation which has a member of the equation as complicated as the expression just mentioned, we suggest you follow the steps outlined in the *procedure for solving linear equations*. For some equations you may see a clever shortcut; however, this procedure will always work. The main idea involved has already been mentioned—*after simplifying both members, isolate the terms involving the variable in one member of the equation and the constant terms in the other member.*

PROCEDURE FOR SOLVING LINEAR EQUATIONS	EXAMPLE Solve for x:

EXAMPLE   Solve for x:

$2(3x - 5) = 7(2x + 5) - 5$

Solution:

1. Simplify each member of the equation to produce an equivalent equation.

$$2(3x - 5) = 7(2x + 5) - 5$$
$$6x - 10 = 14x + 35 - 5$$
$$6x - 10 = 14x + 30$$

2. Using the addition-subtraction principle of equality, isolate the variable terms in one member and the constant terms in the other member.

$$6x - 10 + \mathbf{10} = 14x + 30 + \mathbf{10}$$
$$6x = 14x + 40$$
$$6x - \mathbf{14x} = 14x + 40 - \mathbf{14x}$$
$$-8x = 40$$

3. Using the multiplication-division principle of equality, solve the equivalent equation produced in step 2.

$$\frac{-1}{8}(-8x) = \frac{-1}{8}(40)$$
$$x = -5$$

Example 1   Solve $3(v - 3) = 2(5v - 11) - 1$

$$\begin{aligned}
\textit{Solution:} \quad 3(v - 3) &= 2(5v - 11) - 1 \\
3v - 9 &= 10v - 22 - 1 \\
3v - 9 &= 10v - 23 \\
3v - 9 + \mathbf{9} &= 10v - 23 + \mathbf{9} \\
3v &= 10v - 14 \\
3v - \mathbf{10v} &= 10v - 14 - \mathbf{10v} \\
-7v &= -14 \\
\frac{-7v}{-7} &= \frac{-14}{-7} \\
v &= 2
\end{aligned}$$

Example 2   Solve $11(x - 3) + 4(2x + 1) = 5(3x - 2) + 13$.

$$\begin{aligned}
\textit{Solution:} \quad 11(x - 3) + 4(2x + 1) &= 5(3x - 2) + 13 \\
11x - 33 + 8x + 4 &= 15x - 10 + 13 \\
19x - 29 &= 15x + 3 \\
19x - 29 + \mathbf{29} &= 15x + 3 + \mathbf{29} \\
19x &= 15x + 32 \\
19x - \mathbf{15x} &= 15x + 32 - \mathbf{15x} \\
4x &= 32 \\
\frac{4x}{4} &= \frac{32}{4} \\
x &= 8
\end{aligned}$$

Example 3   Solve $2 - 6(y + 1) = 4(2 - 3y) + 6$ and check the solution.

Solution:
$$2 - 6(y + 1) = 4(2 - 3y) + 6$$
$$2 - 6y - 6 = 8 - 12y + 6$$
$$-6y - 4 = -12y + 14$$
$$-6y - 4 + \mathbf{12y} = -12y + 14 + \mathbf{12y}$$
$$6y - 4 = 14$$
$$6y - 4 + \mathbf{4} = 14 + \mathbf{4}$$
$$6y = 18$$
$$\frac{6y}{\mathbf{6}} = \frac{18}{\mathbf{6}}$$
$$y = 3$$

Check:
$$2 - 6(y + 1) = 4(2 - 3y) + 6$$
$$2 - 6((3) + 1) = 4(2 - 3(3)) + 6$$
$$2 - 6(4) = 4(2 - 9) + 6$$
$$2 - 24 = 4(-7) + 6$$
$$-22 = -28 + 6$$
$$-22 = -22$$

Example 4   Solve $0.475 + 0.2(a - 2.5) = -3.1 - 0.8a$

Solution:
$$0.475 + 0.2(a - 2.5) = -3.1 - 0.8a$$
$$0.475 + 0.2a - 0.5 = -3.1 - 0.8a$$
$$0.2a - 0.025 = -3.1 - 0.8a$$
$$0.2a - 0.025 + \mathbf{0.8a} + \mathbf{0.025} = -3.1 - 0.8a + \mathbf{0.8a} + \mathbf{0.025}$$
$$a = -3.075$$

Example 5   Solve $4(x - 1) + 3(x + 2) = 7(x + 1) - 5$.

Solution:
$$4(x - 1) + 3(x + 2) = 7(x + 1) - 5$$
$$4x - 4 + 3x + 6 = 7x + 7 - 5$$
$$7x + 2 = 7x + 2$$
$$7x + 2 - \mathbf{7x} - \mathbf{2} = 7x + 2 - \mathbf{7x} - \mathbf{2}$$
$$0 = 0$$

Note that the last equation is an identity. Hence, the original equation is also an identity and every real number is a solution.

Example 6   The sum of a number and three times that number is ninety-nine plus the number. Find the number.

Solution: Let n represent the number.

$$n + 3n = 99 + n$$
$$4n = 99 + n$$
$$3n = 99$$
$$n = 33$$

Thus the number is 33.

10.6   Exercises

Solve each of the equations in Exercises 1–20 and then check your solutions.

1. $4(r - 2) = 3(r - 1) + 15$

2. $2(3a - 8) = 5 - 4(a - 1)$

3. $3(2n - 2) = 2(n - 1) - 12$

4. $5(4x - 2) = 9 - 2(7x + 1)$

5. $6(2m - 3) = 7(m - 3) - 2$

6. $6z + 2 = 2(17 - z)$

7. $3(p + 4) = 3(2 - p)$

8. $2(b + 1) = b - 7$

9. $7(v - 4) = v + 2$

10. $4(c + 18) = 2(4c + 18)$

11. $6(y + 1) - 4 = 4(3y - 2)$

12. $2(y - 3) = 4 + (y - 14)$

13. $4(w + 2) = 8(w - 4) + 8$

14. $13(x + 4) = 8(x + 1) - 2$

15. $8s + 11 = 4(s + 2)$

16. $5v - 6(v + 1) = 3(v - 2)$

17. $4(t + 2) = 30 - (t - 3)$

18. $3(w - 6) = 4(w + 1) - 28$

19. $18(y + 1) = 17(y - 1) + 36$

20. $11(y + 3) = 2(5y - 1) + 34$

Solve each of the equations in Exercises 21–60. Translate the word problems into equations and then solve the equations.

21. $12m + 5m - 4 = 3m - 4 + 2m$

22. $-24 - 3y - 2y = 52$

23. Negative three added to two times a number is equal to five. (*Hint:* Let x represent the number.)

24. $5.5a = 90 + 2a - a$

25. $4(x - 3) = 3x - 5$

26. $(t - 6) + (t - 2) + t = -8$

27. A number subtracted from two-thirds of itself is equal to six.

28. $-15(2 - m) - 10m = -35$

29. 2y minus 6 plus 7y equals 8 minus 9y minus 10 plus 18y.

30. $k - (0.5k + 2.6) = 17.6$

31. $111(2t + 3) = 2(108t + 87) + 3$

32. $7y - 4 + 3y = 10y - 4$

33. If x − 3 is increased by x − 2, the result is 0.

34. $5x + (10x - 8) = 9x + (6x + 2)$

35. $-4(x - 2) = 3(x - 7) - 5(4 - x) - 1$

36. $7(b - 2) - 2(3 + b) = 0$

37. $1.5m - (1 - m) = -0.1m - 2.8(1 - 2m)$

38. $13 + 5(2x + 1) = 3(x - 6) + 1$

39. $2(x + 5) - 2(x + 5) = 4$

40. Negative three times the quantity two t minus 5 is equal to three times the quantity one minus t.

41. $2t = 2t$

42. $-2m + 4(2m - 1) = 32$

43. Six more than twice a number is equal to two increased by the number.

44. $2(h + 2) + 2(h + 3) = 27$ **45.** $-2(6 - z) = 3z + 1$

46. $3(x + 5) = (x + 3) + x$ **47.** $4x - 3(x + 2) = 4$

48. Six times a number decreased by five times the number is equal to ten.

49. $62 - 2(2x + 21) = -4$ **50.** $4(4y + 9) = -(9 - y)$

51. $5x + 2(1 - 4x) = 1$

52. $20x + 18 - 5x + 10 = 21x + 10$

53. $1.2z + 0.4(125 - 2z) = 3.5 - 0.8z$

54. $4x - 12 - 3x + 5 = 0$

55. $7(n - 1) - 4(2n + 3) = 2(n + 1) - 3(4 - n)$

56. A number is equal to two times itself. Find the number.

57. $-3(2x - 5) - (-6x + 3) = 12$

58. $3(2x - 5) - (-6x + 3) = 12$

59. $6(x - 9) = 8(x - 6) + 29$

60. $3(m + 1) - 15 = 2m - 10$

10.7   Solving First-Degree Equations Involving Fractions

Since the arithmetic of fractions is more complex than the arithmetic of integers, the solution of a linear equation involving fractions may include an extra step. One way to simplify a fractional equation is to convert it to an equivalent equation which does not involve fractions. The procedure for accomplishing this will now be outlined.

PROCEDURE FOR SOLVING EQUATIONS CONTAINING FRACTIONS	*EXAMPLE* $\dfrac{3x}{7} + \dfrac{4}{5} = \dfrac{3x}{5} + \dfrac{2}{7}$
1. Determine the LCD of all the terms of the equation.	*Solution:* The LCD is 35
2. Multiply both members of the equation by this LCD. (Caution: To accomplish this you must multiply each **term** by the LCD.)	$35\left(\dfrac{3x}{7} + \dfrac{4}{5}\right) = 35\left(\dfrac{3x}{5} + \dfrac{2}{7}\right)$
3. Solve the resulting equivalent equation using the procedure outlined in Section 10.6.	$\overset{5}{\cancel{35}}\left(\dfrac{3x}{\cancel{7}}\right) + \overset{7}{\cancel{35}}\left(\dfrac{4}{\cancel{5}}\right) = \overset{7}{\cancel{35}}\left(\dfrac{3x}{\cancel{5}}\right) + \overset{5}{\cancel{35}}\left(\dfrac{2}{\cancel{7}}\right)$ $15x + 28 = 21x + 10$ $15x + 28 - \mathbf{10} = 21x + 10 - \mathbf{10}$ $15x + 18 = 21x$ $15x + 18 - \mathbf{15x} = 21x - \mathbf{15x}$ $18 = 6x$ $\dfrac{18}{6} = \dfrac{6x}{6}$ $3 = x \quad \text{or} \quad x = 3$

Example 1   Solve $\dfrac{y}{2} + 3 = 5$ and then check the solution.

Solution:

$$\frac{y}{2} + 3 = 5$$

$$2\left(\frac{y}{2} + 3\right) = 2(5)$$

$$2\left(\frac{y}{2}\right) + 2(3) = 10$$

$$y + 6 = 10$$

$$y + 6 - \mathbf{6} = 10 - \mathbf{6}$$

$$y = 4$$

Check:

$$\frac{y}{2} + 3 = 5$$

$$\frac{4}{2} + 3 = 5$$

$$2 + 3 = 5$$

$$5 = 5$$

Example 2   Solve $\frac{x}{3} = \frac{x}{2} - 2$ and check the solution.

Solution: $\frac{x}{3} = \frac{x}{2} - 2$

$$6\left(\frac{x}{3}\right) = 6\left(\frac{x}{2} - 2\right)$$

$$2x = 3x - 12$$

$$2x - 3x = 3x - 12 - 3x$$

$$-x = -12$$

$$x = 12$$

Check: $\frac{x}{3} = \frac{x}{2} - 2$

$$\frac{12}{3} = \frac{12}{2} - 2$$

$$4 = 6 - 2$$

$$4 = 4$$

Example 3   Solve $\frac{5x}{6} + \frac{49}{27} = \frac{12x + 7}{27}$ and check the solution.

Solution: $\frac{5x}{6} + \frac{49}{27} = \frac{12x + 7}{27}$

$$\overset{9}{\cancel{54}}\left(\frac{5x}{\cancel{6}}\right) + \overset{2}{\cancel{54}}\left(\frac{49}{\cancel{27}}\right) = \overset{2}{\cancel{54}}\left(\frac{12x + 7}{\cancel{27}}\right)$$

$$9(5x) + 2(49) = 2(12x + 7)$$

$$45x + 98 = 24x + 14$$

$$45x + 98 - \mathbf{24x} = 24x + 14 - \mathbf{24x}$$

$$21x + 98 - \mathbf{98} = 14 - \mathbf{98}$$

$$21x = -84$$

$$\frac{21x}{21} = \frac{-84}{21}$$

$$x = -4$$

Check: $\frac{5x}{6} + \frac{49}{27} = \frac{12x + 7}{27}$

$$\frac{5(\mathbf{-4})}{6} + \frac{49}{27} = \frac{12(\mathbf{-4}) + 7}{27}$$

$$\frac{-20}{6} + \frac{49}{27} = \frac{-48 + 7}{27}$$

$$\frac{-10}{3} + \frac{49}{27} = -\frac{41}{27}$$

$$\frac{-90}{27} + \frac{49}{27} = -\frac{41}{27}$$

$$\frac{-41}{27} = -\frac{41}{27}$$

Example 4   Solve $\frac{v}{3} + 20 = \frac{v}{4} - \frac{v}{12} - \frac{v}{9}$ and check the solution.

Solution: The denominators are 3, 1, 4, 12, and 9.
The LCD = $2 \cdot 2 \cdot 3 \cdot 3 = 36$.

$$36\left(\frac{v}{3}\right) + 36(20) = 36\left(\frac{v}{4}\right) - 36\left(\frac{v}{12}\right) - 36\left(\frac{v}{9}\right)$$

$$12v + 720 = 9v - 3v - 4v$$

$$12v + 720 = 2v$$

$$12v + 720 - 720 - 2v = 2v - 720 - 2v$$

$$10v = -720$$

$$\frac{10v}{10} = \frac{-720}{10}$$

$$v = -72$$

Check: $\dfrac{v}{3} + 20 = \dfrac{v}{4} - \dfrac{v}{12} - \dfrac{v}{9}$

$$\frac{-72}{3} + 20 = \frac{-72}{4} - \frac{-72}{12} - \frac{-72}{9}$$

$$-24 + 20 = -18 - (-6) - (-8)$$

$$-4 = -18 + 6 + 8$$

$$-4 = -18 + 14$$

$$-4 = -4$$

Example 5   Solve $\dfrac{x}{14} - \dfrac{2}{5} = \dfrac{x}{21} + \dfrac{4x}{175}$ and check the solution.

Solution: LCD = $2 \cdot 3 \cdot 5 \cdot 5 \cdot 7 = 1050$, since $14 = 2 \cdot 7$

$$5 = 5$$
$$21 = 3 \cdot 7$$
$$175 = 5 \cdot 5 \cdot 7$$

$$\overset{75}{\cancel{1050}}\left(\frac{x}{\cancel{14}}\right) - \overset{210}{\cancel{1050}}\left(\frac{2}{\cancel{5}}\right) = \overset{50}{\cancel{1050}}\left(\frac{x}{\cancel{21}}\right) + \overset{6}{\cancel{1050}}\left(\frac{4x}{\cancel{175}}\right)$$

$$75x - 420 = 50x + 24x$$

$$75x - 420 = 74x$$

$$75x - 420 + 420 - 74x = 74x + 420 - 74x$$

$$x = 420$$

Check: $\dfrac{x}{14} - \dfrac{2}{5} = \dfrac{x}{21} + \dfrac{4x}{175}$

$$\frac{420}{14} - \frac{2}{5} = \frac{420}{21} + \frac{4(\overset{12}{\cancel{420}})}{\underset{5}{\cancel{175}}}$$

$$30 - \frac{2}{5} = 20 + \frac{4(12)}{5}$$

$$29\frac{3}{5} = 20 + \frac{48}{5}$$

$$29\frac{3}{5} = 20 + 9\frac{3}{5}$$

$$29\frac{3}{5} = 29\frac{3}{5}$$

10.7 Exercises

Solve each of the equations in Exercises 1–20 and then check your solutions.

1. $\dfrac{w}{7} = 35$

2. $\dfrac{k}{4} = 8$

3. $\dfrac{z}{8} = 4$

4. $\dfrac{n}{3} = \dfrac{1}{2}$

5. $\dfrac{t}{6} = -3$

6. $\dfrac{m}{-2} = 11$

7. $\dfrac{v}{6} = \dfrac{1}{3}$

8. $\dfrac{y}{18} = \dfrac{-2}{3}$

9. $\dfrac{b}{5} = -4$

10. $\dfrac{x}{4} = \dfrac{1}{7}$

11. $\dfrac{x}{2} = \dfrac{x}{3} + 6$

12. $\dfrac{b}{3} = 2 - \dfrac{b}{5}$

13. $\dfrac{z}{7} = 10 - \dfrac{z}{3}$

14. $\dfrac{w}{6} - \dfrac{w}{15} = \dfrac{w}{10}$

15. $\dfrac{v}{55} + \dfrac{v}{33} = \dfrac{v}{15} - 3$

16. $\dfrac{a+6}{5} = \dfrac{a+2}{3}$

17. $\dfrac{t-4}{11} = \dfrac{6t+103}{5} - 1$

18. $\dfrac{2r+6}{4} = \dfrac{3r+9}{5}$

19. $\dfrac{x+2}{7} = \dfrac{x-3}{2}$

20. $\dfrac{3y-3}{6} = \dfrac{4y+1}{15} + 2$

Solve each of the equations in Exercises 21–60. Translate the word problems into equations and then solve the equations.

21. $\dfrac{x}{2} = 15$

22. $\dfrac{-2k}{3} = -4$

23. $9 = \dfrac{3w}{5}$

24. $\dfrac{4}{7}x = \dfrac{3}{4}$

25. $\dfrac{4z}{7} = \dfrac{5z}{14} + 3$

26. $\dfrac{5}{8}t - 17 = 33$

27. Twenty-five less than half a number is equal to one-fourth of the number. Find the number. (*Hint:* Let x represent the number.)

28. $\dfrac{y}{6} - 7 = -1$

29. $\dfrac{t}{3} + 2 = 2 + \dfrac{t}{6} + \dfrac{t}{6}$

30. $\dfrac{8}{5}y - y = 5$

31. One-half of a number is equal to the sum of three-fourths of the number and negative 2. Find the number.

32. $\dfrac{5}{2}k - 6\left(\dfrac{1}{4}k + 7\right) = 18$

33. $\dfrac{1}{12}z + 1 = -2$

34. One-fourth of 3x is equal to 6. Find x.

35. $\dfrac{y}{3} + \dfrac{2}{3} = \dfrac{y}{2} + \dfrac{3}{2}$

36. $\dfrac{3}{4}x = \dfrac{5}{2}x$

37. A number is equal to two-thirds minus itself. Find the number.

38. $\dfrac{1}{12}z + 1 = \dfrac{1}{3}z - 2$

39. $\dfrac{1}{2} - \dfrac{1}{2}c - \dfrac{5}{2} = 4c - 2 - \dfrac{7}{2}c$

40. $\dfrac{9t}{2} - \dfrac{3t}{2} - \dfrac{7t}{2} = \dfrac{3}{2}$

41. $8\dfrac{1}{2}n + 3\dfrac{1}{4}n - 5\dfrac{3}{4}n = 108$

42. $\dfrac{-2(5-z)}{5} = \dfrac{2z+2}{3}$

43. $\dfrac{x}{2} - \dfrac{x}{3} = \dfrac{x}{5} - 1$

44. $-\dfrac{3}{2}(2t - 5) = \dfrac{3}{2}(1 - t)$

45. $\dfrac{x}{3} - \dfrac{1}{6} = \dfrac{x+1}{9}$

46. If x − 3 is divided by 5, the result is 2. Find x.

47. The sum of a number and one-third of the number is equal to the sum of twice the number and ten.

48. $\dfrac{3}{4}w + 1 = \dfrac{5}{6}w - 3$

49. $\dfrac{z}{2} - \dfrac{z}{3} = \dfrac{z}{4} - \dfrac{z}{18} - \dfrac{7}{36}$

50. $\dfrac{y}{3} + 11 = 6 - \dfrac{y}{2}$

51. $\dfrac{24w - 67}{60} = \dfrac{3w - 8}{12}$

52. One-third of x is two more than one-fifth of x. Find x.

53. $\dfrac{2}{3}(h + 2) = \dfrac{5}{6}(h + 3)$

54. $\dfrac{z}{6} = -\dfrac{3}{4}$

55. Two and five-eighths x plus three-eighths is equal to two minus five x.

56. $\dfrac{4x}{7} = \dfrac{1}{7}$

57. $4\dfrac{2}{3}h = 14$

58. $\dfrac{1}{2}h - \dfrac{3}{4}h + 1 = \dfrac{1}{4}h + 10$

59. $\dfrac{x}{5} - \dfrac{4}{3} = -\dfrac{1}{2}$

60. $\dfrac{4x+2}{5} + 2 = 3 - \dfrac{3-7x}{2}$

10.8 Solving Equations for a Specified Variable

Formulas are mathematical statements which usually contain more than one letter or variable. For example, the familiar $A = l \cdot w$ is the formula for the area of a rectangle. It is sometimes convenient to express formulas in different forms. Other forms of the formula $A = l \cdot w$ are $l = \dfrac{A}{w}$ and $w = \dfrac{A}{l}$. If you wish to find the width of a rectangle when its area and length are known, it is most convenient to use the form $w = \dfrac{A}{l}$. The principles of equality can be used to find these forms. Again, the main idea is to isolate in one member of the equation terms involving the variable for which you wish to solve. The variable you are solving for is called the **specified** or **indicated variable**. In the equation

$$w = \frac{A}{l} \qquad \text{w is the specified variable.}$$

A technique which can be used to solve for the specified variable in the types of equations we will consider is given in the *procedure for solving linear equations for a specified variable*.

PROCEDURE FOR SOLVING LINEAR EQUATIONS FOR A SPECIFIED VARIABLE	EXAMPLE
	Solve $2(3a - b) = 7(2a + b) - 5$ for a. *Solution:*
1. Simplify each member of the equation.	$6\underline{a} - 2b = 14\underline{a} + 7b - 5$
2. Use the addition-subtraction principle to isolate in one member of the equation the terms containing the specified variable.	$6\underline{a} - 2b + \mathbf{2b} = 14\underline{a} + 7b - 5 + \mathbf{2b}$ $6\underline{a} = 14\underline{a} + 9b - 5$ $6a - \mathbf{14a} = 14\underline{a} + 9b - 5 - \mathbf{14a}$ $-8\underline{a} = 9b - 5$
3. Solve the equation produced in step 2 for the specified variable by using the multiplication-division principle.	$\dfrac{-8\underline{a}}{-8} = \dfrac{9b - 5}{-8}$ $\underline{a} = \dfrac{9b - 5}{-8}$

Example 1   Solve $a = b + c$ for b.

$$a = \underline{b} + c$$

$$a - c = \underline{b} + c - \mathbf{c} \qquad \text{To isolate b, subtract c from both members.}$$

$$a - c = \underline{b}$$

Thus, $b = a - c$.

Example 2   Solve $a = 2c + b$ for c.

$$a = 2\underline{c} + b$$

$$a - \mathbf{b} = 2\underline{c} + b - \mathbf{b} \qquad \text{To isolate the term containing c, subtract b from both members of the equation.}$$

$$a - b = 2\underline{c}$$

$$\frac{a - b}{2} = \frac{2\underline{c}}{2} \qquad \text{Divide by 2 in order to solve for c.}$$

$$\frac{a - b}{2} = \underline{c}$$

Thus, $c = \dfrac{a - b}{2}$.

Example 3   Solve c = πd for d.

$$c = \pi \underline{d}$$

$$\frac{c}{\pi} = \frac{\pi \underline{d}}{\pi} \qquad \text{Divide both members by } \pi.$$

$$\frac{c}{\pi} = \underline{d}$$

Thus $d = \dfrac{c}{\pi}$.

Example 4   Solve T = πrh + 2πr² for h.

$$T = \pi r \underline{h} + 2\pi r^2 \qquad \text{To isolate the term containing}$$

$$T - 2\pi r^2 = \pi r \underline{h} + 2\pi r^2 - 2\pi r^2 \qquad \substack{\text{h, subtract the term } 2\pi r^2 \text{ from}\\ \text{both members.}}$$

$$T - 2\pi r^2 = \pi r \underline{h}$$

$$\frac{T - 2\pi r^2}{\pi r} = \frac{\pi r \underline{h}}{\pi r} \qquad \text{To solve for h, divide both sides by } \pi r.$$

$$\frac{T - 2\pi r^2}{\pi r} = \underline{h}$$

Thus, $h = \dfrac{T - 2\pi r^2}{\pi r}$.

Example 5   Solve $\dfrac{x}{m} = n$ for x.

$$\frac{\underline{x}}{m} = n$$

$$\mathbf{m} \cdot \frac{\underline{x}}{m} = \mathbf{m} \cdot n \qquad \text{Multiply both members by m.}$$

$$\underline{x} = mn$$

Example 6   Solve $l = a + (n - 1)d$ for n.

First, simplify the right member of the equation.

$$l = a + (\underline{n} - 1)d$$

$$l = a + \underline{n}d - d$$

To isolate the term containing n, subtract a and then add d to both members.

$$l - a = \underline{n}d - d$$

$$l - a + d = \underline{n}d$$

To solve for n, divide both members by d.

$$\frac{l - a + d}{d} = \underline{n}$$

Thus $n = \dfrac{l - a + d}{d}$.

Example 7   If $A = \frac{1}{2}h(a + b)$, $A = 200$, $h = 7$, and $a = 15$, then find b.

First solve the equation for b, and then substitute the values for A, h, and a.

$$A = \frac{1}{2}h(a + \underline{b})$$

$$2A = ha + h\underline{b}$$

$$2A - ha = h\underline{b}$$

$$\frac{2A - ha}{h} = \underline{b}$$

$$\frac{400 - 105}{7} = \underline{b}$$

$$\frac{295}{7} = \underline{b}$$

$$\underline{b} = 42\frac{1}{7}$$

Alternate Solution: First substitute for A, h, and a; and then solve for b.

$$A = \frac{1}{2}h(a + b)$$

$$200 = \frac{1}{2}(7)(15 + b)$$

$$200 = \frac{7}{2}(15 + b)$$

$$\frac{400}{2} = \frac{105}{2} + \frac{7}{2}b$$

$$\frac{400}{2} - \frac{105}{2} = \frac{7}{2}b$$

$$\frac{295}{2} = \frac{7}{2}b$$

$$295 = 7b$$

$$\frac{295}{7} = b \text{ or } b = 42\frac{1}{7}$$

10.8 Exercises

Solve each of the equations in Exercises 1–25 for the indicated variable.

1. $rt = d$ for t

2. $nx = 2d$ for x

3. $2ya = 5a$ for y

4. $x + m = m$ for x

5. $x - m = n$ for x

6. $p = a + b + c$ for b

7. $S = \frac{n}{2}(a + l)$ for a

8. $F = \frac{9}{5}C + 32$ for C

9. $C = \frac{5}{9}(F - 32)$ for F

10. $A = lw$ for w

11. $rst = xyz$ for y

12. $2\pi rt = S$ for t

13. $\frac{w}{r} = -d$ for w

14. $V = lwh$ for h

15. $2u + n = m$ for u

16. $2bx = b$ for x

17. $n - a = b + c$ for n

18. $3(x + a) = b$ for x

19. $A = \frac{1}{2}h(a + b)$ for a   **20.** i = prt for p

21. $A = \frac{1}{2}bh$ for b   **22.** v = k + gt for k

23. $V = \pi r^2 h$ for h   **24.** $l = a + (n - 1)d$ for a

25. $\frac{2n}{b} = a$ for n

Using the given formula, evaluate Exercises 26–35 for the indicated letter.

26. $A = lw$; A = 36, l = 9, w = ?

27. p = a + b + c; p = 30, b = 7, c = 18, a = ?

28. $V = \pi r^2 h$; V = 462, $\pi \doteq \frac{22}{7}$, r = 7, h = ?

29. $A = \frac{1}{2}h(a + b)$; A = 100, h = 4, b = 7, a = ?

30. $A = \frac{1}{2}bh$; A = 50, b = 5, h = ?

31. V = k + gt; V = −100, g = ?, t = 5, k = 60

32. I = PRT; I = 15, P = 300, T = 1, R = ?

33. $V = \frac{4}{3}\pi r^3$; $\pi \doteq 3.14$, r = 12, V = ?

34. $k = \frac{mv^2}{2}$; k = 125,000, v = 100, m = ?

35. $V = lwh$; V = 2000, l = 5, w = 40, h = ?

Important Terms Used in this Chapter	addition–subtraction principle of equality	multiplication–division principle of equality
	conditional equation	root
	contradiction	satisfy
	equation	solution
	equivalent equations	solve by inspection
	first-degree equation	solving an equation
	identity	specified variable
	linear equation	

REVIEW EXERCISES

Answer Exercises 1–20 as true (T) or false (F).

1. 2 = 5 − 3 is an identity.

2. 4x = 8 is a contradiction.

3. Two satisfies the equation 3x = 7.

4. 0x = 7 is a linear equation in the variable x.

5. 7w = 18 is a conditional equation.

6. 7w = 18w − 11w is an identity.

7. Four is a root of 7x = 35.

8. To check a root of an equation, substitute the root for the variable in the equation, and then evaluate each member of the equation.

9. 2 = 3 is a contradiction.

10. The solution of 4x = 20 is 5.

11. Equivalent equations have the same roots.

12. 25 = 12 and 5 = 6 are equivalent.

13. x = 3 and $x = \frac{1}{3}$ are equivalent.

14. 2x + 1 = x and x = 1 are equivalent.

15. 5x = 75 and x = 15 are equivalent.

16. 5V = 9 − 3V is a linear equation in one variable.

17. 2w = 3w is a linear equation in one variable.

18. $r^2 = 5r + 1$ is a linear equation in one variable.

19. 2(z + 3) = 3(z − 2) is a linear equation in one variable.

20. x + y = 7 is a linear equation in one variable.

Solve the equations in Exercises 21–45. Check the solutions to Exercises 21–27.

21. 4y = 36   **22.** 11w − 15 = 18

23. 14x − 15 = 19x   **24.** $\frac{2}{3}x = 10$

25. 12x − 8 = 7x + 2   **26.** 8x + 3 − 5x = 2x + 4

27. 5(x − 3) = 2(x + 4)   **28.** $\frac{x}{6} - \frac{x}{5} = \frac{x}{15} - \frac{x}{10}$

29. $\frac{x + 5}{12} = \frac{x + 8}{15}$ **30.** $\frac{x}{11} - 4 = \frac{x}{7}$

31. $\dfrac{4y - 36}{9} = 4$

32. $\dfrac{11w}{7} - \dfrac{15}{2} = \dfrac{27}{14}$

33. $\dfrac{19x}{12} - \dfrac{15}{4} = \dfrac{14x}{9}$

34. The sum of x and eighteen is the same as the product of three and x. Find x.

35. $\dfrac{5}{7}z + 13 = 58$

36. The sum of 4 times x and 17 is equal to the difference of 50 minus 7 times x. Find x.

37. If a number N is increased by twelve, the result will be four times as large as N. Find N.

38. $3w - 2 + 2w - 9 = -4 + 8$

39. $0.23t + 0.48 = 0.27t - 0.52$

40. $4(3v - 5) + (v - 1) = 7(v - 1)$

41. $4(v - 1) = 4v + 2$

42. w decreased by 240 is equal to seven-thirteenths of w. Find w.

43. $5(u + 3) = 4(u + 4) - (1 - u)$

44. $5x + 19 = 19x + 5$

45. $c - \dfrac{1}{3}c = -12$

Solve the equations in Exercises 46–50 for x.

46. $ax = b$

47. $ax + b = c$

48. $x = y^2 + 4(x - 1)$

49. $x + a = b$

50. $2(x + b) = 3(x - c)$

11

WORD PROBLEMS

PRETEST

1. Translate each of the following phrases into an algebraic form (Objective 1).
   a. Nine more than x
   b. x minus nine is equal to w.
   c. Nine less x is the same as x plus four.
   d. The quotient of x and nine
   e. Two-thirds of x equals seven.
   f. The difference two x minus four
   g. The product of six and y
   h. Twice the quantity y plus three
   i. The square of y is sixteen.
   j. Five times the quantity y minus two

2. In each of the following problems, represent all the quantities in terms of the same variable (Objective 2).

   a. The sum of two numbers is 85 and x represents one of the numbers. Represent the other number in terms of x.

   b. The product of y and another number is 58. Represent the other number in terms of y.

   c. One number is five less than twice another. Represent both numbers using the variable w.

   d. Represent the sum of three consecutive integers using the variable n to represent one of the integers.

   e. The area of a rectangular room is 130 square feet. Represent the length in terms of the width w.

3. Solve the following problems for the numbers described (Objective 3).

   a. Find three consecutive odd integers whose sum is 309.

   b. The sum of three numbers is eight. The second is five more than the first, and the third is seven more than the second. Find these numbers.

4. Solve the following problems and check your solutions (Objective 4).

   a. A tollway machine contained $2,100 worth of quarters, dimes, and nickels. A typical collection will contain 9 times as many dimes as quarters and 5 times as many nickels as quarters. If this collection is typical, how many quarters are in the collection?

   b. A pharmacist must prepare 40 milliliters of a 25% solution. She does not have this particular solution in stock, but she does have both a 40% solution and a 20% solution on hand. How much of each should be mixed to obtain the desired prescription?

5. Solve the following problems and check your solution (Objective 5).

   a. A $500 savings account earned $28.75 interest during a one year period. What was the yearly rate of interest?

b. A family inherited $30,000. They decided to invest some in a savings account (which they might use for some purchases) and the remainder would be invested in a one-year certificate of deposit. The savings account paid 6% annual interest and the certificate paid 8.25%. If their interest for the year was $2385, how much did they invest in each account?

6. Solve the following problems and check your solution (Objective 6).

a. If you observe a flash of lightning and then hear the thunder 10 seconds later, how far away from you did the lightning strike? The speed of sound is approximately 1100 feet per second. (Approximate the distance in miles using 5280 feet = 1 mile.)

b. Two military squads are 60 kilometers apart. They plan to rendezvous at some intermediate point. If one squad hikes 1 kilometer per hour faster than the other squad and they meet in 4 hours, what is the rate of each squad in kilometers per hour?

Mathematics is an important tool for solving problems in many disciplines. Employees in electronics, business, data processing, construction, science, and many other areas require special mathematical skills. The phrase "applied mathematics" means the use of mathematics to solve "real-life" problems. Frequently the goal of applications of mathematics to "real-life" problems is to translate a problem into an equation or inequality. Examples of problems which can be translated for a mathematical solution are given below.

How much carpeting will I need to carpet my den?

I have $1000 to invest. Should I put it in a savings account or should I invest in a certificate of deposit?

My car averages 25 miles per gallon. If the capacity of my gasoline tank is 18 gallons, can I travel 500 miles without filling up?

These problems are not given as nice equations; they are stated verbally. This chapter will give you further practice translating word problems into algebraic statements. We have introduced word problems throughout this text in an attempt to reduce the apprehension that many students have about this topic. Specifically, you have already translated verbal statements to algebraic statements. You also have solved many equations and inequalities. Thus all you need is an opportunity to polish these skills with additional practice.

The emphasis in this chapter is on using algebra to solve word problems. Students with insight and skill can sometimes solve even difficult word problems using only arithmetic. However, the student who learns to work effectively with variables can systematically solve a wide variety of problems.

As we pointed out in Chapter 1, one way to learn to work word problems is to study worked examples and then work similar problems on your own. This chapter is organized to help you do that. We have carefully selected certain types of problems for your practice. The long-range goal is for you to acquire the skills which will allow you to solve an even wider variety of problems.

11.1 Problems Involving One Unknown Quantity

Since arithmetic involves four basic operations, the task of solving a word problem often involves deciding which operation or operations should be used. A great deal of emphasis was placed upon this skill in Unit I. Example 1 will review some frequently-used key words and phrases.

Example 1	*Word or Phrase*	*Algebraic Expression*

a. Two plus x $2 + x$

b. Twice x $2x$

c. x squared x^2

d. Five less than the width, w $w - 5$

e. Quotient of y and four $\dfrac{y}{4}$

f. Two-thirds of n $\dfrac{2}{3}n$ Recall that "of" used in this context indicates multiplication.

g. x plus four is equal to ten. $x + 4 = 10$ Recall that "is equal to," "is the same as," "is," and "equals" are key words which are translated as "=."

h. w is the same as two. $w = 2$

i. y divided by two is sixteen. $\dfrac{y}{2} = 16$

j. Twice z equals z minus eight. $2z = z - 8$

In a word problem you are asked to find the value of some unknown quantity. A **variable** is used to represent this quantity. Even though x is frequently chosen as the variable, any letter can be used. In fact it is often more meaningful to use some other letter such as w for width, l for length, V for volume, A for area, etc.

Example 2	Choose a variable to represent the unknown quantity and then translate the phrase into an algebraic expression.

a. Three more than a certain number $n + 3$

b. A number decreased by nine $n - 9$

c. Hours traveled minus 2 hours for resting $h - 2$

d. The sum of a number and five $n + 5$

e. One-third of his salary $\dfrac{1}{3}s$

Note: The word "of" appears in both Examples d and e. However, it does not indicate multiplication in both examples. Thus you must always carefully consider the context in which the word "of" is used.

The skills reviewed in Examples 1 and 2 form a basis for working many word problems. In applying these skills, there is no one procedure that will work for every problem. However, the following general strategy may be helpful.

First, read the problem quickly in order to get an overview. Then reread it carefully and interpret each word and phrase for its exact mathematical meaning. Next, translate the problem into an equation and solve the equation using the techniques from the previous chapter. Finally, check to see that the answer meets the conditions specified in the problem.

When you follow this procedure, it will be helpful to read the problem with pencil in hand. You should circle key words or jot down key facts as you read. Do not just sit and gaze at the problem—look for words that indicate addition, subtraction, multiplication, division, or equality. Identifying the key pieces of information will start you on your way to finding the solution to the problem.

Example 3   Three more than twice a certain number equals thirty-three. Find this number.

Steps:

1. Choose a variable to represent the unknown quantity.

   Let n = the number.

2. Translate "three <u>more than</u> <u>twice</u> a certain number" as:

   $2n + 3$

3. Form the equality "three more than twice a certain number <u>equals</u> thirty-three."

   $2n + 3 = 33$

4. Solution:

   $2n + 3 = 33$

   $2n = 30$

   $n = 15$

5. *Check:*   The number is 15.

   Twice 15 = 30.

   3 more than 30 = 33.

   Since 3 more than 30 is 33, the conditions in the problem are met.

Example 4   One-half of a number plus three-fourths of the same number is fifteen. What is this number?

Steps:

1. Choose a variable to represent the unknown quantity.

   Let y = the number.

2. Translate "$\frac{1}{2}$ of a number <u>plus</u> $\frac{3}{4}$ of the the number" as:

   $\frac{1}{2}y + \frac{3}{4}y$

3. Form the equality "$\frac{1}{2}$ of a number plus $\frac{3}{4}$ of the number <u>is</u> 15."

   $\frac{1}{2}y + \frac{3}{4}y = 15$

4. Solution:

$$\frac{1}{2}y + \frac{3}{4}y = 15$$

$$4\left(\frac{1}{2}y + \frac{3}{4}y\right) = 4(15)$$

$$4\left(\frac{1}{2}y\right) + 4\left(\frac{3}{4}y\right) = 60$$

$$2y + 3y = 60$$

$$5y = 60$$

$$y = 12$$

5. *Check:* The number is 12.

One-half of the number $= \dfrac{1}{2} \cdot 12 = 6.$

Three-fourths of the number $= \dfrac{3}{4} \cdot 12 = 9.$

The sum, $6 + 9$, is 15, so 12 meets the conditions specified in the problem.

If your solution does not meet the conditions specified in the problem, you may have made an error either in the solution of the equation or in forming the equation. If this happens, check your solution in the equation before trying to form a different equation.

Example 5   Three times a number is the same as five more than the number. What is this number?

Steps:

1. Choose a variable to represent the unknown quantity.   Let n = the number.

2. Translate "3 <u>times</u> a number" as: $3n$
   Translate "5 <u>more than</u> the number" as: $n + 5$

3. Form the equality "3 times a number <u>is</u> <u>the same as</u> 5 more than the number." $3n = n + 5$

4. Solution: $3n = n + 5$

                                              $2n = 5$

                                              $n = \dfrac{5}{2}$

5. *Check:* The number is $\dfrac{5}{2}$.

Three times the number $= 3 \cdot \dfrac{5}{2} = \dfrac{15}{2}.$

Five more than the number $= \dfrac{5}{2} + 5 = \dfrac{5}{2} + \dfrac{10}{2} = \dfrac{15}{2}.$

These are the same, so $\dfrac{5}{2}$ meets the conditions of the problem.

Example 6   The value of a house today is \$4,000 more than three times its value twenty years ago. If the current value of the house is \$100,000, what was its value twenty years ago?

Steps:

1. Choose a variable to represent the unknown quantity.   Let v = value of house twenty years ago.

2. Translate "\$4,000 <u>more than</u> 3 <u>times</u> its value 20 years ago" as:

$3v + 4000$

3. Form the equality "The value of the house today <u>is</u> \$4000 more than three times its value twenty years ago."

$\$100000 = 3v + 4000$

4. Solution:

$$100000 = 3v + 4000$$
$$96000 = 3v$$
$$32000 = v$$

5. *Check:* The value 20 years ago was \$32000.

Three times the value 20 years ago = $3 \cdot 32000 = \$96000$.

\$4000 more than \$96000 = \$100000.

Since this is equal to the present value, \$32000 meets the conditions specified in the problem.

In Examples 3–6, the equation was found by translating a sentence taken directly from the problem. However, in some instances it may be necessary to reword the problem so that the relationship of equality is evident. This is illustrated in Example 7.

Example 7   What number is decreased by four when it is multiplied by three?

Steps:

1. Choose a variable to represent the unknown quantity.

Let n = the number.

2. Translate "number <u>decreased</u> by 4" as:
Translate "number <u>multiplied</u> by 3" as:

$n - 4$
$3n$

3. Form the equality:
By rewording the problem you obtain the following statement: "a number decreased by 4 <u>is the same as</u> the number multiplied by 3."

$n - 4 = 3n$

4. Solution:

$$n - 4 = 3n$$
$$-4 = 2n$$
$$-2 = n$$
$$n = -2$$

5. *Check:* The number is −2.

−2 decreased by $4 = -2 - 4 = -6$.

−2 multiplied by $3 = -6$.

These are the same, so the conditions of the problem are met.

11.1   Exercises

Translate each of the phrases or sentences in Exercises 1–60 into an algebraic expression. If necessary, first choose a variable to represent the unknown quantity.

1. The product of x and 23

2. The product of w and 14

3. The sum of seven z and four

4. Seven divided by t

5. s divided by seven

6. Seven divided into w

7. a divided into seven

8. Eleven less x

9. Eleven less than w

10. Six less than y

11. Six less s

12. Two k minus 1

13. Three y plus 11

14. Three times the quantity x plus 11

15. The sum of four x and 7

16. a plus eight equals seventeen.

17. w minus three is equal to eleven.

18. z divided by two is nine.

19. Twice x is the same as four more than y.

20. Five more than a number

21. Five less than a number

22. A number divided by five

23. Five times a number

24. The quotient of a number and five

25. Seven less than a number

26. Seven more than a number

27. The quotient of a number divided by seven

28. Seven times a number

29. A number decreased by eleven

30. A number increased by eleven

31. Nine more than a number is zero.

32. Five times the sum of a number and three equals thirty.

33. Three times a number is the same as four more than the number.

34. A number squared is thirty-six.

35. Twice a number

36. A number squared

37. Four times a number

38. Twice the quantity x plus 1

39. A number cubed

40. Subtract 13 from x.

41. Subtract S from 7.

42. Twice the quantity x minus 1

43. One-half of a number

44. One-third of a number

45. One-fourth of the quantity x plus 19

46. One-fifth of the quantity x minus 13

47. One-half of a number is seventeen.

48. Two-thirds of a number equals eight.

49. One-seventh of a number is the same as six less than the number.

50. Five more than one-fourth of a number is nine.

51. Double the quantity x plus 5

52. Subtract eight from w.

53. Subtract x from 8.

54. Subtract twelve x from three.

55. Five times the quantity two x plus four

56. Two times the quantity five x plus four

57. Four times the quantity two x plus five

58. Twice the quantity y plus four equals three times the quantity y minus four.

59. Twice the quantity w plus five equals three times the quantity w minus seven.

60. The square of the quantity b plus 5 is the same as b squared plus ten b plus twenty-five.

Use equations to solve the word problems in Exercises 61–85. First choose a variable to represent the unknown quantity and then form an equation from the stated problem.

61. Three times a number decreased by 5 is 7. What is the number?

62. What number equals 106 when it is doubled?

63. Seven more than 3 times a number is 31. What is the number?

64. If a number is increased by 12, its value is tripled. What is the number?

65. Find a number that is decreased by 11 when you form one-half of the number.

66. A motorist bought a new car and his MPG (miles per gallon) rating doubled. If his new car gets 32.8 miles per gallon, how many MPG did his old car get?

67. The current value of a business machine is $100 less than one-half of its cost. If it is worth $450 today, what was its cost?

68. Three times a number is four more than twice the number. Find this number.

69. A chemical calls for twice as much hydrogen as oxygen. How much oxygen will be required if 18 moles of hydrogen are used?

70. A child is 3 inches more than twice her height 7 years ago. If she is currently 51 inches tall, how tall was she 7 years ago?

71. If 3 times a number is decreased by 7, the result is the same as 5 times the number increased by 11. What is the number?

72. Five times the sum of a number and 12 is equal to 4 times the number increased by 6.

73. The current price per share in a computer company is $5 per share more than 3 times its 1960 value. If the current price is $155 per share, what was the 1960 value?

74. If 8 is added to $\frac{1}{7}$ of a number the result is 14. What is this number?

75. A typist can now type 7 words more than triple her beginning rate. If her current rate is 82 words per minute, what was her beginning rate?

76. The width of a picture frame is 3 inches more than twice its length. If the frame is 25 inches wide, how long is it?

77. The sum of $\frac{1}{3}$ of a number and $\frac{1}{5}$ of the number is 16. Find the number.

78. **One half of the sum of a number and 5 is 6 less than the number. What is the number?**

79. If you would double the number of strikeouts a baseball player has, this result would still have to be increased by 18 to equal the number of singles he has hit. If he has 62 singles, how many times has he struck out?

80. Two of the partners in a business decided to sell their interest. Together they received $77,000 for their portion of the business. If one owned one-half interest in the business and the other owned a one-fifth interest, how much is the entire business worth?

81. Six times a certain number is equal to four less than twice the same number. What is the number?

82. What number is three times as large as itself?

83. What number is doubled when it is increased by one?

84. Six less than six times a certain number is four less than four times the same number. Find this number.

85. What number is increased by thirty when it is divided by two?

11.2   Problems Involving More Than One Unknown Quantity

Relating one unknown quantity to another is an underlying concept in solving most word problems. This section contains word problems in which you are asked to find more than one quantity. However, before you attempt to solve problems of this type, you need practice in relating one unknown quantity to another. If two quantities are being sought, choose a variable to represent one quantity and then represent the second quantity in terms of this variable. There will always be some phrase or sentence in the problem that tells you how the two quantities are related. The first four examples will give you practice in representing one quantity in terms of another.

Example 1   Represent each of the following quantities in terms of the variable given.

    a. Let x = a number. How would you represent a second number which is 6 less than this number?

    Solution: x − 6 = second number

    b. Let *l* = length of a rectangle. How would you represent the width if it is 3 more than twice the length?

    Solution: 2*l* + 3 = width

c. Let x = first of four numbers. How would you represent three other numbers if the second is one-half of the first, the third is four less than the first, and the fourth is 6 more than the second?

Solution: $\frac{1}{2}x$ = second number

$x - 4$ = third number

$\frac{1}{2}x + 6$ = fourth number

d. Let N = John's age.
1. How would you represent John's age 3 years from now?
    Solution: $N + 3$

2. How would you represent Tom's age if he is twice as old as John?
    Solution: $2N$

3. How would you represent Harry's age if he is 6 years younger than John?
    Solution: $N - 6$

4. How would you represent Dick's age if he is 5 years older than Tom? (Recall that Tom's age was 2N.)
    Solution: $2N + 5$

5. How would you represent Angela's age if her age equals the combined ages of Harry and Tom?
    Solution: $(N - 6) + 2N = 3N - 6$

6. How would you represent Timothy's age if John is twice as old as Timothy?
    Solution: $\frac{1}{2}N$

e. A board is f feet long. What is its length in inches?
    Solution: $12f$ = length in inches

f. A turkey should cook for h hours. How many minutes should it cook?
    Solution: $60h$ = cooking time in minutes

Many problems in this section will involve the **integers,** the numbers . . . -4, -3, -2, -1, 0, 1, 2, 3, 4, ... Integers which differ by 1 are called **consecutive integers.** For example, 32 and 33 are two consecutive integers. The integers -2, -1, 0, and 1 are four consecutive integers. We will also consider problems involving the even integers, . . . -8, -6, -4, -2, 0, 2, 4, 6, . . . and the odd integers, . . . -5, -3, -1, 1, 3, 5, 7, ... The integers 6 and 8 are two consecutive even integers while 11, 13, and 15 are three consecutive odd integers. Note that consecutive even integers and consecutive odd integers both differ by 2.

Example 2   Represent each of the following quantities in terms of the given variable.

a. Let n = an integer. How would you represent the next consecutive integer?

Solution: Since consecutive integers differ by 1, we will have to add 1 to get the next integer. Therefore, the next consecutive integer = n + 1.

b. Let n = an integer. How would you represent the next 3 consecutive integers?

Solution: Let n = first integer

n + 1 = second integer

n + 2 = third integer

n + 3 = fourth integer

c. Let k = an even integer. How would you represent the next consecutive even integer?

Solution: Since consecutive even integers differ by 2, we will add 2 to get the next even integer. Therefore, the next even integer = k + 2.

d. Let n = an even integer. How would you represent the next two consecutive even integers?

Solution: Let n = given even integer

n + 2 = next even integer

n + 4 = third even integer

e. Let x = an odd integer. How would you represent the next three consecutive odd integers?

Solution: Let x = given odd integer

x + 2 = second odd integer

x + 4 = third odd integer

x + 6 = fourth odd integer

Many word problems give the sum, difference, product, or quotient of two numbers and then ask you to represent one of the numbers in terms of the other. For example, suppose the sum of two numbers is 22 and one of the numbers is 5. Then the other number is 17 since 22 − 5 = 17. We can check this by noting that 5 + 17 = 22. Further examples involving this concept are given below.

Example 3   **a.** The sum of two numbers is 22 and one number is given below. What is the other number?

Given Number	Calculation	Other Number
5	22 − 5	17
8	22 − 8	14
13	22 − 13	9
x	22 − x	22 − x
w	22 − w	22 − w
7z	22 − 7z	22 − 7z

b. The product of two numbers is 24 and one number is given below. What is the other number?

Given Number	Calculation	Other Number
6	$24 \div 6$	4
8	$24 \div 8$	3
$\dfrac{1}{2}$	$24 \div \dfrac{1}{2}$	48
a	$24 \div a$	$\dfrac{24}{a}$
y	$24 \div y$	$\dfrac{24}{y}$

Example 4   Represent each of the following quantities in terms of the given variables.

a. If the sum of two numbers is 30 and x represents one of the numbers, how would you represent the other number?

Solution: $30 - x$ *Check:* $x + (30 - x) = 30$

b. If the product of two numbers is 48 and y represents one of the numbers, how would you represent the other number?

Solution: $\dfrac{48}{y}$ *Check:* $\dfrac{y}{1} \cdot \dfrac{48}{y} = 48$

c. A carpenter plans to cut a 12-foot board into 2 pieces. If *l* represents the length of one piece, how could you represent the length of the other piece? (Ignore the loss due to the cut.)

Solution: If the board is 12 ft long, and a piece *l* ft long is cut off, then *l* feet have been subtracted from the 12-foot board. Thus we let *l* = length of first piece and $12 - l$ = length of second piece.

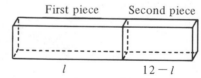

First piece   Second piece

l $12 - l$

Check: length of first piece + length of second piece = total length
$$l \quad + \quad (12 - l) \quad = \quad 12$$

d. The desired area of a rectangular bedroom is 168 square feet. If the room is built w feet wide, how long should it be?

Solution: The product of the length and width is the area. Thus the length is given by $\dfrac{168}{w}$.

Check: (width)(length) = area
$$(w)\left(\dfrac{168}{w}\right) = 168$$

e. A woman plans to invest $6000, part at 6% and the remainder at 8%. If x represents the amount invested at 6%, how would you represent the amount invested at 8%?

Solution: $6000 - x$ *Check:* $x + (6000 - x) = 6000$

Representing one quantity in terms of another is the first step in solving word problems where more than one quantity is being sought. The strategy presented in the previous section can be extended to include this type of problem.

STRATEGY FOR SOLVING WORD PROBLEMS

1. *Determine what you are asked to find.* Quickly read through the problem for an overview, and then carefully reread the problem until you understand what you are asked to find. Look for a question mark or the words "find" or "evaluate."

2. *Choose a variable and specify what it represents.* For example, let x = the number sought.

3. *Represent any other quantities asked for in terms of the variable chosen in step 2.* Look for some phrase or sentence in the problem which tells you how the quantities are related and then translate these phrases into mathematical statements.

4. *Set up the equation.* Reread the problem and identify key phrases which suggest mathematical operations and then translate these phrases into mathematical expressions. Then look for the key word (or words) which indicate equality.

5. *Solve the equation for the variable chosen in step 2.* Find any other quantities asked for, and then check your solution.

Example 5   One number is 2 more than 3 times another number. What are the numbers if their sum is 82?

Steps:

1. The question is: — What are the numbers?

2. Identify the unknown: — Let x = one of the numbers.

3. Represent the other quantity. Translate "2 <u>more than</u> 3 times another number" as: — $3x + 2$ = the other number

4. Set up the equation.
   Translate "their <u>sum</u>" as: — $x + (3x + 2)$
   Translate "sum <u>is</u> 82" as: — $x + (3x + 2) = 82$

5. Solution:
$$x + (3x + 2) = 82$$
$$4x + 2 = 82$$
$$4x = 80$$
$$x = 20$$
$$3x + 2 = 62$$

Check: The numbers are 20 and 62. Note that 2 more than 3 times $20 = 3(20) + 2 = 62$, and $20 + 62 = 82$. Thus the conditions of the problem have been met.

When solving a problem for more than one unknown quantity, make sure that you find all values asked for. For instance, in Example 5 you found that

$x = 20$. This is not the complete answer since you were asked to find two numbers. To find the other number, you substituted 20 for x in the expression $3x + 2$ and got 62. The answer to the problem is a pair of numbers, 20 and 62.

Example 6   Find two consecutive integers whose sum is 75.

Steps:

1. You are asked to: Find two consecutive integers.

2. Identify the unknown: Let x = smaller integer.

3. Represent the other quantity. Key $x + 1$ = larger integer
    phrase: "consecutive integers."

4. Set up the equation.
    Translate "whose <u>sum</u>" as: $x + (x + 1)$
    Translate "sum <u>is</u> 75" as: $x + (x + 1) = 75$

5. Solution: $x + (x + 1) = 75$

$$2x + 1 = 75$$
$$2x = 74$$
$$x = 37$$
$$x + 1 = 38$$

Check: The numbers are 37 and 38. Since they are consecutive and their sum is 75, these answers check.

Example 7   The sum of two consecutive odd integers is 108. Find these integers.

Steps:

1. You are asked to: Find two consecutive odd integers.

2. Identify the unknown: Let x = smaller integer.

3. Represent the other quantity. Key $x + 2$ = larger integer
    phrase: "consecutive odd integers"

4. Set up the equation.
    Translate "<u>sum</u> of" as: $x + (x + 2)$
    Translate "<u>sum</u> <u>is</u> 108" as: $x + (x + 2) = 108$

5. Solution: $x + (x + 2) = 108$

$$2x + 2 = 108$$
$$2x = 106$$
$$x = 53$$
$$x + 2 = 55$$

Check: Since 53 and 55 are consecutive odd integers and since $53 + 55 = 108$, the solution checks.

Example 8   Find three consecutive even integers such that the sum of the first two equals the third.

Steps:

1. You are asked to: Find 3 consecutive even integers.

2. Identify the unknown: Let x = smallest integer.

3. Represent the other quantities. Key $x + 2$ = second integer
   phrase: "consecutive even integers" $x + 4$ = third integer

4. Set up the equation.
   Translate "<u>sum</u> of first two" as: $x + (x + 2)$
   Translate "sum of first two <u>equals</u> the
   third" as: $x + (x + 2) = x + 4$

5. Solution: $x + (x + 2) = x + 4$

$$2x + 2 = x + 4$$
$$2x = x + 2$$
$$x = 2$$
$$x + 2 = 4$$
$$x + 4 = 6$$

Check: The numbers are 2, 4, and 6. Since 2, 4, and 6 are consecutive even integers and since the sum of the first two, $2 + 4$, is equal to the third, the solution checks.

Example 9   The difference of two whole numbers is 9 and their sum is 85. What are these numbers?

Steps:

1. The question is: What are the numbers?

2. Identify the unknown: Let x = one of the numbers.

3. Represent the other quantity. Key $9 + x$ = other number
   phrase: "their difference is 9"

4. Set up the equation.
   Translate "their <u>sum</u>" as: $x + (9 + x)$
   Translate "sum <u>is 85</u>" as: $x + (9 + x) = 85$

5. Solution: $x + (9 + x) = 85$

$$2x + 9 = 85$$
$$2x = 76$$
$$x = 38$$
$$9 + x = 47$$

Check: The numbers are 38 and 47. Their difference is 9 and their sum is 85, so this solution checks.

Frequently a word problem can be solved in more than one way. In Example 9, you are given two different relationships between the numbers you are trying to find. In the solution above, the "difference" relationship was used to represent the second number and the "sum" relationship was used to write the equation. Another approach would be to use the "sum" relationship to represent the second number and the "difference" relationship to write the equation.

Alternate Solution of Example 9

Steps:

1. The question is: What are the numbers?

2. Identify the unknown: Let x = the larger number.

3. Represent the other quantity. Key $85 - x$ = the smaller number.
   phrase: "their <u>sum</u> is 85"

4. Set up the equation.
   Translate "the <u>difference</u>" as: $x - (85 - x)$
   Translate "difference <u>is</u> 9" as: $x - (85 - x) = 9$

5. Solution: $x - (85 - x) = 9$

   $$x - 85 + x = 9$$
   $$2x = 94$$
   $$x = 47$$
   $$85 - x = 38$$

Compare this solution with the first one given. Notice that the equations are different, but the two numbers found are the same.

Example 10   A carpenter is outlining the form which will be used when some concrete is poured in the construction of a house. He has a 16-foot piece of string, all of which is needed to outline the rectangular form. If the length of the form exceeds the width by 3 feet, what are the dimensions of the form?

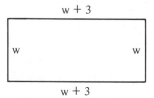

Steps:

1. The question is: What are the dimensions (i.e., length and width) of the form?

2. Identify the unknown: Let w = the width.

3. Represent the other quantity. Key $w + 3$ = the length
   phrase: "length exceeds the width by 3"

4. Set up the equation. From reading the
   problem, we can determine that the perimeter of the form is 16 feet.

Represent the perimeter as: $2w + 2(w + 3)$
The perimeter <u>is</u> 16. $2w + 2(w + 3) = 16$
5. Solution: $2w + 2(w + 3) = 16$

$$2w + 2w + 6 = 16$$
$$4w = 10$$
$$w = \frac{10}{4}$$
$$w = 2.5 \text{ ft}$$
$$w + 3 = 5.5 \text{ ft}$$

Check: The width is 2.5 and the length is 5.5 ft. Since $2(2.5) + 2(5.5) = 5 + 11 = 16$, the solution checks.

11.2   Exercises

In each of Exercises 1-50 represent all the quantities in terms of the same variable.

1. If the sum of two numbers is 56 and x represents one of the numbers, then represent the second number in terms of x.

2. One number is 15 less than 3 times a second number. If n represents the second number, how would you represent the first number?

3. The cost of one suit is $25 more than that of a second suit. If C represents the cost of the second suit, how would you represent the cost of the first suit?

4. The product of two numbers is 105 and the first number is represented by n. How could you represent the other number?

5. A woman made a $35 down payment on a new coat. If the cost is d dollars, what amount does she still owe?

6. Represent the new bank balance in terms of the old balance b after a withdrawal of $150 is made.

7. Represent the new bank balance in terms of the old balance b after a deposit of $879.15 is made.

8. If the sum of two numbers is S and one of the numbers is 8, what is the other number?

9. One number is y. When it is multiplied by a second number the product is 144. Represent the second number in terms of y.

10. Tom, Dick, and Harry were comparing their test scores. If Tom's score was 15 points higher than Harry's score, and Dick's score was 10 points lower than Harry's score, let x represent one of the scores and represent the others in terms of x. (*Hint:* Let x represent Harry's score.)

11. If the sum of two numbers is S and the first number is 11, represent the other number in terms of S.

12. A 16-ounce box of cereal costs c cents. What is the cost per ounce?

13. If gasoline costs c cents per gallon, how many gallons can you purchase for $10?

14. If C represents the current cost of gasoline per gallon, represent the cost if the price doubles.

15. The sum of two numbers is 85 and w represents one number. How would you represent the other?

16. The difference between two numbers is 6. If S represents the smaller number, how could you represent the larger number?

17. The difference between two numbers is 11. If S represents the smaller number, how could you represent the larger number?

18. The difference between two numbers is 6. If *l* represents the larger number, how could you represent the smaller number?

19. The difference between two numbers is 11. If *l* represents the larger number, how could you represent the smaller number?

20. A cook decided to make three pies instead of one. If S represents the amount of sugar required for one pie, how would you represent the amount needed for the three pies?

21. A carpenter wished to divide a board 24 feet long so that one piece would be 5 times as long as the second piece. If S represents the shorter piece, how could you represent the longer piece?

22. A boat was traveling at the rate of m miles per hour in still water. If the current was 5 miles per hour, how would you represent the actual rate of the boat when traveling with the current? How would you represent the actual rate when traveling against the current?

23. Two cars departed in opposite directions. After 3 hours they were 145 miles apart. If one car traveled m miles, what is the distance the other traveled?

24. John's age is y years. If he were three years older, he would be half his father's age. Represent his father's age in terms of y.

25. The first of four numbers is $\frac{3}{4}$ of the third number, the second is four times the first, and the fourth is $\frac{1}{2}$ of the third. Let n represent one of the numbers, and then represent the other numbers in terms of n.

26. If n represents a whole number, how should you represent each of the next two whole numbers?

27. If n represents an odd integer, how could you represent the next three odd integers?

28. The house numbers on one side of Main Street are all consecutive odd numbers. Assume that in one block no numbers are skipped and that the numbers increase as you go from south to north. If n represents the number of a house in the middle of the block, what is the house number of each of the neighboring houses?

29. Two friends started traveling toward each other from towns that are 240 miles apart. They met somewhere between the two towns. If d represents the distance one friend traveled, then represent the distance traveled by the other in terms of d.

30. If a race takes m minutes, how many seconds will this be?

31. If a carpet is f feet long, what is its length in yards?

32. Represent the exterior width of a concrete block basement in terms of its interior width w. Assume w is given in feet and a concrete block is 8 inches wide.

33. A piece of pipe has an inner diameter of d inches. The wall of the pipe is $\frac{1}{4}$ inch thick. What is the outer diameter?

34. A single roll of wallpaper is 192 inches long. How many rolls will be needed for a length of *l* inches?

35. Each revolution of a bicycle wheel moves the bicycle forward f feet. How many revolutions are necessary to travel one mile? (1 mile = 5280 feet)

36. A three-ounce piece of candy contains c calories. How many calories are there per ounce?

37. A student scored 80% on an examination. If there were p points on the exam, how many points did the student receive?

38. If m represents an even integer, how could you represent the next three consecutive even integers?

39. How could you represent the sum of two consecutive integers if m represents one of the integers?

40. A student earned 85 points on a test with the number of possible points represented by p. What fractional portion of the points did the student earn?

41. A man sold three-fifths of his shares in a computer company. If S represents the number of shares he originally owned, represent the number he sold and the number he still owns.

42. The area of a rectangle is 244 square feet. If the width is w, represent the length in terms of w.

43. Represent the sum of three consecutive odd integers in terms of the first integer n.

44. Golden rectangles are renowned in art and architecture for their eye-pleasing proportions. The length of a golden rectangle is approximately 1.618 times its width. Express the perimeter of a golden rectangle in terms of its width w.

45. The perimeter of a rectangle is 120 inches and the length is *l*. Represent the width in terms of *l*.

46. Two integers are consecutive multiples of 5. If K represents one number, how could you represent the other? (*Hint:* Multiples of 5, such as 5, 10, 15, etc., differ by 5.)

47. How could you represent 3 consecutive multiples of 7? Let n represent the first. (*Hint:* Multiples of 7, such as 7, 14, 21, etc., differ by 7.)

48. Represent the sum of two consecutive multiples of 9 in terms of the first integer m.

49. A trust fund invested $100,000. Part of the fund was invested in stocks and the remainder in bonds. If b represents the amount invested in bonds, how could you represent the amount invested in stocks?

50. Two hikers are 12 miles apart. They plan to meet at some intermediate point. If one hikes x miles, how far must the other hike?

Use equations to solve the word problems in Exercises 51-80. First represent all unknowns in terms of a single variable.

51. Find two consecutive integers whose sum is:
    a. 91   b. 75   c. −125   d. −41

52. Find two consecutive even integers whose sum is:
    a. 82   b. −206   c. −74   d. 190

53. Find two consecutive odd integers whose sum is:
    a. 112   b. −224   c. −84   d. 628

54. Find three consecutive integers whose sum is:
    a. 51   b. 255   c. −198   d. −1230

55. Find three consecutive even integers whose sum is:
a. 228   b. 342   c. −1332   d. −1968

56. Find three consecutive odd integers whose sum is:
a. 339   b. 129   c. −75   d. −273

57. Find four consecutive even integers whose sum is 4.

58. Find four consecutive odd integers whose sum is 104.

59. Find four consecutive odd integers whose sum is 8.

60. Find four consecutive even integers whose sum is 356.

61. One number is 5 more than 2 times the other. What are the numbers if their sum is 44?

62. One number is 7 more than 5 times the other. What are the numbers if their sum is 61?

63. One number is 6 less than 3 times the other. What are the numbers if their sum is 22?

64. One number is 9 less than 4 times the other. What are the numbers if their sum is 31?

65. The first of two numbers is 7 less than the second. Find these numbers if their sum is 33.

66. The difference of two numbers is 12 and their sum is 68. Find these numbers.

67. The first of two numbers is 25 more than the other and their sum is 11. Find these numbers.

68. One number is 19 more than another and their sum is 253. Find these numbers.

69. The sum of three numbers is 40. If the second is one-third the first and the third is 5 more than the second, what are the numbers?

70. When a number is decreased by eighteen, the result is fourteen. What is the number?

71. The sum of three numbers is 71. The second number is 8 less than the first, and the third is 3 more than twice the first. Find the numbers.

72. The sum of 2 more than three times a number and 5 less than four times the number is 133. What is this number?

73. A homeowner has 170 feet of fencing which he plans to use to form a rectangular playing area for his child. His wife suggests that the length of the play area should be 15 feet longer than the width. What are the dimensions of the play area that could be formed if all the fencing is used? (Assume that any gate will not affect the calculations.)

74. The first of four whole numbers is twice the third, the second is four less than the first, and the fourth is six more than the second. If their sum is 82, find the numbers.

75. A chemistry experiment requires two parts of chemical A for one part of chemical B and three parts of chemical C for one part of chemical B. If 36 milliliters are required for the mixture, how much of each chemical should be used?

76. The second of four whole numbers is four more than the first number. The third is twice the fourth, and the fourth is two less than the first. If their sum is 93, find the numbers.

77. There were $\frac{2}{3}$ as many B's as A's on a math test. If 30 students made either an A or a B, how many made A's? How many made B's?

78. The difference of two whole numbers is 12 and their sum is 90. What are the numbers?

79. The temperature Fahrenheit is nine-fifths of the temperature Centigrade plus 32°. Normal body temperature is 98.6° Fahrenheit. Calculate normal body temperature on the Centigrade scale.

80. Use Exercise 79 to convert the temperature of a warm 86° Fahrenheit reading to the corresponding Centigrade reading.

11.3   Mixture Problems

We will now investigate a variety of mixture problems. Although the problems have distinct features, in each problem one quantity can be considered as a mixture of other quantities. For example:

1. A collection of coins can be considered as a mixture of pennies, nickels, dimes, and quarters.

2. Homeowners seed their lawns with a mixture of grasses in order to obtain the best of each variety.

3. Stores sell candy or nuts as a mixture of different varieties since customers find this more appealing than a single variety.

4. Farmers mix chemicals when planting in order to fertilize and to treat for insects and weeds at the same time.

5. Pharmacists mix solutions since some drugs are too strong to be taken in pure form.

Even though these problems come from various disciplines, they can all be worked by the same method. This is one of the properties which makes mathematics such a powerful tool—the ability to use one technique to solve seemingly unrelated problems.

Study the worked examples in this section before you try the exercises. You may even want to cover the answer once you have read through an example and then see if you can work the example by yourself. If any particular exercise causes you difficulty, you may want to refer to a similar example in the text.

We will continue to use the basic strategy outlined in the previous section. That is, we will (1) read and reread, (2) identify the unknown, (3) translate key phrases, (4) form the equation, and (5) solve and check. For mixture problems, the equality is based upon the principle shown below.

> Amount in first + Amount in second = Amount in mixture

Example 1   A paperboy starts on his collections. At most stops he has to make change. From past experience he knows that he needs three times as many pennies as nickels. He starts with $4.00 worth of pennies and nickels as part of his change. If he takes three times as many pennies as nickels, how many does he take of each?

Steps:

1. The question is: "How many pennies and nickels does he take?"
2. We identify the number of nickels as: Let n = number of nickels.
3. We translate <u>3 times</u> as many pennies as nickels as: 3n = number of pennies.
4. The equality is: Value of pennies + Value of nickels = Total value of coins.

    Observe the value of   1 nickel = 1($0.05) = 0.05
                              2 nickels = 2($0.05) = 0.10
                              3 nickels = 3($0.05) = 0.15

    Thus the value of   n nickels = n($0.05) = 0.05n

    Likewise, the value of 3n pennies = 3n($0.01) = 0.03n

    Equation: Value of pennies + Value of nickels = Total value
                              $0.03n + 0.05n = 4.00$

5. Solution: $0.03n + 0.05n = 4.00$

                              $0.08n = 4.00$

                              $8n = 400$

                              $n = 50$

                              $3n = 150$

    Thus, he has 150 pennies and 50 nickels.

    Check: a)   150 pennies is 3 times the number of nickels, 50.

                  b) $1.50   (Value of pennies)
                       +2.50   (Value of nickels)
                       $4.00   (Total value)

Example 2   A homeowner purchased a mixture of two types of grass seed: fine perennial bluegrass and a rapid growing annual rye. The bluegrass cost $1.20 per pound and the rye was $0.95 per pound. If she purchased 20 pounds of the mixture for $23.00, how many pounds of each type of seed did she buy?

Steps:

1. The question is: "How many pounds each of rye and bluegrass?"

2. We identify the number of pounds of rye as: Let r = number lbs of rye.

3. Since sum of rye and bluegrass is 20 lbs: $20 - r$ = number lbs bluegrass.

4. Rye costs $0.95 per pound; thus cost of rye is: 0.95r.
   Bluegrass costs $1.20 per pound; thus cost of bluegrass is: $1.20(20 - r)$.

   Equality: Value of rye + Value of bluegrass = Total value

   $$0.95r + 1.20(20 - r) = 23.00$$

5. Solution: $0.95r + 1.20(20 - r) = 23.00$

   $$95r + 120(20 - r) = 2300 \text{ (multiply by 100)}$$
   $$95r + 2400 - 120r = 2300$$
   $$-25r + 2400 = 2300$$
   $$-25r = -100$$
   $$r = 4$$
   $$20 - r = 16$$

Thus she has 4 pounds of rye and 16 pounds of bluegrass.

Check: a)   4 pounds + 16 pounds = 20 total pounds.

b) $\underline{\begin{aligned} 4(\$0.95) &= \$\ 3.80 \quad \text{(Value of rye)} \\ +16(\$1.20) \quad &+19.20 \quad \text{(Value of bluegrass)} \end{aligned}}$
                 $\$23.00$ (Total value)

Example 3   A nurse must administer 5 ounces of a 12% solution of medicine. In stock are a 25% solution and a 5% solution of this medicine. How many ounces of each should she mix to obtain the 5 ounces of a 12% solution?

Steps:

1. The question is: "How many ounces each of 25% solution and 5% solution?"

2. We identify the number of ounces of 25% solution as: Let x = number of ounces of 25% solution.

3. The two solutions total 5 ounces. Thus if there are x ounces of one solution, the other amount is represented by $5 - x$. Thus, $5 - x$ = number of ounces of 5% solution.

4. Equality: Medicine in 25% solution + Medicine in 5% solution = Total medicine in 12% solution.

   Observe that the amount of medicine in the 25% solution is 25% of x; and that the amount of medicine in the 5% solution is 5% of $(5 - x)$; and that the amount of medicine in the 12% mixture is 12% of 5 ounces.

   $$0.25x = \text{amount of medicine in 25\% solution}$$
   $$0.05(5 - x) = \text{amount of medicine in 12\% solution}$$
   $$(0.12)(5) = \text{amount of medicine in 12\% mixture}$$

Equality: $(0.25)(x) + (0.05)(5 - x) = (0.12)(5)$

5. Solution: Sometimes it is convenient to summarize the information in a table as shown below.

	Ounces of solution	Percent	Ounces of medicine
5% solution	$5 - x$	5%	$(0.05)(5 - x)$
25% solution	x	25%	$(0.25)(x)$
Total mixture	5	12%	$(0.12)(5)$

$$0.25x + (0.05)(5 - x) = (0.12)(5)$$
$$25x + 5(5 - x) = (12)(5)$$
$$25x + 25 - 5x = 60$$
$$20x + 25 = 60$$
$$20x = 35$$
$$x = 1.75$$
$$\text{Thus, } 5 - x = 3.25$$

She should mix 1.75 ounces of the 25% solution with 3.25 ounces of the 5% solution.

Check: a) 1.75 ounces + 3.25 ounces = 5 total ounces

   b)   (0.05)(3.25) = 0.1625   (ounces of medicine in 5% solution)
       +(0.25)(1.75) = 0.4375   (ounces of medicine in 25% solution)
        (0.12)(5)   0.6000   (ounces of medicine in 12% mixture)

11.3   Exercises

Solve each of Exercises 1–20 and check your solution.

1. A supermarket usually needs three times as many five-dollar bills as ten-dollar bills in order to transact its weekly business. On one trip to the bank the manager planned to pick up $5,000 worth of fives and tens. How many of each bill should she get?

2. A boy collected seventy-five coins consisting of nickels and dimes from his allowance. If the coins are worth $5.95, how many of each has he collected?

3. A vending machine takes quarters and dimes but gives change in nickels and dimes. The selections chosen by the customers usually result in four times as many nickels as dimes being used in change. How many of each coin should be placed in the change hoppers by the vendor, if he is putting $45 in change in the machine?

4. At the end of her shift, a toll booth operator removed four times as many dimes as quarters from the coin box. If the total was $104, how many dimes and quarters were there?

5. Each week a shopping mall donates the coins from its fountain to a local charity. Past statistics indicate that the charity should expect twice as many quarters as half dollars, ten times as many dimes as quarters, six times as many nickels as dimes, and three times as many pennies as nickels. If one week's collection was $37.80, how many half dollars were collected?

6. Tickets to a school play sold for $2 for each adult and $0.75 for each child. A total of 350 people paid $450 to see the play. How many adult tickets were sold?

7. A softball team budgeted $120 for the purchase of new bats. Wooden bats were $11 each and aluminum bats were $23 each. They purchased 6 bats and spent $114. How many of each type of bat did they purchase?

8. A homeowner purchased a mixture of grass seed composed of fine perennial bluegrass and a rapid growing annual rye. The blue grass cost $1.25 per pound and the rye was $.95 per pound. If she purchased 25 pounds of the mixture for $29.15, how many pounds of each did she buy?

9. A newspaper carrier had 239 customers on his route. Some of the customers took the Sunday edition which was 25¢ extra, and the rest took only the weekday editions for $2.00 per week. On his weekly report the carrier recorded total sales of $488.25, but he failed to record how many customers took the

Sunday edition. The circulation manager completed the report. How many took the Sunday edition?

10. A man made two investments totaling $25,000. He made a profit of 5% on the first investment and 2% on the second investment. If his total profit was $650, how much was each investment?

11. A nurse must administer 4 ounces of a 15% solution of medicine. In stock are a 10% solution and a 50% solution of this medicine. How many ounces of each should she mix to obtain the 4 ounces of a 15% solution?

12. A druggist needs 20 milliliters of a 30% solution. To obtain this he mixes an 80% stock of the solution with a dilutant (0% solution). How many milliliters of the stock and how many milliliters of dilutant should be used?

13. A hospital needs 82.5 liters of a 20% disinfectant solution. How many liters of a 60% and a 15% solution could be mixed to obtain this 20% solution?

14. How many gallons each of a 20% nitric acid solution and a 45% nitric acid solution must be used to make 6 gallons of a 30% nitric acid solution?

15. A goldsmith has two alloys that are respectively 50% and 80% pure gold. How many grains of each must he use to make 300 grains of an alloy that is 72% pure gold?

16. A farmer is mixing a herbicide from a container of a 90% solution with pure water (0% solution). How much solution and how much water are needed to fill a 225-gallon tank with 2% solution?

17. A farmer is mixing an insecticide from a container of a 100% solution with pure water (0% solution). How much solution and how much water are needed to fill a 500-gallon tank with 1% solution?

18. A theater will seat 500 people. On a sell-out night the theater grossed $1945 from tickets. The adult tickets were $5 each and the children's tickets were $2 each. How many adults were there that night?

19. An insurance agent receives a 12% fee on each new policy payment and a 3% fee for each renewal payment. If the agent received $1350 on total payments of $26,250, how much were the payments for new policies?

20. One paint contains 18 parts of pigment per gallon while another contains 3 parts of pigment per gallon. How much of each should be mixed to fill a 30-gallon barrel with paint containing 5.5 parts of pigment per gallon?

11.4   Investment Problems

The formula presented in this section is one that almost everyone should know. Most of us borrow money at least once in our lives, particularly to buy either a car or a house. Those who don't have to borrow are probably using this formula when they invest their money.

The problems we are dealing with assume simple yearly interest. Charge card corporations and finance companies often use more complicated methods. This is one reason the government passed the Truth in Lending law which requires a statement of the simple interest rate.

Terminology:

P = Principal = the amount of money borrowed (or invested)
R = Rate   = the rate of interest as a percent (for us, simple interest)
T = Time   = the time in years the principal is borrowed (or invested)
I = Interest = the fee charged for the use of the principal

Formula: I = PRT

Interest equals principal times rate times the time (in years).

Example 1   Find the yearly interest on $800 which is invested at 7% interest.

Solution: I = PRT where P = $800, R = 7% = 0.07, and T = 1 year.

$$I = (800)(0.07)(1)$$
$$I = \$56.00$$

Example 2   A father wishes to establish a fund that will pay his child $45 per month interest. How much should he invest in an account that pays monthly interest at a rate of 9% per year?

Solution: $I = PRT$ where $I = \$45$, $R = 9\%/\text{year} = 0.09$, and $T = 1 \text{ month} = \dfrac{1}{12} \text{ year}$.

$$45 = P(0.09)\left(\frac{1}{12}\right)$$

$$12(45) = 0.09P$$

$$\frac{(12)(45)}{0.09} = P$$

$$P = \frac{(12)(\overset{5}{\cancel{45}})(100)}{\underset{1}{\cancel{9}}}$$

$$P = \$6000$$

Check: $P \cdot R \cdot T = (6000)(0.09)\left(\dfrac{1}{12}\right) = 45$

Example 3   If an investment of $12,000 earned $720.00 interest in one year, what was the rate of interest?

Solution: $I = PRT$ where $I = \$720.00$, $P = \$12,000$, and $T = 1 \text{ year}$.

$$720.00 = (12,000)(R)(1)$$

$$\frac{720}{12000} = R$$

$$0.06 = R$$

Thus $R = 6\%$.

Example 4   A city invested some of its early tax returns in certificates of deposit. The interest on one of the $50,000 certificates was $1,312.50 when the certificate matured. If the yearly rate of interest on the certificate was 10.5 percent, how many months had the city held this certificate?

Solution: $I = PRT$ where $I = \$1,312.50$, $P = \$50,000$, $R = 10.5\% = 0.1050$, and $T = \text{Time in years}$.

$$1312.50 = (50000)(.105)T$$

$$1312.50 = (50)(105)T$$

$$T = \frac{1312.50}{(50)(105)}$$

$$T = \frac{1}{4} \text{ year}$$

$$T = \frac{1}{4}(12 \text{ months})$$

$$T = 3 \text{ months}$$

Example 5   Members of a family had $20,000 to invest. They decided to invest a portion in a one-year certificate which paid 8.5% and the rest in their savings account which paid 6.25%. The advantage of the savings account was that this money was more accessible if they needed it. However, they did not withdraw any during the year and at the end of the year their interest from these investments was $1587.50. How much did they invest in each account?

Solution: We will use the method from the previous section on mixtures as well as the formula I = PRT.

Interest on Certificate + Interest on Savings Account = Total Interest

Principal invested in Certificate + Principal invested in Savings Account = Total Principal Invested

This information is concisely summarized by the following table.

	Principal	Rate	Time	Interest
Savings Account	P	0.0625	1 year	P(0.0625)(1)
Certificate	20000 − P	0.0850	1 year	(20000 − P)(0.0850)1
Total	$20,000			$1587.50

$$(20000 - P)(0.0850) + 0.0625P = 1587.50$$
$$(20,000 - P)(850) + 625P = 15,875,000 \text{ (multiplied by 10000)}$$
$$17,000,000 - 850P + 625P = 15,875,000$$
$$-225P = -1,125,000$$
$$P = \$5,000$$
$$20,000 - P = \$15,000$$

The family invested $15,000 in the certificate and $5,000 in the savings account.

Check: (15,000)(0.0850)(1) + (5,000)(0.0625) = $1275 + $312.50 = $1587.50

11.4   Exercises

Solve each of Exercises 1-20 and check your solution.

1. Determine the interest on $8560 at 7% for 6 months.

2. A customer owes $300 on his credit card bill for a full month. Assume the interest rate is equivalent to 18% simple yearly interest. How much interest is due at the end of the month?

3. A customer owes $540 on his credit card bill for a full month. Assume the interest rate is equivalent to 18% simple yearly interest. How much interest is due at the end of the month?

4. How much interest will be earned on a $6\frac{1}{4}$% savings account if interest is paid on $275 at the end of a six-month period?

5. A bank has an 8% account that will pay monthly interest by mail. How much would a widow have to invest in such an account in order to receive a monthly check of $150?

6. A family can afford to pay at most $75 each month in interest. If they borrow money at 10.5%, how much can they afford to borrow?

7. A family is planning a fund for their daughter's college education. They wish to establish an 8.5% account that will pay $4,000 interest at the end of each year. How much principal is needed for this purpose?

8. A bank has a 7.5% account that will pay monthly interest by mail. How much would a person have to invest in such an account in order to receive a monthly check of $200?

9. If a $400 savings account earned $29 interest at the end of the year, what was the yearly interest rate?

10. If an investment of $500 earned $40 interest in one year, what was the rate of interest?

11. If an investment of $700 earned $31.50 interest in six months, what was the yearly rate of interest?

12. If a $600 savings account earned $46.50 interest at the end of the year, what was the yearly interest rate?

13. A loan shark charged $15 interest for loaning $100 for one month. What was the yearly rate of interest on this loan?

14. An insurance company pays interest on a portion of the premiums it collects. At the end of a year a policy-holder received an interest check for $6.75. If the principal used to calculate this amount was $225.00, what rate of interest is the insurance company using?

15. For how many months must $750 be left in a 7% account if the interest to be paid at the end of this period is $35?

16. At the end of the first interest period a bank mailed an interest check for $7.50. If there was $1,000 in the 9% account, how many months were in the interest period?

17. At the end of the first interest period a bank mailed an interest check for $24.00. If there was $1,200 in this 8% account, how many months were in the interest period?

18. How many months must $3,500 be left in a 7.5% account if the interest to be paid at the end of this period is $196.88. (Give answer to nearest month.)

19. A family invested $4500, part of it at 8% and the rest at 10% annual interest. If the interest from these investments was $400 for the year, how much was invested at each rate?

20. A school district planned to invest $200,000 of its building funds in two kinds of 6-month certificates of deposit. If the rates are 8.5% and 9%, how much should be invested at each rate in order to earn $8,812.50 in 6 months?

11.5 Motion Problems

In all word problems it is important to read carefully and interpret the problem carefully. In a difficult word problem it is especially important to do this part by part rather than try to solve the problem in one giant step. For many problems it is helpful to form an over-all mental picture of the problem. Often a simple sketch will clarify the relationship between the quantities involved and may be the key to gaining an over-all grasp of the problem.

As a means of illustrating problems where simple sketches are helpful, this section will investigate problems dealing with distance, rate, and time. The basic formula is:

$$\text{Distance} = \text{Rate} \times \text{Time} \quad \text{or} \quad D = RT$$

In simple problems which use this formula, you will be given two of the quantities involved and asked to find the third one. It may be helpful to note that all of the following are equivalent:

$$D = RT \qquad T = \frac{D}{R} \qquad R = \frac{D}{T}$$

To use this formula correctly you must first express all the values in compatible units. If the rate is expressed in miles per hour, then time must be expressed in hours and distance in miles. If the time were given in minutes, it would be necessary to convert this time to the corresponding number of hours.

Example 1   If you travel at 50 mph (miles per hour) for three hours, how far will you travel?

Solution: $D = RT$

$D = (50)(3)$

$D = 150$ miles

Example 2   If a train travels at 80 mph, how long will it take the train to travel 500 miles?

Solution: $D = RT$

$$500 = 80T$$

$$T = \frac{500}{80}$$

$$T = 6\frac{1}{4} \text{ hr or 6 hours 15 minutes} \qquad \left(\frac{1}{4} \text{ of 60 minutes} = 15 \text{ minutes}\right)$$

Most distance, rate, and time problems involve two objects (or one object under two distinct circumstances) and thus two separate applications of the formula $D = RT$. To simplify the step-by-step analysis for these problems, we suggest that you use the following procedure.

> 1. Make a sketch which describes the over-all situation.
> 2. Make a table (as illustrated below) which concisely summarizes all the given information.
> 3. Decide what the equality is based upon.

In many of these problems the equality is based upon distance; however, the equality may also be based upon either rate or time. Read the problem over carefully and make a sketch before you decide what the equality is based upon.

Example 3   How long will it take a man running at 12 mph to overtake his wife who runs at 10 mph if he departs 15 minutes after she does?

Steps:

1. The question is: "What is the time for the man?"

2. We identify husband's time as: T = time traveled by husband.

3. And his wife left 15 minutes or $\frac{1}{4}$ hour earlier: $T + \frac{1}{4}$ = time traveled by wife.

4. The equality is: Distance traveled by husband = Distance traveled by wife.

Observe: $D = RT$, so husband's distance $D = (12)(T)$

$$\text{wife's distance } D = (10)\left(T + \frac{1}{4}\right).$$

The table below summarizes this information.

	Distance	Rate	Time
Husband	$12T$	12	T
Wife	$10\left(T + \frac{1}{4}\right)$	10	$T + \frac{1}{4}$

5. Solution:

$$12T = 10\left(T + \frac{1}{4}\right)$$

$$12T = 10T + 10\left(\frac{1}{4}\right)$$

$$2T = \frac{10}{4}$$

$$T = \frac{5}{4}$$

$$T = 1\frac{1}{4} \text{ hours} = 1 \text{ hour } 15 \text{ min}$$

It will take the man 1 hour and 15 minutes to overtake his wife.

Check: Man $D = 12\left(1\frac{1}{4}\right) = 12\left(\frac{5}{4}\right) = 15$ miles

Woman $D = 10\left(1\frac{1}{4} + \frac{1}{4}\right) = 10\left(\frac{6}{4}\right) = 10\left(\frac{3}{2}\right) = 15$ miles

They ran the same distance; thus he has overtaken her.

Example 4   Two friends leave two towns at the same time and start traveling toward each other in autos. One averages 40 mph and the other 50 mph. How far does each travel before meeting if the towns are 270 miles apart? (Note: they each travel the same length of time.)

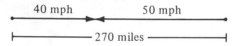

Steps:

Let   D = distance traveled by 40-mph car.
270 − D = distance traveled by 50-mph car.
(The total distance traveled by the two cars is 270 miles.)

Equality: They travel the same time. Time of 40-mph car = Time of 50-mph car.

Since D = RT, $T = \dfrac{D}{R}$.

Thus time of 40-mph car is: $T = \dfrac{D}{40}$

and time of 50-mph car is: $T = \dfrac{270 - D}{50}$

The equality based upon time is: $\dfrac{D}{40} = \dfrac{270 - D}{50}$

	Distance	Rate	Time
40-mph car	D	40	$\dfrac{D}{40}$
50-mph car	270 − D	50	$\dfrac{270 - D}{50}$

Solution:
$$\frac{D}{40} = \frac{270 - D}{50}$$

$$200\left(\frac{D}{40}\right) = 200\left(\frac{270 - D}{50}\right)$$

$$5D = 4(270 - D)$$

$$5D = 1080 - 4D$$

$$9D = 1080$$

$$D = \frac{1080}{9} = 120 \text{ miles}$$

$$270 - D = 150 \text{ miles}$$

The faster car travels 150 miles and the slower car travels 120 miles (a total of 270 miles).

Check: Faster $\dfrac{150}{50} = 3$ hr

Slower $\dfrac{120}{40} = 3$ hr

They would both travel 3 hours.

Example 5   Two truckers wave as they depart from a truck stop and travel in opposite directions on an interstate highway. One truck (exceeding the legal limit) travels 8 mph faster than the other. If the trucks are 224 miles apart after 2 hours, how fast is each traveling?

Steps:

Let R = rate of slower truck
R + 8 = rate of faster truck

Fill in the table.

	Distance	Rate	Time
Slower truck	?	R	2
Faster truck	?	R + 8	2

Note that each traveled 2 hours, so both the rate and the time have been filled in.

Using the formula D = RT, we can represent the distance each traveled. The distance of each is represented in the following table.

	Distance	Rate	Time
Slower truck	2R	R	2
Faster truck	2(R + 8)	R + 8	2

Equality: From the sketch we note that the total distance traveled by the two trucks is 224 miles.

Distance of slower truck + Distance of faster truck = 224

$$2R + 2(R + 8) = 224$$

Solution:
$$2R + 2(R + 8) = 224$$
$$2R + 2R + 16 = 224$$
$$4R + 16 = 224$$
$$4R = 208$$
$$R = 52 \text{ mph}$$
$$R + 8 = 60 \text{ mph}$$

The slower truck is traveling at 52 mph (which is within the 55 mph limit) while the faster truck is exceeding the limit by traveling at 60 mph.

Check: $2(52) + 2(60) = 104 + 120 = 224$ miles.

11.5 Exercises

Solve each of Exercises 1-12 and check your solution.

1. If a plane travels over the ground at 458 mph, how far will it go in 3 hours?

2. A bicyclist pedals at a rate of 32 mph. How far will she go in 15 minutes? (*Hint:* Convert minutes to hours.)

3. The speed of sound is about 1100 feet per second. If you observe a flash of lightning and then hear the thunder 5 seconds later, how far away from you did the lightning strike? Is this less than or more than a mile? (A mile is 5280 feet.)

4. A soldier saw the flash of an enemy artillery piece on a hillside in the distance. Seven seconds later he heard the muzzle blast. Approximate the distance to the artillery piece using the fact that sound travels at about 1100 feet per second.

5. A military jet traveled 2800 miles from New York to Los Angeles in 2 hours. How fast was the jet going?

6. A tornado traveled 75 miles in $2\frac{1}{2}$ hours. How fast was the tornado moving?

7. The winner of the Boston Marathon won the race with a time of 2 hours and a few minutes. If we approximate his distance as 26 miles and his time as 2 hours, what is an approximation of his rate for the marathon?

8. The winner of the 1923 Indianapolis 500 had a time of $5\frac{1}{2}$ hours. How fast did Tommy Milton, the winner, average for the 500 miles?

9. A scientific probe is sent to Saturn. When the probe is 67,068,000 miles from Earth a signal is sent from the probe to Earth. If the signal travels at 186,300 miles per second (roughly the speed of light), how many seconds elapse before the signal is received on Earth? How many minutes is this?

10. How many hours will it take a jet traveling at 580 mph to travel 725 miles from Atlanta to Chicago?

11. How much time will it take a race car averaging 125 mph to run a 400-mile race?

12. A pitcher throws a baseball at 132 ft per second. How many seconds will it take the ball to travel 66 feet?

For each of Exercises 13-20 make a sketch of the situation and fill in a table (as shown below) before attempting to answer the question.

	Distance	Rate	Time
First case			
Second case			

13. One canoe departs 20 minutes after another. The first canoe travels at 6 mph while the second travels at 10 mph. Determine the time the second must travel in order to overtake the first. (*Hint:* Convert minutes to hours.)

14. Two commuters both live 36 miles from work. However, the route one must take is so congested that it takes her 45 minutes longer to go one way. The second commuter has a better route and drives twice as fast as the first. How much time does it take each to drive one way?

15. Two military squads are 26 miles apart. They plan to rendezvous at some intermediate point. If one squad hikes at $\frac{2}{3}$ mph faster than the other and they meet in 3 hours, what is the rate of each squad?

16. Two cars start toward each other at the same time from two cities 360 miles apart. If one car averages 40 miles an hour and the other 50 miles an hour, how much time will elapse before they meet?

17. One train is 30 mph faster than another. If they depart a terminal at the same time and travel in opposite directions, they will be 800 miles apart after 4 hours. How fast is each traveling?

18. Two police cars depart from the same point traveling in opposite directions looking for a suspect in a stolen truck. One car averages 6 mph faster than the other car. Find the speed of each if they are 176 miles apart at the end of 55 minutes.

19. A man jogged to his friend's house at a rate of 6 mph. The trip took 20 minutes. How long will it take him to return home if he can maintain a rate of 8 mph?

20. A jet and a tanker are 1050 miles apart. The jet radios to set up a rendezvous for refueling in 1 hour and 15 minutes. If the tanker flies at 265 mph, how fast must the jet fly?

Important Terms Used in this Chapter	consecutive integers	odd integers
	even numbers	positive integers
	distance	principal
	integers	rate of interest
	interest	rate of travel
		time

Important Formulas Used in this Chapter	I = PRT	D = RT

REVIEW EXERCISES

Translate each of the verbal expressions in Exercises 1-10 into an algebraic expression.

1. The product of nine and w equals eleven.

2. Three-fourths of y

3. Twice the sum of 3x and 4

4. z minus five is equal to eleven.

5. The sum of x and y is three.

6. Five less than three n

7. The difference z minus seven

8. The quotient of nine divided into w

9. The quotient of nine divided by w

10. Four times y is the same as y cubed.

In each of Exercises 11-15 represent all the quantities in terms of the same variable.

11. The sum of x and another number is 80. Represent the other number in terms of x.

12. The area of a rectangular room is 168 square feet. Represent the length in terms of the width w.

13. The difference of two numbers is 15. If s represents the smaller number, how could you represent the larger number?

14. Represent the sum of two consecutive even integers in terms of the first integer n.

15. The perimeter of a rectangle is 200 meters. Represent the length in terms of the width w.

Solve Exercises 16-30 and check your solution.

16. Find two consecutive integers whose sum is 107.

17. Find two consecutive even integers whose sum is 74.

18. Find two consecutive odd integers whose sum is 180.

19. Find three consecutive odd integers whose sum is 183.

20. The sum of three numbers is 57. The second number is twice the first number, and the third number is five more than the first number. Find these numbers.

21. A grocer mixed 50 pounds of protein patties consisting of beef and soybean meal. The beef costs the grocer $1.10 per pound and the soybean meal costs $0.60 per pound. If the 50 pounds cost the grocer $50.00, how many pounds of soybean meal were mixed with the beef?

22. A repairman purchased 40 transistors to replenish his stock. Some of the transistors cost $1.15 each and the rest cost $1.85 each. If the total cost was $56.50, how many of each did he buy?

23. How many months are required for a $760 note to earn interest of $52.73 when the simple rate is 9.25%? (Round to the nearest month.)

24. During the year a church received the interest from a $50,000 investment. Part of the investment was in 8.75% government bonds and the rest was invested in a 7% bank account. If the total interest received was $4,235, how much was invested in each account?

25. If a bullet travels at 400 meters per second, how far will it travel in 5 seconds?

26. A plane took 3 hours to fly 1275 miles. What was the rate of the plane?

27. A NASA scientific package sent a message speeding to Earth at 186,300 miles per second. If this probe is 2,235,600 miles from Earth, how much time will be required before Earth receives the signal?

28. What number is twice its own value?

29. What number is decreased by 3 when it is doubled?

30. Two snowplows meet at the salt warehouse, load up with salt, and depart simultaneously in opposite directions. The combined route they cover is 108 miles. However, one must drive 4 mph faster in order to cover his portion of the route. After 2 hours both have completed their portions of the route and reverse their directions to clear the other lane. What is the rate of each?

12

APPLICATION OF EQUATIONS

PRETEST

1. Rewrite each of the following ratios as a fraction reduced to simplest form (Objective 1).

   a. $\frac{12}{15}$ b. 4:12

   c. Ratio of 6 to 3   d. 14:21

   e. 18:27   f. 10:15

   g. Ratio of 16 to 24   h. 8 of 12

   i. 25:15   j. 36:27

2. Determine the answer to each of the following questions by using ratios (Objective 2).

   a. Which of the following items is the best buy?
      1. a 3-ounce box of cream cheese for 37¢
      2. an 8-ounce box of cream cheese for 63¢
      3. a 10-ounce box of cream cheese for 79¢

   b. In 1931, three National League ball players finished the season in a near tie for the batting championship. Chick Hafey had 157 hits in 450 at bats, Billy Terry had 213 hits in 611 at bats, and Jim Bottomley had 133 hits in 382 at bats. Who won the championship and what was his batting average?

3. Determine the unknown quantity, that is, solve for x, in each of the following problems (Objective 3).

   a. $\frac{x}{12} = \frac{3}{6}$ b. 5:19 = 15:x

   c. 3:x = 6:14   d. $\frac{22}{11} = \frac{x}{9}$

   e. A photograph 5 inches wide and 7 inches long is enlarged so that the width is 8 inches. What is the length of the enlargement?

4. Solve the following problems (Objective 4).

   a. A boy who weighed 150 pounds at the start of the school year weighed 168 at the end of the year. What was the percent increase in his weight during the year?

   b. A factory which produces light bulbs finds that 3.5% of the bulbs are defective. How many defectives would be expected in a shipment of 48,000 bulbs?

   c. A financial advisor suggests that a couple invest no more than 225% of their yearly income in a home. How much must they earn in order to afford a $51,750 house?

Assume that all measurements in Exercises 5–7 are given in inches.

5. Determine the perimeter of each of the following figures (Objective 5).

a.

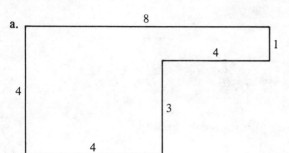

b.

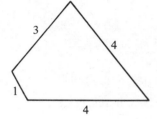

c.

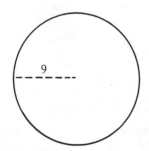

d.

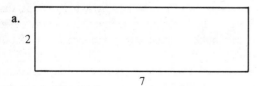

6. Determine the area of each of the following plane figures (Objective 6).

a.

b.

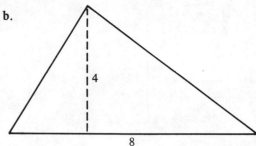

c.

d.

7. Determine the volume of each of the following solids (Objective 7).

a.

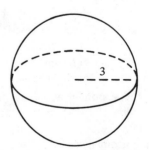

b.

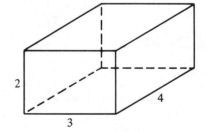

c.

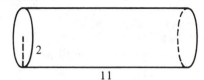
2
11

d.

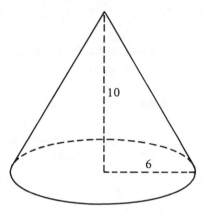
10
6

8. Use the Pythagorean formula to solve the following problems (Objective 8).

a. A construction company is burying a pipe diagonally across a rectangular lot. Calculate the length of pipe necessary to cross the lot shown below.

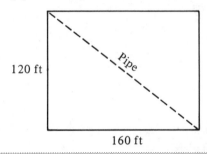
120 ft
Pipe
160 ft

b. A homeowner was concerned that his basement wall might be moving due to heavy rains in the spring. He made the measurements shown below. Is the wall perpendicular to the floor?

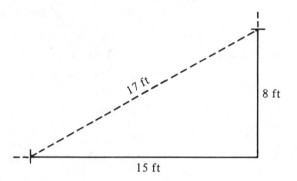
17 ft
8 ft
15 ft

9. Identify each of the following real numbers as either rational or irrational (Objective 9).

a. $\frac{1}{3}$ b. 0.6666666...

c. $\pi = 3.1415927...$ d. $0.53\overline{12}$

e. 1.01001000100001000001...

10. Convert each of the following decimals to a fraction in simplest form (Objective 10).

a. 0.375   b. $2.\overline{4}$ c. $5.\overline{27}$

This chapter will present a variety of topics: ratio and proportion, percent, area, volume, the Pythagorean theorem, and repeating decimals. The title "Applications of Equations" indicates that problems involving these diverse topics can all be worked by using equations. Although this chapter covers only a few applications, it illustrates the use of skills which you developed in Chapter 10.

One difference between this chapter and others in this text is that all but the first two sections of this chapter are independent of each other. You should not do the work on proportions (Section 12.2) without first covering ratios (Section 12.1). However, you may select any other section in the chapter without doing the remaining sections.

12.1 Ratio

There are various ways in which numbers may be compared. For example, when we say that nine is four more than five, we are comparing nine and five.

If we say that ten is five times two, we are comparing ten and two. Numbers can also be compared by dividing one number by the other. This particular type of comparison occurs frequently in a variety of situations and is called a ratio. That is, the **ratio of x to y** is the quotient $x \div y$. This ratio is frequently written as x:y and read as "the ratio of x to y." Since $x \div y = \dfrac{x}{y}$, ratios are also frequently written as fractions.

You are already familiar with the concept of ratio since batting averages, gas mileage, and cost per unit are obtained by dividing one quantity by another quantity. However, there are many other applications involving ratios. For example, you have heard or read expressions such as "9 out of 10 people surveyed used Brand X" or "Retail sales are up 2% over May of last year" or "Store A sells pencils 6 for a dollar." All of these expressions involve the comparison of a pair of numbers which can be thought of as a ratio. In other words 9 out of 10 is the ratio of 9 to 10, $2\% = \dfrac{2}{100}$ or the ratio of 2 to 100, and 6 pencils for a dollar is the ratio of 6 pencils to one dollar. The following examples illustrate some of the applications of ratios.

Example 1   In a class of 25 students there are 10 girls and 15 boys.

a. Find the ratio of girls to boys.

There are 10 girls to 15 boys or $\dfrac{10 \text{ girls}}{15 \text{ boys}}$.

Since $\dfrac{10}{15} = \dfrac{2}{3}$, we could also say the ratio of girls to boys is 2:3 or $\dfrac{2}{3}$. Thus, for every 2 girls in the class there are 3 boys.

b. Find the ratio of boys to girls.

There are 15 boys to 10 girls or $\dfrac{15 \text{ boys}}{10 \text{ girls}}$.

The ratio is $\dfrac{15}{10} = \dfrac{3}{2}$. This can also be indicated as 3:2 or $\dfrac{3}{2}$:1 or $1\dfrac{1}{2}$:1. Thus there are $1\dfrac{1}{2}$ boys for each girl.

c. Find the ratio of girls to total students.

This ratio is $\dfrac{10 \text{ girls}}{25 \text{ students}}$. The ratio $\dfrac{10}{25} = \dfrac{2}{5}$. Thus $\dfrac{2}{5}$ of the students are girls.

Note that in each case the ratio was expressed in fractional form and then the fraction was reduced to lowest terms by dividing the numerator and denominator by the same nonzero number.

Example 2   What is the ratio of 3 feet to 4 yards?

If the quantities being compared are not expressed in the same unit, they should be changed to the same unit if possible. If this is not possible, the units must be stated with the ratio. Since 3 feet = 1 yard, the ratio is 1 yard to 4 yards or 1:4. Also note that if we convert 4 yards to feet, the ratio is $\dfrac{3 \text{ feet}}{12 \text{ feet}}$ which equals $\dfrac{1}{4}$.

Example 3   In a regular deck of 52 playing cards, what is the ratio of diamonds to all cards in the deck?

Since 13 of the 52 cards are diamonds, this ratio is $\frac{13}{52}$ or $\frac{1}{4}$.

Example 4   A baseball player's batting average is defined as the ratio of the hits he makes to the number of times he is officially at bat. This ratio is usually expressed as a decimal to the nearest thousandth. If a player had 4 hits in 15 official times at bat, find his batting average.

The ratio is $\frac{4}{15}$.

$$15 \overline{)\begin{array}{c} 0.2666 \\ 4.0000 \end{array}}$$
$$\begin{array}{r} 3\,0 \\ \hline 1\,00 \\ 90 \\ \hline 100 \\ 90 \\ \hline 100 \\ 90 \\ \hline 10 \end{array}$$

$\frac{4}{15} \doteq 0.267$

Thus, his batting average is 0.267.

Example 5   A factory which produces nails finds that the ratio of defective nails produced to total nails produced is 1.5 to 100. What does this mean?

Note that this ratio does not give the actual number of nails involved since it is impossible to produce 1.5 nails. This ratio means that the average number of defective nails is 1.5 per 100. Thus, in 1000 nails, you would expect 15 defective nails. This follows since $\frac{1.5}{100} = \frac{1.5 \times 10}{100 \times 10} = \frac{15}{1000}$.

In some problems, such as those involving unit pricing, it is convenient to express a ratio as a fraction with a denominator of one. This can be accomplished by dividing the numerator and the denominator by the denominator. Consider the fraction $\frac{3}{2}$. Since $\frac{3}{2} = \frac{3 \div 2}{2 \div 2} = \frac{1.5}{1}$, the ratio of 3 to 2 could also be expressed as the ratio of 1.5 to 1. Instances where this interpretation is useful are given in Examples 6, 7, and 8.

Example 6   a.   A 17-ounce can of peas cost 37¢. Find the cost per ounce.

The ratio of cost to weight is 37¢ to 17 ounces. Since $\frac{37}{17} = \frac{37 \div 17}{17 \div 17} \doteq \frac{2.2}{1}$, the ratio is 2.2¢ to 1 ounce. Thus, the cost is 2.2¢ per ounce. Note that since the quantities are not in the same unit, the units must be stated in order to give meaning to the ratio.

b.   Find the unit price of a 32-ounce can of peas which cost 73¢.

The unit price is the price per ounce. Since $\frac{73}{32} = \frac{73 \div 32}{32 \div 32} \doteq \frac{2.3}{1}$, the price is 2.3¢ per ounce.

c.   In **a** and **b** above, which can is the better buy?

The 17-ounce can is the better buy because the unit price is less.

Example 7   A 48-ounce container of cranberry juice costs 98¢ while a 32-ounce container costs 78¢. Which is the better buy?

The unit cost for the larger container can be found by considering the ratio $\frac{98}{48}$.

Since $\frac{98}{48} = \frac{98 \div 48}{48 \div 48} \doteq \frac{2.0}{1}$, the cost is approximately 2¢ per ounce. For the smaller container, $\frac{78}{32} = \frac{78 \div 32}{32 \div 32} \doteq \frac{2.4}{1}$. Thus, the price per ounce is approximately 2.4¢. The 48-ounce container is a slightly better buy.

Example 8   A car traveled 260 miles on 13 gallons of gas. Find the ratio of miles to gallons and express this ratio with a denominator of 1.

The ratio of miles to gallons is $\frac{260 \text{ miles}}{13 \text{ gallons}}$. Since $\frac{260}{13} = \frac{260 \div 13}{13 \div 13} = \frac{20}{1}$, the ratio is 20 miles to 1 gallon of gas. Thus, the car has averaged 20 miles per gallon of gas.

Since ratios are fractions, they are sometimes expressed as decimals or percents. In fact, percents can be considered as a special kind of ratio—one in which the denominator is 100.

Example 9   Express 23% as a ratio.

$$23\% = 23 \cdot \frac{1}{100} = \frac{23}{100}$$

Thus, 23% is the ratio of 23 to 100.

Example 10   Express the ratio 1:5 as a percent.

$$1{:}5 = \frac{1}{5} = 0.20 = 20\%$$

Example 11   Two basketball teams had the following records:

	Won	Played
Team I	25	35
Team II	27	39

Which team had the better record?

Team I's ratio of games won to games played is $\frac{25}{35} = \frac{5}{7}$; that is, Team I won $\frac{5}{7}$ of their games. Team II's ratio of games won to games played is $\frac{27}{39} = \frac{9}{13}$, so they won $\frac{9}{13}$ of their games. Since $\frac{5}{7} \doteq 0.71 = 71\%$ and $\frac{9}{13} \doteq 0.69 = 69\%$, Team I won a larger percent of the games they played, so they have a better record.

12.1 Exercises

Rewrite each of the ratios in Exercises 1-6 as a fraction.

1. 2:5

2. 3:7

3. Ratio of 6 to 11

4. Ratio of 9 to 13

5. 12:17

6. 18:23

Rewrite each of the ratios in Exercises 7-12 using the colon notation.

7. Ratio of 5 to 7

8. Ratio of 17 to 25

9. $\frac{2}{3}$

10. $\frac{11}{19}$

11. Ratio of 4 to 9

12. Ratio of 27 to 88

Translate each of the ratios in Exercises 13-18 into a statement of the form "ratio of x to y."

13. $\frac{5}{8}$

14. $\frac{3}{16}$

15. $\frac{7}{3}$

16. 9:14

17. 6:13

18. 5:6

Rewrite each of the percents in Exercises 19-24 as a ratio.

19. 25%

20. 30%

21. 50%

22. 75%

23. 17%

24. 19%

Rewrite each of the ratios in Exercises 25-30 as a percent.

25. 1:4

26. 1:8

27. 3:4

28. 3:10

29. 2:1

30. 1:2

Rewrite Exercises 31-40 as ratios in fractional form with the fractions reduced to simplest form.

31. Ratio of 24 to 18

32. 21:24

33. 12:36

34. Ratio of 13 to 39

35. Ratio of 18 to 24

36. 14:21

37. 15:25

38. Ratio of 35 to 63

39. 25%

40. 50%

Rewrite Exercises 41-50 as ratios in fractional form with a denominator of 1.

41. 14:7

42. Ratio of 32 to 8

43. 19:38

44. $\frac{2}{3}:\frac{1}{3}$

45. 144:48

46. Ratio of 100 to 5

47. Ratio of 12 to 2

48. 3:6

49. Ratio of 4 to 8

50. 6:3

Find the unit cost of each of the items in Exercises 51-56.

51. 6 bottles of soda for $1.28; cost per bottle = _____

52. A case of 24 bottles of beer costs $6.24; cost per bottle = _____

53. A 20-pound bag of fertilizer costs $9.20; cost per pound = _____

54. A box of 100 envelopes costs $.89; cost per envelope = _____

55. An 84-ounce box of detergent costs $2.94; cost per ounce = _____

56. A 5-pound bag of flour costs $1.15; cost per pound = _____

57. In 1963, Stan Musial had 86 hits in 337 official at bats. Find his batting average for this year, his last in the major leagues.

58. Which of the following represents the lowest price per ounce for rice?

    12 ounces for 40¢
    14 ounces for 45¢
    1 pound 12 ounces for 85¢
    2 pounds for 99¢

59. In a regular deck of 52 cards, what is the ratio of jacks to all cards in the deck?

60. A car travels 342 miles on 13.2 gallons of gas. Find, to the nearest tenth of a mile, the number of miles the car averaged per gallon of gas.

61. Soft drinks can be purchased in the following quantities at the indicated prices.

    32 oz for 37¢ *1.14*
    48 oz for 59¢ *1.23*
    72 oz for 99¢ *1.38*

Which is the best buy?

62. Three ounces of lemon juice are used in making a 10-ounce glass of lemonade. Find the ratio of lemon juice to water in the lemonade.

63. If the rear gear of a bicycle has 80 teeth and the drive gear has 15 teeth, express the ratio of the teeth in the rear gear to those in the drive gear.

64. During the quarter a student has accumulated 252 points out of a possible 400 points. What percent of the points has the student earned?

65. A student's grade point average (GPA) is defined as the ratio of his total grade points to the total number of hours he has attempted. During one quarter a student taking 15 hours earned 57 grade points. Find his GPA.

12.2 Proportion

An equation which states that two ratios are equal is called a proportion if it is a true statement. Thus $\frac{2}{3} = \frac{4}{6}$ is a proportion. It can also be written as 2:3 = 4:6.

Equations such as $\frac{5}{6} = \frac{10}{x}$ and $\frac{2}{t} = \frac{5}{u}$, which contain variables, are also proportions if the variables are replaced with values which make the equation true.

Thus, a true equation of the form $\frac{a}{b} = \frac{c}{d}$ is a **proportion**. It can be written as a:b = c:d and is read "a is to b as c is to d." The four numbers a, b, c, and d are called the **terms of the proportion**—a is the first term, b is the second term, c is the third term, and d is the fourth term. The first and fourth terms are called the **extremes** and the second and third terms are called the **means**. The terms are said to be **in proportion** or **proportional**. A statement such as $\frac{1}{2} = \frac{3}{4}$ is not a proportion since the statement is false.

The topics of ratio and proportion are important because they are simple yet powerful tools which can be used to solve many problems. For example, the usefulness of these tools is shown by the fact that many mathematics items on civil service tests and other employment tests are merely problems involving ratio and proportion.

Example 1   In each of the following proportions, identify the means and the extremes.

	Extremes	*Means*
a. $\frac{4}{6} = \frac{2}{3}$	4, 3	6, 2
b. $\frac{a}{b} = \frac{c}{d}$	a, d	b, c
c. $2\frac{1}{2}:4 = 5:x$	$2\frac{1}{2}$, x	4, 5
d. 4:6 = 2:3	4, 3	6, 2

Since a proportion is a special type of equation, the principles which apply to equations also apply to proportions. Thus, if one term of the proportion is not known, you can use the properties of equations to solve for the unknown term in the proportion.

Example 2   Find the value of x which will make $\frac{2}{3} = \frac{x}{7}$ a true statement.

$$\frac{2}{3} = \frac{x}{7}$$

$$3 \cdot 7 \cdot \frac{2}{3} = 3 \cdot 7 \cdot \frac{x}{7}$$

$$7 \cdot 2 = 3x$$

$$14 = 3x$$

$$x = \frac{14}{3}$$

Note that 14 is the product of the extremes, and that 3x is the product of the means.

This example suggests an important property of proportions which is stated below.

PROPERTY OF PROPORTIONS

In a proportion, the product of the extremes is equal to the product of the means, that is, $\frac{a}{b} = \frac{c}{d}$ if and only if ad = bc.

Note that since ad = bc, it is also true that bc = ad. Thus, the property of proportions could also be stated as "the product of the means is equal to the product of the extremes." It follows from this property that if ad ≠ bc, then $\frac{a}{b} \neq \frac{c}{d}$. For example, $\frac{1}{2} \neq \frac{3}{4}$ since $1 \cdot 4 \neq 2 \cdot 3$.

Example 3 Use the property of proportions to determine if the following equations are proportions.

a. $\frac{4}{6} = \frac{2}{5}$

The product of the means (6 times 2) is not equal to the product of the extremes (4 times 5). Thus, $\frac{4}{6} \neq \frac{2}{5}$. Since $\frac{4}{6} = \frac{2}{5}$ is not a true statement, this is not a proportion.

b. 4:6 = 2:3

The product of the means (6 and 2) is 12. The product of the extremes (4 and 3) is 12. Since these products are equal, the statement is a proportion.

Example 4   Are the terms 2, 3, 4, and 9 in proportion?

For 2, 3, 4, and 9 to be in proportion the product of the extremes (2 and 9) must equal the product of the means (3 and 4). Since $2 \cdot 9 \neq 3 \cdot 4$, the terms are not in proportion.

The property of proportions gives us a simple way of determining when two fractions are equivalent. For example, $\frac{2}{3} \neq \frac{4}{5}$ since $\frac{2}{3} = \frac{10}{15}$ and $\frac{4}{5} = \frac{12}{15}$. Now we can determine that $\frac{2}{3} \neq \frac{4}{5}$ by observing that the product of the means $(3 \cdot 4 = 12)$ is not equal to the product of the extremes $(2 \cdot 5 = 10)$.

Example 5   Determine which of the following pairs of fractions are equivalent.

a. $\frac{2}{7}, \frac{3}{11}$

$\frac{2}{7} \neq \frac{3}{11}$ since $2 \cdot 11 \neq 3 \cdot 7$

b. $\frac{-50}{130}, \frac{35}{-91}$

$\frac{-50}{130} = \frac{35}{-91}$ since $(-50)(-91) = 4550$ and $(35)(130) = 4550$

When solving a proportion for an unknown term, you may use the multiplication property of equations. However, the work may be simplified by using the property of proportions.

Example 6   Form an equation, and solve for the value of x that will satisfy the following proportions.

a. $2\frac{1}{2}:x = 4:5$

$$\frac{2\frac{1}{2}}{x} = \frac{4}{5}$$

$$4 \cdot x = 2\frac{1}{2} \cdot 5$$

$$4x = \frac{5}{2} \cdot \frac{5}{1}$$

$$4x = \frac{25}{2}$$

$$\frac{1}{4} \cdot 4x = \frac{25}{2} \cdot \frac{1}{4}$$

$$x = \frac{25}{8}$$

b. 9 is to 10 as 27 is to x.

$$\frac{9}{10} = \frac{27}{x}$$

$$9x = 10 \cdot 27$$

$$9x = 270$$

$$\frac{9x}{9} = \frac{270}{9}$$

$$x = 30$$

c. $\dfrac{x}{4} = \dfrac{5}{10}$ d. $\dfrac{2x}{x+7} = \dfrac{1}{4}$

$$4 \cdot 5 = 10 \cdot x$$ $$4 \cdot 2x = 1(x + 7)$$
$$20 = 10x$$ $$8x = x + 7$$
$$2 = x$$ $$7x = 7$$
$$x = 2$$ $$x = 1$$

Proportions are particularly useful in solving many types of word problems. The first step is to decide on a ratio which uses the numbers which are to be compared. Consider the following examples.

Example 7   In a class the ratio of girls to boys is 3:2. If there are 15 girls, how many boys are there in the class?

Let b represent the number of boys. The ratio of girls to boys would be 15 to b. This should be the same as 3:2, the ratio of girls to boys; that is,

$$\frac{15 \text{ girls}}{b \text{ boys}} = \frac{3 \text{ girls}}{2 \text{ boys}}$$

$$\frac{15}{b} = \frac{3}{2}$$

$$3b = 30$$

$$b = 10$$

Thus, there are 10 boys in the class.

Example 8   A man earns \$532 in 2 months. At this rate, how much would he earn in 5 months?

Let x represent the amount he will earn in 5 months. Then his ratio of earnings to months can be represented as x dollars to 5 months or as 532 dollars to 2 months. Thus,

$$\frac{x \text{ dollars}}{5 \text{ months}} = \frac{532 \text{ dollars}}{2 \text{ months}}$$

$$\frac{x}{5} = \frac{532}{2}$$

$$5 \cdot 532 = 2x$$

$$2660 = 2x$$

$$1330 = x$$

$$x = 1330$$

Therefore, his earnings would be \$1330 for 5 months.

Example 9   Each inch on a map represents a distance of 115 miles. What distance corresponds to 6.5 inches on the map?

Let x be the number of miles represented by 6.5 inches. Then the ratio 1 inch : 115 miles must be the same as the ratio 6.5 inches : x miles. Thus,

$$\frac{1 \text{ inch}}{115 \text{ miles}} = \frac{6.5 \text{ inches}}{x \text{ miles}}$$

$$\frac{1}{115} = \frac{6.5}{x}$$

$$115 \cdot 6.5 = 1 \cdot x$$

$$747.5 = x$$

$$x = 747.5$$

Thus, 6.5 inches on the map represent 747.5 miles.

Example 10   In a basketball game the ratio of field goals made to field goals attempted was 5 to 8. If twenty-five field goals were made, how many were attempted?

Let x be the number of goals attempted, then the ratio of goals made to goals attempted can be represented by either 25:x or 5:8. Thus,

$$\frac{5}{8} = \frac{25}{x}$$

$$8 \cdot 25 = 5 \cdot x$$

$$200 = 5x$$

$$40 = x$$

$$x = 40$$

Therefore, 40 goals were attempted.

12.2 Exercises

In each of the proportions in Exercises 1–6, identify the means and the extremes.

1. $\frac{12}{13} = \frac{60}{65}$

2. $4:2.5 = 8:5$

3. $m:s = t:a$

4. z is to y as 2 is to z

5. 2x is to 13 as y is to 32

6. $\frac{100}{200} = \frac{5000}{x+4}$

Determine which of the equations in Exercises 7–12 are proportions.

7. $\frac{2}{5} = \frac{4}{25}$

8. $3:\frac{1}{2} = 4:\frac{2}{3}$

9. $\frac{3}{4} = \frac{15}{20}$

10. $\frac{2}{5} = \frac{8}{9}$

11. $\frac{1}{2} = \frac{50}{100}$

12. $4:6 = 6:9$

Determine which of the terms in Exercises 13–18 are in proportion.

13. $\frac{1}{2}$, 3, 4, and 2

14. 32, 8, 8, and 2

15. 2, 4, 6, and 8

16. 2, 4, 6, and 12

17. 4, 6, 8, and 12

18. 1, 2, 3, and 7

Find the value of the variable that will satisfy each of the proportions in Exercises 19–32.

19. $\frac{x}{20} = \frac{3}{5}$

20. $5:6 = 6:3c$

21. $\frac{5}{3} = \frac{x}{6}$

22. $\frac{12}{x} = \frac{9}{18}$

23. $a:3 = 4:6$

24. $5:11 = 15:y$

25. $3.3:\frac{1}{2}y = 4:5$

26. $\frac{2\frac{1}{2}}{f} = \frac{3}{7}$

27. $\dfrac{7}{0.5} = \dfrac{14}{a-4}$

28. $s:2 = 5:2$

29. $(t-1):5 = t:2$

30. $\dfrac{3}{4} = \dfrac{z+2}{z}$

31. $\dfrac{x}{2x-3} = \dfrac{3}{5}$

32. $3x:4 = (x-5):1$

Determine which of the pairs of fractions in Exercises 33–42 are equivalent.

33. $\dfrac{2}{3}, \dfrac{14}{21}$

34. $\dfrac{4}{8}, \dfrac{5}{10}$

35. $\dfrac{10}{6}, \dfrac{15}{9}$

36. $\dfrac{5}{15}, \dfrac{4}{12}$

37. $\dfrac{5}{6}, \dfrac{27}{29}$

38. $\dfrac{8}{33}, \dfrac{56}{77}$

39. $\dfrac{2x}{3y}, \dfrac{4x}{6y}$

40. $\dfrac{1148}{738}, \dfrac{98}{63}$

41. $\dfrac{14a}{33b}, \dfrac{35a}{55b}$

42. $\dfrac{121}{33}, \dfrac{22}{6}$

Use proportions to solve Exercises 43–65.

43. A recipe for 50 people uses 3 cups of sugar. How many cups of sugar are needed for 75 people? $\Big($*Hint:* Consider the proportion $\dfrac{3}{50} = \dfrac{c}{75}.\Big)$

44. The lengths of the corresponding sides of the two triangles shown below are proportional, that is, the ratio of 6 to 4 is the same as the ratio of x to 12. Find x.

6 in
4 in
x
12 in

45. The ratio of students to teachers at a school is 33:1. If the school has 85 teachers, how many students are there?

46. A dieter lost 2 pounds in three weeks. If he continues to lose at this rate, how many weeks (to the nearest week) will it take him to lose 25 pounds?

47. A car averages 34 miles per gallon of gas. How many gallons (to the nearest tenth) would be required for a trip of 800 miles?

48. Under a microscope an insect antenna which measures $1\frac{1}{2}$ millimeters long is magnified to 15 millimeters. Find the length to which an object 1 millimeter long would be magnified under this microscope.

49. If 6 centimeters on a map represents 300 kilometers, how much does 1 centimeter represent?

50. A recipe for six adults calls for $\dfrac{3}{4}$ tsp of salt. If this recipe is used to cook for four people, how much salt is required?

51. A factory which produces light bulbs finds that 2.5 out of 100 bulbs are defective. In a group of 10,000 bulbs, how many would be expected to be defective?

52. A bartender has a 32-oz liquor bottle from which he dispenses drinks containing $\dfrac{4}{5}$ oz of liquor. How many drinks can be made from this bottle? (*Hint:* Let d represent the number of drinks. Consider the proportion $\dfrac{4}{5}:1 = 32:d$.)

53. A family is making a trip on which they will travel 5200 miles. In 3 days they traveled 1950 miles. If they continue at this rate, how many more days will be needed to complete the trip?

54. The lengths of the corresponding sides of the two rectangles shown below are proportional. Find the length of the side labeled x.

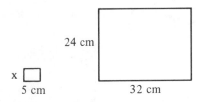

24 cm
x
5 cm
32 cm

55. A photograph is 7 centimeters wide and 10 centimeters long. What would be the width of the photograph if it is enlarged so that the length is 25 centimeters?

56. The slope of a roof is the ratio of the rise to the run. If the roof rises 4 feet over a run of 16 feet, what is the slope?

57. If 8 pieces of taffy sell for 10¢, how much would 20 pieces cost?

58. If a man earned $75 on a $937.50 investment, how much would he need to invest to earn $180 at this same rate of return?

59. The slope of a cable is the ratio of the rise to the run. A man wished to determine the height of a television tower. The distance (or run) from the bottom of the cable to the bottom of the tower was measured to be 40 feet. He then measured a rise of 8 feet and found a run of 6 feet. What is the height of the tower? (*Hint:* The slope of the cable is always the same, so the ratio of 8 to 6 should be the same as the ratio of the height of the tower to 40 feet.)

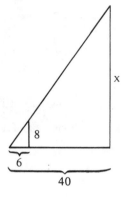

60. A farmer paid $984 to have 1200 bales of hay baled and stored in his barn. What would it cost to bale and store 2700 bales of hay?

61. If 118 bricks are used in constructing a 2-foot section of a wall, how many bricks will be needed for a similar wall 24 feet long?

62. If 11 cubic feet of sand is needed to make 32 cubic feet of concrete, how much sand is needed to make 224 cubic feet of concrete?

63. If 3 cubic feet of gravel is contained in 6 cubic feet of concrete, how much concrete can be made using 270 cubic feet of gravel?

64. Show that if $\frac{a}{b} = \frac{c}{d}$, then $ad = bc$. (*Hint:* Start by multiplying both sides of the proportion by bd.)

65. Show that if $ad = bc$, then $\frac{a}{b} = \frac{c}{d}$. (*Hint:* Start by dividing both sides of $ad = bc$ by b.)

12.3 Percent Problems

There are three common types of percent problems. In each of these problems there are three quantities involved: the base (B), the rate of percent (R), and the amount (A). The variable that you are solving for determines which type of problem it is. Fortunately all of these problems are easily worked using the following percentage formula.

PERCENTAGE FORMULA

$$A = RB$$

The **A**mount equals the **R**ate of percent times the **B**ase.

- -

Example: 15   equals 25% of 60.

          Amount   Rate   Base

The first task you must do to use this formula properly is to identify correctly the three quantities involved in the formula. Rewording the given problem to fit the standard format given in the box can help with this identification.

In Example 1 both the rate and the base are known and the amount is then calculated. Note the different wordings used in this type of percent problem and in the types illustrated in the next two examples.

Example 1 a. What   is 25% of 200?

Amount   Rate   Base

$$A = (0.25)(200)$$
$$A = 50$$

Answer: 50 is 25% of 200.

b. 10% of 80 is equal to what number?
    Rewording: **What number** is 10% of 80?

Amount   Rate   Base

$$A = (0.10)(80)$$
$$A = 8$$

Answer: 10% of 80 is equal to 8.

A second type of percent problem is the type which gives the amount and the base and asks for the rate.

Example 2 a. 50   is **what percent** of 200?

Amount   Rate   Base

$$50 = R(200)$$
$$\frac{50}{200} = R$$
$$\frac{1}{4} = R$$
$$R = 0.25 \text{ or } 25\%$$

Answer: 50 is 25% of 200.

b. What percent of 60 is 12?
    Rewording: **12**   is **what percent** of 60?

Amount   Rate   Base

$$12 = R(60)$$
$$\frac{12}{60} = R \qquad \frac{4}{30} \text{ or } \frac{1}{5}$$
$$\frac{1}{5} = R$$
$$R = 0.20 \text{ or } 20\%$$

Answer: 12 is 20% of 60.

A third type of percent problem gives the amount and the rate and asks for the base.

Example 3   a. 15   is 50% of **what number?**

Amount   Rate   Base

$$15 = 0.50B$$

$$\frac{15}{0.50} = B$$

$$B = 30$$

Answer: 15 is 50% of 30.

b. 40% of what quantity is 80?

Rewording: **80**   is **40%** of what **quantity?**

Amount   Rate   Base

$$80 = 0.40B$$

$$\frac{80}{0.40} = B$$

$$B = 200$$

Answer: 40% of 200 is 80.

Example 4   The state sales tax rate in Illinois is 5%. If you purchase a new car which costs $6,500, how much sales tax will you owe?

Rewording: What is 5% of $6,500?

$$A = 0.05(6500)$$

$$A = 325$$

Answer: The Illinois sales tax will be $325.

Example 5   In 1965, Willie Mays had 177 hits in 558 at bats. What was his batting average, that is, his rate of hits?

Rewording: 177 is what percent of 558?

$$177 = R(558)$$

$$\frac{177}{558} = R$$

$$R \doteq 0.317 \text{ or } 31.7\%$$

Answer: His batting average for 1965 was 0.317.

Example 6   A family had \$3,500 for the down payment on a house. If the minimum down payment must be 10% of the cost of the house, what is the highest price they could afford for a house?

Rewording: 3500 is 10% of what number?

$$3500 = 0.10B$$

$$\frac{3500}{0.10} = B$$

$$B = 35{,}000$$

Answer: They could afford a \$35,000 house.

Example 7   A \$40,000 house is rented for \$320 per month. What percent of the value of the house is the monthly rental?

Rewording: 320 is what percent of 40,000?

$$320 = R(40{,}000)$$

$$\frac{320}{40{,}000} = R$$

$$R = 0.008$$

$$R = 0.8\%$$

Answer: The monthly rental is 0.8% of the \$40,000 value.

Example 8   A baseball team won 54% of the first 150 games of its schedule. How many games had the team won?

Rewording: What number of games is 54% of 150 games?

$$A = 0.54(150)$$

$$A = 81$$

Answer: The team won 81 of the first 150 games.

Example 9   A car contains 12% more aluminum this year than the same model did last year. If the car contains 30 more pounds of aluminum this year, how much aluminum did last year's model contain?

Rewording: 30 is 12% of what number?

$$30 = 0.12B$$

$$\frac{30}{0.12} = B$$

$$B = 250$$

Answer: The old model contained 250 pounds of aluminum.

12.3 Exercises

Using the formula $A = RB$, fill in the missing values in the table below for Exercises 1–20.

	A *Amount*	R *Rate of Percent*	B *Base*
1.	———	6%	500
2.	———	7%	300
3.	40	8%	———
4.	450	9%	———
5.	60	———	1200
6.	70	———	500
7.	———	82%	340
8.	———	95%	490
9.	———	0.5%	600
10.	———	0.25%	800
11.	31	25%	———
12.	40	12.5%	———
13.	900	30%	———
14.	2.4	———	60
15.	0.16	———	32
16.	157.5	———	1,500
17.	2550	———	30,000
18.	57	9.5%	———
19.	19	2%	———
20.	1.55	2.5%	———

Solve the percent problems in Exercises 21–50 for the unknown value. If necessary, first restate the problem in the standard form.

21. What number is 20% of 35?

22. 15% of 40 is equal to what number?

23. 30 is what percent of 150?

24. What percent of 80 is 40?

25. What percent of 40 is 80?

26. 15 is 60% of what number?

27. 45% of 200 is equal to what number?

28. 16 is what percent of 4?

29. 50 is 20% of what number?

30. 30% of a number is 15. What is the number?

31. If a student answered 72 of 96 questions correctly, what percent did the student answer correctly?

32. In a college algebra class, 20% of the 30 students received C's. How many students received a C?

33. The sales tax rate on a consumer item is 8%. If you purchase $20 worth of this item, how much sales tax will you pay?

34. A patient received 4.05 grams of medicine. The drug was injected into the patient along with a dilutant. If the drug is 15% of the injected mixture, how much of the drug was injected?

35. If a family must pay 24% of its $32,000 income in federal income taxes, what is the income tax?

36. In a class survey, 24 of 40 students wanted their exam on Monday. What percent favored the Monday test date?

37. If the sales tax on a $950 minicomputer is $38, what is the sales tax rate?

38. In a bottle factory, the manufacturing process produces 2% defectives. If the factory produces 35,000 bottles per day, how many defectives are produced each day?

39. An inventor claims his product will boost a car's mileage by 20%. If this claim is true and a car receives a 5-mile per gallon boost, how many miles per gallon did the car average originally?

40. The price of a $40 blouse is increased by 4%. What is the price increase?

41. A union contract called for a 6% increase in the hourly rate of pay. If the current rate is $7.50, what is the increase in the hourly wage?

42. A homeowner's monthly expenses were $500. $80 of these expenses were for energy. What percent of the expenses were for energy?

43. In 1956, Mickey Mantle had 188 hits in 533 at bats. What was his batting average for the year?

44. A box of cereal increased in price from $1.05 to $1.26. What was the rate of increase for this $.21 price change?

45. An employee receiving a 5% pay increase said that he was grossing only $60 more per month. Determine from his comment his gross monthly salary before the raise.

46. Mrs. Dees pays her cook $500 per month. From this amount the cook pays $76.85 for social security and other taxes. What percent of the salary goes for taxes? (Round to the nearest percent.)

47. A share of stock on the New York exchange was selling for $32.50 per share. If the price dropped $7.50 per share, what percent of the original price was the decrease? (Round to the nearest percent.)

48. The population of Oklahoma in 1960 was 2,328,711. From 1960 to 1970 there was about a 9.9% increase in the population. What was the population in 1970? (Round to the nearest person.)

49. Only 55% of the season tickets for the home-town baseball team have been sold. If there are 540 tickets still available, how many have been sold? (*Hint:* What percent of the tickets are still available?)

50. A sale on watches advertised $\frac{1}{4}$ off. If a watch originally sold for $69.00, what will be the new price? (*Hint:* First solve for the amount the price is marked down.)

12.4 Geometry—Length and Area

Correct measurements of length and area play an important part in the building of homes, cars, and equipment. Before we purchase many items we must first decide the size of the item we want or how much of the product we desire. For example, your TV may be a 19-inch set (the length of the diagonal of the picture tube) and it may rest upon 48 square yards of carpeting.

Many distinct units are used as standards to measure length and width. For example, to measure length we use inches, feet, yards, rods, and miles. Our country, since the Metric Conversion Act of 1975, is slowly replacing all these various units with one metric unit, the meter. However, the actual process of measurement is the same with all units. A selected unit, such as the inch, is repetitively marked off on a ruler or tape and then this instrument is used to count the number of units contained in an object.

In the following examples many of the figures shown are polygons. **Polygons** are closed figures whose sides are straight lines. The **perimeter** of a polygon is the distance around the figure and may be calculated by finding the sum of the lengths of the sides. The only figure shown below which is not a polygon is a circle. The distance around a circle is called the **circumference**.

Example 1   Find the perimeter of each of the following figures.

a.

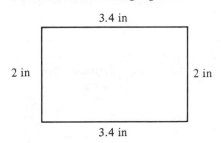

3.4 in + 2 in + 3.4 in + 2 in = 10.8 in

b.

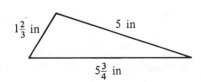

$$1\frac{2}{3} \text{ in} + 5\frac{3}{4} \text{ in} + 5 \text{ in} = 1\frac{8}{12} \text{ in} + 5\frac{9}{12} \text{ in} + 5 \text{ in} = 11\frac{17}{12} \text{ in} = 12\frac{5}{12} \text{ in}$$

c.

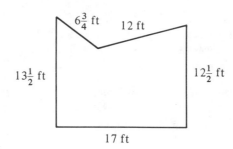

$$13\frac{1}{2} \text{ ft} + 6\frac{3}{4} \text{ ft} + 12 \text{ ft} + 12\frac{1}{2} \text{ ft} + 17 \text{ ft} = 61\frac{3}{4} \text{ ft}$$

C = πd is the formula which is used to find the circumference of a circle. C represents the circumference, π is an irrational number which is approximately 3.14, and d represents the diameter.

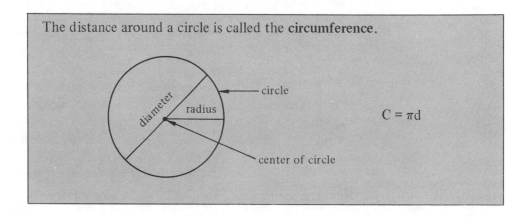

The distance around a circle is called the **circumference**.

C = πd

Example 2   a. Find the circumference of a circle which has a diameter of 10 units. (Use π ≐ 3.14.)

   C = πd

   ≐ (3.14)(10 units)

   ≐ 31.4 units

Thus the circumference is approximately 31.4 units.

b. Find the circumference of a circle of radius 3 meters. (Since the diameter is twice as long as the radius, the diameter is 6 meters.)

   C = πd

   C ≐ (3.14)(6 meters)

   C ≐ 18.84 meters

Next to length, the type of measurement we use most frequently is area. When we calculate the area of a region, we determine the amount of space occupied by this plane figure. Many items are measured by the square unit (a measure of area): carpeting by the square yard, construction costs by the square foot, lot size by the square foot, and farms by the acre. To measure the area of a geometric region you must find the number of unit squares in the region.

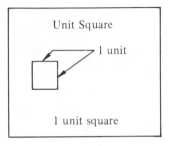

Example 3 The following regions are marked off into unit squares. Count the number of unit squares contained in each region.

a.

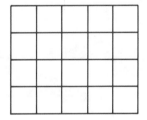

Three unit squares are contained in the figure, so the area of the figure is 3 square units.

b.

The area is 20 square units.

Example 4 The following region is marked off into unit squares. Find the area of each of the shaded regions.

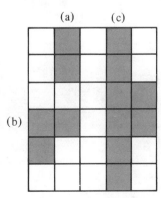

a. 2 square units
b. 3 square units
c. 8 square units

Example 5   Estimate the area of each of the shaded figures.

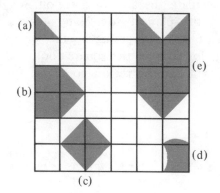

a. $\frac{1}{2}$ square unit

b. 3 square units

c. 2 square units

d. 1 square unit

e. 6 square units

Example 6   Find the length, width, and area of each of the following rectangles. Assume the sides of each square are 1 foot in length.

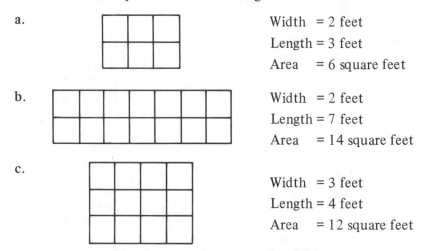

a.   Width = 2 feet
      Length = 3 feet
      Area   = 6 square feet

b.   Width = 2 feet
      Length = 7 feet
      Area   = 14 square feet

c.   Width = 3 feet
      Length = 4 feet
      Area   = 12 square feet

Obviously, it is inconvenient to divide a rectangle into squares and count these squares. A much faster method is to compute the area based upon the width and length. In Example 6, note that in part **a**, (3)(2) = 6; in part **b**, (7)(2) = 14; and in part **c**, (4)(3) = 12. In general, the area of a rectangle is found by multiplying the length times the width—provided the dimensions are in the same units.

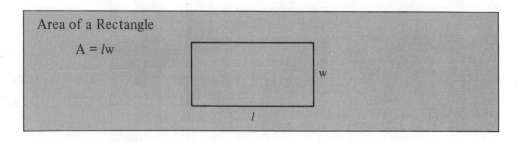

Area of a Rectangle

$A = l\text{w}$

Example 7   Determine the area of each of the following shaded triangular regions. Assume the sides of each square are 1 foot in length.

a.

Width = 2 ft
Length = 4 ft
Area   = 4 sq ft

b.

Width = 2 ft
Length = 7 ft
Area   = 7 sq ft

c.

Width = 3 ft
Length = 4 ft
Area   = 6 sq ft

A general formula based upon these observations is given below. Also included is the formula for the area of a circle.

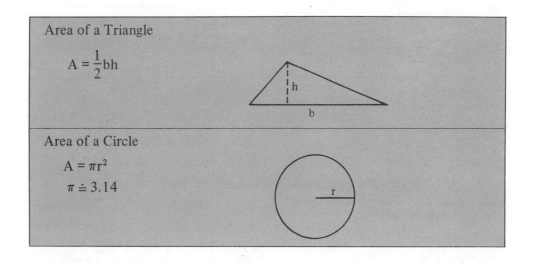

Area of a Triangle

$$A = \frac{1}{2}bh$$

Area of a Circle

$$A = \pi r^2$$

$$\pi \doteq 3.14$$

Example 8   Calculate the area of each of the following figures. Assume all units are given in inches.

a.

$A = lw$
$A = (11 \text{ inches})(2 \text{ inches})$
$A = 22 \text{ inches}^2$

The area of the rectangle is 22 square inches.

b.

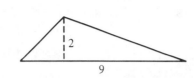

$A = \frac{1}{2}bh$

$A = \frac{1}{2}(9 \text{ inches})(2 \text{ inches})$

$A = 9 \text{ inches}^2$

The area of the triangle is 9 square inches.

c.

$A = \pi r^2$

$A = \pi(4 \text{ inches})^2$

$A = 16\pi \text{ inches}^2$

$A \doteq 16(3.14) \text{ inches}^2$

$A \doteq 50.24 \text{ inches}^2$

The area of the circle is approximately 50 square inches.

A common method used to calculate the area of some unfamiliar geometric figures is first to partition the region into rectangles, triangles, and circles. The area of each of these portions is then calculated using the appropriate formula and the results are summed to find the total area of the original figure.

Example 9   Calculate the area enclosed by the following figure.

This figure is a rectangular region with semicircles at each end. Thus the total area is equal to the area of a rectangle plus the area of 1 circle (2 semicircles).

Total Area = Area of Rectangle + Area of Circle*

$A = (2)(5) + (\pi)(1)^2$

$A = 10 + \pi$

$A \doteq 10 + 3.14$

$A \doteq 13.14$

The total area is approximately 13.14 square inches.

Note: If the diameter of the circle is 2, its radius is 1.

12.4 Exercises

Calculate the perimeter of each of the regions shown in Exercises 1-6. Assume that all lengths are in meters. Approximate π by 3.14.

1.

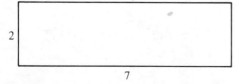

2.

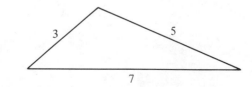

3.

4.

5.

6.

Calculate the area of each of the shaded areas in Exercises 7–12. Assume that each unit square is one square inch.

7.

8.

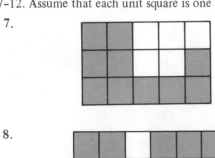
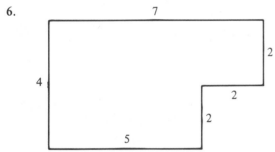

9.

10.

11.

12.

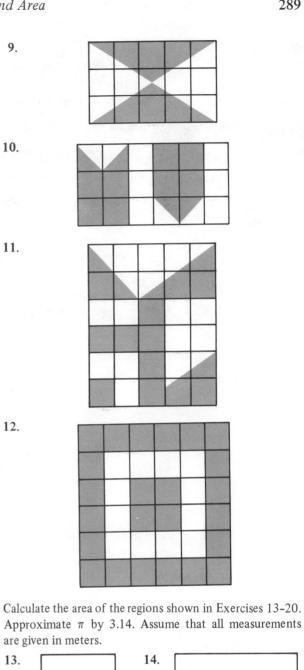

Calculate the area of the regions shown in Exercises 13–20. Approximate π by 3.14. Assume that all measurements are given in meters.

13.

14.

15.

16.

17.

18.

19.

20.

12.5 Geometry—Volume

The **volume** of a closed three-dimensional figure is a measure of its capacity. Volumes are given in cubic units since the capacity is found by determining the number of unit cubes contained in the figure.

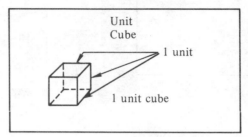

Many common items are described by giving their volume. Car manufacturers are fond of novel ways of illustrating their cargo space in TV commercials. We also describe engine sizes by their displacement and buy freezers and refrigerators based upon their capacity.

Example 1   Find the volume of each of the following figures. Assume that the sides of each cube are 1 inch in length.

a.

Width   = 1 inch
Length = 3 inches
Height = 1 inch
Volume = 3 cubic inches

b.

Width   = 2 inches
Length = 3 inches
Height = 1 inch
Volume = 6 inches³

c.

Width   = 3 inches
Length = 4 inches
Height = 2 inches
Volume = 24 inches³

The general formulas for calculating the volumes of parallelipipeds (boxes), cylinders, cones, and spheres are given below.

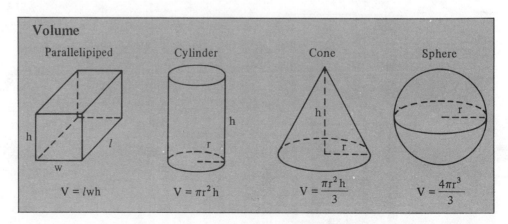

Volume

Parallelipiped $V = lwh$

Cylinder $V = \pi r^2 h$

Cone $V = \dfrac{\pi r^2 h}{3}$

Sphere $V = \dfrac{4\pi r^3}{3}$

Example 2   Calculate the volume of each of the following figures. Assume that all units are given in meters.

a.

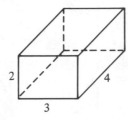

$V = l \cdot w \cdot h$

$V = (4 \text{ meters})(3 \text{ meters})(2 \text{ meters})$

$V = 24 \text{ meters}^3$

The volume is 24 cubic meters.

b.

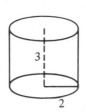

$V = \pi r^2 h$

$V = \pi (2 \text{ meters})^2 (3 \text{ meters})$

$V = 12\pi \text{ meters}^3$

$V \doteq 37.68 \text{ meters}^3$

The volume is approximately 37.68 cubic meters.

c.

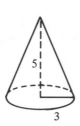

$V = \dfrac{\pi r^2 h}{3}$

$V = \dfrac{\pi (3 \text{ meters})^2 (5 \text{ meters})}{3}$

$V = 15\pi \text{ meters}^3$

$V \doteq 47.1 \text{ meters}^3$

The volume is approximately 47.1 cubic meters.

d.

$V = \dfrac{4\pi r^3}{3}$

$V = \dfrac{4\pi (6 \text{ meters})^3}{3}$

$V = 288\pi \text{ meters}^3$

$V \doteq 904.32 \text{ meters}^3$

The volume is approximately 904.32 cubic meters.

12.5   Exercises

Determine the volume of each of the solids in Exercises 1–6, assuming that each unit cube is one cubic foot.

1.

2.

3.

4.

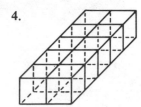

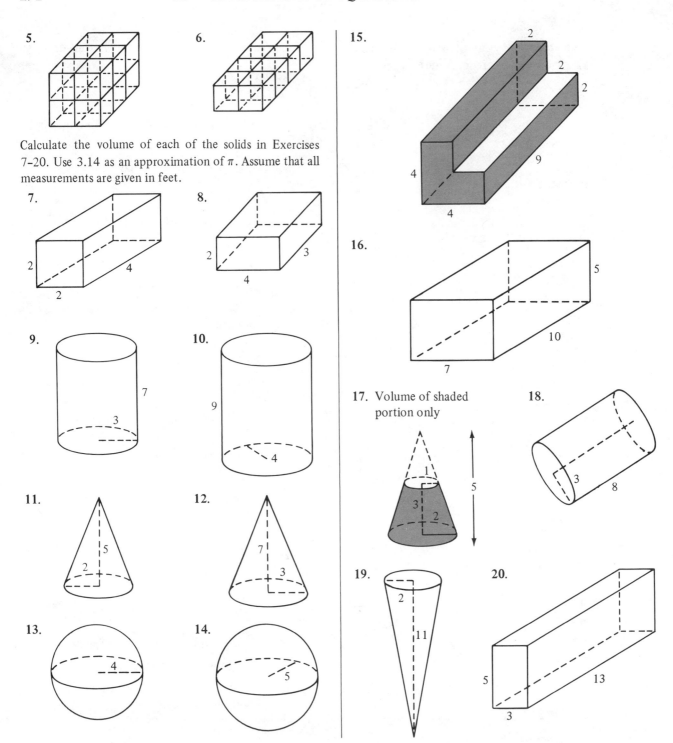

5.

6.

Calculate the volume of each of the solids in Exercises 7–20. Use 3.14 as an approximation of π. Assume that all measurements are given in feet.

7.

8.

9.

10.

11.

12.

13.

14.

15.

16.

17. Volume of shaded portion only

18.

19.

20.

12.6   The Pythagorean Theorem

One of the most well-known formulas in the history of mathematics is the formula credited to Pythagoras. The reason this formula is so familiar is that it is used daily by mathematicians, engineers, surveyors, machinists, carpenters, and others solving everyday problems. The reason this theorem is so important is that it states an important relationship between the sides of a right triangle.

A **right triangle** is a triangle containing a 90° angle; that is, two sides of the triangle are **perpendicular**.

Right triangle:

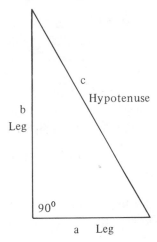

Sides a and b are perpendicular.

The sides a and b that are perpendicular are called **legs** of the triangle and the longest side c is referred to as the **hypotenuse**. The Pythagorean theorem states that in a right triangle the sum of the squares of the legs is equal to the square of the hypotenuse.* In fact, it is also true that if the sum of the squares of the legs equals the square of the hypotenuse, then the triangle is a right triangle. These two results are easier to remember by stating the formula below.

Pythagorean Formula

A triangle is a right triangle if and only if
$$a^2 + b^2 = c^2$$

Right triangle

Caution: This formula works only for right triangles.

Example 1   Obtain a ruler and a piece of paper. If the sides of the paper are perpendicular, make a mark on one side 3 inches from one corner and another mark 4 inches from the corner on the other side. Now measure the length between the marks. By the Pythagorean formula this hypotenuse must be 5 inches long.

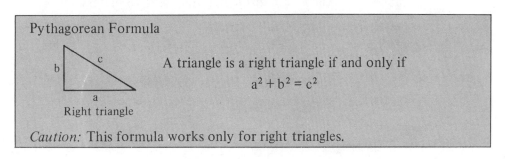

$$a^2 + b^2 = c^2$$
$$3^2 + 4^2 = 5^2$$
$$9 + 16 = 25$$
$$25 = 25$$

*There are many separate proofs of this famous theorem. The book *The Pythagorean Proposition* by Elisha S. Loomis contains 370 proofs of the Pythagorean Theorem. President James Garfield is also credited with a proof of this theorem.

The Pythagorean formula is frequently used to calculate the length of one side of a right triangle when the other two sides are known. In this case, an alternate form of this formula is used.

Alternate forms of $a^2 + b^2 = c^2$

$$a^2 = c^2 - b^2 \text{ or } a = \sqrt{c^2 - b^2}$$
$$b^2 = c^2 - a^2 \text{ or } b = \sqrt{c^2 - a^2}$$
$$c^2 = a^2 + b^2 \text{ or } c = \sqrt{a^2 + b^2}$$

To approximate the lengths of the sides you may want to review the material on square roots in Section 5.2.

Example 2   Find the length of the hypotenuse of a right triangle with legs of 5 meters and 12 meters.

$$c = \sqrt{a^2 + b^2}$$
$$c = \sqrt{(12)^2 + (5)^2}$$
$$c = \sqrt{144 + 25}$$
$$c = \sqrt{169}$$
$$c = 13$$

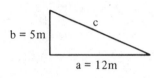

The length of the hypotenuse is 13 meters.

Example 3   The hypotenuse of a right triangle is 25 feet and one leg is 24 feet. How long is the other leg?

Let b = length of other leg. Then

$$b = \sqrt{c^2 - a^2}$$
$$b = \sqrt{(25)^2 - (24)^2}$$
$$b = \sqrt{625 - 576}$$
$$b = \sqrt{49}$$
$$b = 7$$

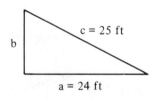

The length of the other leg is 7 feet.

Example 4   Approximate the length of the hypotenuse of a right triangle when one leg is measured as 9.0 centimeters and the other leg is measured as 14.0 centimeters.

Let c = length of hypotenuse.

$$c = \sqrt{a^2 + b^2}$$
$$c = \sqrt{(14)^2 + (9)^2}$$
$$c = \sqrt{196 + 81}$$
$$c = \sqrt{277}$$
$$c = 16.6 \text{ (approximation to the nearest tenth)}$$

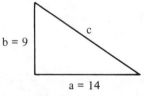

The length of the hypotenuse is approximately 16.6 centimeters.

Example 5   A carpenter is framing a house and is trying to "square" the corner of the house. To determine if the walls are "square" (perpendicular) he makes the measurements shown in the figure below. Are the walls "square"?

If the triangle is a right triangle, then c^2 must equal $a^2 + b^2$. Let $a = 15'$, $b = 8'$, and $c = 17'$.

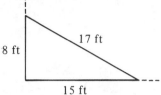

$$a^2 + b^2 = c^2$$
$$(15)^2 + (8)^2 = (17)^2$$
$$225 + 64 = 289$$
$$289 = 289$$

Yes, the walls are "square" since $a^2 + b^2 = c^2$; that is, the walls are perpendicular.

Example 6   A triangle has sides of 4, 7, and 10. Is this a right triangle?

Let a and b represent the shorter sides and c be the longest side. To be a right triangle, c^2 must equal $a^2 + b^2$. Let $a = 7$, $b = 4$, and $c = 10$.

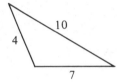

$$a^2 + b^2 = 7^2 + 4^2 \qquad c^2 = 10^2$$
$$= 49 + 16 \qquad\quad = 100$$
$$= 65$$

Since $65 \neq 100$, the triangle is not a right triangle.

Example 7   A cable is used as a brace to support two sides of a metal bin. Find the length of cable needed in the figure below.

Let c = length of the cable. Then

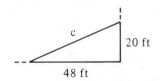

$$c = \sqrt{a^2 + b^2}$$
$$c = \sqrt{(48)^2 + (20)^2}$$
$$c = \sqrt{2304 + 400}$$
$$c = \sqrt{2704}$$
$$c = 52$$

The cable must be 52 feet long.

12.6 Exercises

Use the Pythagorean theorem to calculate the lengths of the missing sides of each of the right triangles in Exercises 1–9.

1.

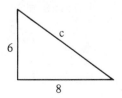

2.

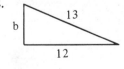

3.

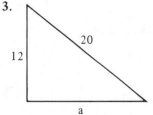

4.

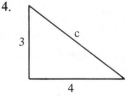

5.
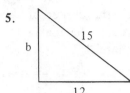

6.
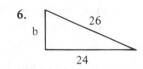

7.
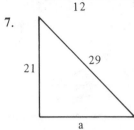

8.
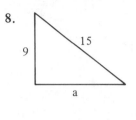

9.
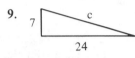

Use the Pythagorean formula to determine whether the triangles whose sides are given in Exercises 10–17 are right triangles.

10. 6 inches, 8 inches, 10 inches

11. 5 cm, 12 cm, 13 cm

12. 2 m, 3 m, 6 m

13. 4 inches, 5 inches, 6 inches

14. 10 ft, 14 ft, 16 ft

15. 12 yd, 16 yd, 20 yd

16. 7 cm, 24 cm, 25 cm

17. 6 ft, 7 ft, 8 ft

Use the Pythagorean formula to solve the applied problems in Exercises 18–25. Refer to the sketches to select a, b, and c properly.

18. A jogger left home and jogged 3 miles west and 4 miles north. The jogger rested and calculated her distance from home "as the crow flies." What was this distance?

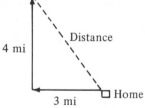

19. An airplane left an airport and flew 15 miles east and then 8 miles south. At this point, how far from the airport is the airplane?

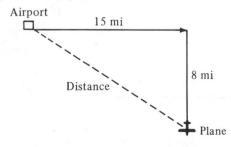

20. A rafter is cut to span a length of 24 feet while the roof rises 7 feet. What length rafter will be needed just to span this distance?

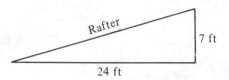

21. A diagonal brace is needed to reinforce a rectangular gate. The gate is three feet wide and four feet high. How long is the diagonal brace?

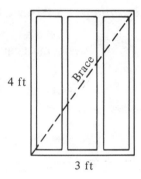

22. A 17-foot ladder leans against a wall. If the ladder is 8 feet from the base of the wall, how far is it from the bottom of the wall to the top of the ladder?

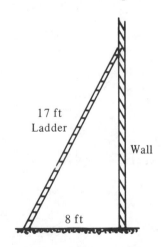

23. A swimming pool has a bottom which slopes from the middle of the pool to the deep end. Using the figure below, calculate the depth of the drop-off, b.

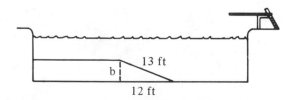

24. A carpenter is trying to "square" the corner of a building. He makes a mark 6 feet along one wall and another mark 8 feet along the other wall. What must the distance between the marks be if the walls are "square"?

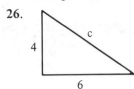

25. An electrician plans to run wiring diagonally across a rectangular room. The room is 12 feet long and 16 feet wide. What is the length of the wiring that crosses this room?

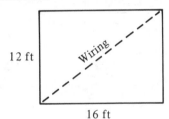

Approximate to the nearest tenth the lengths of the missing sides of the right triangles in Exercises 26-30.

26.

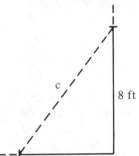

27.

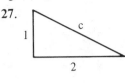

28.

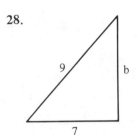

29.

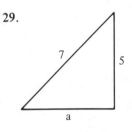

30.

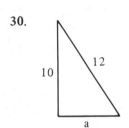

12.7   Terminating, Repeating, and Nonrepeating Decimals

One of the beauties of mathematics is its ability to obtain information about abstractions using symbols to represent these abstractions. For example, the number line is a visualization we frequently use to illustrate the set of real numbers and to show the order relation between numbers. Sometimes we use many different symbols to represent the same idea. For example, the number one-half can be symbolized by $\frac{2}{4}$, $\frac{6}{12}$, 0.5, $\sqrt{\frac{1}{4}}$, 50%, or $\frac{1}{2}$. Frequently, however, one way of symbolizing an idea leads to a different approach which may result in a better understanding of that idea.

Although rational numbers are defined in terms of ratios of integers, we may be able to gain a better understanding of both rational numbers and irrational numbers by examining rational numbers in decimal form rather than in fractional form. In order to accomplish this objective, we will need to form equations and solve them. Recall that fractions can be expressed as decimals by dividing the numerator by the denominator.

Example 1   a. $\dfrac{1}{8} = 0.125$ b. $\dfrac{2}{3} = 0.6666...$

Since

$$\begin{array}{r} 0.125 \\ 8\overline{)1.000} \\ \underline{8} \\ 20 \\ \underline{16} \\ 40 \\ \underline{40} \end{array}$$

Since

$$\begin{array}{r} 0.6666 \\ 3\overline{)2.0000} \\ \underline{1\,8} \\ 20 \\ \underline{18} \\ 20 \\ \underline{18} \\ 20 \\ \underline{18} \\ 2 \end{array}$$

Note that the division in the first example is exact while the division in the second example is not exact. In the number 0.6666... the three dots following the last digit indicate that the 6s continue without end. Since 0.125 has a last digit or terminal digit, we call 0.125 a **terminating decimal**. The decimal 0.6666... does not have a last or terminal digit so we call 0.6666... a **nonterminating decimal**.

Example 2   a. 2.4683 is a terminating decimal.

b. 7.864444... is a nonterminating decimal.

c. 123 is a terminating decimal (consider 123 = 123.)

d. 487.5367777... is a nonterminating decimal.

e. 4.80000... is a terminating decimal. Since the three dots indicate that only zeros are continuing without end, 4.80000... is the same as 4.8 which terminates.

f. 0.07838383... is a nonterminating decimal.

g. 81.364873209... is a nonterminating decimal.

Another notation that is frequently used to show that the 6s in 0.6666... continue infinitely is to write $0.\overline{6}$ for this number. The bar over the 6 means that this digit repeats infinitely. Similarly, the number 0.07838383... can be denoted by $0.07\overline{83}$, where the bar extends over both the digit 8 and the digit 3. This means that this pair of digits continues to repeat. In this text, a repeating decimal will be indicated either by the bar or by writing the repeating digits three times followed by three dots.

Example 3   a. $1.6555... = 1.6\overline{5}$

b. $4.0696969... = 4.0\overline{69}$

c. $0.173173173... = 0.\overline{173}$

d. $0.142857142857142857... = 0.\overline{142857}$

The bar can be used only when a digit or sequence of digits repeats infinitely. A decimal which has an infinitely repeating sequence of digits is called a **repeating decimal**. For some nonterminating decimals there is never a pattern of digits which repeats itself. These decimals are called **nonrepeating decimals**. Thus a nonrepeating decimal is nonterminating.

Example 4   a. $0.\overline{3}$ is a repeating decimal since the digit 3 repeats infinitely.

b. $109.4\overline{72}$ is a repeating decimal since the sequence of digits 72 repeats endlessly.

c. 1.41421356237... is a nonrepeating decimal since no sequence of digits repeats. The three dots mean that the number has an infinite series of digits, but it does not indicate what the next digit is since no pattern has repeated.

d. 0.00417417417... is a repeating decimal where the sequence of digits 417 repeats again and again.

e. 3.14159... is a nonrepeating decimal.

Now that we are familiar with all possible types of decimals (terminating, repeating, and nonrepeating), we can return to the problem of converting a rational number to a decimal. If a rational number $\dfrac{n}{m}$ is converted to a decimal, the result is either a terminating decimal or a repeating decimal. Several examples will be given to illustrate this important fact, and then a discussion will be given to explain why a rational number must be either a terminating or a repeating decimal.

Example 5   a. $\dfrac{5}{6} = 0.8\overline{3}$

Since

```
      0.8333
   6)5.0000
     4 8
     ────
       20
       18
     ────
       20
       18
     ────
       20
       18
     ────
        2
```

b. $\dfrac{31}{250} = 0.124$

Since

```
        0.124
   250)31.000
       25 0
       ────
        600
        500
       ────
       1000
       1000
       ────
```

c. $\dfrac{142}{99} = 1.\overline{43}$

Since

$$
\begin{array}{r}
1.4343 \\
99\overline{)142.0000} \\
\underline{99} \\
430 \\
\underline{396} \\
340 \\
\underline{297} \\
430 \\
\underline{396} \\
340 \\
\underline{297} \\
43
\end{array}
$$

d. $\dfrac{104}{333} = 0.\overline{312}$

Since

$$
\begin{array}{r}
0.312312 \\
333\overline{)104.000000} \\
\underline{99\ 9} \\
4\ 10 \\
\underline{3\ 33} \\
770 \\
\underline{666} \\
1040 \\
\underline{999} \\
410 \\
\underline{333} \\
770 \\
\underline{666} \\
104
\end{array}
$$

e. $\dfrac{1}{7} = 0.\overline{142857}$ Since

$$
\begin{array}{r}
0.1428571428571 \\
7\overline{)1.0000000000000} \\
\underline{7} \\
30 \\
\underline{28} \\
20 \\
\underline{14} \\
60 \\
\underline{56} \\
40 \\
\underline{35} \\
50 \\
\underline{49} \\
10 \\
\underline{7} \\
30 \\
\underline{28} \\
20 \\
\underline{14} \\
60 \\
\underline{56} \\
40 \\
\underline{35} \\
50 \\
\underline{49} \\
10 \\
\underline{7} \\
3
\end{array}
$$

A rational number may result in a terminating decimal since the division may be exact. The preceding example has shown, however, that for some rational numbers the division process may never terminate. By examining $\frac{1}{7} = 0.\overline{142857}$, we can gain some insight as to why a rational number must be a repeating decimal if it does not terminate. When we divide, we only allow remainders smaller than the divisor. Thus, the only nonzero remainders possible when we divide by 7 are 1, 2, 3, 4, 5, and 6. This means that by the 6th step in the division process all remainders have already been used. Since some remainder must now repeat, the quotient will repeat the sequence of digits obtained when 7 was earlier divided into that number. Regardless of the denominator of a fraction, the number of possible nonzero remainders is finite. Thus, either the division eventually terminates, or it must repeat itself when the remainders start to repeat. This important result is summarized in the box that follows.

> The rational number $\frac{n}{m}$ can be rewritten either as a terminating decimal or as a decimal which repeats in a sequence of fewer than m digits.

Example 6   Convert $\frac{2}{13}$ to a decimal.

The possible nonzero remainders are 1, 2, 3, 4, 5, 6, 7, 8, 9, 10, 11, and 12.

```
        0.15384615
   13/2.00000000
       1 3
       ───
        70
        65
        ──
        50
        39
        ──
        110
        104
        ───
         60
         52
         ──
         80
         78
         ──
      →  20   (Remainder of 2 is a repeat of first digit.)
         13
         ──
         70
         65
         ──
          5
```

Note that the remainder 2 is a repeat of an earlier digit listed above and, thus, the sequence of digits in the quotient now begins to repeat. Observe that all possible remainders did not occur before a remainder repeated.

Thus, $\frac{2}{13} = 0.\overline{153846}$.

Example 7   Convert $\frac{5}{6}$ to a decimal.

Before dividing, note that the division must eventually be exact or repeat since the only possible nonzero remainders are 1, 2, 3, 4, and 5.

$$
\begin{array}{r}
0.833 \\
6\overline{)5.00} \\
\underline{4\,8} \\
20 \\
\underline{18} \\
20 \\
\underline{18} \\
2
\end{array}
\quad \text{(remainder 2 repeats)}
$$

Example 8   Convert $\frac{218}{7}$ to a decimal.

$$
\begin{array}{r}
31.142857 \\
7\overline{)218.000000} \\
\underline{21} \\
8 \\
\underline{7} \\
10 \\
\underline{7} \\
30 \\
\underline{28} \\
20 \\
\underline{14} \\
60 \\
\underline{56} \\
40 \\
\underline{35} \\
50 \\
\underline{49} \\
1
\end{array}
\quad \text{(remainder 1 repeats)}
$$

Alternate Solution:

$$\frac{218}{7} = 31\frac{1}{7}$$

$$\frac{1}{7} = 0.\overline{142857} \quad \text{(\textit{Note}: See Example 5e.)}$$

Note that since the denominator is 7, the decimal must have a sequence of 6 or fewer digits repeating. However, since the fraction is improper, some digits occur before the start of the sequence which repeats.

Thus, $\frac{218}{7} = 31.\overline{142857}$.

Example 9   Convert $\frac{1}{17}$ to a decimal.

$$
\begin{array}{r}
.0588235294\overset{\downarrow}{1}1764\overset{\downarrow}{7}05 \\
17\overline{)1.000000000000000000} \\
\underline{85} \\
150 \\
\underline{136} \\
140 \\
\underline{136} \\
40 \\
\underline{34} \\
60 \\
\underline{51} \\
90 \\
\underline{85} \\
50 \\
\underline{34} \\
160 \\
\underline{153} \\
70 \\
\underline{68} \\
20 \\
\underline{17} \\
30 \\
\underline{17} \\
130 \\
\underline{119} \\
110 \\
\underline{102} \\
80 \\
\underline{68} \\
120 \\
\underline{119} \\
10 \\
\underline{0} \\
100 \\
\underline{85} \\
15
\end{array}
$$

(Remainder 1 is a repeat of first digit.)

Note that the recurrence of a digit in the quotient does *not* mean the entire sequence of digits is beginning to repeat. Be sure to continue dividing until a remainder repeats. A repeating remainder is what signals that the sequence of digits in the quotient is going to repeat.

Thus, $\frac{1}{17} = 0.\overline{0588235294117647}$

Since the denominator of $\frac{1}{17}$ is 17, we know that the decimal form should repeat in a sequence of 16 or fewer digits. In fact, $0.\overline{0588235294117647}$ does repeat a sequence of 16 digits.

In Chapter 4 we converted terminating decimals to fractions. We will now give a procedure for converting repeating decimals to fractions. The result of these two pieces of information is summarized below.

> Every terminating decimal and repeating decimal can be rewritten as a rational number in the form $\frac{n}{m}$.

Thus, all rational numbers are terminating or repeating decimals and all terminating and repeating decimals are rational numbers. From this fact we can derive a new description of the irrational numbers. The irrational numbers must be those real numbers which are nonrepeating decimals. Irrational numbers are very important and useful, but they are somewhat more difficult to work with. One reason for this is that an irrational cannot be written as a fraction, so in decimal form there is no repeating pattern which we can use to represent that irrational number.

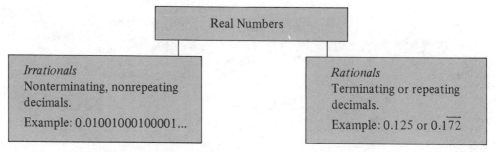

Real Numbers	
Irrationals Nonterminating, nonrepeating decimals. Example: 0.01001000100001...	*Rationals* Terminating or repeating decimals. Example: 0.125 or 0.1$\overline{72}$

After making these observations, let us return to the conversion of decimals to fractions. Before examining the repeating case, first recall how to convert terminating decimals to fractions by examining the following examples.

Example 10   Convert 0.125 to a fraction.

$$0.125 = \frac{125}{1000}$$
$$= \frac{1}{8}$$

Thus, $0.125 = \frac{1}{8}$.

Example 11   Convert 12.308 to a fraction.

$$12.308 = \frac{12{,}308}{1000}$$
$$= \frac{3077}{250}$$

Thus, $12.308 = \frac{3077}{250}$.

We are able to write down the fractional equivalent of a terminating decimal by determining the power of ten which should be in the denominator. Since repeating decimals never terminate, there is no power of ten which will be the correct denominator. Thus, we will need a new technique for converting repeating decimals to fractions. Linear equations can be used to make these conversions.

Example 12   Convert $0.\overline{3}$ to a fraction.

Hint 1: $0.\overline{3} = 0.333... = 0.3333... = 0.33333...$ etc.
  The three dots indicate that the digit 3 repeats infinitely. The number of 3s that we actually write to illustrate this is arbitrary.

Hint 2: Equals subtracted from equals are equal.

Hint 3: $3.000... = 3$
  Although 3.000... has an infinite number of zeros repeating, we consider 3.000... a terminating decimal since 3.000... = 3.

Let N = 0.333..., then

    10N = 3.33... (multiply both sides by 10)

or   10N = 3.333... (see *Hint 1*)

Thus,

$$
\begin{array}{ll}
10N = 3.333... & \\
-(N = .333...) & \\
\hline
9N = 3.000... & \text{(see *Hint 2*)} \\
9N = 3 & \text{(see *Hint 3*)} \\
\end{array}
$$

$$N = \frac{3}{9}$$

$$N = \frac{1}{3}$$

Since N = 0.333... and $N = \frac{1}{3}$, we conclude that $0.333... = \frac{1}{3}$.

Example 13   Convert $0.\overline{15}$ to a fraction.

Let N = 0.151515..., then

    100N = 15.151515...

Note after multiplying by 100, we can align the digits with those in the first equation so that when we subtract, we will be subtracting each digit from itself.

$$
\begin{array}{l}
100N = 15.151515... \\
-(N = 0.151515...) \\
\hline
99N = 15.000000... \\
99N = 15 \\
\end{array}
$$

$$N = \frac{15}{99}$$

$$= \frac{5}{33}$$

Thus, $0.\overline{15} = \frac{5}{33}$.

Example 14   Convert 1.2444... to a fraction.

Let N = 1.2444..., then

$$10N = 12.444...$$

$$\begin{aligned} 10N &= 12.4444... \\ -(N &= 1.2444...) \\ \hline 9N &= 11.2000... \end{aligned}$$

$$9N = 11.2$$

Note that since 11.2 is not a whole number, we multiply both sides of the equation by ten to obtain an equivalent equation with a whole number on the right side of the equation.

$$90N = 112$$

$$N = \frac{112}{90}$$

$$= \frac{56}{45}$$

Thus, $1.2444... = \dfrac{56}{45}$.

The steps illustrated in these examples are summarized in the following *procedure for converting a repeating decimal to a fraction.*

PROCEDURE FOR CONVERTING A REPEATING DECIMAL TO A FRACTION	EXAMPLE
	Convert $0.\overline{432}$ to a fraction. *Solution:*
1. Let N equal the decimal given.	1. Let N = 0.4323232...
2. Determine the sequence of digits which is repeating.	2. The sequence of digits 32 is repeating.
3. Multiply both sides of the equation by a power of ten containing as many zeroes as there are digits in the sequence of digits which repeats.	3. $\quad (100)N = (100)(.4323232...)$ $\quad\quad 100N = 43.23232...$
4. Form a new equation by subtracting the left member of the equation in step 1 from the left member in step 3 and the right member in step 1 from the right member in step 3.	4. $\quad 100N = 43.23232...$ $\quad\quad\underline{-(N = 0.43232...)}$ $\quad\quad 99N = 42.80000....$
5. Solve the resulting equation for N. Express N as a fraction in lowest terms.	5. $\quad\quad 99N = 42.8$ $\quad\quad 990N = 428$ $\quad\quad\quad N = \dfrac{428}{990} = \dfrac{214}{495}$
	Thus, $0.\overline{432} = \dfrac{214}{495}$

Example 15 Convert $0.\overline{6}$ to a fraction.

Let N = 0.666..., then

$$10N = 6.666...$$
$$\underline{-(N = 0.666...)}$$
$$9N = 6.000...$$
$$9N = 6$$
$$N = \frac{6}{9}$$
$$N = \frac{2}{3}$$

Thus, $0.\overline{6} = \frac{2}{3}$.

Example 16 Convert 0.777... to a fraction.

Let x = 0.777..., then

$$10x = 7.777...$$
$$\underline{-(x = 0.777...)}$$
$$9x = 7.000...$$
$$9x = 7$$
$$x = \frac{7}{9}$$

Thus, $0.777... = \frac{7}{9}$.

Example 17 Convert $0.\overline{428571}$ to a fraction.

Let $q = 0.\overline{428571}$, then

$$1,000,000q = 428,571.428571...$$
$$\underline{-(q = \qquad\quad 0.428571...)}$$
$$999,999q = 428,571.000000...$$
$$999,999q = 428,571$$
$$q = \frac{428,571}{999,999}$$
$$q = \frac{3 \cdot \cancel{142857}}{7 \cdot \cancel{142857}}$$
$$q = \frac{3}{7}$$

Thus, $0.\overline{428571} = \frac{3}{7}$.

Example 18   Convert 0.042121... to a fraction.

Let m = 0.042121..., then

$$100m = 4.212121...$$
$$\underline{-(m = 0.042121...)}$$
$$99m = 4.170000...$$
$$99m = 4.17$$
$$9900m = 417$$
$$m = \frac{417}{9900} = \frac{\cancel{3} \cdot 139}{\cancel{3} \cdot 3300}$$
$$m = \frac{139}{3300}$$

Thus, $0.042121... = \dfrac{139}{3300}$.

Example 19   Convert $8.4\overline{713}$ to a fraction.

Let N = 8.4713713..., then

$$1000N = 8471.3713713...$$
$$\underline{-(N = \qquad 8.4713713...)}$$
$$999N = 8462.9000000...$$
$$999N = 8462.9$$
$$9990N = 84629$$
$$N = \frac{84629}{9990}$$

Thus, $8.4\overline{713713}... = \dfrac{84{,}629}{9{,}990}$.

Example 20   Show that 0.999999... is the same as 1.

Let N = 0.999..., then

$$10N = 9.999...$$
$$\underline{-(N = \ .999...)}$$
$$9N = 9.000...$$
$$9N = 9$$
$$N = 1$$

Thus, 0.999... = 1.

Comment: This fact astounds many students the first time they encounter it since the numbers "look" different. Of course, we have seen the same value in different forms before. For example, $\frac{1}{2} = \frac{2}{4} = 0.5 = 50\%$ etc. If you still believe they are different, then try to determine the amount by which they differ. Do they differ by 0.1, 0.01, 0.001, or ...?

Example 21   Classify each of the following as either a rational number or an irrational number.

a. 0.1875   is a terminating decimal; therefore, it is a rational number.

b. 0.18757575...   is a repeating decimal; therefore, it is a rational number.

c. 0.03040506070809011012...   is a nonrepeating decimal; therefore, it is an irrational number.

12.7   Exercises

Classify each of the decimals in Exercises 1–20 as either terminating, repeating, or nonrepeating.

1. 0.478
2. 132
3. $403.5\overline{7}$
4. 7.13922
5. 139.77777
6. $23.\overline{95}$
7. 67
8. 43.999...
9. 1.4789732571...
10. $1{,}073.03\overline{56175}$
11. 0.272727...
12. 412.4576324879...
13. $0.\overline{428571}$
14. $9.130\overline{0}$
15. 82.40000...
16. 0.012345678900112233...
17. $17.8\overline{0}$
18. 3.141592654...
19. 2.718281...
20. 481.76000...

Classify each of the numbers in Exercises 21–30 as either a rational number or an irrational number.

21. $\frac{2}{3}$
22. 7.84
23. $0.06\overline{7}$
24. 0.030030003...
25. 405.2
26. 37
27. 0.45146147148...
28. $0.247\overline{3}$
29. 0.898989...
30. $\frac{5}{7}$

Express each of the fractions in Exercises 31–50 as a decimal. For repeating decimals be sure to indicate the entire sequence of digits which is repeating.

31. $\frac{7}{8}$
32. $\frac{2}{9}$
33. $\frac{5}{18}$
34. $\frac{1}{3}$
35. $\frac{4}{7}$
36. $\frac{3}{16}$
37. $\frac{6}{35}$
38. $\frac{6}{7}$
39. $\frac{2}{13}$
40. $\frac{1}{33}$
41. $\frac{14}{9}$
42. $\frac{8}{35}$
43. $\frac{7}{16}$
44. $\frac{5}{8}$
45. $\frac{5}{33}$
46. $\frac{19}{13}$
47. $\frac{3}{17}$
48. $\frac{34}{17}$
49. $\frac{159}{70}$
50. $\frac{13}{30}$

Convert each of the decimals in Exercises 51–70 to a fraction.

51. 0.3125
52. 0.444
53. $0.\overline{13}$
54. 0.875
55. 13.02111...
56. $0.6\overline{5}$
57. $0.\overline{714285}$
58. $8.\overline{3}$
59. $0.\overline{384615}$
60. $0.\overline{857142}$
61. $6.\overline{6}$
62. 1.08
63. $3.\overline{9}$
64. $0.\overline{8}$
65. 12.35
66. $0.\overline{538461}$
67. 4.999...
68. 7.99999...
69. 15.01333...
70. $8.\overline{9}$

Important Terms
Used in this Chapter

area	perimeter
circle	perpendicular
circumference	polygon
cone	proportion
cube	proportional
cylinder	Pythagorean theorem
extremes	rate of percent
hypotenuse	ratio
in proportion	rectangle
leg	repeating decimal
length	right triangle
means	sphere
nonrepeating decimal	square
nonterminating decimal	terminating decimal
parallelepiped	terms of a proportion
percent	triangle
percentage	volume

Important Formulas
Used in this Chapter

Perimeter

Perimeter of a polygon: add lengths of all sides

Circumference of a circle: $C = \pi d$

Area

Circle: $A = \pi r^2$

Rectangle: $A = lw$

Triangle: $A = \dfrac{bh}{2}$

Pythagorean Theorem: $a^2 + b^2 = c^2$
Alternate Forms: $c = \sqrt{a^2 + b^2}$
$b = \sqrt{c^2 - a^2}$
$a = \sqrt{c^2 - b^2}$

Volume

Cone: $V = \dfrac{\pi r^2 h}{3}$

Cylinder: $V = \pi r^2 h$

Parallelepiped: $V = lwh$

Sphere: $V = \dfrac{4\pi r^3}{3}$

Percent

$A = RB$

Amount = Rate of Percent $\times$ Base

REVIEW EXERCISES

Rewrite each of the ratios in Exercises 1-10 as a fraction in simplest form.

1. 75:100   **2.** 30%

3. 18:35   **4.** Ratio of 24 to 40

5. 77:88   **6.** 25/35

7. Ratio of 187 to 209   **8.** 100/50

9. 500%   **10.** 12:2

11. Which of the following is the best buy?
  a. A box of 150 facial tissues which costs 70¢.
  b. A box of 200 facial tissues which costs 90¢.
  c. A box of 125 facial tissues which costs 55¢.

12. In 1972, Roberto Clemente had 118 hits in 378 at bats. In 1973, Rod Carew had 203 hits in 580 at bats. Which of these represents the better batting average?

Identify the means and the extremes of the proportions in Exercises 13–16.

13. 3:11 = 15:55   **14.** $\dfrac{x}{7} = \dfrac{13}{35}$

15. 8:3 = 56:21   **16.** $\dfrac{w}{z} = \dfrac{a}{b}$

Solve each of the proportions in Exercises 17-20 for the unknown term.

17. $\dfrac{12}{x} = \dfrac{72}{80}$ **18.** $\dfrac{35}{21} = \dfrac{y}{3}$

19. x:10 = 11:2   **20.** 66:154 = 3:a

Using the property of proportions, determine which of the pairs of fractions in Exercises 21–24 are equivalent.

21. $\dfrac{35}{14}, \dfrac{45}{18}$ **22.** $\dfrac{11}{33}, \dfrac{5}{15}$

23. $\dfrac{6}{21}, \dfrac{9}{28}$ **24.** $\dfrac{30}{24}, \dfrac{40}{44}$

Use proportions to solve Exercises 25 and 26.

25. If 186 bricks are used in constructing a 3-foot section of a wall, how many bricks will be needed for a similar wall 22 feet long?

26. On a map, 1.5 inches represents 100 miles. How many miles are represented by 5 inches?

Use the formula A = RB to solve Exercises 27-29.

In 1978, form 5695 of the Internal Revenue Service contained a statement which said "Enter 15% of line 2 (but do not enter more than $300)." Use this statement in Exercises 27, 28, and 29.

27. If line 2 is $470, what should be entered on line 3 to comply with this instruction?

28. If 15% of line 2 is the same as the $300 limitation, what is the amount on line 2?

29. If the amount on line 2 is $2500, you must enter $300 on line 3. In this particular case, what percent of line 2 are you really allowed to enter?

Calculate the perimeter of each of the plane regions in Exercises 30-33. Assume that all lengths are given in meters.

30.   **31.**

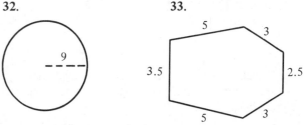

32.   **33.**

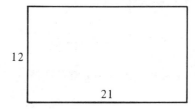

Calculate the area of each of the regions in Exercises 34-37. Assume that all lengths are given in feet.

34.

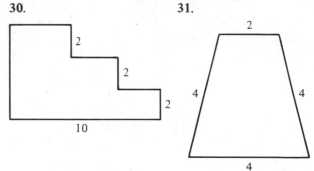

35.

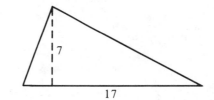

36.

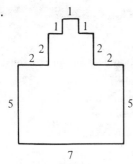

37.

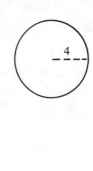

Calculate the volume of each of the solids in Exercises 38–40. Assume that all measurements are given in meters.

38.   **39.**   **40.**

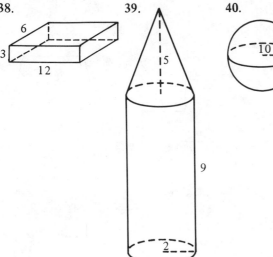

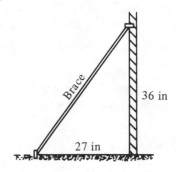

41. Calculate the length of one leg of a right triangle if the hypotenuse is 29 cm long and the other leg is 21 cm long.

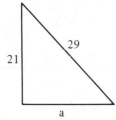

42. A park employee is outlining a new softball diamond. In order to insure that the left and right field lines are perpendicular, he places a mark along each line. The mark down the left field line is 24 feet from home plate and the mark down the right field line is 32 feet. How many feet should be between the marks?

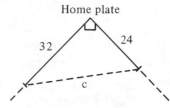

43. Calculate the length of the brace needed in the figure below.

Identify each of the real numbers in Exercises 44–47 as either rational or irrational.

44. $\frac{2}{5}$

45. 0.222...

46. .121221222122221...

47. $0.\overline{15}$

Convert each of the decimals in Exercises 48–50 to a fraction in reduced form.

48. .875   **49.** $2.\overline{5}$ **50.** $1.\overline{54}$

13

FIRST-DEGREE INEQUALITIES IN ONE VARIABLE

OBJECTIVES

Upon completion of this chapter you should be able to:

1. Translate algebraic statements involving the order relations into verbal statements. — 13.1

2. Translate verbal statements involving the order relations into algebraic statements. — 13.1

3. Determine the order relation between any pair of rational numbers. — 13.1

4. Recognize an absolute inequality and a contradiction. — 13.2, 13.5

5. Draw the graph of the solution of a linear inequality. — 13.2–13.5

6. Solve linear inequalities using the multiplication–division principle for inequalities, and check values from the solution set. — 13.3

7. Solve linear inequalities using the addition–subtraction principle for inequalities, and check values from the solution set. — 13.4

8. Solve more general forms of linear inequalities, and check values from the solution set. — 13.5

PRETEST

1. Translate each of the following into verbal statements (Objective 1).
   a. $2 < 10$
   b. $5 > -10$
   c. $6 \geqslant 6$
   d. $7 \leqslant 20$

2. Translate each of the following into algebraic form (Objective 2).
   a. Nine is less than or equal to twenty.
   b. Seven is greater than five.
   c. Eleven is greater than or equal to eleven.
   d. Negative nine is less than zero.

3. Answer the following true (T) or false (F) (Objective 3).
   a. $-6 < 3$
   b. $3 > 3$
   c. $1.25 \leqslant 1.3$
   d. $\frac{4}{7} < \frac{5}{9}$
   e. $-\frac{2}{3} > \frac{5}{9}$
   f. $0 > -15$

4. Identify each of the following as a conditional inequality, an absolute inequality, or a contradiction. Give the solution for each inequality (Objective 4).
   a. $x > x + 1$
   b. $2x + 1 > x - 5$
   c. $2(x + 6) < 2x + 6$
   d. $10x < x$
   e. $x + 4 \geqslant x + 4$
   f. $-3x + 4 < 2x + 4$

5. Draw the graph of each of the following inequalities (Objective 5).
   a. $x > 5$
   b. $x \leqslant 10$
   c. $x \geqslant 0$
   d. $x > 1.5$
   e. $x < -3$
   f. $x < -3\frac{2}{3}$

6. Solve each of the following inequalities. Draw the graph of the solution and check one value from each solution set (Objective 6).
   a. $3x > 30$
   b. $\frac{-x}{3} \geqslant 30$

c. $\frac{2}{3}x \geqslant -30$ d. $-\frac{3}{2}x < 30$

7. Solve each of the following inequalities. Draw the graph of the solution and check one value from each solution set (Objective 7).

   a. $30x + 15 < 20x - 15$ b. $-1.2x - 5 \geqslant 0.8x - 7$
   c. $3x + 4 > 3x + 2$ d. $5x - 8 < 8x + 12$

8. Solve each of the following inequalities. Draw the graph of the solution and check one value from each solution set (Objective 8).

   a. $2x + 3(x - 5) - 20 < -6x + 2(x - 6) + 4$
   b. $\frac{x}{2} + \frac{x}{3} + \frac{x}{4} > \frac{1}{2} + \frac{1}{3} + \frac{1}{4}$
   c. $6x + 4 - 9x \leqslant 2x + 7 + 6x$
   d. $\frac{2}{3}x + 5 \geqslant \frac{x}{7} + 10 - x + 2$

In previous chapters we have examined many statements of equality and we have solved a variety of conditional equations. Many problems that need to be solved can be represented as equations, but some problems are best represented by inequalities. For example, a manager may insist that expenses be less than a certain amount. Likewise, an engineer may insist upon a steel beam more than 2 inches thick. In this chapter we will concentrate on conditional inequalities and their solutions.

13.1 The Order Relations

We have already used the symbol $\neq$ to show that two mathematical expressions do not have the same value. The symbol $\neq$ was read *is not equal to* or *does not equal.* Four other symbols of inequality will now be explained:

   1. $x > y$ is read x **is greater than** y
   2. $x < y$ is read x **is less than** y
   3. $x \geqslant y$ is read x **is greater than or equal to** y
   4. $x \leqslant y$ is read x **is less than or equal to** y

The inequalities $x < y$, $x > y$, $x \geqslant y$, and $x \leqslant y$ are also called **order relations** since they refer to the order in which we write down the coordinates of the points on the real number line.

   1. $x > y$ if the graph of x is to the right of the graph of y.
   2. $x < y$ if the graph of x is to the left of the graph of y.
   3. $x \geqslant y$ if x and y represent the same point or if x lies to the right of y.
   4. $x \leqslant y$ if x and y represent the same point or if x lies to the left of y.

Example 1   a. $5 > 3$ is read *five is greater than three.* On the number line, 5 is to the right of 3.

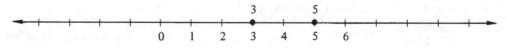

Note that since this is equivalent to saying that three is to the left of five, this inequality is equivalent to the inequality $3 < 5$.

b. $1.5 < 2$ is read *one and five tenths is less than two.* On the number line, 1.5 is to the left of 2.

Note that since 2 is to the right of 1.5, this also means that $2 > 1.5$.

c. $x < x + 1$ is read *x is less than x + 1.* On the number line, x is to the left of x + 1.

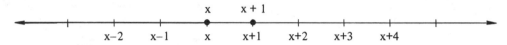

d. $2 \geqslant 2$ is read *two is greater than or equal to two.* This statement is true if 2 is greater than 2 or if 2 is equal to 2. Since 2 does equal 2, these values represent the same point on the number line, and this statement is true.

e. $w > w - 2$ is read *w is greater than w - 2.* On the number line, w is located two units to the right of w minus two.

f. $3 \leqslant 3$ is read *3 is less than or equal to 3.* This statement is true since three does equal three. Both members of the inequality represent the same point on the number line.

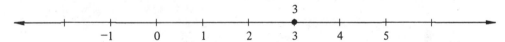

g. $-2 \geqslant -5$ is read *-2 is greater than or equal to -5.* This statement is true since -2 is greater than -5. On the number line, negative two is to the right of negative five.

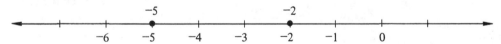

h. $0.714 < 0.718$ is read *0.714 is less than 0.718.* On the number line, 0.714 is 0.004 of a unit to the left of 0.718.

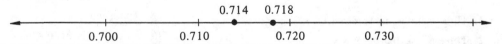

While the number line provides an excellent visualization of the order property of numbers, sometimes we must examine fractions in an alternate form before we can determine how they are positioned on the number line.

Consider the two methods described below for determining which is greater, $\frac{2}{3}$ or $\frac{3}{5}$. The first method is based on finding the LCD, and in the second method, a decimal approximation of each fraction is found.

Method A.   LCD = 15

$$\frac{2}{3} = \frac{2 \cdot 5}{3 \cdot 5} = \frac{10}{15}$$

$$\frac{3}{5} = \frac{3 \cdot 3}{5 \cdot 3} = \frac{9}{15}$$

Since $10 > 9$, $\frac{10}{15} > \frac{9}{15}$. Thus $\frac{2}{3} > \frac{3}{5}$.

Method B.

$$\frac{2}{3} \doteq 0.67 \quad \text{(rounded to nearest hundredth)}$$

$$\frac{3}{5} = 0.60$$

Since $0.67 > 0.60$, $\frac{2}{3} > \frac{3}{5}$.

Example 2   Determine the order relation between the following pairs of numbers.

a. $\frac{4}{9}$ and $\frac{1}{2}$

$$\frac{4}{9} \doteq 0.44 \quad \text{(rounded to nearest hundredth)}$$

$$\frac{1}{2} = 0.50$$

Thus $\frac{4}{9} < \frac{1}{2}$.

b. $\frac{22}{33}$ and $\frac{34}{51}$

$$\frac{22}{33} = \frac{2}{3} \text{ and } \frac{34}{51} = \frac{2}{3}$$

Therefore $\frac{22}{33} = \frac{34}{51}$.

c. $\frac{19}{26}$ and $\frac{28}{39}$

LCD = 78

$$\frac{19}{26} = \frac{57}{78}, \frac{28}{39} = \frac{56}{78}$$

Since $57 > 56$, $\frac{19}{26} > \frac{28}{39}$.

d. $\frac{47}{59}$ and $\frac{135}{147}$

$$\frac{47}{59} \doteq 0.80 \text{ and } \frac{135}{147} \doteq 0.92$$

(rounded to nearest hundredth)

Hence $\frac{47}{59} < \frac{135}{147}$.

13.1 Exercises

Write the expressions in Exercises 1-12 in algebraic form.

1. Three is not equal to seven.
2. Three is less than seven.
3. Three is less than or equal to seven.
4. x plus one is greater than x.
5. Eighteen and three tenths is greater than or equal to eighteen and three tenths.
6. Four is the same as two squared.
7. Twelve is less than twenty.
8. Ten is unequal to five.
9. x plus ten is greater than x.
10. Eight is less than or equal to eight.
11. Sixteen is the same as four squared.
12. Seven-tenths is greater than or equal to four-tenths.

Write a verbal expression for each of the Exercises 13-24.

13. $7 \geqslant 7$
14. $-5 < -3$

15. $\frac{1}{2} > \frac{1}{3}$

16. $x \neq y$

17. $-17 \leqslant -17$

18. $2(x+1) = 5x$

19. $-12 < -5$

20. $9 \geqslant 7$

21. $y \neq z$

22. $\frac{3}{4} > \frac{1}{2}$

23. $3(x+1) = 3x+3$

24. $-6 \leqslant -6$

Determine the order relation between the pairs of numbers in Exercises 25-50.

25. 2, 6

26. −4, −3

27. −2, −6

28. −2, 6

29. 2, −6

30. 0, 2

31. 0, −2

32. 0, 6

33. 0, −6

34. $0, \frac{1}{2}$

35. $0, -\frac{1}{2}$

36. $\frac{1}{2}, -\frac{1}{2}$

37. $\frac{2}{7}, \frac{3}{11}$

38. −0.419, −0.418

39. $\frac{-11}{23}, \frac{-13}{27}$

40. 149.3, 149.03

41. −1, 1

42. 3, −2

43. −8, −5

44. 4, 10

45. −0.63, −0.61

46. $\frac{1}{9}, \frac{2}{17}$

47. 13.5, 13.05

48. $\frac{-9}{20}, \frac{-11}{41}$

49. 2, −4

50. −2, 2

13.2 Conditional Inequalities

Most of the inequalities in the previous section involved only constants in both members of the inequality. Even those inequalities which contained variables, such as $x < x + 1$, were of a very special type. Inequalities which are true for all values of the variable involved are called **absolute inequalities.** Since the inequality $x < x + 1$ is true for each real number substituted for x, this is an absolute inequality. Inequalities which are not true for any value of the variable are called **contradictions.** Since the inequality $x > x + 1$ is not true for any value of x, this is a contradiction.

This section will investigate conditional inequalities and these inequalities will be solved in the next two sections. A **conditional inequality** is an inequality which is true for some, but not all, values of the variable involved. For example, $3x > 6$ is true if x is replaced by 3, 5, or 17 but is false if x is replaced by −7, 0, or 2. Hence, $3x > 6$ is a conditional inequality.

The **solution set** of an inequality is the set or collection of all values which make the inequality a true statement. The number line is an excellent way of visualizing the solution set, which is often an infinite set. The graph of the solution set for $x > 4$ can be drawn as follows.

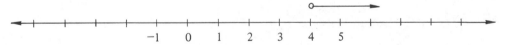

The *open circle* above 4 indicates that 4 is not an element or member of the solution set. That is, 4 does *not make* the inequality a true statement. A *solid dot* would be used instead of the open circle if 4 were included in the solution set. Note that the values to the left of 4 are not in the solution set. The *arrow* to the right of the open circle indicates that all values located to the right of 4 on the number line are in the solution set. If the arrow had pointed to the left, this would indicate that the numbers less than 4 were in the solution set.

As a test of the accuracy of our graph, let us take a value that is included in the graph of the solution set, such as 5. Substituting 5 into the inequality $x > 4$, we obtain $5 > 4$, which is a true statement. This does not verify that our graph is perfect. However, it does test the value 5, and it verifies that we have pointed the arrow in the proper direction.

Although 5 is one solution for the inequality $x > 4$, note that 5 is not the only solution. The solution set for an inequality must contain all the numbers which make the statement true.

Example 1   Sketch the graph of the solution set of each of the following inequalities.

a. $x \leqslant 2$

b. $y > -3$

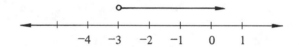

c. $x < -1$

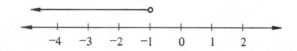

d. $x \geqslant \dfrac{1}{2}$

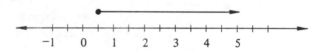

e. $1.67 > x$ (*Hint:* Rewrite as $x < 1.67$.)

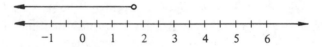

f. $-2.5 \leqslant x$ (*Hint:* Rewrite as $x \geqslant -2.5$.)

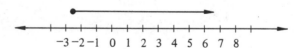

Example 2   a. Sketch the graph of the solution set of $x < -4$ and use the graph to determine which of the following are solutions.

$$5, -4, -6, -2, -5$$

The graph of the solution set of $x < -4$ is sketched as follows:

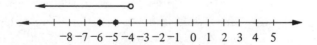

-6 and -5 are solutions; 5, -4, and -2 are not solutions.

b. Determine which of the following are solutions of $x \geq 5$ and sketch the graph of the solution set.

$$1, 2, 3, 4, 5, 6, 7\frac{1}{2}$$

The graph of the solution set of $x \geq 5$ is sketched as follows:

5, 6, and $7\frac{1}{2}$ are solutions; 1, 2, 3, and 4 are not solutions.

Example 3   List 10 elements from the solution set of $x \geq 10$.

$$10, 10\frac{1}{2}, 11, 15, 20, 100, 500, 501, 508.3, 1000$$

Any number greater than or equal to ten could be listed.

13.2   Exercises

Sketch the graphs of the inequalities in Exercises 1-10.

1. $x \geq 3$ **2.** $x < -4$ **3.** $-3 < x$

4. $-0.718 \geq x$ **5.** $x < 4.1$ **6.** $x \leq -5\frac{2}{3}$

7. $4\frac{1}{2} > x$ **8.** $x \geq 300$ **9.** $589 \leq x$

10. $x < 750$

Write an inequality for the graphs in Exercises 11-18. Use x as the unknown.

11.

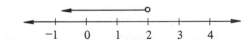

12.

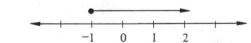

13.

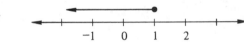

14.

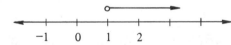

15.

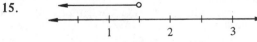

16.

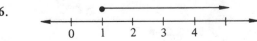

17.

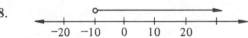

18.

In Exercises 19-23, answer true (T) or false (F).

19. -4 is a solution for $x < 3$.

20. The only solutions for $x > 1$ are 2, 4, and 6.

21. 0 is a solution for $x > 0$.

22. 2, 4, and 6 are solutions for $x > 1$.

23. The only solutions for $x > 0$ are 1, 2, 3, 4, etc.

24. Which of the following are solutions for $x < 10$?
2, 4, 6, 8, 10, 12

25. Which of the following are solutions for $x > -5$?
$-10, -5, 0, 5, 10$

26. Which of the following are solutions for $x > 0$?
$-1, -2, -3, -4, -5$

27. List 4 elements from the solution set of $x > 10$.

28. List 6 elements from the solution set of $x \leq -4$.

29. List 5 numbers that are not in the solution set of $x < 100$.

30. List 3 numbers that are not in the solution set of $x \geq -12$.

13.3 Solving Conditional Inequalities Using the Multiplication–Division Principle

Solving conditional inequalities is nearly identical to solving conditional equations. The basic idea of isolating the variable in one member of the inequality and the constants in the other member is the same main idea we used to solve equations. Given an inequality, we will systematically form equivalent inequalities until only the variable remains on one side of the inequality. Inequalities are **equivalent** when they have the same solution set. There are two major principles that are used to produce equivalent inequalities. The first is the multiplication–division principle.

If you multiply both sides of an equation by any nonzero number, you will obtain an equivalent equation. If you multiply both sides of an inequality by a nonzero number, the resulting inequality may not be equivalent to the original inequality. Therefore, you must exercise caution when multiplying or dividing both sides of an inequality by the same number. The following three graphs suggest how you may obtain an equivalent inequality.

Multiply by 2

When both sides of the inequality $2 < 3$ are multiplied by 2, we obtain 4 and 6. If we compare the points corresponding to 2 and 3 and then 4 and 6, we see that multiplying by 2 seems to stretch out the space between the two but preserves the original order of the points. Thus, $2 < 3$ and $4 < 6$.

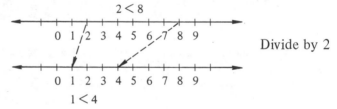

Divide by 2

Dividing by a positive 2 seems to shrink the space between the numbers on the number line, but the original order of the points is preserved.

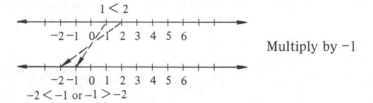

Multiply by −1

When both members of the inequality $1 < 2$ are multiplied by −1, however, we obtain −1 and −2. If we compare the points corresponding to 1 and 2 and then −1 and −2, we see that multiplying by −1 reversed the order of the original points. In other words, $1 < 2$ but $-1 > -2$. That is, we start with a number line looking like this:

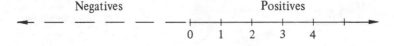

After multiplying by −1, the number line is reversed to the following:

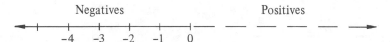

Consider the *multiplication–division principle for inequalities* which summarizes what was suggested by these graphs.

> **THE MULTIPLICATION–DIVISION PRINCIPLE FOR INEQUALITIES**
>
> a. If both members of an inequality are multiplied or divided by a positive number, the result is an equivalent inequality with the order preserved.
>
> b. If both members of an inequality are multiplied or divided by a negative number *and the order is reversed*, the result is an equivalent inequality.

Example 1   Solve the inequality $3x \leqslant 9$, and sketch the graph of the solution set.

$$3x \leqslant 9$$

$$\frac{3x}{3} \leqslant \frac{9}{3} \qquad \text{Division by positive 3 preserves the order}$$

$$x \leqslant 3$$

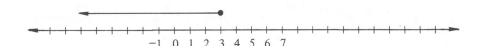

Set builder notation is sometimes used to indicate solution sets. To describe the solution set in this example we could write $\{x \mid x \leqslant 3\}$, read as "the set of all values of x such that x is less than or equal to 3."

Example 2   Solve the following inequalities, draw graphs of the solution sets, and test one value from each solution set.

a. $-2x > 1$

$$-2x > 1$$

$$\frac{-2x}{-2} < \frac{1}{-2} \qquad \text{Division by negative 2 reverses the order}$$

$$x < -\frac{1}{2}$$

Test the value of −3: $-2(-3) > 1$
$$6 > 1$$
Since this is a true statement, −3 is in the solution set.

b. $1.17x \geqslant -0.234$

$$1.17x \geqslant -0.234$$

$$\frac{1.17x}{1.17} \geqslant \frac{-0.234}{1.17} \qquad \text{Division by positive 1.17 preserves the order}$$

$$x \geqslant -0.2$$

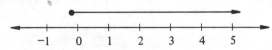

Test the value 0: $1.17(0) \geqslant -0.234$

$$0 \geqslant -0.234$$

Since this is a true statement, 0 is in the solution set.

c. $-\frac{14}{15} \geqslant -\frac{2}{3}x$

$$-\frac{14}{15} \geqslant -\frac{2}{3}x$$

$$\left(-\frac{3}{2}\right)\left(-\frac{14}{15}\right) \leqslant \left(-\frac{3}{2}\right)\left(-\frac{2}{3}x\right) \qquad \begin{array}{l}\text{Multiplication by a negative value}\\ \text{reverses the order}\end{array}$$

$$\frac{7}{5} \leqslant x$$

Thus, $x \geqslant \frac{7}{5}$.

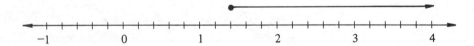

Test the value 3: $-\frac{14}{15} \geqslant -\frac{2}{3}(3)$

$$-\frac{14}{15} \geqslant -2$$

Since this is a true statement, 3 is in the solution set.

Example 3   Write the following statements as inequalities and solve.

a. Two-thirds of m is less than 16.

$$\frac{2}{3}m < 16$$

$$\frac{3}{2} \cdot \frac{2}{3}m < \frac{3}{2} \cdot \frac{16}{1} \qquad \begin{array}{l}\text{Multiplication by a positive value preserves the}\\ \text{order}\end{array}$$

$$m < 24$$

b. The opposite of five times a certain number is greater than thirty-five.

$$-5n > 35$$

$$\frac{-5n}{-5} < \frac{35}{-5} \qquad \text{Division by a negative value reverses the order}$$

$$n < -7$$

13.3 Exercises

Solve each of the inequalities in Exercises 1–10 and sketch the graph of the solution set. Check one value from the solution set.

1. $2x \geqslant 4$ **2.** $-2x < 4$

3. $-3x > 6$ **4.** $-3x < 6$

5. $\frac{1}{2}x > 4$ **6.** $-\frac{1}{2}x > 4$

7. $-\frac{1}{3}x \leqslant 1$ **8.** $-\frac{1}{3}x > 1$

9. $\frac{-3}{4}x < 3$ **10.** $\frac{-4}{3}x \leqslant 4$

Solve each of the inequalities in Exercises 11–50.

11. $7x \geqslant 4$ **12.** $1.1x < -1.21$

13. Three times x is less than sixty-three.

14. $-4x > 8$ **15.** $-10 > -5x$

16. $\frac{4}{7}x \leqslant \frac{-8}{21}$

17. Seven-ninths of a certain number is greater than forty-nine fiftieths. (Let n represent the number.)

18. $-4\frac{1}{3}x > \frac{26}{9}$ **19.** $0.417x < -8.34$

20. Seven-thousandths of a number is greater than or equal to forty-nine ten-thousandths.

21. $-168x \leqslant 84$ **22.** $-3\frac{5}{9} > -5\frac{1}{3}x$

23. $2 \leqslant -x$ **24.** $2x < 30$

25. $0.6x > 3$

26. Ninety-eight is less than x divided by 2.

27. $-5x \leqslant -15$ **28.** $12 < -\frac{2}{3}x$

29. $2 \leqslant 3x$

30. Forty-eight ten-thousandths is less than or equal to forty-eight times x.

31. $8x < 0$ **32.** $7x \leqslant 45$

33. $45y < 5$

34. Three-fourths x is less than 39.

35. $\frac{x}{0.013} > -5.4$ **36.** $\frac{-2w}{7} < -10$

37. $\frac{-5h}{3} \leqslant 25$

38. Five-sevenths of a certain number is greater than or equal to twenty-five. (Let n represent the number.)

39. $-3\frac{1}{2}x > \frac{18}{7}$ **40.** $4\frac{3}{7}x < 2\frac{20}{21}$

41. $-x \geqslant 4$

42. Twelve-thousandths of a number is greater than or equal to forty-eight ten-thousandths.

43. $\frac{x}{1.3} \leqslant 17$ **44.** $\frac{-11x}{4} < 0$

45. $5x < 30$ **46.** $0.6x > 4$

47. Fifty-six is less than x divided by two.

48. $-8x \leqslant -56$ **49.** $\frac{-2}{5}x > 12$

50. $5x > 3$

13.4 Solving Conditional Inequalities Using the Addition–Subtraction Principle

The two graphs which follow will illustrate the second principle for producing equivalent inequalities, the addition–subtraction principle. This principle can be used to isolate the variable terms in one member of the inequalities in exactly the same way the addition–subtraction principle for equalities is used.

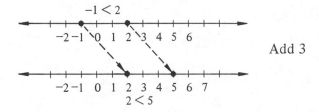

By adding 3 to both sides of the inequality $-1 < 2$, the points -1 and 2 have both been repositioned 3 units to the right on the number line. The important thing to note is that both points were shifted so that their position relative to each other is the same. On the graph, −1 is to the left of 2. After the points are repositioned three units to the right, 2 is to the left of 5.

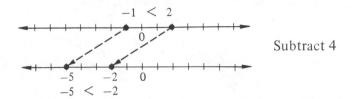

Subtract 4

In this graph, both points are shifted 4 units to the left. However, the relative position between the points is maintained. In particular, since −1 is to the left of 2, −5 is to the left of −2. In other words, the order is preserved.

> ### THE ADDITION–SUBTRACTION PRINCIPLE FOR INEQUALITIES
> If the same number is added to or subtracted from both members of an inequality, the result is an equivalent inequality with the order preserved.

Example 1 Solve the following inequalities, sketch the graph of the solution set of each, and check one value from the solution set.

a. $x + 3 \leqslant 7$

$$x + 3 \leqslant 7$$
$$x + 3 - 3 \leqslant 7 - 3 \qquad \text{Subtracting 3 preserves the order}$$
$$x \leqslant 4$$

Test the value 2: $2 + 3 < 7$
$$5 < 7$$
Since this is a true statement, 2 is in the solution set.

b. $3x + 11 > 2x + 8$

$$3x + 11 > 2x + 8$$
$$3x + 11 - 11 > 2x + 8 - 11 \qquad \text{Subtracting 11 preserves the order}$$
$$3x > 2x - 3$$
$$3x - 2x > 2x - 3 - 2x \qquad \text{Subtracting 2x preserves the order}$$
$$x > -3$$

Test the value −2: $3(-2) + 11 > 2(-2) + 8$
$$-6 + 11 > -4 + 8$$
$$5 > 4$$
Since this is a true statement, −2 is a solution.

In the following examples both principles for solving inequalities are used. We generally use the *addition-subtraction principle* first. This enables us to isolate the variables on one side of the inequality and the constants on the other side. Then we use the *multiplication-division principle* to obtain a coefficient of 1 for the variable.

Example 2 Solve the following inequalities, sketch the graph of the solution set, and check one value from the solution set.

a. $2x + 3 < 15$

$$2x + 3 < 15$$
$$2x + 3 - 3 < 15 - 3 \qquad \text{Subtracting 3 preserves the order}$$
$$2x < 12$$
$$\frac{2x}{2} < \frac{12}{2} \qquad \text{Dividing by positive 2 preserves the order}$$
$$x < 6$$

Test the value 0: $2(0) + 3 < 15$
$$3 < 15$$
Since this is true, 0 is a solution.

b. $6x - 2 \geqslant 8x + 12$

$$6x - 2 \geqslant 8x + 12$$
$$6x - 2 - 8x \geqslant 8x + 12 - 8x \qquad \text{Subtracting 8x preserves the order}$$
$$-2x - 2 \geqslant +12$$
$$-2x - 2 + 2 \geqslant 12 + 2 \qquad \text{Adding 2 preserves the order}$$
$$-2x \geqslant 14$$
$$\frac{-2x}{-2} \leqslant \frac{14}{-2} \qquad \text{Dividing by negative 2 reverses the order}$$
$$x \leqslant -7$$

Alternate Solution: The solution can be shortened as follows if the addition and division are done mentally.

$$6x - 2 \geqslant 8x + 12$$
$$-2x \geqslant 14$$
$$x \leqslant -7$$

Test the value -8: $6(-8) - 2 \geqslant 8(-8) + 12$
$$-48 - 2 \geqslant -64 + 12$$
$$-50 \geqslant -52$$
Since this is a true statement, -8 is a solution.

c. $5x - 17 \geqslant 6x - 18$

$$5x - 17 \geqslant 6x - 18$$
$$5x - 17 + \mathbf{18} \geqslant 6x - 18 + \mathbf{18} \qquad \text{Adding 18 preserves the order}$$
$$5x + 1 \geqslant 6x$$
$$5x + 1 - \mathbf{5x} \geqslant 6x - \mathbf{5x} \qquad \text{Subtracting 5x preserves the order}$$
$$1 \geqslant x$$
$$x \leqslant 1 \qquad \text{These last two inequalities express the same thing.}$$

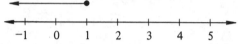

Alternate Solution:

$$5x - 17 \geqslant 6x - 18$$
$$5x - 17 - \mathbf{6x} \geqslant 6x - 18 - \mathbf{6x} \qquad \text{Subtracting 6x preserves the order}$$
$$-x - 17 \geqslant -18$$
$$-x - 17 + \mathbf{17} \geqslant -18 + \mathbf{17} \qquad \text{Adding 17 preserves the order}$$
$$-x \geqslant -1$$
$$(-x)(\mathbf{-1}) \leqslant (-1)(\mathbf{-1}) \qquad \text{Multiplying by } -1 \text{ reverses the order}$$
$$x \leqslant 1$$

Test the value 0: $5(\mathbf{0}) - 17 \geqslant 6(\mathbf{0}) - 18$
$$-17 \geqslant -18$$
Since this is a true statement, 0 is in the solution set.

d. $4x - 20 \leqslant 9x + 18$

$$4x - 20 \leqslant 9x + 18$$
$$4x - 20 - \mathbf{9x} \leqslant 9x + 18 - \mathbf{9x} \qquad \text{Subtracting 9x preserves the order}$$
$$-5x - 20 \leqslant 18$$
$$-5x - 20 + \mathbf{20} \leqslant 18 + \mathbf{20} \qquad \text{Adding 20 preserves the order}$$
$$-5x \leqslant 38$$
$$\frac{-5x}{\mathbf{-5}} \geqslant \frac{38}{\mathbf{-5}} \qquad \text{Dividing by } -5 \text{ reverses the order}$$
$$x \geqslant \frac{-38}{5} \text{ or } -7\frac{3}{5}$$

Alternate Solution:

$$4x - 20 \leqslant 9x + 18$$

$$4x - 20 - \mathbf{18} \leqslant 9x + 18 - \mathbf{18} \qquad \text{Subtracting 18 preserves the order}$$

$$4x - 38 \leqslant 9x$$

$$4x - 38 - \mathbf{4x} \leqslant 9x - \mathbf{4x} \qquad \text{Subtracting 4x preserves the order}$$

$$-38 \leqslant 5x$$

$$\frac{-38}{5} \leqslant \frac{5x}{5} \qquad \qquad \text{Dividing by 5 preserves the order}$$

$$\frac{-38}{5} \leqslant x$$

$$x \geqslant -\frac{38}{5} \text{ or } -7\frac{3}{5} \qquad \text{The last two inequalities express the same thing.}$$

Test the value -7: $4(-7) - 20 \leqslant 9(-7) + 18$

$$-28 - 20 \leqslant -63 + 18$$

$$-48 \leqslant -45$$

Since this is a true statement, -7 is in the solution set.

13.4 Exercises

Solve each of the inequalities in Exercises 1–10 and sketch the graph of the solution set. Check one value from the solution set.

1. $x + 7 > 11$ 2. $x - 7 < -11$

3. $x - 7 < -3$ 4. $x - 3 < -7$

5. $x + 6 \geqslant -2$ 6. $x - 2 \geqslant -6$

7. $2x + 3 < x + 4$ 8. $x - 4 < 2x - 3$

9. $5x - 3 \geqslant 4x - 9$ 10. $4x + 9 \geqslant 5x + 3$

Solve each of the inequalities in Exercises 11–50.

11. $x + 7 \leqslant 9$ 12. $2x - 3 > x - 4$

13. $5x - 8 < 4x - 8$ 14. $5x - 7 \leqslant 6x - 9$

15. $5x \leqslant 21 - 2x$ 16. $0.45x < 0.04x + 0.533$

17. $9x + 7 \geqslant 5x - 13$ 18. $23x - 14 > 18x + 56$

19. $5x + 21 \geqslant 13 - 3x$

20. $3.11x - 2.19 \geqslant 2.11x - 1.19$

21. If seven is added to x, the result is more than twelve.

22. If nine is subtracted from x, the result is less than twenty.

23. $0.5x + 3 \leqslant 0.2x - 5$ 24. $-8x + 5 < -10x + 5$

25. $-\frac{2}{3}x - \frac{1}{2} \leqslant \frac{4}{5}x$ (Multiply both numbers by LCD.)

26. Three x plus seven is greater than four x minus 12.

27. Three-tenths of x is greater than the sum of seven-tenths x and seven.

28. $6.4x - 0.54 \leqslant 0.4x - 0.46$

29. $4x - 9 > 6x - 9$ 30. $25x - 12 \geqslant 30x + 15$

31. $x - 7 \leqslant 10$ 32. $5x + 11 > 6$

33. $-2x - 7 < 11$ 34. $1.23x - 0.668 \geqslant 8.065$

35. $\frac{3}{5}x + 1 > 22$ 36. $17x > 15 + 2x$

37. $3x + 5 < 6x + 5$ 38. $29x < 4x - 0.50$

39. $83x \geqslant -57x + 13$ 40. $-17x < -2x + 45$

41. $5x + 3 \leqslant 2x + 45$ 42. $8x - 12 > 6x - 12$

43. $\frac{1}{2}x > \frac{3}{4}x - 2$

44. If twenty is added to 3x, the result is more than fifty-six.

45. If eighteen is subtracted from 4x, the result is less than negative eighteen and one-tenth.

46. $75x - 420 \leqslant 50x + 24$ 47. $12x + 720 < 9x - 34$

48. $9w - 27 \leqslant 8w + 24$

49. If sixteen is subtracted from 5x, the result is no more than the sum of 3x and ten. (*No more than* means the same thing as *less than or equal to*.)

50. If 5x is subtracted from sixteen the result is no less than the sum of 3x and ten. (*No less than* means the same thing as *greater than or equal to*.)

13.5 Solving More General Forms of Linear Inequalities

In Sections 13.3 and 13.4, you probably noticed that the procedure for solving linear inequalities is nearly identical to the procedure for solving linear equalities. The only exception is that you must determine at each step whether the order is preserved or reversed. As you study the examples in this section and work the exercises that follow, note how similar the procedure is to that outlined for solving linear equations.

Example 1 Solve the following inequality, and sketch the graph of the solution set:
$$5(3x - 2) + 13 < 11(x - 3) + 4(2x + 1)$$

$15x - 10 + 13 < 11x - 33 + 8x + 4$	Remove parentheses.
$15x + 3 < 19x - 29$	Combine like terms.
$15x + 3 - 3 < 19x - 29 - 3$	Subtract 3 (preserves the order).
$15x < 19x - 32$	
$15x - \mathbf{19x} < 19x - 32 - \mathbf{19x}$	Subtract 19x (preserves the order).
$-4x < -32$	
$\dfrac{-4x}{-4} > \dfrac{-32}{-4}$	Divide by −4 (reverses the order)
$x > 8$	

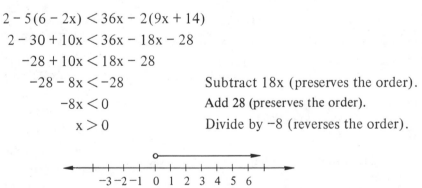

Example 2 Solve $2 - 5(6 - 2x) < 36x - 2(9x + 14)$, and sketch the graph of the solution set.

$2 - 5(6 - 2x) < 36x - 2(9x + 14)$	
$2 - 30 + 10x < 36x - 18x - 28$	
$-28 + 10x < 18x - 28$	
$-28 - 8x < -28$	Subtract 18x (preserves the order).
$-8x < 0$	Add 28 (preserves the order).
$x > 0$	Divide by −8 (reverses the order).

Example 3 If four is added to a certain number and the result is multiplied by three, this quantity is at least ten times the number diminished by eight. Find the number.

Let n represent the number. Since "at least" means the same thing as "greater than or equal to," this statement may be written algebraically as

$$3(n + 4) \geqslant 10n - 8$$
$$3n + 12 \geqslant 10n - 8$$
$$-7n \geqslant -20 \qquad \text{Subtract 10n (preserves the order).}$$
$$n \leqslant 2\frac{6}{7} \qquad \text{Divide by } -7 \text{ (reverses the order).}$$

The required number could be any number less than or equal to $2\frac{6}{7}$.

When we discussed linear equations we considered two special cases:

1. An equation with no solution—a contradiction.
2. An equation that has every real number as a solution—an identity.

The following examples illustrate the corresponding cases for inequalities.

Example 4 $3x + 4 < 3x + 1$

$$3x + 4 - 3x < 3x + 1 - 3x$$
$$4 < 1$$

Since $4 < 1$ is not true, the original inequality is a contradiction, and there is no value for x which will make it a true statement. Thus the inequality has no solution.

Example 5 $3x + 4 > 3x + 1$

$$3x + 4 - 3x > 3x + 1 - 3x$$
$$4 > 1$$

Since $4 > 1$ is true, any number that is substituted in the original inequality will make it a true statement. Thus, the original statement is an absolute inequality, and the solution set is the set of real numbers.

13.5 Exercises

Solve each of the inequalities in Exercises 1–10 and sketch the graph of the solution set. Check one value from the solution set.

1. $3(2x - 5) < 5x - 18$
2. $7x + 14 > 2(3x + 8)$
3. $-2(5x + 2) \geqslant 1 - 9x$
4. $-5(2x - 2) \leqslant 7 - 11x$
5. $4(2x - 3) \leqslant 2(3x - 8)$
6. $5(3x - 6) \geqslant 3(6x - 4)$
7. $2(3x - 4) > 3(x - 6) + 1$
8. $2(11x - 3) < 5(3x + 2) - 20$
9. $5(x - 4) + 6 < 6(x + 4) - 30$
10. $7(x - 3) + 6 > 12(x + 4) + 2$

Solve each of the inequalities in Exercises 11–50 and sketch the graph of the solution set.

11. $5x \geqslant 21 - 2x$
12. $0.04x + 0.533 > 0.45x$
13. Two x plus 4 is greater than 16.
14. $2(3x - 5) \leqslant 7(2x + 5) - 5$

15. $4(2 - 3x) + 6 > 2 - 6(x + 1)$

16. Two-thirds of x is less than x minus 5.

17. Two times a certain number is greater than twelve more than twice this number. (Let n represent the number.)

18. Four times the quantity x minus three is no more than 6x plus 15. ("No more than" means the same thing as "less than or equal to.")

19. $-4(x - 3) < 5 - 4x$

20. $\frac{3x}{7} + \frac{4}{5} > \frac{3x}{5} + \frac{2}{7}$ (*Hint:* Multiply both members of the inequality by the LCD.)

21. Nine more than two-thirds of a number is at least five more than one half the same number. ("At least" means the same as "greater than or equal to.")

22. Nine less than 2x is greater than two times the quantity x minus 12. (Check at least one value.)

23. $\frac{-x}{2} < \frac{1}{4}$ 24. $9 > \frac{3w}{5}$

25. If 4 is added to x, the result is no more than 10.

26. Two times a certain number is greater than the sum of twice the number and 10.

27. $\frac{x - 3}{10} \geqslant \frac{x - 2}{3}$ 28. $3\frac{2}{9}x \geqslant 4\frac{1}{6}x - (x - 6)$

29. If 12 is subtracted from x, the result is at most 20. ("At most" means the same as "less than or equal to.")

30. Three less than four times x is at least two more than three times x. ("At least" means the same as "is greater than or equal to.")

31. Three times a certain number is less than two times the same number. (Let n represent the number.)

32. Two-sevenths of x is greater than $\frac{8}{9}$.

33. $6x + 15 < 3(2x + 10)$ 34. $8 > 6w + 8$

35. 0.48 times x is no more than 12.

36. The opposite of 5x is less than 25.

37. Six less than 10x is less than 4 more than 12x.

38. Five more than 2x is at least three more than x.

39. $8(x + 5) \geqslant 4x + 2(x - 10)$

40. $\frac{2}{3}x + \frac{1}{2} \leqslant \frac{7}{9} - \frac{1}{3}$

41. $3(x - 2) < 4(x + 1) - 5$

42. $4(x - 3) + 7 \geqslant 2x + 3(x - 2)$

43. $17(x + 1) \geqslant 16(x + 1)$

44. $5(x + 3) > 4(x - 1) + 17$

45. $4(x - 1) + 3(x + 2) \leqslant 7(x + 1) - 5$

46. $-4(x - 2) < 3(x - 7) - 5(4 - x) - 1$

47. The product obtained when four is multiplied by the quantity 2x plus three is less than the sum of 8x and twelve.

48. $7(n - 1) - 4(2n + 3) > 2(n + 1) - 3(4 - n)$

49. $\frac{x}{3} - \frac{1}{6} < \frac{x + 1}{9}$ 50. $\frac{x - 3}{11} \geqslant \frac{x - 2}{3}$

Important Terms Used in this Chapter	absolute inequality
	addition–subtraction principle
	  for inequalities
	conditional inequalities
	contradiction
	equivalent inequalities
	is greater than
		is greater than or equal to
		is less than
		is less than or equal to
		multiplication–division principle
		  for inequalities
		order relations
		solution set for an inequality

Important Symbols Used in this Chapter	$<$ indicates one quantity is less than another $>$ indicates one quantity is greater than another $\neq$ indicates one quantity is not equal to another $\leqslant$ indicates one quantity is less than or equal to another quantity $\geqslant$ indicates one quantity is greater than or equal to another quantity

REVIEW EXERCISES

Answer Exercises 1-20 as true (T) or false (F).

1. $-2 > -2$
2. $-2 \geqslant -2$
3. $2 \geqslant -2$
4. $2 > -2$
5. $0 < -2$
6. $0 < 2$ is read 0 is less than 2.
7. $0 \leqslant 2$ is read 0 is less than 2.
8. $0.8176 > 0.817$
9. $\frac{-3}{17} > \frac{-4}{17}$
10. $\frac{19}{43} < \frac{23}{57}$

11. The graph of $x > 3$ is

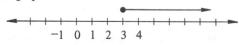

12. The graph of $x < -2$ is

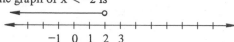

13. The graph of $x \geqslant 4$ is

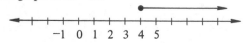

14. The graph of $x < \frac{1}{2}$ is

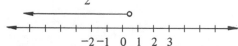

15. The graph of $5 < x$ is

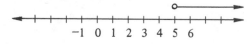

16. $x > x - 1$ is an absolute inequality.
17. $3x > 6$ is a contradiction.
18. Adding *any* number to both members of an inequality preserves the order relation.
19. Subtracting *any* number from both members of an inequality preserves the order relation.
20. Multiplying *any* number times both members of an inequality preserves the order relation.

Solve the inequalities in Exercises 21-35 and sketch the graph of the solution set. Check one value from the solution set.

21. $x + 5 < 6$
22. $x > 2x - 3$
23. The sum of x and 8 is less than 25. Find x.
24. Five-eighths of a certain number is more than 50. Find the number.
25. $3x + 5 \leqslant 2x + 6$
26. $4x < 8$
27. $-5x \geqslant 15$
28. $-\frac{1}{3}x < 1$
29. If 3x is added to 5x, the result is no more than 8x plus ten.
30. The sum of a number and two is less than the sum of the same number and one.
31. $11x - 15 > 18$
32. $4(3x - 5) + (x - 1) \leqslant 7(x - 1)$
33. $4(x - 1) < 4x + 2$
34. $5(x + 3) \geqslant 4(x + 4) - (1 - x)$
35. Four times the quantity $x - 4$ is no more than three times the quantity $x + 10$.

GRAPHS AND SYSTEMS OF EQUATIONS

14

GRAPHS

PRETEST

1. The following pictograph shows the amount of personal income which was derived from direct government payments from 1945–1975 (Objective 1).

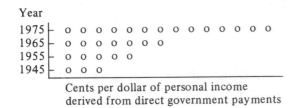

Cents per dollar of personal income derived from direct government payments

a. How many cents per dollar of personal income came from government payments in 1955?

b. What is the apparent trend in amount of personal income derived from government payments?

c. What was the contribution of the government for each $1000 of personal income in 1975?

The following bar graph shows the tar (mg/cig) present in 9 brands of cigarettes (Objective 1).

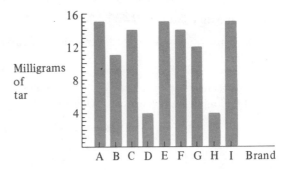

d. Which brand(s) had the lowest amount of tar?

e. How many milligrams of tar did each brand F cigarette contain?

f. Which brand(s) contained exactly 12 milligrams of tar per cigarette?

335

A family had a monthly income of $1200. The following circle graph shows the percent of this family's monthly income which was spent on various items (Objective 1).

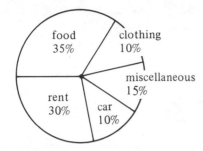

g. On which item was the largest proportion of the family income spent?

h. How much money was spent on car expenses each month?

i. How much was spent on clothing by this family during one year?

The following line graph shows the monthly car expenses for a family (Objective 1).

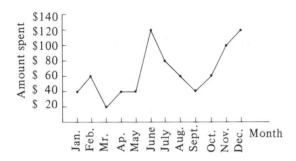

j. What was the lowest monthly expense of the year?

k. Which month did it occur?

l. In how many months were the monthly expenses less than $50?

2. a. Draw a rectangular coordinate system, and plot and label the following ordered pairs (Objective 2):

$(2, 4)$, $\left(-\frac{1}{2}, 0\right)$, $(-7, -5)$, $(0, 4)$, $(-2, 4)$, $(-2, -4)$, $(2, -4)$, $\left(4\frac{1}{2}, \frac{7}{3}\right)$.

b. Identify the quadrant in which each of the points in part a is located. If a point is not in a quadrant, specify the axis on which it is located (Objective 2).

c. Give the coordinates of each of the points indicated in the rectangular coordinate system below (Objective 2).

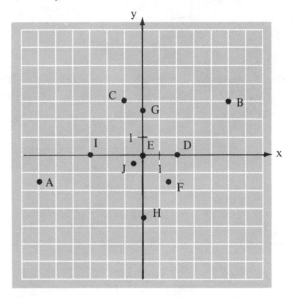

3. Draw the graph of each of the following equations (Objective 3).

a. $x + y = 5$ **b.** $-2x = 9$

c. $2x = 15 - 3y$ **d.** $y = -\frac{1}{2}x - 1$

4. Write the following equations in slope-intercept form (Objective 4).

a. $2x + y = 5$ **b.** $3x - 4y = 9$

c. $x - 3y = 8$ **d.** $2x + 3y = 15$

5. Give the slope of the equations in question 4 (Objective 5).

6. Draw the graph of each of the following inequalities (Objective 6).

a. $x - y \leqslant 7$ **b.** $3x \geqslant 5y$

c. $5x + 7y > 11$ **d.** $y < -2$

Graphs are pictures or diagrams that show the relationship between quantities that can be counted or measured. Newspapers, magazines, and textbooks use many different kinds of graphs to present data to their readers. The most common are bar graphs, pictographs, circle graphs, and line graphs. Our primary interest will be interpreting these graphs.

Graphs that are specifically used to show mathematical relationships are usually drawn on a coordinate system which uses two perpendicular lines as axes. Solution sets of equations and inequalities can be represented on a coordinate system as points, lines, or regions on the graph.

14.1 Interpreting Graphs

A **bar graph** is a very common graph that can be easily interpreted. The essential feature of the bar graph is a number scale against which the length of the bars may be compared. The bars may be horizontal or vertical. The following examples demonstrate some of the ways bar graphs are used.

Example 1   The bar graph that follows shows the average class size per year for a large metropolitan high school.

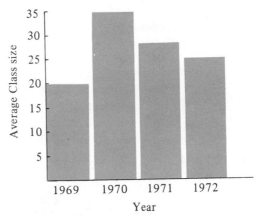

Use the graph to answer the following questions:

a. What was the average class size in 1972?   25

b. During which year was the average class size largest?   1970

c. Which year were there more students per class, 1969 or 1971?   1971

Example 2   A public opinion poll asked students to compare the current student activities program with the program of the previous year. The response is interpreted by the following graph.

Student Activities Program

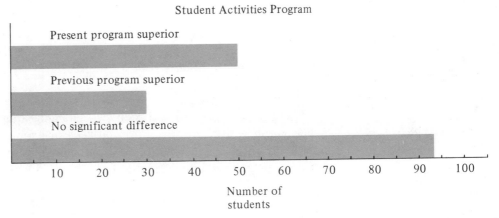

Use the graph to answer the following questions.

a. How many students thought the present program was superior?   50

b. How many students thought the previous program was superior?   30

c. How many students thought there was no significant difference?
(The answer here must be estimated.)   93

A **pictograph** is actually a variation of a bar graph. For example, the picto-graph below conveys the same information as the bar graph in Example 2, but stick figures are used to represent the students.

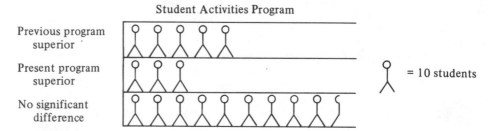

When parts of a quantity are to be compared, a **circle graph** is often used. The circle represents the whole quantity, and the pie-like wedges of various sizes represent the parts. The wedges are proportional to the parts they repre-sent with larger wedges representing larger parts. Some geometry is needed to construct a circle graph in the proper proportion; however, the interpretation is fairly simple. Percents are used to label circle graphs since percents can easily be interpreted to represent a part of the whole.

Example 3   The graph in Figure 14.1 shows how the income from a small business was dis-persed. Use the graph to answer the following questions.

Figure 14.1

a. What percent of the income was spent for supplies and wages
combined?   55%

b. What percent of the income was spent for operating expenses?   33%

c. What percent was profit?   12%

The entire graph represents 100%.

When the relationship between two changing quantities is being compared, a **line graph** is usually more informative. For example, if we are interested in comparing the fall enrollments at a junior college for five consecutive years, a broken line graph like the one that follows could be used.

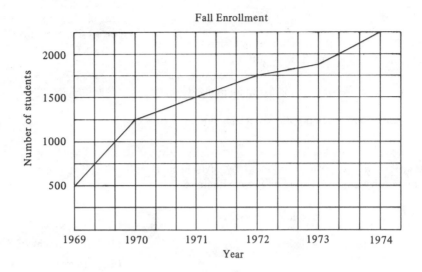

Fall Enrollment

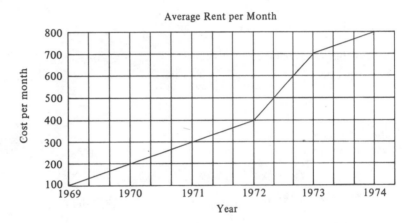

Example 4   The graph below shows the cost for cottage rental in a vacation resort for six consecutive years.

Use the graph to answer the following questions.

a. What was the average monthly rent in 1971? $300

b. What was the average monthly rent in 1974? $800

c. What was the increase from 1972 to 1973? $300

d. What was the percent of increase from 1972 to 1973? (What % of $400 is $300?)   75%

Although different types of graphs may be used to represent the same information, one type is often more appropriate than another. This is an important consideration for anyone who is constructing graphs. However, our primary concern in this chapter is interpreting graphs rather than constructing them.

14.1 Exercises

1. Find examples of a bar graph, a pictograph, a circle graph, and a line graph in a current newspaper or magazine. If the graphs cannot be clipped out, make a copy and be prepared to discuss them in class.

2. Make up at least three questions pertaining to each of the graphs you selected in Exercise 1.

The graph below shows the enrollment by division in a small community college. Use it to answer Exercises 3-5.

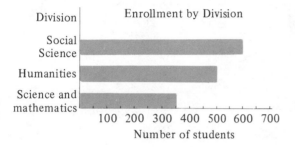

3. How many students are enrolled in each division?

4. Which division had the greatest enrollment?

5. How many more students were enrolled in humanities than in science and mathematics?

The graph that follows shows the number of votes each candidate received in an election for mayor. Use it to answer Exercises 6-9.

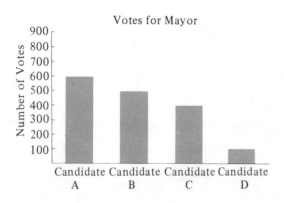

6. How many votes did each candidate receive?

7. Which candidates received less than 500 votes?

8. Which candidate received the fewest votes?

9. How many more votes than candidate D did candidate A receive?

This circle graph shows the relationship between the number of A's, B's, and C's in a class where no student made below a C grade. Use it to answer Exercises 10-12.

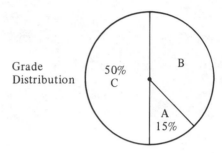

10. What percent of the class made an A?

11. What percent of the class made a B?

12. If there were 40 students in the class, how many made an A? How many made a B? How many made a C?

The pictograph below represents the profits made by an investment company. Use it to answer Exercises 13-15.

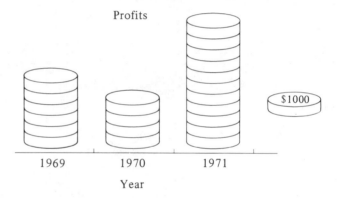

13. What was the profit in 1969?

14. How much more profit was made in 1971 than was made in 1969?

15. What was the percent of decrease from 1969 to 1970?

The following graph shows the total production for each of six months at a parts factory. Use it to answer Exercises 16-19.

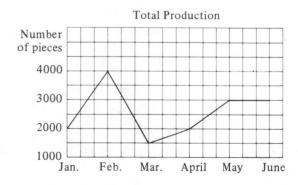

16. What was the production in May?

17. Which month had the greatest production?

18. What was the difference between the highest production and the lowest production for these months?

19. Was the production the same for any two months?

This circle graph shows the relationship between the number of boys and the number of girls in a typing class. Use it to answer Exercises 20–23.

Typing Class Enrollment

20. What percent of the class was girls?

21. What percent was boys?

22. If there were 30 in the class, how many girls were there?

23. If there were 30 in the class, how many boys were there?

The graph below shows the sales record for the top four salesmen in a large auto dealership. Use it to answer Exercises 24 and 25.

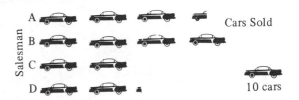

24. Which salesman sold the most cars? How many did he sell?

25. Which salesman sold the least number of cars? What was the difference in the number of cars sold by the top salesman and the number of cars sold by the salesman who was fourth from the top?

14.2   Graphs of Points on a Rectangular Coordinate System

You should recall from Chapter 5 how to plot points on a number line.

Example 1   Represent the graphs of -5, $-3\frac{1}{2}$, and 2.

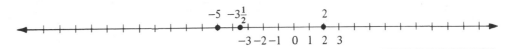

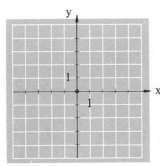

Figure 14.2

The graph shown in Example 1 has only one dimension. If two number lines are drawn so that they intersect at the **origin** and are perpendicular, then they form a **rectangular coordinate system**. (See Figure 14.2.)

The horizontal number line is called the **x-axis** and the vertical number line is called the **y-axis**. The coordinates above the origin on the y-axis are positive and the coordinates below the origin are negative.

On the graph, the positive side of each axis is indicated with an arrow. The origin is labeled and the scale indicated. This graph enables us to locate or **plot points** on a plane which has two dimensions.

Points on the plane correspond to **ordered pairs** of real numbers. Examples of ordered pairs are $(4, 3)$, $(-4, 2)$, $\left(6\frac{1}{2}, 10\right)$, $(5, 0)$, and $(0, 5)$. Note that parentheses are used to enclose the ordered pairs, and a comma is used between the numbers. The numbers in the ordered pair are called the **coordinates** of the

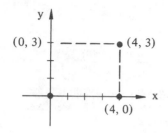

Figure 14.3

point. To locate the point on the graph corresponding to (4, 3), follow these steps (see Figure 14.3):

1. Locate +4 on the x-axis (4 is called the x-value).

2. Locate +3 on the y-axis (3 is called the y-value).

3. Complete a rectangle using these two points and the origin as three of the corners.

4. Indicate the fourth corner (4, 3).

The first number in the ordered pair is the x-value and is called the **x-coordinate**. The second number is the y-value and is called the **y-coordinate**. The x-value is also called the **abscissa** and the y-value the **ordinate**. With practice you should be able to locate or *plot* points without actually completing the rectangle.

Example 2   **a.** Draw a rectangular coordinate system and plot the points corresponding to the following ordered pairs: $(-4, 2)$, $\left(6\frac{1}{2}, -4\right)$, $(-5, -3)$, $(5, 0)$, $(5, 3)$, $(5, -3)$, and $(0, -3)$.

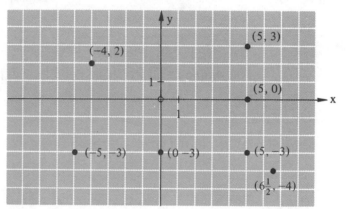

b. Draw a rectangular coordinate system and locate 4 points which have an abscissa of -3. Give the coordinates of each point.

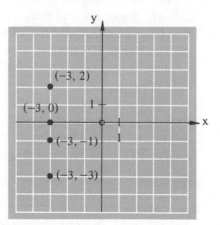

The axes divide the plane into four regions which are called **quadrants**. They are numbered counter-clockwise from I through IV starting with the part of the graph that is to the right of the y-axis and above the x-axis. The sign pattern for the coordinates is the same within each quadrant, and points on either axis have at least one 0 coordinate.

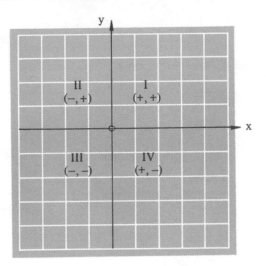

14.2 Exercises

1. What are the coordinates of the points A–G in the following graph?

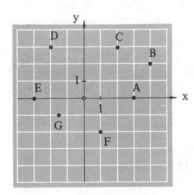

2. Give the quadrant in which each of the points in Exercise 1 is located, or the axis on which it lies.

3. Draw a rectangular coordinate system and plot the points with the following coordinates: $(-3, -7)$, $\left(6\frac{1}{2}, 5\right)$, $(-4, 3)$, $(-6, 2)$, $(0, -5)$, $(5, 0)$, $(-3, -2)$, $(0, 4)$.

Draw a rectangular coordinate system and plot the points which meet the conditions given in Exercises 4–6.

4. Three points with abscissa of -5. Give the coordinates.

5. Two points that are in the first quadrant. Give the coordinates.

6. Two points that are on the y-axis. Give the coordinates.

7. Give the quadrant in which each of the points A–G in the following graph is located.

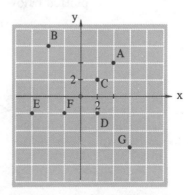

8. What are the coordinates of each of the points A–G in Exercise 7?

9. What are the coordinates of points A–F in the following graph?

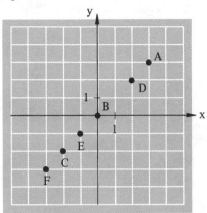

10. Draw a rectangular coordinate system and plot the points with the following coordinates: $(-2, -8)$, $\left(6, 3\frac{1}{2}\right)$, $\left(-5\frac{1}{2}, 2\right)$, $\left(6\frac{1}{3}, 0\right)$, $(0, -5)$, $(0, 0)$, $(4, -5)$, $(-3, 5)$.

11. Give the coordinates of the points A, B, and C in the following graph.

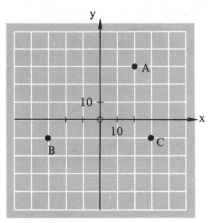

12. Give the quadrant (or axis) in which each of the points in Exercise 11 is located.

Draw a rectangular coordinate system and plot the points that meet the conditions given in Exercises 13–15.

13. Three points with ordinate 5. (Give the coordinates.)

14. Two points in the third quadrant. (Give the coordinates.)

15. Two points that are on the x-axis. (Give the coordinates.)

14.3   Graphs of First-Degree Equations in Two Variables

In Chapters 10 and 13 we studied first-degree equations and inequalities in one variable. We will now consider first-degree equations in two variables. For example, $x + y = 10$ is a first-degree equation in the variables x and y. A solution for a first-degree equation in two variables is an ordered pair of numbers. The first number is a value of x and the second number is a value of y. When these values are substituted for x and y in the equation, a true statement is obtained. One solution of $x + y = 10$ is $(6, 4)$ since $6 + 4 = 10$. Other solutions would be $(0, 10)$, $(10, 0)$, $(4, 6)$, $(9, 1)$, and $(1, 9)$. In fact, this equation has an infinite number of solutions. The points corresponding to some of the solutions are plotted below.

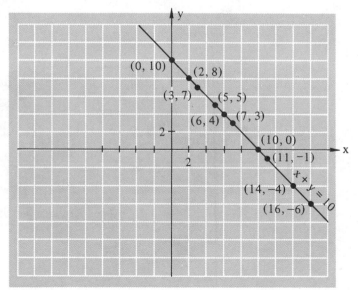

Note that these points lie on a straight line. The line is called the graph of the equation. The graph of any ordered pair that is a solution of the equation will lie on the line. Also, any ordered pair associated with a point on the line will satisfy the equation. Because the graph is a straight line, a first-degree equation in two variables is called a **linear equation**.

Example 1   Find three ordered pairs which are solutions of $x - y = 10$. Plot these points and connect them with a straight line.

(14, 4), (12, 2), and (11, 1) are all solutions for $x - y = 10$, since

$$14 - 4 = 10, \quad 12 - 2 = 10, \quad \text{and} \quad 11 - 1 = 10.$$

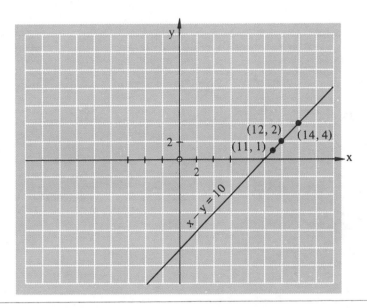

To draw the graph of a linear equation in two variables, you must find and plot at least two ordered pairs which are solutions of the equation. To check your work, it is wise to plot a third point. In a very simple equation such as the one above, the points can often be found by inspection. In other cases a simple procedure can be followed to obtain the desired ordered pairs.

Consider the equation $2x + 3y = 12$. The best procedure here is to choose a convenient value for one variable and find the corresponding value for the second variable. In other words, follow these steps:

1. Choose a convenient value for x.
2. Substitute this value of x in the equation and find the corresponding value of y.

An alternative procedure would be to follow these steps:

1. Choose a convenient value for y.
2. Substitute this value of y in the equation and find the corresponding value of x.

It is important to realize that if you choose to consider the x-value first, you can use any value for x. However, once you set an x-value, there can be only one corresponding y-value. If you choose the y-value first, you must then calculate the corresponding value of x. With practice, you can learn to choose *convenient values.* For example, you may wish to avoid choosing values which would require plotting points with extremely large values of x and y, points that are very close together, or points with fractional coordinates that are difficult to place accurately on the coordinate system.

The point where the line crosses the x-axis is called the **x-intercept** and the point where the line crosses the y-axis is called the **y-intercept**. To find the x-intercept, set y equal to 0, and solve for x. Similarly, by letting x equal 0, it is possible to find the y-intercept. The intercepts are generally very easy to find and are extremely important points to plot.

Example 2   Draw the graph of $2x + 3y = 12$.

If $x = 0$,

$$2(0) + 3y = 12$$
$$3y = 12$$
$$y = 4$$

Thus, $(0, 4)$ is a solution. (This point is also called the y-intercept.)

If $y = 0$,

$$2x + 3(0) = 12$$
$$2x = 12$$
$$x = 6$$

Thus, $(6, 0)$ is a solution. (This point is also called the x-intercept.)

If $x = 1$,

$$2(1) + 3y = 12$$
$$2 + 3y = 12$$
$$3y = 10$$
$$y = 3\frac{1}{3}$$

Thus, $\left(1, 3\frac{1}{3}\right)$ is a solution.

If we plot these three ordered pairs and join them with a straight line, the result is the graph of the equation $2x + 3y = 12$. The equation is written on the line, and each point is labeled with its coordinates.

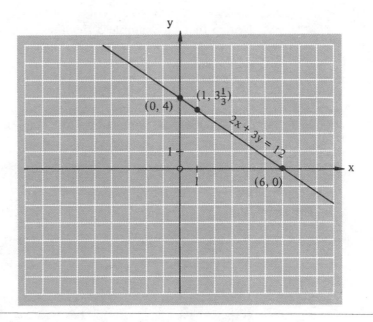

Note that all of the x-values chosen in Example 2 were integers, but one of the corresponding y-values was not.

Even though integers are usually chosen for the initial substitution, it is not necessary to have only integral coordinates. Fractional coordinates can be plotted by estimating. The information in Example 2 can be summarized in a table as follows:

x	y
0	4
6	0
1	$3\frac{1}{3}$

Example 3   Draw the graph of x = 6.

Since there is no y in the equation, y can have any value, but x must always equal 6. The equation can be thought of as x + 0y = 6. For example, (6, 4), (6, −2), (6, 0) are all solutions for the equation x = 6.

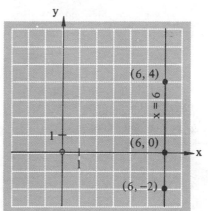

Notice that in Example 3 the line $x = 6$ is parallel to the y-axis. Any time x is equal to a constant, the graph will be parallel to the y-axis. (Recall that two lines are **parallel** if they never intersect.)

Example 4   **a.** Examine the following graphs of $x = -2$, $x = 6$, and $x = -4$.

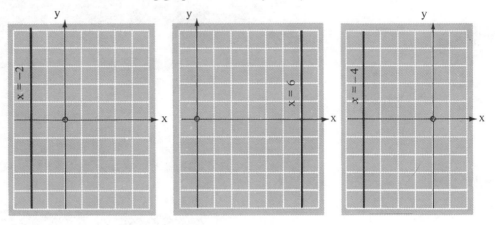

b. Draw the graphs of $y = +4\frac{1}{2}$ and $y = -4\frac{1}{2}$.

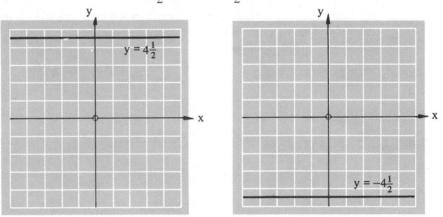

As Examples 3 and 4 illustrate, *the graph of an equation where y equals a constant is a line parallel to the x-axis, and the graph of an equation where x equals a constant is a line parallel to the y-axis.*

14.3 Exercises

1. Complete the following table for the equation $2x - 5y = 13$. Plot the resulting points, and draw a line through the points.

x	y
0	
	0
1	

2. Complete the following table for the equation $x = 5$. Plot the resulting points, and draw the graph of the equation.

x	y
	-6
	3
	$5\frac{1}{2}$

3. Complete the following table for the equation y = −3. Plot the resulting points, and draw the graph of the equation.

x	y
0	
−5	
5	

4. Given the equation 3x + 7y = 49, find the ordinate when the abscissa is 0.

5. Given the equation 5x − 2y = 10, find the ordinate when the abscissa is 2.

6. Given the equation 5x = 6y + 3, find the abscissa when the ordinate is 2.

Find the x-intercept and the y-intercept of each of the equations in Exercises 7–16.

7. 15x + y = 5   **8.** 2x − y = 10
9. 3x = 2y   **10.** x + 3y = 6
11. x − 2y = 10   **12.** 3x + 4y = 16
13. 3x − 4y = 16   **14.** 2x + 5y = 10
15. 3x + 5y = 12   **16.** 6x − 4y = 10

Plot at least 3 ordered pairs from the solution set of each of the equations in Exercises 17–26 and then draw the graph of each equation.

17. x + y = 8   **18.** x = 7
19. y − x = 7   **20.** 2x + y = 10
21. 3x − 2y = 12   **22.** x = y

23. 6x + 2y = 9   **24.** 5x − 3y = 15
25. 2x = −8   **26.** 5x − 3y = 12

What is the equation of each of the lines in Exercises 27 and 28?

27. **28.**

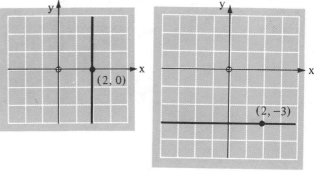

What is the equation of each of the lines in Exercises 29 and 30? (*Hint:* For Exercise 30, write the coordinates of several points, and observe the relationship between x and y.)

29. **30.**

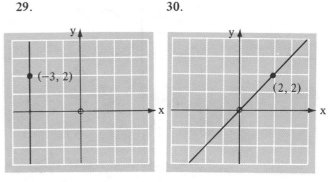

14.4   Special Forms of Linear Equations

Equations, like algebraic expressions, can be written in different forms. In previous sections, we have used a form where both variable terms were in the left member of the equation and the constant term was isolated in the right member: ax + by = c. Another form places all nonzero terms in the left member, which leaves zero for the right member of the equation: ax + by − c = 0. For example, 2x + y = 16 could be written as 2x + y − 16 = 0.

There are several standard forms that can be used to draw the graphs of equations. These forms are so important that they have special names. The **slope-intercept** form is one of the most useful. When an equation is expressed in this form, it is possible to identify the slope of the line and the y-intercept directly from the equation.

The **slope** of a line is the ratio of the *rise* to the *run*.

Example 1

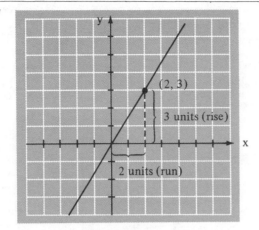

The slope of this line is $\dfrac{\text{rise}}{\text{run}}$ or $\dfrac{3}{2}$.

Example 2

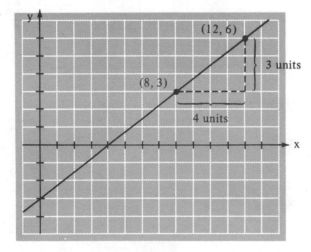

The slope of this line is $\dfrac{3}{4}$.

Example 3

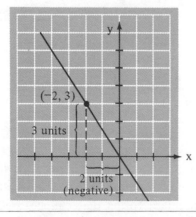

The slope of this line is $\dfrac{3}{-2}$ or $-\dfrac{3}{2}$.

The slope-intercept form is **y = mx + b**. To write an equation in this form, solve the equation for y and let the x term preceed the constant in the right member. When the equation is in slope-intercept form, m (the coefficient of the x term) is the slope and b is the y-intercept.

Example 4   Write the following equations in slope-intercept form, and find the slope and the y-intercept for each equation.

a. $2x + y = 6$

$$2x + y = 6$$
$$y = -2x + 6$$

The slope is $\dfrac{-2}{1}$, and the y-intercept is $(0, 6)$.

b. $3x - 2y = 6$

$$3x - 2y = 6$$
$$-2y = -3x + 6$$
$$y = \dfrac{3}{2}x - 3$$

The slope is $\dfrac{3}{2}$ and the y-intercept is $(0, -3)$.

The slope-intercept form can also be used to draw the graph of a linear equation.

PROCEDURE FOR DRAWING THE GRAPH OF A LINEAR EQUATION USING THE SLOPE-INTERCEPT FORM	*EXAMPLE* Use slope-intercept form to draw the graph of $3x - 2y = 6$.
1. Write the equation in slope-intercept form.	1. $3x - 2y = 6$ $$y = \dfrac{3}{2}x - 3$$
2. Identify the slope and the y-intercept, and let a = the numerator of the slope and b = the denominator of the slope.	2. The slope is $\dfrac{3}{2}$; a = 3, b = 2. The y-intercept is $(0, -3)$.
3. Plot the y-intercept. a. If the slope is *positive*, start at the y-intercept and count *down a* units on the graph. From this point count *b* units to the *left*. Identify the coordinates of the point located.	3. a. 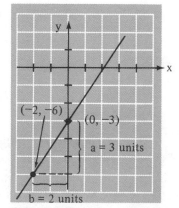
b. If the slope is *negative*, start at the y-intercept and count *down a* units on the graph. From this point count *b* units to the *right*. Identify the coordinates of the point located.	b. See Example 5 for negative slope.
4. Draw the line containing the y-intercept and the point found in step 3.	4. Join $(0, -3)$ and $(-2, -6)$.

Example 5   Use the slope-intercept form to draw the graph of $3x + 2y = 6$.

$$3x + 2y = 6$$
$$y = -\frac{3}{2}x + 3$$

The slope is $-\frac{3}{2}$, and the y-intercept is $(0, 3)$.

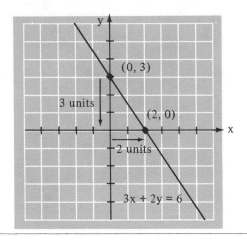

Another form for a linear equation is called the **point-slope form.** $y - b = m(x - a)$ is the equation of a line which passes through the **point (a, b)** and has **slope m.** Equations are not usually written in this form, but when they are it is possible to find the slope and the coordinate of a point through which the graph will pass. Also, an equation written in this form can be changed to one of the other forms if this is desirable.

Example 6   Give the slope and the coordinates of one point on the graph of each of the following equations.

a. $y - 3 = \frac{2}{3}(x - 4)$

   Solution: The slope $= \frac{2}{3}$ and one point is $(4, 3)$.

b. $y + 6 = 4(x - 3)$

   Solution: The slope $= \frac{4}{1}$ and one point is $(3, -6)$.

Example 7   Write the following equations in the form $ax + by = c$, and then in the form $y = mx + b$.

a. $y - 3 = 6(x - 2)$

$$y - 3 = 6(x - 2)$$
$$y - 3 = 6x - 12$$
$$-6x + y = -9$$
$$6x - y = 9 \qquad (ax + by = c)$$
$$-y = -6x + 9$$
$$y = 6x - 9 \quad (y = mx + b)$$

b. $y - 4 = \dfrac{3}{2}(x + 5)$

$$y - 4 = \dfrac{3}{2}(x + 5)$$

$$2(y - 4) = 3(x + 5)$$

$$2y - 8 = 3x + 15$$

$$-3x + 2y = 23$$

$$3x - 2y = -23 \qquad (ax + by = c)$$

$$-2y = -3x - 23$$

$$y = \dfrac{3}{2}x + \dfrac{23}{2} \qquad (y = mx + b)$$

14.4   Exercises

Write the equations in Exercises 1–6 in the form $ax + by = c$.

1. $9x + 6 = 3y$

2. $2y = 6x - 12$

3. $6y + 2x = 12$

4. $3x = 12y - 12$

5. $5x - 2y - 10 = 0$

6. $5x + 3y - 8 = 0$

Write the equations in Exercises 7–12 in the form $ax + by - c = 0$.

7. $2x + 6 = 3y$

8. $2y = 6x - 12$

9. $6y + 2x = 12$

10. $3x = 12y - 12$

11. $5x + 10 - 2y = 0$

12. $5x + 3y = 8$

Write the equations in Exercises 13–22 in the slope-intercept form: $y = mx + b$.

13. $3x + y = 12$

14. $5x - y = 10$

15. $6x + 2y = 10$

16. $2y - 6x = 12$

17. $5x - 2y = 30$

18. $6x + 4y = 12$

19. $9x - 2y = 15$

20. $3x - y = 10$

21. $6x - 4y = 13$

22. $5x + 3y = 16$

Use the slope-intercept form to find the slope and the y-intercept of the equations in Exercises 23–26.

23. $9x - y = 15$

24. $3x - y = 10$

25. $6x - 4y = 13$

26. $5x + 3y = 15$

Use the slope-intercept form to draw the graphs of the equations in Exercises 27–32.

27. $3x + y = 12$

28. $5x - y = 4$

29. $6x + 4y = 12$

30. $2x + 3y = 9$

31. $5x - 2y = 10$

32. $3x - 5y = 10$

Find the slope and the coordinates of one point through which the graphs of the equations in Exercises 33–35 would pass.

33. $y - 2 = 3(x - 5)$

34. $y - 1 = 6(x - 2)$

35. $y + 3 = 5(x - 3)$

14.5   Graphs of First-Degree Inequalities in Two Variables

The solution of an inequality is related to the solution of the corresponding equation. This can be illustrated for first-degree equations in one unknown by comparing the solutions given in Example 1.

Example 1   Solve $x + 6 = 10$ and $x + 6 < 10$.

$$x + 6 = 10$$
$$x = 4$$

$$x + 6 < 10$$
$$x < 4$$

Notice that the solution for $x + 6 < 10$ is the set of points to the left of $x = 4$ which is the solution for $x + 6 = 10$.

In general, *the graph of the solution for a linear inequality in one variable is the set of points either to the right or to the left of the point of equality.* The point of equality may or may not be in the solution.

Similarly, the set of points which satisfy a first-degree inequality in two variables lies on one side of the *line* which is the graph of the equation obtained when the symbol of inequality is replaced with a symbol of equality. The line may or may not be in the solution.

Example 2   Draw the graph of $x + y > 10$.

First draw the graph of $x + y = 10$.

x	y
5	5
2	8
3	7

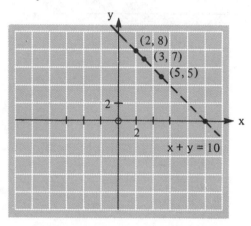

Then determine which side of the line contains the solutions for the inequality and shade that region.

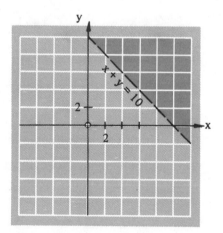

The line divides the plane into two regions. *All the points in one of these regions satisfy the inequality and the points in the other region do not satisfy the inequality.* The points *on* the line satisfy the equation. By choosing a point in one region and substituting its coordinates in the inequality, we can easily determine the side of the line which corresponds to the solution set.

For example, let's test the origin for the problem in Example 2. The coordinates are (0, 0). If we substitute these coordinates in $x + y > 10$, we have $0 + 0 > 10$ or $0 > 10$. Since this is not a true statement, (0, 0) is not a solution for $x + y > 10$. It follows then that the points on the opposite side of the line are solutions for the inequality.

It is customary to shade the region which corresponds to the solution of the inequality. A dotted line is used for the graph of the equation since the points on the line are not solutions for the inequality.

PROCEDURE FOR DRAWING THE GRAPHS OF FIRST-DEGREE INEQUALITIES IN TWO VARIABLES

1. Replace the inequality symbol with the symbol of equality.

2. Draw the graph of the resulting equation. Use a dotted line for the graph of the equation if there is no symbol of equality in the original problem, or use a solid line if the problem contains both a symbol of inequality and one of equality.

3. Determine which region corresponds to the solution set of the inequality by testing the coordinates of a point on one side of the line.

4. Shade the region identified in step 3.

EXAMPLE   Draw the graph of $2x - y < 10$.
Solution:

$2x - y = 10$

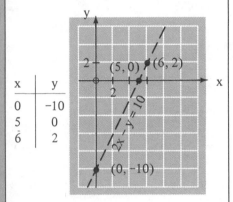

x	y
0	-10
5	0
6	2

Test (0, -4) in $2x - y < 10$.

$$2(0) - (-4) < 10$$
$$4 < 10$$

Since (0, -4) is in the solution set, we can shade the solution set as shown below.

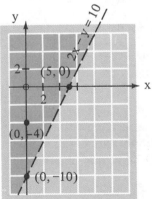

In this case the line representing the graph of the equation is not a part of the solution set for the inequality.

Example 3   Draw the graph of $5x \leqslant 2y$.

First draw the graph of $5x = 2y$.

x	y
0	0
2	5
-2	-5

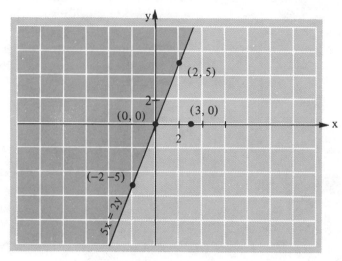

Test $(3, 0)$:

$$5(3) \leqslant 2(0)$$
$$15 \leqslant 0$$

Since this is not a true statement, $(3, 0)$ is not a solution for the inequality. The region on the opposite side of the line should be shaded and the graph of the equation is a solid line since it is in the solution set.

14.5 Exercises

1. Is $(4, 3)$ a solution for $3x - 12y < 15$?
2. Is $(5, 4)$ a solution for $y < 4$?
3. Is $(5, 4)$ a solution for $x \leqslant 5$?
4. Is $(5, 4)$ a solution for $3x > 2y$?
5. Is $(5, 4)$ a solution for $x \leqslant y$?

Draw the graphs of the inequalities in Exercises 6-25.

6. $x + y < 5$
7. $2x - y \geqslant 10$
8. $3x \geqslant 2y$
9. $x + 3y < 6$
10. $x - 2y \leqslant 10$
11. $3x + 4y \geqslant 16$
12. $3x - 4y > 16$
13. $2x + 5y \geqslant 10$
14. $3x < 12$
15. $6x - 4y < 10$
16. $y < 4x - 3$
17. $y < 2x$
18. $3x \geqslant 4y$
19. $3x - y \geqslant 0$
20. $y \leqslant \frac{1}{2}x + 6$
21. $y \geqslant -2$
22. $y - x < 10$
23. $3x - y \geqslant 0$
24. $4x + 2y \geqslant 5$
25. $2x - 3y \leqslant 4$

Important Terms Used in this Chapter	abscissa	pictograph
	axis	plot points
	bar graph	point-slope form
	circle graph	quadrant
	coordinates	rectangular coordinate system
	graph	slope
	ordered pair	slope-intercept form
	ordinate	x-intercept
	origin	y-intercept
	parallel	

REVIEW EXERCISES

The following graph was used to compare the annual sales of the five salesmen for a company that manufactured farm machinery. Use it for Exercises 1-6.

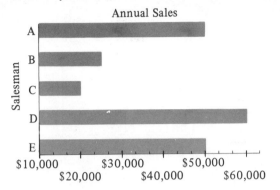

1. Rank the 5 salesmen from highest to lowest in sales.

2. What was the difference between the sales made by the top salesman and the sales made by the second ranking salesman?

3. Which salesman made the least sales?

4. What was the total amount sold by the five salesmen?

5. How much did the top man sell?

6. What percent of the total sales did the top salesman make? (Round to the nearest percent.)

The following circle graph represents the distribution of the school population in a school district that is composed of elementary, middle, senior high, and vocational schools. Use it for Exercises 7-10.

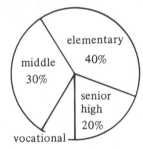

7. What percent of the school population was in senior high school?

8. What was the ratio of the number of students in vocational schools to the number of students in elementary schools?

9. If there are 30,000 students in the system, how many are in middle school?

10. What type of school had the smallest enrollment?

11. Draw a rectangular coordinate system and plot the following points: $(-1, -1)$, $(-2, -2)$, $(3, 3)$, $(4, 4)$.

12. Give the quadrant for each of the points plotted in Exercise 11.

13. Draw a rectangular coordinate system. Plot 3 points that have an abscissa of -5, and 3 points that have an ordinate of 4.

14. Give the coordinates of the points plotted in the following rectangular coordinate system.

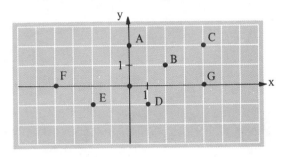

15. Give the quadrant or axis for each of the points plotted in Exercise 14.

Draw the graphs of the equations in Exercises 16-21.

16. $2x + y = 12$ 17. $x + y = -4$

18. $2x = 5y$ 19. $x = -8$

20. $3y = 9$ 21. $2x - 5y = 9$

Write each equation in Exercises 22-25 in slope-intercept form, and draw the graph of each.

22. $2x + y = 3$ 23. $8x - 12y = 24$

24. $x + 3y = 6$ 25. $4x - y = 16$

Draw the graph of each of the inequalities in Exercises 26-30.

26. $4x + 3y > 12$ 27. $2x - 3y \leqslant -6$

28. $y \leqslant 4$ 29. $x < y$

30. $5x + 7y < 11$

FIRST-DEGREE EQUATIONS IN TWO VARIABLES

OBJECTIVES

Upon completion of this chapter you should be able to:

1. Solve systems of linear equations by using graphs.   15.1
2. Solve systems of linear equations by the addition method.   15.2
3. Solve systems of linear equations by substitution.   15.3
4. Recognize systems that have no solution and systems that have an infinite number of solutions.   15.1, 15.2, 15.3
5. Choose the most appropriate method for solving a system of linear equations when the method is not specified.   15.1, 15.2, 15.3
6. Use systems of linear equations to solve word problems.   15.2, 15.3

PRETEST

1. Solve each of the following systems by using graphs (Objective 1).

   a. $-x - y = -6$
   $-3x + y = -2$

   b. $x = y$
   $3x + 2y = 15$

   c. $3x - 6y = 18$
   $x - 2y = 6$

   d. $x = 6$
   $3x - y = 9$

2. Solve each of the following systems by the addition-subtraction method (Objective 2).

   a. $-x - y = -6$
   $-3x + y = -2$

   b. $x = y$
   $3x + 2y = 15$

   c. $3x - 6y = 18$
   $x - 2y = 6$

   d. $x = 6$
   $3x - 4y = 9$

3. Solve each of the following systems by the substitution method (Objective 3).

   a. $-x - y = -6$
   $-3x + y = -2$

   b. $x = y$
   $3x + 2y = 15$

   c. $3x - 6y = 18$
   $x - 2y = 6$

   d. $x = 6$
   $3x - 4y = 9$

4. Indicate whether each of the following systems has a unique solution, no solution, or an infinite number of solutions (Objective 4).

a.

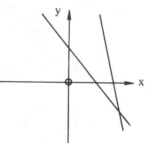

b.

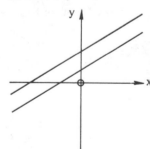

c.

d. $x = 4$
$y = -2$

e. $x + y = 3$
$2x + 2y = 6$

f. $x + y = 3$
$x + y = 10$

5. Solve each of the following by any appropriate method (Objective 5).

   a. $x + 12y = 34$
      $7x + 9y = 13$

   b. $2x + 5y = 22$
      $2x - 6y = 0$

   c. $5x - 2y = 0$
      $x + y = 0$

   d. $2x + 3y = 3$
      $-4x - 6y = -6$

   e. $5x - 2y = 10$
      $x = 6$

   f. $2x + 3y = 4$
      $12x + 18y = 4$

6. Set up a system of equations to solve the following problem (Objective 6): Find two numbers if their sum is sixteen and twice the first number minus the second number is five.

Many of the practical problems in mathematics deal with relationships between quantities. The businessman is interested in the relationship between his sales and his profit. The engineer is interested in the relationship between the amount of dirt which must be excavated in a building project and the number of truck loads that will be needed to haul away this dirt. To solve problems dealing with relationships involving several quantities, mathematicians have devised techniques for setting up systems of equations to represent these relationships and finding a common solution for the equations. When two or more equations are considered simultaneously, the equations are referred to as a **system of equations.** In advanced mathematics we can deal with many variables and many equations at the same time. However, in this text we will limit our discussion to systems of two first-degree equations in two variables.

15.1   Solving Systems of Linear Equations by Using Graphs

A **solution** of a system of equations is an ordered pair that is a solution of each equation in the system. Although a single linear equation in two variables has an infinite number of solutions, there may be only one solution for a pair of equations. An intuitive way to show this is to draw the graphs of the equations. Although the use of graphs is not the most efficient method for finding a solution, we will consider this method first because it illustrates visually how solutions of the two equations are related. Recall that the graph of a first-degree equation in two variables is a straight line. Since there are three ways in which two lines in the same plane can be related, it follows that there are three types of solutions for a system of two linear equations.

1. Their graphs may intersect in one and only one point as shown in Figure 15.1. In this case, the graphs have only one point in common. The ordered pair corresponding to the point of intersection is the solution of the system.

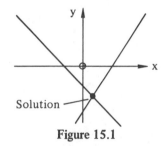

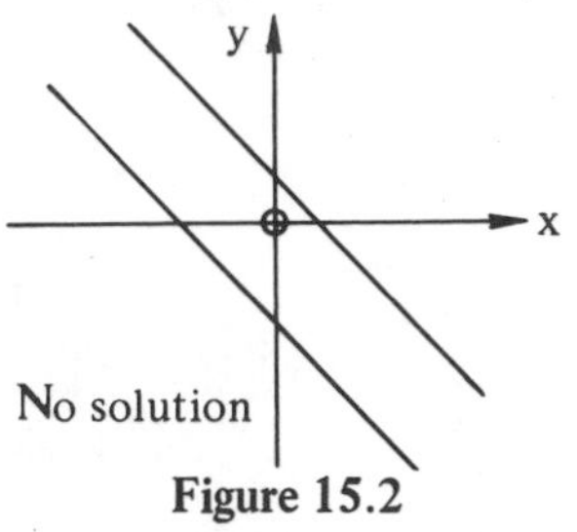

Figure 15.1

2. Their graphs may be parallel lines as shown in Figure 15.2. These equations have no common solution since their graphs will never intersect.

Figure 15.2

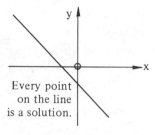

Every point
on the line
is a solution.

Figure 15.3

3. Their graphs may be the same line as shown in Figure 15.3. These equations are equivalent; that is, any ordered pair that is a solution for one is a solution for the other. Thus, the solution is an infinite number of ordered pairs and corresponds to the line.

To solve a system of equations by using graphs, draw the graphs of the two equations on the same coordinate system. If the lines intersect, the solution will be the ordered pair which corresponds to the point of intersection. If the lines are parallel, the system has no solution. If the lines coincide, every point on the line is a solution of the system.

Example 1   Solve: $x + 2y = 10$
$3x - y = 9$

Draw the graph of $x + 2y = 10$.   Draw the graph of $3x - y = 9$.

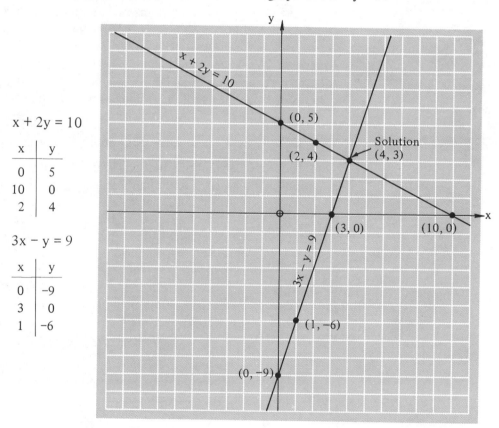

$x + 2y = 10$

x	y
0	5
10	0
2	4

$3x - y = 9$

x	y
0	-9
3	0
1	-6

Since the lines intersect at $(4, 3)$, this is the common solution. It should satisfy both equations.

Check: $x + 2y = 10$ $3x - y = 9$
$4 + 6 = 10$ $12 - 3 = 9$
$10 = 10$ $9 = 9$

Example 2   Solve: $x + 2y = 4$
$2x + 4y = 16$

Draw the graph of x + 2y = 4.   Draw the graph of 2x + 4y = 16.

x + 2y = 4

x	y
0	2
4	0
2	1

2x + 4y = 16

x	y
0	4
8	0
2	3

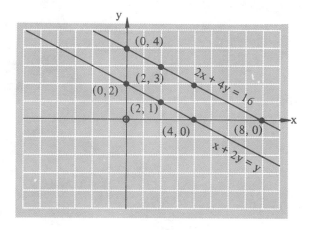

Since the graphs do not intersect, the equations have no solution in common, and the system has no solution.

If the equations in Example 2 are written in slope-intercept form, it is obvious that they have the same slope but different y-intercepts.

$$\text{If } x + 2y = 4, \quad y = -\frac{1}{2}x + 2$$

$$\text{If } 2x + 4y = 16, \quad y = -\frac{1}{2}x + 4$$

If two equations have the same slope and different y-intercepts, their graphs will be parallel lines and the equations will have no common solution.

Example 3   Solve: 3x − 4y = 12
                        6x − 8y = 24

Draw the graph of 3x − 4y = 12.   Draw the graph of 6x − 8y = 24.

3x − 4y = 12

x	y
0	−3
4	0
2	$-1\frac{1}{2}$

6x − 8y = 24

x	y
0	−3
4	0
6	$1\frac{1}{2}$

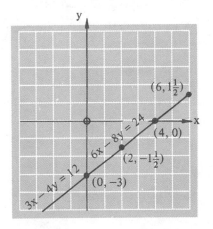

Since the points for both graphs lie on the same line, any solution for one equation is also a solution for the other equation; that is, every point on the line is a solution.

Example 4   Solve: $5x - y = 10$
$2x + y = 6$

Draw the graph of $5x - y = 10$.   Draw the graph of $2x + y = 6$.

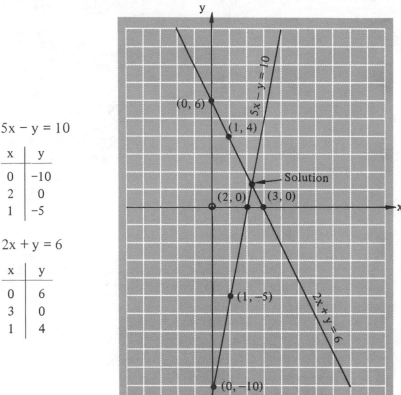

$5x - y = 10$

x	y
0	−10
2	0
1	−5

$2x + y = 6$

x	y
0	6
3	0
1	4

When the lines intersect between units on the graph, as in Example 4, an approximation for the coordinates of the point must be used because it is difficult to read the exact solution. For this reason, an algebraic method is usually preferred since the exact answer is obtained and it is generally less time consuming. We will discuss two algebraic methods for solving systems of equations in Sections 15.2 and 15.3.

15.1 Exercises

Solve the systems in Exercises 1-20 by using graphs and check your solution in both equations.

1. $x + y = 6$
$x - y = 6$

2. $x + 2y = 6$
$-x + 3y = 4$

3. $x + 3y = 6$
$2x + 4y = 4$

4. $2x - y = 7$
$3x + 2y = 14$

5. $5x + 10y = 10$
$-x - 2y = -3$

6. $2x + 3y = 0$
$3x - 4y = 17$

7. $x = 6$
$2x + y = 4$

8. $2x + y = 12$
$-x + 4y = 3$

9. $2x + y = 0$
$5x - y = 7$

10. $5x + 10y = 10$
$x + 2y = 2$

11. $2x + y = 10$
$3x - y = 5$

12. $y = 2x + 4$
$y = -4$

13. $3x + y = 7$
$x + 2y = 4$

14. $2x + y = 1$
$4x + 2y = 4$

15. $x + y = 9$
$2x - y = 6$

16. $x + 2y = 8$
$3x + y = 9$

17. $2x + y = 6$
$6x + 3y = 18$

18. $4x + 2y = 14$
$x - 3y = -14$

19. $x + y = 6$
$2x - y = 6$

20. $2x + y = 11$
$x + 2y = 13$

15.2 Solving Systems of Equations by the Addition Method

The addition method for solving systems of equations is an algebraic method which is also called the addition–subtraction method or the elimination method. In this method two equations are added (or subtracted) in order to eliminate one of the variables.

Example 1   Solve the following system: $2x + y = 10$
$$3x - y = 5$$

1. Add the two equations to produce a single equation in one variable.

    $$\begin{aligned} 2x + y &= 10 \\ \underline{3x - y} &= \underline{5} \\ 5x &= 15 \end{aligned}$$

2. Solve this equation.

    $$x = 3$$

3. Substitute the value obtained in step 2 into one of the original equations.

    $$2(3) + y = 10$$

4. Solve this equation for the second variable.

    $$6 + y = 10$$
    $$y = 4$$

5. Express the solution as an ordered pair.

    The solution is $(3, 4)$.

6. Check the solution in both equations.

    $$\begin{array}{ll} 2(3) + 4 = 10 & 3(3) - 4 = 5 \\ 6 + 4 = 10 & 9 - 4 = 5 \\ 10 = 10 & 5 = 5 \end{array}$$

If adding the two original equations does not eliminate one of the variables, the multiplication property of equality can be used to obtain appropriate coefficients to accomplish this. Observe in each of the following examples that the multiplier is chosen so that one of the variables will have numerical coefficients in each equation with the same absolute value but opposite signs. In some cases the multiplication principle may be applied to only one equation, but in other problems both equations must be multiplied by an appropriate constant so that the addition will eliminate one of the variables.

Example 2   Solve the following system: $2x - y = 7$
$$3x + 2y = 14$$

1. Multiply both members of the first equation by 2.

    $$2(2x - y) = (7)2$$
    $$4x - 2y = 14$$

2. Add the resulting equation to the second equation in the system.

    $$\begin{aligned} 4x - 2y &= 14 \\ \underline{3x + 2y} &= \underline{14} \\ 7x &= 28 \end{aligned}$$

3. Solve the resulting equation for x.

$$x = 4$$

4. Substitute this value into one of the original equations.

$$2(4) - y = 7$$

5. Solve this equation for y.

$$8 - y = 7$$
$$-y = -1$$
$$y = 1$$

6. The resulting ordered pair is the solution of the system.

(4, 1) is the solution.

7. Check the solution in both equations.

$$2(4) - (1) = 7 \qquad 3(4) + 2(1) = 14$$
$$8 - 1 = 7 \qquad 12 + 2 = 14$$
$$7 = 7 \qquad 14 = 14$$

In this text, systems to be solved by addition are written with the variable terms in the left member and the constant terms in the right member. The equations should be in this form before using the procedure for solving systems of equations by the addition method.

PROCEDURE FOR SOLVING SYSTEMS OF EQUATIONS BY THE ADDITION METHOD	EXAMPLE $\quad 2x + 3y = 0$ $\qquad\qquad\qquad 3x - 4y = 17$
1. If necessary, use the multiplication property of equality to obtain coefficients for one of the variables which have the same absolute value and opposite signs.	Multiply both members of the first equation by 4. Multiply both members of the second equation by 3.
2. Add the two equations.	$8x + 12y = \ 0$ $\underline{9x - 12y = 51}$ $17x \qquad\ = 51$
3. Solve the equation found in step 2.	$x = 3$
4. Substitute this value into one of the original equations.	$2(3) + 3y = 0$
5. Solve the equation found in step 4.	$6 + 3y = 0$ $3y = -6$ $y = -2$
6. The ordered pair obtained in steps 3 and 5 is the desired solution.	(3, -2) is the solution.
7. Check the solution in both equations.	$2(3) + 3(-2) = 0$ $6 - 6 = 0$ $0 = 0$ $3(3) - 4(-2) = 17$ $9 + 8 = 17$ $17 = 17$

Example 3   Solve the following system: $2x + 5y = 5$
$5x - 3y = -3$

Multiply both members of the first equation by 5. Multiply both members of the second equation by -2.

$$10x + 25y = 25$$
$$\underline{-10x + 6y = 6}$$

Add the resulting equations. Solve for y.

$$31y = 31$$
$$y = 1$$

Substitute into one of the original equations, and solve for x.

$$2x + 5y = 5$$
$$2x + 5(1) = 5$$
$$2x = 0$$
$$x = 0$$

Solution: (0, 1)
Check: $2(0) + 5(1) = 5$ $5(0) - 3(1) = -3$
$5 = 5$ $-3 = -3$

Example 4   Solve the following system: $5x + 10y = 10$
$-x - 2y = -3$

$$5x + 10y = 10$$
$$\underline{-5x - 10y = -15}$$
$$0 = -5 \quad \text{Contradiction}$$

No solution

We have followed the procedure for solving by addition in Example 4, but instead of eliminating one of the variables, we have eliminated them both. Since the resulting statement is a contradiction, the system has no solution. (Note that in Exercises 15.1, Exercise 5 asks for the solution of this system by using graphs.)

Example 5   Solve the following system: $5x + 10y = 10$
$x + 2y = 2$

$$5x + 10y = 10$$
$$\underline{5x - 10y = -10}$$
$$0 = 0 \quad \text{Identity}$$

Infinite number of solutions

The result in Example 5 is similar to the one obtained in Example 4 with one very significant difference. The statement obtained when the two equations are added is a true statement. This indicates that the two equations are equivalent and therefore have the same solution set. Thus, the system has an infinite number of solutions. (For a similar example see Exercises 15.1, Exercise 10.)

Example 6   Solve the following system: $x - 2y = 6$
$$3x + y = 3$$

$$
\begin{array}{ll}
x - 2y = 6 & -3x + 6y = -18 \\
6x + 2y = 6 & 3x + y = 3 \\
\hline
7x = 12 & 7y = -15 \\
\end{array}
$$

$$x = \frac{12}{7} \qquad y = \frac{-15}{7}$$

Solution: $\left(\dfrac{12}{7}, \dfrac{-15}{7}\right)$

Check: $\dfrac{12}{7} - 2\left(\dfrac{-15}{7}\right) = 6$

$$\frac{12}{7} + \frac{30}{7} = 6$$

$$6 = 6$$

$$3\left(\frac{12}{7}\right) + \left(\frac{-15}{7}\right) = 3$$

$$\frac{36}{7} - \frac{15}{7} = 3$$

$$3 = 3$$

In the solution to Example 6, instead of substituting to find the second variable, we repeated the addition process. This alternate method may be helpful when the first coordinate found is a fraction.

Word problems which require that two quantities be found are particularly suitable to solution by systems. Since two variables are used, two equations should be formed from the given relationships.

Example 7   Set up a system of equations to solve the following problem. The sum of two numbers is 26, and their difference is 2. Find the numbers.

Let x = first number and y = second number.

$$
\begin{array}{l}
x + y = 26 \\
x - y = 2 \\
\hline
2x = 28 \\
\end{array}
$$

$$x = 14 \qquad \text{(first number)}$$

$$14 + y = 26$$

$$y = 12 \qquad \text{(second number)}$$

Check:

Sum	Difference
14	14
12	−12
26	2

15.2 Exercises

Solve the systems in Exercises 1–20 by the addition method. Check your solutions in both equations.

1. $x + y = 6$
$x - y = 6$

2. $x + 2y = 6$
$-x + 3y = 4$

3. $5x - y = 10$
$2x + y = 4$

4. $2x + y = 0$
$5x - y = 7$

5. $3x - 6y = 12$
$x - 2y = 4$

6. $x + y = 21$
$-2x + 2y = -6$

7. $3x + 2y = 15$
$4x - y = 9$

8. $2x + 3y = -8$
$5x + 4y = -20$

9. $x + 2y = 7$
$x + y = 2$

10. $2x + y = 7$
$2x - y = 7$

11. $5x + 4y = 12$
$2x - 5y = 8$

12. $6x + 5y = 9$
$-4x + 3y = 5$

13. $3x - 6y = 12$
$-x + 2y = 6$

14. $3x - 6y = -9$
$2x + 2y = 6$

15. $-x + y = 0$
$x + y = 0$

16. $6x + 2y = 26$
$3x - 3y = 9$

17. $5x - y = 10$
$2x + 2y = 14$

18. $-2x + 3y = -5$
$3x - 2y = 0$

19. $2x - y = -10$
$x - y = -5$

20. $2x - y = 12$
$8x - 4y = 2$

21. Anthony is six years younger than Paul, and their combined age is the same as Jennifer's. If Jennifer is 24, how old are Anthony and Paul?

22. Timothy has 19 coins in his pocket. He has only nickels and dimes. If there are 5 more dimes than nickels, how many nickels and how many dimes does he have?

23. Timothy has only nickels and dimes in his pocket. When he last counted he had 5 more dimes than nickels. He has forgotten the exact number of coins, but he recalls that the value was $3.80. How many dimes does he have?

24. A man made two investments totaling $25,000. He made a profit of 5% on the first investment and 2% on the second investment. If his total profit was $650, how much was each investment?

25. The first of two even integers is 10 less than the second. Find the two even integers if their sum is 18.

15.3 Solving Systems of Equations by Substitution

If at least one of the variables in one of the equations of a linear system has a coefficient of 1, then the system may be solved easily by the substitution method.

In order to use this method it is necessary to solve one of the equations for one of the variables in terms of the other variable. This procedure was discussed in Section 10.8.

Example 1   Solve: $3x + 5y = 11$
$2x - y = 2$

Solve $2x - y = 2$ for y in terms of x.

$$2x - y = 2$$
$$-y = 2 - 2x$$
$$y = -2 + 2x$$
$$y = 2x - 2$$

Substitute the value of y into the first equation.

$$3x + 5(2x - 2) = 11$$
$$3x + 10x - 10 = 11$$
$$13x - 10 = 11$$
$$13x = 21$$
$$x = \frac{21}{13}$$

Substitute this value of x into the equation which has been solved for y.

$$y = 2x - 2$$

$$y = 2\left(\frac{21}{13}\right) - 2$$

$$y = \frac{42}{13} - \frac{26}{13}$$

$$y = \frac{16}{13}$$

Solution: $\left(\frac{21}{13}, \frac{16}{13}\right)$

Check:

$$3\left(\frac{21}{13}\right) + 5\left(\frac{16}{13}\right) = 11 \qquad 2\left(\frac{21}{13}\right) - \frac{16}{13} = 2$$

$$\frac{63}{13} + \frac{80}{13} = 11 \qquad \frac{42}{13} - \frac{16}{13} = 2$$

$$\frac{143}{13} = 11 \qquad \frac{26}{13} = 2$$

$$11 = 11 \qquad 2 = 2$$

PROCEDURE FOR SOLVING LINEAR SYSTEMS BY SUBSTITUTION	*EXAMPLE* Solve: $x + 3y = 6$ $\qquad 2x + 4y = 4$
	Solution:
1. Solve one of the equations for one of the variables in terms of the other variable.	Solve $x + 3y = 6$ for x in terms of y. $\qquad x = 6 - 3y$
2. Substitute the expression obtained in step 1 into the other equation, thus eliminating one of the variables.	Substitute the expression $6 - 3y$ for x in the second equation. $2x + 4y = 4$ becomes $$2(6 - 3y) + 4y = 4$$
3. Solve the resulting equation.	$$12 - 6y + 4y = 4$$ $$-2y = -8$$ $$y = 4$$
4. Substitute the value obtained in step 3 into the equation obtained in step 1 to find the value of the other variable.	From step 1 $$x = 6 - 3y$$ $$x = 6 - 3(4)$$ $$x = 6 - 12$$ $$x = -6$$
5. The ordered pair obtained in steps 3 and 4 is the solution.	$(-6, 4)$ is the solution.
6. Check the solution in both equations.	$$-6 + 3(4) = 6$$ $$-6 + 12 = 6$$ $$6 = 6$$ $$2(-6) + 4(4) = 4$$ $$-12 + 16 = 4$$ $$4 = 4$$

Example 2   Solve: $6x - 2y = 12$
$3x - y = 6$

Solve for y in terms of x:

$$3x - y = 6$$
$$-y = 6 - 3x$$
$$y = 3x - 6$$

Substituting, we have

$$6x - 2(3x - 6) = 12$$
$$6x - 6x + 12 = 12$$
$$12 = 12 \quad \text{Identity}$$

There are an infinite number of solutions.

Since both variables were eliminated in Example 2 and the resulting statement is true, the two equations are equivalent. Therefore, the system has an infinite number of solutions. In other words, every solution for one equation is also a solution for the other.

Example 3   Solve: $x = 6$
$y = 2x - 5$

$$x = 6$$
$$y = 2(6) - 5$$
$$y = 7$$

Solution: $(6, 7)$

Check: $x = 6$ $y = 2x - 5$
$6 = 6$ $7 = 2(6) - 5$
$7 = 12 - 5$
$7 = 7$

Example 4   Solve: $6x - 2y = 12$
$3x - y = 10$

Solve for y in terms of x:

$$3x - y = 10$$
$$-y = 10 - 3x$$
$$y = 3x - 10$$

Substituting, we have

$$6x - 2(3x - 10) = 12$$
$$6x - 6x + 20 = 12$$
$$20 = 12 \quad \text{Contradiction}$$

No solution

Since both variables were eliminated in Example 4 and the resulting statement is a contradiction, the system has no solution.

Example 5 Anthony has saved six times as much as Mitch. If the sum of their savings is $497, how much has each saved?

Let x = Anthony's savings and y = Mitch's savings.

$$x = 6y$$
$$x + y = 497$$

By substitution,

$$x = 6y$$
$$6y + y = 497$$
$$7y = 497$$
$$y = \$71 \qquad \text{(Mitch's savings)}$$
$$x = 6(71)$$
$$= \$426 \qquad \text{(Anthony's savings)}$$

Check: The sum of their savings is $497.

$$\begin{array}{r} 71 \\ +426 \\ \hline 497 \end{array}$$

15.3 Exercises

Solve the systems in Exercises 1–20 by substitution and check the solution in both equations.

1. $x + y = -9$
    $2x - y = 3$

2. $x + 2y = 6$
    $-x + 3y = 4$

3. $3x - 6y = 12$
    $x - 2y = 4$

4. $2x - 5y = 9$
    $x = 3$

5. $y = 3x - 1$
    $x + 2y = 2$

6. $3x + 2y = 15$
    $4x - y = 9$

7. $x + y = 21$
    $-2x + 2y = -6$

8. $3x - 6y = 12$
    $-x + 2y = 6$

9. $5x - 4y = 3$
    $y = 4$

10. $5x - y = 10$
     $2x + y = 4$

11. $2x + y = 0$
     $5x - y = 7$

12. $3x + y = 8$
     $y = 2x - 2$

13. $3x - y = 4$
     $x + 2y = 6$

14. $6x + 3y = 12$
     $2x + y = 4$

15. $x + 2y = 8$
     $3x + y = 9$

16. $x = 6 - 5y$
     $2x + 9y = 4$

17. $6x - 2y = 20$
     $-3x + y = 8$

18. $4x + 2y = 14$
     $x - 3y = -14$

19. $2x + y = 11$
     $x + 2y = 13$

20. $2x - 5y = 18$
     $x = 12$

21. A carpenter wishes to divide a 32-foot board so that one piece is three times as long as a second piece. How long should each piece be?

22. There were $\frac{2}{3}$ as many B's as A's on a math test. If 30 students made either an A or a B, how many made A's? How many made B's?

23. The first of two whole numbers is five times the second. If their sum is 18, what are the two numbers?

24. Deborah earns $1\frac{1}{2}$ times as much money as Diane. How much does each make if their combined weekly earnings are $350?

25. Two kinds of candy were combined to make 50 pounds of the mixture. The less expensive candy cost 37¢ per pound and the more expensive cost 45¢ per pound. If the total mixture was worth 40¢ per pound, how many pounds were there of each kind of candy?

Important Terms	addition method	substitution method
Used in this Chapter	solution	system of equations

REVIEW EXERCISES

Solve the systems in Exercises 1 and 2 by using graphs.

1. $x + y = 2$
   $2x - y = 4$

2. $x - 2y = 4$
   $2x + y = 3$

Solve the systems in Exercises 3-6 by the addition method.

3. $y - x = 12$
   $x - 2y = 12$

4. $3x + y = 1$
   $-x - 2y = -3$

5. $4x - 9y = 12$
   $4x + 12y = -2$

6. $5x - y = 1$
   $6x + 2y = 14$

Solve the systems in Exercises 7-10 by substitution.

7. $5x + y = 8$
   $3x - 2y = 10$

8. $3x + y = 1$
   $-x - 2y = -3$

9. $10x + 2y = 12$
   $5x + y = 8$

10. $x - y = 0$
    $3x + 2y = 15$

Solve Exercises 11-16 by any appropriate method.

11. $x - 4y = -11$
    $5x - 3y = -4$

12. $2x + 3y = 3$
    $-x - 5y = -4$

13. $6x - 8y = 14$
    $3x - 4y = 7$

14. $5x - 3y = 23$
    $6x + 4y = 20$

15. $y = 2x + 9$
    $3x + y = -1$

16. $4x + y = 1$
    $-4x + 5y = 5$

17. Solve the following system algebraically and verify your solution using a graph. $5x - 2y = 10$
    $x = 2$

Use systems of equations to solve Exercises 18-20.

18. A farmer planted 6 more acres of corn than beans on his 160-acre farm. How many acres did he plant of corn and how many of beans?

19. Andrew's lunch cost 25¢ less than Tom's. The total check for both lunches was $2.75. How much did Tom and Andrew each spend for lunch?

20. The profit from a small business doubled from the first to the second year of operation. If the profit for the first two years was $27,000, what was the profit for each of the first two years?

INTRODUCTION TO INTERMEDIATE ALGEBRA

16

FACTORING POLYNOMIALS

PRETEST

1. Find the greatest common factor of each of the following sets of numbers (Objective 1).
   a. $30, 45, 75$ b. $53, 97$

   Find the greatest common factor of each of the following polynomials (Objective 1).
   c. $18xyz + 9xy + 3yz$ d. $30a^4b^6 - 45a^2b^4 + 75ab^2$
   e. $4a(3a - 4) - 3(3a - 4)$

2. Factor out the greatest common factor of each of the following polynomials (Objective 2).
   a. $18xyz + 9xy + 3yz$ b. $30a^4b^6 - 45a^2b^4 + 75ab^2$

3. Multiply the following binomials by inspection (Objective 3).
   a. $(4x - 5)(x + 5)$ b. $(4x + 5y)(x - 5y)$
   c. $(4x - 5)(x - 5)$ d. $(x - 4)(x + 3)$

4. Find the following products (Objective 4).
   a. $(4x + 5y)(4x - 5y)$ b. $(4x - 5y)(4x - 5y)$
   c. $(4x + 5y)^2$

5. Factor (Objective 5).
   a. $x^2 - 6x - 16$ b. $a^2 + 3ab + 2b^2$
   c. $x^2 + 12x + 35$ d. $6n^2 - n - 7$

6. Factor (Objective 6).
   a. $n^2 - m^2$ b. $9x^2 + 30x + 25$
   c. $x^3 - 27$ d. $8t^3 + 64s^3$

7. Factor (Objective 7).
   a. $8xy + 12x + 14y + 21$ b. $xy - 2y + x^2 - 4$
   c. $a^2 + 2a + 1 - b^2$

8. Factor completely (Objective 8).
   a. $100x^2 - 4y^2$ b. $12a^2 + 60a - 72$
   c. $8 - 4x + 2x^2$ d. $3x^4 - 81x$
   e. $y^2 - 10y + 25$ f. $x^2 + 4$
   g. $12xy - 8x + 27y - 18$ h. $x^3 + 64$
   i. $x^2 - x - 2$ j. $2x(x^2 + 6) - 7(x^2 + 6)$

You should recall from your earlier work with fractions the importance of factoring whole numbers. Factorizations enable us to reduce fractions. Factoring was also essential in finding the LCD of two fractions so that the fractions could be added or subtracted.

Before you continue your work in algebra, you should master some techniques for factoring polynomials. In particular, factoring polynomials is necessary for simplifying rational expressions and for adding rational expressions. It is also extremely useful in solving some second-degree equations.

16.1 The Greatest Common Factor of a Polynomial

One of the first techniques that we will examine is factoring a polynomial to find the greatest common factor. We will begin by defining what is meant by a common factor.

A **common factor** of a set of whole numbers is a natural number that will divide each of these whole numbers. For example, the whole numbers 8 and 12 have common factors 1, 2, and 4. The **greatest common factor** (GCF) of a set of whole numbers is the largest number that will divide each of the whole numbers in the set. For example, the GCF of 8 and 12 is 4 since 4 is the largest number that will divide both 8 and 12. Similarly, you may examine 30 and 18 and find that their GCF is 6.

For many sets of whole numbers you may be able to determine the GCF by inspection; however, for larger numbers you may need a procedure for calculating the GCF. Fortunately, all of the skills which you will need to find a GCF were covered when the concept of LCD (least common denominator) was first discussed in Chapter 2.

RULE FOR FINDING THE GCF OF A SET OF WHOLE NUMBERS	*EXAMPLE*
	Find the GCF of 168, 180, and 300.
	Solution:
1. Obtain the prime factorization of each of the whole numbers for which the GCF is sought.	$168 = 2 \cdot 2 \cdot 2 \cdot 3 \cdot 7$
	$180 = 2 \cdot 2 \cdot 3 \cdot 3 \cdot 5$
	$300 = 2 \cdot 2 \cdot 3 \cdot 5 \cdot 5$
2. Examine the factorizations and list each *common* factor the minimum number of times it occurs in any one of these factorizations.	The factors which are common are 2 and 3.
	2, 2 (the minimum number of times 2 is used as a factor is twice.)
	3 (the minimum number of times 3 is used as a factor is once.)
3. The GCF is the product of the factors listed in step 2.	$GCF = 2 \cdot 2 \cdot 3 = 12$
	Note that neither 5 nor 7 is used as a factor in forming the GCF since they are not factors of each of the given numbers (that is, they are not common factors). Also notice that the GCF 12 does indeed divide 168, 180, and 300.

Example 1 Find the GCF of the following whole numbers.

a. 30 and 18

$$18 = 2 \cdot 3 \cdot 3$$
$$30 = 2 \cdot 3 \cdot 5$$
$$GCF = 2 \cdot 3 = 6$$

Note that 6 does divide both 18 and 30.

b. 55 and 91

$$55 = 5 \cdot 11$$
$$91 = 7 \cdot 13$$
$$GCF = 1$$

Note that 1 is a factor of every number. By examining the prime factorizations, you see that 55 and 91 do not have any other common factor. Therefore, the GCF is 1. (*Comment:* 1 was not listed in the above factorizations because we chose to list only prime factors. One is not a prime and therefore was not listed. If, for convenience, you choose to list the factor 1, this is acceptable.)

c. 42, 70, 63, and 77

$$42 = 2 \cdot 3 \cdot 7$$
$$70 = 2 \cdot 5 \cdot 7$$
$$63 = 3 \cdot 3 \cdot 7$$
$$77 = 7 \cdot 11$$
$$GCF = 7$$

Before considering the GCF of a polynomial it might be wise for us to recall exactly what we mean by a polynomial. A polynomial is an expression which is formed by adding, subtracting, or multiplying constants and variables. This implies that only natural numbers may appear as exponents of variables and that no variables may appear in a denominator. Since this is an introduction to factoring, we shall only consider those polynomials which have integers as numerical coefficients and constant terms.

Example 2 a. $3y^2 - 3y - 4$ is a polynomial with the integers 3 and -3 as numerical coefficients and -4 as the constant term.

b. $x^{1/2} - 4t$ is not a polynomial since the exponent 1/2 is not a natural number.

c. $\frac{1}{2}x^2y - 3xy^2$ is a polynomial, but we will not be factoring it in this material since the numerical coefficient $\frac{1}{2}$ is not an integer.

Now we are ready to describe the greatest common factor of a polynomial and show a method for calculating this GCF. The GCF of a polynomial is determined by examining each term of the polynomial. A common factor of the polynomial is a factor that will divide each term of the polynomial. The **greatest common monomial factor** (GCF) **of a polynomial** is that common factor which has (1) the largest numerical coefficient and (2) the largest exponent on each of the variables. A procedure for finding this monomial is given in the *rule for finding the GCF of a polynomial.*

RULE FOR FINDING THE GCF OF A POLYNOMIAL	EXAMPLE
	Find the GCF of $12x^2y^2 - 20xy^4$. *Solution:*
1. Find the coefficient by determining the GCF of the absolute values of the numerical coefficients.	The numerical coefficients are 12 and -20. The GCF of 12 and 20 is 4.
2. For those variables which occur in every term of the polynomial, list the lowest power which occurs.	x —the lowest power of x occurring is the 1st power. y^2 —the lowest power of y occurring is the 2nd power.
3. The GCF is the product of the coefficient found in step 1 and the variables listed in step 2.	$GCF = 4xy^2$

Example 3   Find the GCF of the following polynomials.

a. $24x^3y^2z^3 - 16x^2y^3z^2 + 40xy^3z^5$

Coefficient: The numerical coefficients are 24, -16, and 40. The GCF of 24, 16, and 40 is 8.

Variables: x (lowest power of x)
　　　　　y^2 (lowest power of y)
　　　　　z^2 (lowest power of z)

　　　$GCF = 8xy^2z^2$

b. $-33a^2b^3c^2 + 77bc^3 + 44c^2$

Coefficient: The numerical coefficients are -33, 77, and 44. The GCF of 33, 77, and 44 is 11.

Variables: not a (a does not occur in every term.)
　　　　　not b (b does not occur in every term.)
　　　　　Therefore, a and b will not occur in the GCF.
　　　　　c^2 (lowest power of c occurring)

　　　$GCF = 11c^2$

c. $26x^2y - 39xz + 65y^2z$

Coefficient: The GCF of 26, 39, and 65 is 13.

Variables: Neither x nor y nor z occurs in every term. Therefore, none of these will occur in the GCF.

　　　$GCF = 13$

Sometimes polynomials are written with symbols of grouping. For polynomials written in this form, the terms sometimes have a common factor which is a polynomial with two or more terms. This is illustrated by the following examples.

Example 4 Find the GCF of each of the following polynomials.

a. $2x(a + b) + y(a + b)$

$(a + b)$ is the GCF since it is the only common factor of both terms.

b. $(x + 2y)a + (x + 2y)3b$

GCF is $(x + 2y)$.

c. $2a(x - y) + 3b(x - y) - 4c(x - y)$

The GCF is $(x - y)$.

d. $2x^3(x + y) + 4x^2(x + y) + 6x(x + y)$

$(x + y)$ is a factor of each term. $2x$ is also a factor of each term. Therefore, $2x(x + y)$ is the GCF.

16.1 Exercises

Find the GCF in Exercises 1–30.

1. 10 and 14
2. 6 and 9
3. 16 and 36
4. 18 and 36
5. 15 and 25
6. 60 and 70
7. $5a^2$ and $25a^3$
8. $6ab^2c$ and $12a^2b^3c$
9. $22a^2b^3c^4$ and $33a^2b^2c^2$
10. m^3n and $49m^2n^2$
11. 33 and 121
12. 56 and 98
13. 128 and 288
14. 16, 18, and 24
15. 140, 210, and 350
16. 31 and 43
17. 34 and 55
18. 15, 35, and 21
19. 14, 20, 35, and 98
20. 6, 10, 14, 22, and 26
21. −33 and 121
22. 33 and −121
23. −33 and −121
24. −16, −18, and 24
25. −33 and 77
26. −17 and 71
27. −15, −35, and −21
28. −323 and 95
29. −51, −85, and 119
30. −66, −88, −110, and −242

Find the GCF of the polynomials in Exercises 31–70.

31. $x^2 + 2x^2 + x^5$
32. $3m^2 - 6m^3 + 9m$
33. $15a^3 + 10a^4 - 30a^6$
34. $8a^3 - 16a^2 + 4a$
35. $28b^2 - 14b^3 + 34b^6$
36. $7a^2 - 14a^3 + 28a$
37. $a^5 - a^{10} + a^{15}$
38. $m^{10} + m^5 - m$
39. $3a^2 - 3a^3 + 3$
40. $7a^5 + 7a^3 - 7$
41. $9x^2 - 18xy - 15y^2$
42. $-x^3y^3 - x^2y + xy^5$

43. $12a^2b^3c - 20ab^2c^2$
44. $4x^3 - 12x^2y^2 - 8xy^3 + 6y^4$
45. $140x^3y^5z^2 - 210x^3y^5z^4 - 350x^{10}y^2z^{11}$
46. $-55x^3y^5z^7 - 91x^2y^6z^9$
47. $24a^3b^2c - 36a^2b^2c^2 + 96ab^3c^3$
48. $111abc - 11bc$
49. $-18x^2yz - 12xy^2z$
50. $4m^3 - 2m^2 + 6m$
51. $-12k^4 - 3h^2k^2 + 33k^3$
52. $6bx^2 - 12bx - 18xy + 6y$
53. $12a^2b^2 - 14a^2b^3 - 10ab^3$
54. $-13x^2y^2 + 17xy^2 - 19xy^3z$
55. $-27x^3 + 8y^3$
56. $2x^5 - 3x^4y + 5x^3y^2 - 7x^2y^3 + 11xy^4$
57. $10x^5 - 15x^4y + 25x^3y^2 - 35x^2y^3 + 55xy^4$
58. $-55a^3x^3 + 35z^2x^4 - 40z^2x^5$
59. $-121h^3k + 143h^2k^4 - 187h^2k^6$
60. $-4500x^2 + 450y^2$
61. $3x(a - b) + 7y(a - b)$
62. $a(a + b) - b(a + b)$
63. $(x + y)(7z) - (x + y)(8w)$
64. $(a - b)(7x) - (a - b)(4y)$
65. $2x(a + 3b) - y(a + 3b)$
66. $a(3x - 4y) + 2b(3x - 4y) + 2c(3x - 4y)$
67. $(2a - 5b)x + (2a - 5b)(5y) + (2a - 5b)(7z)$
68. $9x^3(a - b) + 21x^2(a - b) - 12x(a - b)$
69. $42x^2y(3a - b) - 66xy^2(3a - b)$
70. $x(2a - 3b + c) - 4y(2a - 3b + c)$

16.2 Factoring a Polynomial Using the GCF

To factor a polynomial, you must be able to recognize or calculate at least one factor other than 1 or −1. Usually the greatest common factor of a polynomial is the easiest factor to recognize. We suggest that you always determine the GCF of a polynomial if there is one and factor it out before looking for other factors.

The distributive property, which was first covered in Section 8.3, provides the justification for factoring out the GCF of a polynomial.

DISTRIBUTIVE PROPERTY
$a(b + c) = ab + ac$ or $ab + ac = a(b + c)$

Using the first form of the distributive property we are able to multiply products such as $6xy^2(x − 2y)$.

$$6xy^2(x − 2y) = 6xy^2(x) − 6xy^2(2y) = 6x^2y^2 − 12xy^3$$

In this chapter we will use the second form of the distributive property to factor polynomials such as $6x^2y^2 − 12xy^3$.

$$6x^2y^2 − 12xy^3 = 6xy^2(x) − 6xy^2(2y) = 6xy^2(x − 2y)$$

Note that the GCF of $6x^2y^2 − 12xy^3$ is $6xy^2$ and that the distributive property justifies removing this factor from each term. Consider the steps for accomplishing this described in the *procedure for factoring out the GCF of a polynomial.*

PROCEDURE FOR FACTORING OUT THE GCF OF A POLYNOMIAL	*EXAMPLE* Factor $3x^4 − 6x^3 + 27x^2$. *Solution:*
1. Find the GCF of the polynomial.	The GCF is $3x^2$.
2. Divide the polynomial by the GCF. (This should be done mentally.)	$\dfrac{3x^4 − 6x^3 + 27x^2}{3x^2} = x^2 − 2x + 9$
3. The polynomial is the product of the GCF and the quotient calculated in step 2.	$3x^4 − 6x^3 + 27x^2 = 3x^2(x^2 − 2x + 9)$

Generally, with practice you should be able to write down the answer without showing any intermediate steps. The work is shown here to illustrate the technique.

Example 1   Factor out the GCF of each of the following polynomials.

a. $x^3 − x^2 + x$

  GCF = x

  $$\frac{x^3 − x^2 + x}{x} = x^2 − x + 1$$

Thus, $x^3 − x^2 + x = x(x^2 − x + 1)$

b. $4a^2 + 6a^3 - 10a^{10}$

GCF = $2a^2$

$$\frac{4a^2 + 6a^3 - 10a^{10}}{2a^2} = 2 + 3a - 5a^8$$

Thus, $4a^2 + 6a^3 - 10a^{10} = 2a^2(2 + 3a - 5a^8)$

c. $3x^2 - 3x^4 + 3y^2$

GCF = 3

$$\frac{3x^2 - 3x^4 + 3y^6}{3} = x^2 - x^4 + y^6$$

Thus, $3x^2 - 3x^4 + 3x^6 = 3(x^2 - x^4 + y^6)$

Example 2   Factor out the GCF of each of the following polynomials.

a. $35x^5y^3 - 49x^3y^4$

GCF = $7x^3y^3$

$$\frac{35x^5y^3 - 49x^3y^4}{7x^3y^3} = 5x^2 - 7y$$

Thus, $35x^5y^3 - 49x^3y^4 = 7x^3y^3(5x^2 - 7y)$.

b. $24x^3y^2z^3 - 16x^2y^3z^2 + 40xy^3z^5$

GCF = $8xy^2z^2$

$$\frac{24x^3y^2z^3 - 16x^2y^3z^2 + 40xy^3z^5}{8xy^2z^2} = 3x^2z - 2xy + 5yz^3$$

Thus, $24x^3y^2z^3 - 16x^2y^3z^2 + 40xy^3z^5 = 8xy^2z^2(3x^2z - 2xy + 5yz^3)$.

c. $-26x^3y^2 + 39x^2y^3 - 65xy^4$

GCF = $13xy^2$

$$\frac{-26x^3y^2 + 39x^2y^3 - 65xy^4}{13xy^2} = -2x^2 + 3xy - 5y^2$$

Thus, $-26x^3y^2 + 39x^2y^3 - 65xy^4 = 13xy^2(-2x^2 + 3xy - 5y^2)$

When the coefficient of the first term of the second factor is negative, we sometimes rewrite the factorization to make the coefficient positive. To accomplish this, factor -1 out of the second factor. Thus in Example 2c,

$$-2x^2 + 3xy - 5y^2 = -1(2x^2 - 3xy + 5y^2)$$

or

$$-26x^3y^2 + 39x^2y^3 - 65xy^4 = 13xy^2(-2x^2 + 3xy - 5y^2)$$
$$= 13xy^2(-1)(2x^2 - 3xy + 5y^2)$$
$$= -13xy^2(2x^2 - 3xy + 5y^2)$$

The same result can be accomplished by factoring out $-13xy^2$.

Example 3   Factor −1 out of the following polynomials.

    a. $a - 3 = -1(-a + 3)$ or $-(3 - a)$

    b. $2a + b = -1(-2a - b)$ or $-(-2a - b)$

    c. $-x + y = -1(x - y)$ or $-(x - y)$

       The greatest common factor may be a polynomial of more than one term. This is illustrated in Example 4.

Example 4   Factor out the greatest common factor of each of the following polynomials.

    a. $2x(a + b) + y(a + b)$

       $(a + b)$ is a common factor of both terms.

$$\frac{2x(a + b) + y(a + b)}{(a + b)} = \frac{2x(a + b)}{(a + b)} + \frac{y(a + b)}{(a + b)} = 2x + y$$

       Thus, $2x(a + b) + y(a + b) = (2x + y)(a + b)$.

       Check:

       $(2x + y)(a + b) = 2x(a + b) + y(a + b)$ by the distributive property.

    b. $(x + 2y)a + (x + 2y)(3b)$

       The common binomial factor is $(x + 2y)$.

       Thus, $(x + 2y)a + (x + 2y)(3b) = (x + 2y)(a + 3b)$

    c. $2a(x - y) + 3b(x - y) - 4c(x - y)$

       The GCF is $(x - y)$.

       Thus, $2a(x - y) + 3b(x - y) - 4c(x - y) = (2a + 3b - 4c)(x - y)$

    d. $2x^3(x + y) + 4x^2(x + y) + 6x(x + y)$

       The GCF is $2x(x + y)$.

$$\frac{2x^3(x + y) + 4x^2(x + y) + 6x(x + y)}{2x(x + y)} = x^2 + 2x + 3$$

       Thus, $2x^3(x + 6) + 4x^2(x + y) + 6x(x + y) = (x^2 + 2x + 3)(2x)(x + y)$

$$= 2x(x + y)(x^2 + 2x + 3)$$

16.2 Exercises

Factor out the GCF in each of the polynomials in Exercises 1–40.

  1. $7x - 7y$

  3. $x^3 - x^2$

  5. $4x + 6x^2$

  7. $3m^4 + 3m^2$

  9. $22x^6 - 33x^3$

11. $13x - 26y$

13. $x^3y^3 - x^2y^4$

  2. $24a - 16b$

  4. $m^4 + m^6$

  6. $15a^6 - 20a^4$

  8. $8m^5 + 4m^{10}$

10. $18m^3 + 9m$

12. $15x^3 + 45y^3$

14. $m^4n^2 + 3m^2n^6$

15. $49x^2 - 63xy + 14y$

16. $63x^2y - 72xy^2 + 99$

17. $5x^2y - 7xy^2$

18. $9a^3b - 6a^2b - 7ab^2$

19. $3m^4 - 9m^3 + 6m^2$

20. $12x^5 - 18x^4 + 12x^3$

21. $12a^2b^3c - 20ab^2c$

22. $35x^3y^3 - 49x^2y^4 + 77xy^5$

23. $140x^2y^5 - 210x^3y^5 - 350x^{10}y^5$

24. $-55x^3y^4z^7 - 121x^2y^6z^9 + 33xy^5z^{11}$

25. $4x^3y - 12x^2y^2 - 6xy^3$

26. $111abcd - 11abcde$

27. $3x^3 - 6x^2y + 3xy^2$ 28. $5xz^4 - 5xy^4$ 35. $3x(2a - 3b + c) - 2y(2a - 3b + c)$

29. $-12k^4 - 3h^2k^2 + 33k^3$ 30. $-27x^3y + 8y^4$ 36. $(a + 11b)(17x) - (a + 11b)(9y) + (a + 11b)(4z)$

31. $2x(a + 3b) - y(a + 3b)$ 37. $x(5a - 4b - 3c - d) - 2y(5a - 4b - 3c - d)$

32. $(x + y)(7z) - (x + y)(11w)$ 38. $(14x - 3y + z)a - (14x - 3y + z)(2b)$

33. $3a(19x - y) + 5b(19x - y) - 7c(19x - y)$ 39. $15x^3y(2a - 7c) - 21x^2y^2(2a - 7c)$

34. $-2a(x + 3y) - 4c(x + 3y)$ 40. $(117x + 31y)(54a^3b^2c) - 90a^2b^3c^3(117x + 31y)$

16.3 Multiplying Binomials by Inspection

Skillful factoring requires that you be able to recognize some factors by inspection. For example, you should be able to factor 35 as $5 \cdot 7$ without going through a division procedure. Of course, you are not expected to factor 1147 by inspection. Similarly, you should soon be able to factor $x^2 - y^2$ by inspection, but you will not be expected to factor $15x^2 - 2xy + 24y^2$ by inspection.

To obtain this skill of recognizing certain factors by inspection, you need to be able to multiply two binomials mentally. Without pencil and paper you should be able to determine the product of 5 times 24 even though you probably do not have this product memorized. Similarly, it will be quite helpful if you can calculate the product of $(2x - y)$ times $(x - 3y)$ without pencil and paper. In order to develop this skill, we must first review the procedure for multiplying polynomials.

In Section 8.4 the distributive property was used to multiply two polynomials. Both a horizontal and vertical format were used.

$$\textit{Horizontal format:} \quad (x - 3y)(2x - y) = x(2x - y) - 3y(2x - y)$$
$$= 2x^2 - xy - 6xy + 3y^2$$
$$= 2x^2 - 7xy + 3y^2$$

$$
\begin{array}{r}
\textit{Vertical format:} \quad 2x - \ y \\
x - 3y \\
\hline
2x^2 - \ xy \\
- 6xy + 3y^2 \\
\hline
2x^2 - 7xy + 3y^2
\end{array}
$$

The steps are exactly the same in both cases. Note that in the final result the first term, $2x^2$, is the product of the first terms from each of the binomial factors (x and 2x). The second term, $-7xy$, is the sum of the cross products ($-xy$ and $-6xy$). The third term, $3y^2$, is the product of the last terms from each of the binomial factors ($-3y$ and $-y$). The cross products are obtained in the horizontal format by multiplying the outside terms (x and $-y$) and the inside terms ($-3y$ and 2x). In the vertical format the cross products are found by multiplying the terms indicated by the arrows in the following figure. (The name "cross product" comes from this format.)

By examining the problem in horizontal format we can write down the product without recopying the problem or writing down any intermediate steps. This is illustrated by Example 1.

Example 1   Find the product $(x - 5y)(2x + y)$.

$$(x - 5y)(2x + y)$$

$2x^2$ is the product of the first terms.

$$(x - 5y)(2x + y)$$

$xy - 10xy = -9xy$. Thus $-9xy$ is the sum of the cross products.

$$(x - 5y)(2x + y)$$

$-5y^2$ is the product of the last terms.
Thus, $(x - 5y)(2x + y) = 2x^2 - 9xy - 5y^2$.

The best way to illustrate these steps in a text is to show them. However, the idea is for you to practice until you can write down the last step shown without showing the intermediate work.

PROCEDURE FOR MULTIPLYING BINOMIALS BY INSPECTION

EXAMPLE
$(3x + y)(x - 4y)$
Solution:

Find the product by

$(3x + y)(x - 4y) = 3x^2 - 11xy - 4y^2$

a. writing down the term which is the product of the first terms of the binomials

b. writing down the term which is the sum of the cross products

c. writing down the term which is the product of the last terms of the binomials

---

Example 2   Determine the following products using the horizontal format. Show the intermediate steps.

a. $(3x + 2y)(5x + 7y)$

$(3x + 2y)(5x + 7y)$ $15x^2$ (product of first terms)

$(3x + 2y)(5x + 7y)$ $31xy$ (the sum of the cross products $21xy$ and $10xy$)

$(3x + 2y)(5x + 7y)$ $14y^2$ (product of last terms)

Thus, $(3x + 2y)(5x + 7y) = 15x^2 + 31xy + 14y^2$.

b. $(-2a + 11b)(3a + 10b)$

$(-2a + 11b)(3a + 10b)$　　$-6a^2$ (product of first terms)

$(-2a + 11b)(3a + 10b)$　　13ab (the sum of the cross products $-20ab$ and 33ab)

$(-2a + 11b)(3a + 10b)$　　$110b^2$ (product of last terms)

Thus, $(-2a + 11b)(3a + 10b) = -6a^2 + 13ab + 110b^2$.

Example 3　Write the following products by inspection without showing any intermediate steps.

a. $(x - 7)(x - 8) = x^2 - 15x + 56$

b. $(3a - 3b)(4a + 2b) = 12a^2 - 6ab - 6b^2$

c. $(6w + 5z)(9w - 7z) = 54w^2 + 3wz - 35z^2$

16.3　Exercises

Fill in the middle term in the products in Exercises 1–10.

1. $(a - 1)(a + 3) = a^2 + \boxed{} - 3$

2. $(x + 5)(x + 2) = x^2 + \boxed{} + 10$

3. $(a - 4)(a - 3) = a^2 + \boxed{} + 12$

4. $(x - 6)(x + 2) = x^2 + \boxed{} - 12$

5. $(2a - 1)(a + 3) = 2a^2 + \boxed{} - 3$

6. $(3x + 5)(x + 2) = 3x^2 + \boxed{} + 10$

7. $(a - 4)(5a - 3) = 5a^2 + \boxed{} + 12$

8. $(x - 6)(3x + 2) = 3x^2 + \boxed{} - 12$

9. $(2x + 1)(3x - 4) = 6x^2 + \boxed{} - 4$

10. $(5x - 6)(2x - 3) = 10x^2 + \boxed{} + 18$

Determine the products in Exercises 11–50 by inspection.

11. $(x - y)(2x + y)$

12. $(x - y)(x + y)$

13. $(3w - z)(3w - z)$

14. $(5a - b)(2a + b)$

15. $(11x - 4y)(3x + y)$

16. $(5a + 9)(7a - 2)$

17. $(2s - 7t)(3s - 6t)$

18. $(h - 2k)(h - 2k)$

19. $(2a + 3b)(2a - 3b)$

20. $(21a + b)(a - 2b)$

21. $(7w - 4z)(3w + 5z)$

22. $(x + y)^2$

23. $(5a - 7b)(5a + 7b)$

24. $(-10s + t)(-s + t)$

25. $(4x - 7)(3x - 5)$

26. $(x - 8y)(x + 8y)$

27. $(6a - b)(a - 6b)$

28. $(2s - 11t)(3s - 5t)$

29. $(2s + 11t)^2$

30. $(2s - 11t)^2$

31. $(8x - 10)(7x - 3)$

32. $(11a - 9)(11a + 9)$

33. $(a - 9)(a + 8)$

34. $(9a - 1)(8a + 1)$

35. $(2x - y)^2$

36. $(11c - d)(10c + d)$

37. $(xy - 3)(xy + 4)$

38. $(2ab + 1)(3ab + 1)$

39. $(ab - c)(ab + 3c)$

40. $(2xy - z)(3xy - z)$

41. $(wx - yz)^2$

42. $(wx + yz)^2$

43. $(2abc - d)(2abc + d)$

44. $(x^2 - 2)(x^2 + 3)$

45. $(5a^2 - b^2)(2a^2 + b^2)$

46. $(7ax - 3)(ax + 4)$

47. $(2ax^2 - y)(3ax^2 + y)$

48. $(10ab^2c - 1)(10ab^2c + 1)$

49. $(x^3 + 3y^2)(x^3 - 3y^2)$

50. $(w^4 - 3z)(w^4 - 2z)$

16.4 Special Products

After practicing the problems in the preceding section, you should be able to multiply many polynomials by inspection. Certain products occur so frequently, though, that it is useful to single them out to be memorized. If you can recognize these products and write the answer, you will then be able to factor many polynomials by inspection.

Two special products which you should recognize are given in this section. As you study the examples and work the exercises, concentrate upon recognizing the type of problem. Once you know the type you should be able to write down the product quickly.

The first special product we will consider is the square of a binomial. The two cases to be considered are

1. The square of a sum

$$(a + b)^2 = (a + b)(a + b)$$
$$= a^2 + 2ab + b^2$$

2. The square of a difference

$$(a - b)^2 = (a - b)(a - b)$$
$$= a^2 - 2ab + b^2$$

These results are described verbally as follows:

THE SQUARE OF A BINOMIAL

1. The square of the sum of two numbers is the square of the first number plus twice the product of the two numbers plus the square of the second number.

$$(a + b)^2 = a^2 + 2ab + b^2$$

2. The square of the difference of two numbers is the square of the first number minus twice the product of the two numbers plus the square of the second number.

$$(a - b)^2 = a^2 - 2ab + b^2$$

Example 1   Each of the following is the square of a binomial. Write the product.

a. $(x + 3)^2 = x^2 + 6x + 9$

square of first term x
plus twice the product of 3 and x
plus square of the last term 3

b. $(w + 4)^2 = w^2 + 8x + 16$

square of first term w
plus twice the product of 4 and w
plus square of last term 4

c. $(3w + 5z)^2 = (3w)^2 + 2(15wz) + (5z)^2$
                 $= 9w^2 + 30wz + 25z^2$

d. $(7s + 9t)^2 = (7s)^2 + 2(63st) + (9t)^2$
                 $= 49s^2 + 126st + 81t^2$

e. $(z - 1)^2 \;=\; z^2 \;-\; 2z \;+\; 1$

square of first term z ⏐
minus twice the product of 1 and z ⏐
plus square of the last term 1 ⏐

f. $(x - 8)^2 \;=\; x^2 \;-\; 16x + 64$

square of first term x ⏐
minus twice the product of 8 and x ⏐
plus square of the last term 8 ⏐

g. $(6m - 5n)^2 = (6m)^2 - 2(30mn) + (5n)^2$
                 $= 36m^2 - 60mn + 25n^2$

h. $(11w - 4x)^2 = (11w)^2 - 2(44wx) + (4x)^2$
                 $= 121w^2 - 88wx + 16x^2$

Another product that appears frequently is the sum of two terms multiplied by the difference of these terms. This is examined in both the horizontal and vertical formats as follows:

$$(a + b)(a - b) = a^2 - b^2$$

$$\begin{array}{r} a + b \\ a - b \\ \hline a^2 + ab \\ - ab - b^2 \\ \hline a^2 \qquad - b^2 \end{array}$$

Note that this product contains only two terms. As the vertical format clearly illustrates, the cross product is zero since these terms are of opposite sign.

> ### *A SUM TIMES A DIFFERENCE*
>
> The product of the sum of two numbers times their difference equals the difference of their squares.
>
> $$(a + b)(a - b) = a^2 - b^2$$

Example 2   Each of the following is the product of a sum times a difference. Write the product.

a. $(x + 2)(x - 2) \;=\; x^2 \;-\; 4$

square of first number x ⏐
minus the square of second number 2

b. $(w + 5)(w - 5) = w^2 - 25$

square of first number w ⎯⎯┘   ↑
minus the square of the second number 5

c. $(2s + 11)(2s - 11) = (2s)^2 - (11)^2$
           $= 4s^2 - 121$

d. $(7x - 9y)(7x + 9y) = (7x)^2 - (9y)^2$
           $= 49x^2 - 81y^2$

16.4 Exercises

Each of Exercises 1–20 is the square of a binomial. Write the product.

1. $(x + y)^2$
2. $(w + z)^2$
3. $(2a + b)^2$
4. $(c + 3d)^2$
5. $(x + 11)^2$
6. $(4x + 3y)^2$
7. $(7x + 5y)^2$
8. $(9x + 8y)^2$
9. $(3ab + 2c)^2$
10. $(6ab + 5cd)^2$
11. $(w - z)^2$
12. $(x - y)^2$
13. $(x - 3y)^2$
14. $(2x - a)^2$
15. $(4x - 3y)^2$
16. $(7x - 10)^2$
17. $(4w - 11)^2$
18. $(2a - 12b)^2$
19. $(2x - 9)^2$
20. $(7st - w)^2$

Each of Exercises 21–30 is the product of a sum times a difference. Write the product.

21. $(s + t)(s - t)$
22. $(a - b)(a + b)$
23. $(x - y)(x + y)$
24. $(2x - y)(2x + y)$
25. $(7x - 5)(7x + 5)$
26. $(9a + b)(9a - b)$
27. $(5xy - z)(5xy + z)$
28. $(x^2 - y^2)(x^2 + y^2)$
29. $(s^3 + t^3)(s^3 - t^3)$
30. $(4xy^2 - 3w)(4xy^2 + 3w)$

For each of Exercises 31–40 first determine if it is one of the special products discussed in this section, and then write the product.

31. $(x - 7)(x + 7)$
32. $(a - 6)^2$
33. $(y - z)(y - z)$
34. $(w + 5)^2$
35. $(x + 2y)(x + 2y)$
36. $(2a - b)^2$
37. $(2x + 3)(2x - 3)$
38. $(7x - 2y)^2$
39. $(8y - 3z)^2$
40. $(2x - 5y)(2x + 5y)$

Find the products in Exercises 41–60 by inspection.

41. $(4w + 3)^2$
42. $(5a + 6b)(5a + 6b)$
43. $(9k - 5n)^2$
44. $(4ab + 6)^2$
45. $(7a - 8b)(7a + 8b)$
46. $(x + y)(2x + 3y)$
47. $(5a + 3)(2a - 5)$
48. $(7a - 8b)(7a - 8b)$
49. $(20x + 3y)^2$
50. $(9ab - c)(9ab + 2c)$
51. $(8w - 1)(2w - 6)$
52. $(3xy + z)^2$
53. $(5wz^2 - 1)^2$
54. $(6xy + 7z)(6xy - 7z)$
55. $(x^2 + 5)(x^2 - 5)$
56. $(ab + 3c)(2ab - 4c)$
57. $(m - 3n)(m + 6n)$
58. $(ab^2 - c)^2$
59. $(8yz^2 + w^3)^2$
60. $(a^3 - b)(a^3 + b)$

16.5 Factoring Second-Degree Trinomials by Trial and Error

If a whole number cannot be factored as a product of smaller whole numbers, it is called prime. We have stated that we will factor only polynomials with integral coefficients. If we also consider as factors only polynomials with integral coefficients, a polynomial is **prime over the integers** if the GCF of the polynomial is +1 and if it cannot be written as the product of polynomials of lower degree. A polynomial which is not prime is said to be **factorable**.

Example 1   a. $x^2 - 1$ is factorable since it can be written as the product of polynomials of lower degree: $x^2 - 1 = (x - 1)(x + 1)$. Note that each factor is a prime polynomial.

b. $3x - 6$ is factorable, since the GCF is 3, so $3x - 6 = 3(x - 2)$.

c. $x^2 - 5$ is prime over the integers. You could factor $x^2 - 5$ as $(x + \sqrt{5})(x - \sqrt{5})$, but this would not be a factorization over the integers since $\sqrt{5}$ and $-\sqrt{5}$ are not integers.

d. $x^2 + 1$ is prime over the integers.

We will now develop a technique which will enable you to determine whether a second-degree trinomial is prime or factorable. This technique is analogous to a method used to factor whole numbers. While you should be able to factor 35 at sight as $5 \cdot 7$, you probably would not know if 1081 is prime or composite. Nonetheless, you were introduced in Section 2.2 to a systematic procedure for determining the prime factors of a whole number. This procedure, even though systematic, could be described as trial and error since it is possible that you may have to try many divisors before you eventually find one that works.

The procedure for factoring trinomials as the product of two binomials is likewise systematic, and it also involves trial and error. Let us begin by considering trinomials of the form $ax^2 + bx + c$ where a, b, and c are constants and x is the variable. The ax^2 term is called the second-degree term or the **quadratic term**; bx is the first-degree term or the **linear term**; and c is the **constant term**. The trinomial $ax^2 + bx + c$ is referred to as a **quadratic trinomial**.

For example, $6x^2 - 7x - 5$ is a quadratic trinomial with a = 6, b = −7, and c = −5. The quadratic term is $6x^2$; the linear term is −7x; and the constant term is −5. The easiest quadratic trinomials to factor are those in which a = 1. Consider, for example, $x^2 - 5x - 14$. In this trinomial a = 1, b = −5, and c = −14.

Example 2   Verify that $x^2 - 5x - 14$ factors as $(x - 7)(x + 2)$.

Although you should be able to multiply these binomials by inspection, the vertical format is shown here so that the steps can be described.

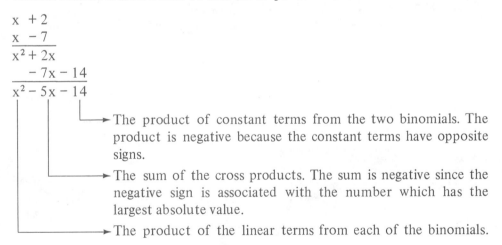

Example 3   Find the following products.

a. $(x + 7)(x + 2)$

$\qquad (x + 7)(x + 2) = x^2 + 9x + 14$

b. $(x - 7)(x - 2)$

$\qquad (x - 7)(x - 2) = x^2 - 9x + 14$

Observation: If the constant term of the trinomial is positive, then the constant terms of the binomial factors must be of the same sign. The sign of both of those constant terms will be the same as the sign of the linear term of the trinomial.

Example 4   Find the following products.

a. $(x + 7)(x - 2)$

$$(x + 7)(x - 2) = x^2 + 5x - 14$$

b. $(x - 7)(x + 2)$

$$(x - 7)(x + 2) = x^2 - 5x - 14$$

Observation: If the constant term of the trinomial is negative, then the constant terms of the binomial factors must be opposite in sign. The sign of the linear term of the trinomial is the same as the constant term which has the largest absolute value.

Example 5   Factor $x^2 - 10x + 24$.

The constant term, 24, is positive; thus, the constant terms of the factors must be of the same sign. Since the linear term, $-10x$, has a negative coefficient, the sign of both of these constant terms is negative. Thus, $x^2 - 10x + 24$ looks like $(x - ?)(x - ?)$. What are the constant terms? (*Hint:* Their product is 24. This allows several of the following possibilities, but only one of these gives the correct linear term. The arrow shows which is the correct term.)

Possible Factors	Resulting Linear Term
$(x - 1)(x - 24)$	$-25x$
$(x - 2)(x - 12)$	$-14x$
$(x - 3)(x - 8)$	$-11x$
$(x - 4)(x - 6)$	$\longrightarrow -10x$

The only possibility which yields the correct linear term is $(x - 4)(x - 6)$. Hence, $x^2 - 10x + 24 = (x - 4)(x - 6)$.

Example 6   Factor $x^2 + 7x - 60$.

-60 is negative so the constant terms are of opposite sign. The linear term $7x$ has a positive coefficient; thus, the positive constant will be larger in absolute value than the negative constant. Examining the factors of 60, we list the possibilities as follows:

Possible Factors	Resulting Linear Term
$(x - 1)(x + 60)$	$59x$
$(x - 2)(x + 30)$	$28x$
$(x - 3)(x + 20)$	$17x$

$$(x - 4)(x + 15) \qquad\qquad 11x$$
$$(x - 5)(x + 12) \qquad\quad \longrightarrow 7x$$
$$(x - 6)(x + 10) \qquad\qquad 4x$$

The correct linear term for this problem is 7x; hence,
$x^2 + 7x - 60 = (x - 5)(x + 12)$.

Example 7 Factor $x^2 - 16x - 36$.

The constant terms must be of opposite sign since -36 is negative. The constant term that is larger in absolute value must be negative since $-16x$ has a negative coefficient. All factors of 36 that fulfill these requirements are listed as possibilities.

Possible Factors	*Resulting Linear Term*
$(x + 1)(x - 36)$	$-35x$
$(x + 2)(x - 18)$	$\longrightarrow -16x$
$(x + 3)(x - 12)$	$-9x$
$(x + 4)(x - 9)$	$-5x$

Since $-16x$ is the correct linear term, $x^2 - 16x - 36 = (x + 2)(x - 18)$.

Example 8 Factor $x^2 + 9x + 12$.

Since 12 is positive and the linear term $9x$ has a positive coefficient, both binomial factors must have positive constant terms. The complete list of possibilities is found by examining the factors of 12.

Possible Factors	*Resulting Linear Term*
$(x + 1)(x + 12)$	$13x$
$(x + 2)(x + 6)$	$8x$
$(x + 3)(x + 4)$	$7x$

Since none of these possibilities works, $x^2 + 9x + 12$ is not factorable over the integers. In other words, $x^2 + 9x + 12$ is prime over the integers.

With practice you should not actually have to list all possibilities in order to factor a trinomial. Rather, you should be able to think of the possibilities and arrive at the correct factorization. The major point is for you to be systematic. There is some trial and error, but there is no need to make wild guesses. By eliminating the possibilities in an orderly manner, you can also be certain when a second-degree trinomial is prime.

Any trinomial of the form $ax^2 + bx + c$ which is not prime can be factored in a similar manner to that of the trinomials in the preceding examples, even if a is not 1.

Example 9   Factor $6x^2 - 7x - 5$.

The constant terms of the binomial factors must be of opposite sign since −5 is negative. Considering only the constant terms, the factors are either $(? + 1)(? - 5)$ or $(? - 1)(? + 5)$. The product of the first terms must be $6x^2$, so we list all such possibilities. (Note that $6x^2$ can be obtained from two different combinations.)

Possible Factors	Resulting Linear Terms
$(6x + 1)(x - 5)$	$-29x$
$(6x - 1)(x + 5)$	$29x$
$(3x + 1)(2x - 5)$	$-13x$
$(3x - 1)(2x + 5)$	$13x$
$(2x + 1)(3x - 5)$	$\longrightarrow -7x$
$(2x - 1)(3x + 5)$	$7x$
$(x + 1)(6x - 5)$	x
$(x - 1)(6x + 5)$	$-x$

Since the correct linear term is $-7x$, the factorization is $6x^2 - 7x - 5 = (2x + 1)(3x - 5)$.

Obviously, when the quadratic term and the constant term of the trinomial have several factors, the number of possibilities becomes quite large. Nonetheless, such a list provides a starting place and is better than random guessing.

Example 10   Factor $12x^2 - 27x + 15$.

Since 15 is positive, the factors of this constant must have the same sign. The linear term, $-27x$, is negative, so both of the factors of 15 must be negative. The product of the first terms must be $12x^2$. The correct factorization is $12x^2 - 27x + 15 = (4x - 5)(3x - 3)$.

Example 11   Factor $26x^2 + 19x - 7$.

The constant terms must be opposite in sign. Why? The product of the first terms must be $26x^2$. Thus $26x^2 + 19x - 7 = (26x - 7)(x + 1)$.

Example 12   Factor $21x^2 - 30x + 10$.

The product of the constant terms must be 10 and the constant terms must be negative. Why? The product of the first terms must be $21x^2$. All such possibilities will be examined.

Possible Factors	Resulting Linear Term
$(21x - 10)(x - 1)$	$-31x$
$(21x - 1)(x - 10)$	$-211x$
$(21x - 5)(x - 2)$	$-47x$
$(21x - 2)(x - 5)$	$-107x$

$(7x - 10)(3x - 1)$	$-37x$
$(7x - 1)(3x - 10)$	$-73x$
$(7x - 5)(3x - 2)$	$-29x$
$(7x - 2)(3x - 5)$	$-41x$

Since none of these possibilities yields the correct linear factor, the polynomial is prime over the integers.

Example 13   Factor $4a^2 + 29ab + 25b^2$.

Although this trinomial is not of the form $ax^2 + bx + c$, we will use the same technique to factor it. Since the last term, $25b^2$, has a positive coefficient and the middle term, $29ab$, is positive, both binomial factors must have positive coefficients for their second terms. The product of the first terms must be $4a^2$, and the product of the last terms must be $25b^2$. The correct choice is $4a^2 + 29ab + 25b^2 = (4a + 25b)(a + b)$.

Example 14   Factor $15s^2 - 29st + 8t^2$.

$$15s^2 - 29st + 8t^2 = (5s - 8t)(3s - t)$$

Check:

$$
\begin{array}{r}
5s - 8t \\
3s - t \\
\hline
15s^2 - 24st \\
-5st + 8t^2 \\
\hline
15s^2 - 29st + 8t^2
\end{array}
$$

This factoring technique can also be applied to binomials if we consider the coefficient of the linear term to be 0. This is illustrated in Examples 15 and 16.

Example 15   Factor $x^2 - 9$.

Since -9 is negative, the constant terms must be of opposite sign.

Possible Factors	*Resulting Linear Term*
$(x + 9)(x - 1)$	$8x$
$(x - 9)(x + 1)$	$-8x$
$(x + 3)(x - 3)$	$\longrightarrow\ 0x$

The linear term is $0x$; thus, $x^2 - 9 = (x + 3)(x - 3)$.

Example 16   Factor $x^2 + 1$.

The constant terms must be of the same sign.

Possible Factors	*Resulting Linear Term*
$(x - 1)(x - 1)$	$-2x$
$(x + 1)(x + 1)$	$2x$

Since none of the possible factors gives a linear term of 0, $x^2 + 1$ is prime over the integers.

16.5 Exercises

Factor each of the trinomials in Exercises 1-75 as the product of two binomials.

1. $x^2 - 5x - 6$

2. $x^2 - x - 6$

3. $x^2 + x - 6$

4. $x^2 + 5x - 6$

5. $x^2 + 7x + 6$

6. $x^2 + 5x + 6$

7. $x^2 - 5x + 6$

8. $x^2 - 7x + 6$

9. $w^2 - w - 30$

10. $a^2 + 12a + 35$

11. $s^2 + 18s + 77$

12. $y^2 + 2y - 48$

13. $b^2 + 9b - 36$

14. $z^2 + 29z + 100$

15. $t^2 + 2t - 48$

16. $x^2 - 5x - 36$

17. $c^2 - 25c + 100$

18. $w^2 - 11w + 28$

19. $r^2 + 2r - 35$

20. $p^2 + 2p - 99$

21. $x^2 + 3x + 2$

22. $y^2 - 2y - 3$

23. $p^2 - 6p - 7$

24. $x^2y^2 + 5xy + 4$

25. $x^2 - 9x - 36$

26. $2x^2 + 5x + 3$

27. $5x^2 + 12x + 7$

28. $11x^2 + 6x - 5$

29. $7y^2 - 6y - 13$

30. $3w^2 - 5w + 2$

31. $9c^2 + c - 8$

32. $4z^2 - 9z + 5$

33. $3a^2 - 8a + 5$

34. $5x^2 + 14x - 3$

35. $2y^2 - 15y + 7$

36. $7w^2 + 69w - 10$

37. $7b^2 + 34b - 5$

38. $33z^2 - 100z + 3$

39. $11t^2 + 100t + 9$

40. $14k^2 - 69k - 5$

41. $10a^2 - 69a - 7$

42. $12w^2 + 49w + 4$

43. $a^2 - ab - 6b^2$

44. $3y^2 - 16yz + 5z^2$

45. $x^2 + xy - 12y^2$

46. $18s^2 + 25st - 3t^2$

47. $p^2 + 9pg + 14g^2$

48. $4x^2 + 20xy + 9y^2$

49. $6c^2 + 35cd + 11d^2$

50. $25m^2 + 20mn - 21n^2$

51. $15x^2 - 26xy + 8y^2$

52. $45a^2 + 79ab + 14b^2$

53. $18s^2 + 15st - 7t^2$

54. $63y^2 - 31yz - 10z^2$

55. $42j^2 - 5jk - 25k^2$

56. $24c^2 - 94cd + 35d^2$

57. $25x^2 + 4xy - 21y^2$

58. $70a^2 + 9ab - 9b^2$

59. $45f^2 - 79fg + 14g^2$

60. $12w^2 + 64wx + 77x^2$

61. $63p^2 + 27pq - 10q^2$

62. $24x^2 - 58xy + 35y^2$

63. $25a^2 - 68ab - 21b^2$

64. $70k^2 - 99kj - 9j^2$

65. $45m^2 + 73mn + 14n^2$

66. $12a^2 + 65ab + 77b^2$

67. $24w^2 - 74wx + 35x^2$

68. $28s^2 + 64st - 15t^2$

69. $28m^2 - 23mn - 15n^2$

70. $30x^2 - 47xy + 7y^2$

71. $99x^2y^2 + 59xyz - 14z^2$

72. $26a^2b^2c^2 - 85abcd - 21d^2$

73. $15s^2 + 31st + 14t^2$

74. $a^2 - 18a + 77$

75. $15b^2 + 32b + 16$

Some of the polynomials in Exercises 76–100 are factorable; some are prime. Factor those which are factorable, and label the others as prime over the integers.

76. $x^2 - 3$

77. $x^2 - 25$

78. $x^2 - 49$

79. $x^2 - 21$

80. $x^2 - 40$

81. $x^2 + 4$

82. $x^2 + 9$

83. $x^2 + 15$

84. $x^2 + 8x + 16$

85. $x^2 + 22x + 121$

86. $x^2 + 8x + 6$

87. $x^2 + 8x + 7$

88. $x^2 - 8x + 7$

89. $x^2 - 6x - 7$

90. $x^2 + 6x - 7$

91. $x^2 + 5x - 7$

92. $3x^2 - 16x - 5$

93. $3x^2 - 14x - 5$

94. $3x^2 + 8x - 5$

95. $3x^2 - 2x - 5$

96. $a^2 + 10a + 28$

97. $a^2 - 3a - 28$

98. $25y^2 - 4yz - 21z^2$

99. $24s^2 + 17st + 35t^2$

100. $24s^2 + 31st + 35t^2$

16.6 Factoring Polynomials by Recognizing Special Forms

Although you can use the trial and error technique to determine the factors of a quadratic trinomial, it is very important that you learn to recognize some special forms that can be factored on sight. However, before you can master this section you should have the special products from Section 16.4 memorized to the extent that you can recognize the results of these products on sight. As you study the special forms of polynomials, pay particular attention to the distinctions made between them so that you will be able to identify them quickly.

> *TRINOMIALS WHICH ARE THE SQUARES OF BINOMIALS*
> $a^2 + 2ab + b^2 = (a + b)^2$
> $a^2 - 2ab + b^2 = (a - b)^2$

Notice that in these special forms both the first and last terms must be perfect squares. The middle term must be twice the product of the numbers which are squared to obtain the first and last terms of the trinomial. Such trinomials are frequently called **perfect square trinomials**. It is possible to factor them by trial and error, but it is less time-consuming to recognize them as special products.

Example 1 Each of the following trinomials is a perfect square. Show the numbers which must be squared to produce the first and last terms and then factor the trinomial.

a. $49x^2 - 14x + 1 = (7x)^2 - 2(7x) + (1)^2$
$$= (7x - 1)^2$$

b. $100s^2 + 60st + 9t^2 = (10s)^2 + 2(30st) + (3t)^2$
$$= (10s + 3t)^2$$

c. $w^2 - 6w + 9 = (w)^2 - 2(3w) + (3)^2$
$$= (w - 3)^2$$

d. $36z^2 + 60z + 25 = (6z)^2 + 2(30z) + (5)^2$
$$= (6z + 5)^2$$

Example 2 Determine which of the following trinomials are squares of a binomial.

a. $18x^2 - 18x + 1$

Since the first term $(18x^2)$ is not a perfect square, the trinomial cannot be the square of a binomial.

b. $25x^2 + 60xy + 36y^2$

This trinomial equals $(5x)^2 + 2(30xy) + (6y)^2$; thus, it is a perfect square. It factors as $(5x + 6y)^2$.

c. $64x^2 - 32x + 2$

Since the last term, 2, is not a perfect square, this trinomial is not a perfect square.

d. $81x^2 + 100y$

This is a binomial, not a trinomial; therefore, it cannot be the square of a binomial.

e. $4w^2 + 5w + 1$

Both the first term and the last term are perfect squares; however, the middle term is not twice the product of the numbers which are squared to obtain the first and last terms. The square of $2w + 1$ is $4w^2 + 4w + 1$, not $4w^2 + 5w + 1$. Therefore, $4w^2 + 5w + 1$ is not a perfect square.

f. $4x^2 - 12xy - 9y^2$

The last term is negative; hence, it cannot be a perfect square. Therefore, the trinomial is not the square of a binomial. For a trinomial to be a perfect square, both the first and last terms must have positive coefficients.

Certain binomials are also considered to be special forms. They are special not only because they occur frequently, but also because the product of a binomial factor and another binomial or trinomial factor normally results in a polynomial with more than two terms. Three special forms are illustrated below.

$$
\begin{array}{lll}
a\ +b & a^2+ab\ +b^2 & a^2-ab\ +b^2 \\
\underline{a\ -b} & \underline{a\ -b} & \underline{a\ +b} \\
a^2+ab & a^3+a^2b+ab^2 & a^3-a^2b+ab^2 \\
\underline{-ab-b^2} & \underline{-a^2b-ab^2-b^3} & \underline{+a^2b-ab^2+b^3} \\
a^2-b^2 & a^3-b^3 & a^3+b^3
\end{array}
$$

Before giving some hints for working with these special forms, we will display them for future reference.

BINOMIALS WHICH ARE SPECIAL FORMS

The difference of two squares. $a^2 - b^2 = (a - b)(a + b)$

The difference of two cubes. $a^3 - b^3 = (a - b)(a^2 + ab + b^2)$

The sum of two cubes. $a^3 + b^3 = (a + b)(a^2 - ab + b^2)$

Notice that the difference of the squares of two numbers $(a^2 - b^2)$ factors as their sum $(a + b)$ times their difference $(a - b)$. The key to determining the factors is to find what quantities are being squared. Then the next step is merely to form the sum and the difference of these quantities.

Example 3   Each of the following binomials is the difference of two squares. Factor these binomials.

a. $v^2 - w^2 = (v + w)(v - w)$

b. $x^2 - 9y^2 = (x)^2 - (3y)^2 = (x + 3y)(x - 3y)$

c. $49a^2 - 36b^2 = (7a)^2 - (6b)^2 = (7a + 6b)(7a - 6b)$

d. $121w^2 - 25 = (11w)^2 - (5)^2 = (11w + 5)(11w - 5)$

Example 4   Determine which of the following are the difference of two squares.

a. $x^2 - xy - y^2$

This is a trinomial, not a binomial. Therefore, it is not the difference of two squares.

b. $3x^2 - 4$

The first term is not a perfect square. This binomial is not the difference of two squares.

c. $169x^6 - 64$

This binomial equals $(13x^3)^2 - (8)^2$. Thus, it is the difference of two squares and factors as $(13x^3 + 8)(13x^3 - 8)$.

d. $25x^2 + 36y^2$

This binomial is the sum of two squares, not the difference of two squares.

e. $100x^2 - 48y^2$

The last term is not a perfect square. This binomial is not the difference of two squares.

Since $a^3 + b^3 = (a + b)(a^2 - ab + b^2)$, we say that the sum of two cubes $(a^3 + b^3)$ factors as a binomial times a trinomial. Note that the binomial factor is the sum of the numbers (a and b) which are cubed. The first term in the trinomial factor is the square of the first number; the middle term is the opposite of the product of the two numbers; while the last term is the square of the second number. The key to factoring this sum is to find the quantities which are being cubed. The importance of these values is illustrated by the grouping symbols in the examples which follow.

Similarly, since $a^3 - b^3 = (a - b)(a^2 + ab + b^2)$, the difference of the cubes of two numbers $(a^3 - b^3)$ factors as their difference $(a - b)$ times the trinomial $(a^2 + ab + b^2)$ whose first term is the square of the first number, whose middle term is the product of the two numbers, and whose last term is the square of the second number.

Example 5   Each of the following binomials is either the sum or the difference of two cubes. Factor these binomials.

a. $s^3 + t^3 = (s + t)(s^2 - st + t^2)$

b. $w^3 - z^3 = (w - z)(w^2 + wz + z^2)$

c. $27x^3 - 1 = (3x)^3 - (1)^3$
$$= [3x - 1][(3x)^2 + (3x)(1) + (1)^2]$$
$$= (3x - 1)(9x^2 + 3x + 1)$$

d. $z^3 + 64 = (z)^3 + (4)^3 = [z + 4][(z)^2 - (z)(4) + (4)^2]$
$$= (z + 4)(z^2 - 4z + 16)$$

e. $y^6 - 1000 = (y^2)^3 - (10)^3 = [y^2 - 10][(y^2)^2 + (y^2)(10) + (10)^2]$
$$= (y^2 - 10)(y^4 + 10y^2 + 100)$$

Example 6   Determine which of the following binomials are the sum or difference of two cubes.

a. $s^6 - t^3$

Since this binomial equals $(s^2)^3 - (t)^3$, it is the difference of two cubes and factors as $(s^2 - t)(s^4 + s^2t + t^2)$.

b. $x^3 + 25$

This is not the sum of two cubes since 25 is not a perfect cube.

   c. $a^3 + 27a^2 + 1$

This is a trinomial, not a binomial, and is, therefore, not the sum or difference of two cubes.

   d. $27x^3 + y^3z^9$

Since this binomial equals $(3x)^3 + (yz^3)^3$, it is the sum of two cubes and factors as $(3x + yz^3)(9x^2 - 3xyz^3 + y^2z^6)$.

   e. $s^7 - t^3$

This is not the difference of two cubes since s^7 is not a perfect cube.

16.6   Exercises

Each of the trinomials in Exercises 1–8 is the square of a binomial. Factor these trinomials.

  1. $x^2 - 2x + 1$ **2.** $y^2 + 2y + 1$

  3. $a^2 + 6a + 9$ **4.** $z^2 - 10z + 25$

  5. $w^2 - 14w + 49$ **6.** $x^2 - 4xy + 4y^2$

  7. $x^2 + 16xy + 64y^2$ **8.** $4w^2 + 12wz + 9z^2$

Each of the binomials in Exercises 9–16 is the difference of two squares. Factor these binomials.

  9. $w^2 - z^2$ **10.** $s^2 - t^2$

 11. $9a^2 - b^2$ **12.** $x^2 - 16y^2$

 13. $s^2 - 81$ **14.** $w^2 - 169$

 15. $144 - x^2$ **16.** $25a^2 - 36b^2$

Each of the binomials in Exercises 17–24 is the sum or difference of two cubes. Factor these binomials.

 17. $x^3 - y^3$ **18.** $c^3 + d^3$

 19. $w^3 + z^3$ **20.** $n^3 - m^3$

 21. $27a^3 + 1$ **22.** $8x^3 - 1$

 23. $x^3 - 8$ **24.** $y^3 - 1000$

Factor each of the polynomials in Exercises 25–32 after determining which special form is given.

 25. $x^2 - 9$ **26.** $s^2 + 4s + 4$

 27. $a^3 - 1$ **28.** $4x^2 - y^2$

 29. $a^2 - 6a + 9$ **30.** $b^3 + 8$

 31. $c^3 - d^3$ **32.** $9a^2 - 49b^2$

In Exercises 33–62, factor the polynomials by one of the following methods:

  a. greatest common factor

  b. trial and error

  c. as the difference of two squares

  d. as the sum of two cubes

  e. as the difference of two cubes

  f. as a perfect square trinomial

If a polynomial cannot be factored by one of these methods, indicate that it is prime over the integers.

 33. $9x^2 - 12xy + 4y^2$ **34.** $25s^4 - 20s^2t + 4t^2$

 35. $x^2 + 3x - 18$ **36.** $64s^3 - t^3$

 37. $x^3 + 125y^3$ **38.** $8n^3 + 27m^3$

 39. $4x^2 + 18x - 8$ **40.** $s^4 + 6s^2 + 9$

 41. $w^6 - 22w^3 + 121$ **42.** $49n^2 - 64m^2$

 43. $16s^2 - 121t^2$ **44.** $4a^2b^2 - 49$

 45. $9x^2y^6 - 1$ **46.** $64x^2 - 16x + 1$

 47. $49s^2 + 28s + 4$ **48.** $3x^2 + 4x - 15$

 49. $5x^2 + 5y^2$ **50.** $2x^2 + 16x + 32$

 51. $216x^3 + y^3$ **52.** $30x^2 - y^2$

 53. $y^4 - 4y^2 + 4$ **54.** $36w^2 - 24w - 4$

 55. $x^3 + 64$ **56.** $w^2 - 8$

 57. $x^2 + y^2$ **58.** $9x^2 - 25y^2$

 59. $k^3 + j^3$ **60.** $x^3 - y^2$

 61. $9x^2 - 30xy + 25y^2$ **62.** $9x^2 + 30xy + 25y^2$

 63. Verify that $(a^3 - b^3)$ factors as $(a - b)(a^2 + ab + b^2)$ by multiplying these factors.

 64. Verify that $144x^2 + 168xy + 49y^2$ factors as $(12x + 7y)^2$ by squaring this binomial.

 65. Verify that $(125s^3 - 64t^3)$ factors as $(5s - 4t) \times (25s^2 + 20st + 16t^2)$ by multiplying these factors.

16.7   Factoring by Grouping

The techniques of factoring discussed in the previous sections are very useful for factoring binomials and trinomials. Many polynomials with four or more terms can be factored if some of the terms are grouped together so that these groups of terms are factorable. This technique of factoring, called **factoring by**

grouping, is illustrated in the following examples. Example 1 shows how to factor by grouping pairs of terms together, 2-2, so that factoring these groups yields a common binomial factor.

Example 1   Factor the following polynomials by the method of grouping 2-2 to obtain a common binomial factor.

a. $xz - yz + xw - yw$

Although the four terms do not have a common factor, we observe that the first two terms have z as a common factor and the last two terms have w as a common factor. Thus, we group the first two terms together and the last two terms together as shown below.

$$xz - yz + xw - yw = (xz - yz) + (xw - yw)$$
$$= (x - y)z + (x - y)w$$
$$= (x - y)(z + w)$$

Note that $(x - y)$ is a factor of both groups. Thus,

$$xz - yz + xw - yw = (x - y)(z + w)$$

b. $3ac - bc + 6a - 2b = (3ac - bc) + (6a - 2b)$
$$= c(3a - b) + 2(3a - b)$$
$$= (3a - b)(c + 2)$$

Note that $(3a - b)$ is a factor of both groups.

c. $2xy + 3xw - 10y - 15w = (2xy + 3xw) - (10y + 15w)$
$$= x(2y + 3w) - 5(2y + 3w)$$
$$= (x - 5)(2y + 3w)$$

Example 2 illustrates grouping three terms which form a perfect square trinomial so that the difference of two squares is obtained. The terms may be grouped 3-1 or 1-3.

Example 2   Factor the following polynomials by the method of grouping 3-1 or 1-3 so that the difference of two squares is obtained.

a. Factor $x^2 - 14x + 49 - 16y^2$.

Group 3-1.

$$x^2 - 14x + 49 - 16y^2 = (x^2 - 14x + 49) - (16y^2)$$
$$= (x - 7)^2 - (4y)^2$$
$$= [(x - 7) + 4y][(x - 7) - 4y]$$
$$= (x + 4y - 7)(x - 4y - 7)$$

b. Factor $z^2 - a^2 + 4ab - 4b^2$.

Group 1-3.

$$z^2 - a^2 + 4ab - 4b^2 = z^2 - (a^2 - 4ab + 4b^2)$$
$$= z^2 - (a - 2b)^2$$
$$= [z - (a - 2b)][z + (a - 2b)]$$
$$= (z - a + 2b)(z + a - 2b)$$

It would be difficult to classify all of the types of grouping which may be used in factoring. Example 3 presents some of the more complex methods.

Example 3 Factor by grouping.

a. $x^2 - y^2 + x - y$

Group 2-2.

$$x^2 - y^2 + x - y = (x^2 - y^2) + (x - y)$$
$$= (x - y)(x + y) + (x - y)(1)$$
$$= (x - y)(x + y + 1)$$

b. $s^3 + t^3 + s + t$

Group 2-2.

$$s^3 + t^3 + s + t = (s^3 + t^3) + (s + t)$$
$$= (s + t)(s^2 - st + t^2) + (s + t)(1)$$
$$= (s + t)(s^2 - st + t^2 + 1)$$

c. $ay^2 + 2ay + y + a + 1$

Group 3-2.

$$ay^2 + 2ay + y + a + 1 = (ay^2 + 2ay + a) + (y + 1)$$
$$= a(y^2 + 2y + 1) + (y + 1)$$
$$= a(y + 1)(y + 1) + (y + 1)(1)$$
$$= (y + 1)[a(y + 1) + 1]$$
$$= (y + 1)[ay + a + 1]$$

Since the terms of a polynomial can be grouped in different ways, it is possible that some groupings will lead to a factorization while other groupings will prove useless. This is illustrated by Example 4.

Example 4 Factor $b^3 + x + x^3 + b$.

If we group the terms as $(b^3 + b) + (x^3 + x)$, then b is a common factor of the first group and x is a common factor of the second group:

$$(b^3 + b) + (x^3 + x) = b(b^2 + 1) + x(x^2 + 1)$$

However, the two groups which result do not have a common factor. Since grouping the polynomial in this way was not fruitful, let us reorder the terms so that the perfect cubes are grouped together:

$$b^3 + x + x^3 + b = (b^3 + x^3) + (x + b)$$
$$= (x^3 + b^3) + (x + b)$$
$$= (x + b)(x^2 - xb + b^2) + (x + b)(1)$$
$$= (x + b)(x^2 - xb + b^2 + 1).$$

Thus, $b^3 + x + x^3 + b = (x + b)(x^2 - xb + b^2 + 1)$.

Practice will improve your ability to group terms together for factorization. Do not be surprised if some of your preliminary groupings are not successful. As you work on the exercises in this section look for groups which are factorable by the techniques discussed in the previous sections. In particular, try to identify groups which share a common factor or groups which are special forms.

16.7 Exercises

Factor the polynomials in Exercises 1–30 using the technique of grouping. (Hints on grouping are given for some of the problems.)

1. $ac + bc + ad + bd$

2. $xy + xz + 2y + 2z$

3. $3a - 6b + 5ac - 10bc$

4. $x^2 - xy + 5x - 5y$

5. $6a^2 + 3ab + 2a + b$

6. $ac + bc + a + b$

7. $ab + bc - ad - cd$

8. $xy + xz - 2y - 2z$

9. $3a - 6b - 5ac + 10bc$

10. $x^2 - xy - 5x + 5y$

11. $ax - a + bx - b$

12. $ab + 5a + b + 5$

13. $sx - 7s + x - 7$

14. $4a^2 + 12a + 9 - 16b^2$ (3-1)

15. $kx - k + jx - j$

16. $3xt - st + 15x - 5s$

17. $9b^2 - 24b + 16 - a^2$ (3-1)

18. $x^3 - x^2 + 7x - 7$

19. $sx^2 + k + s + kx^2$

20. $16x^2 - a^2 - 2a - 1$ (1-3)

21. $x^2 - y^2 + 2y - 1$ (1-3)

22. $x^3 - y^3 + x - y$ (2-2)

23. $3x + 3y + x^3 + y^3$ (2-2)

24. $x^2 - 5xy - 6y^2 + x - 6y$ (3-2)

25. $s^3 + 11s^2 + s + 11$

26. $az^3 + bz^3 + aw^2 + bw^2$

27. $a^2 + 2a + 1 + ab + b$ (3-2)

28. $x^3 + y^3 + x^2 - y^2$ (2-2)

29. $ax - ay - az + bx - by - bz$ (3-3)

30. $kz - kw^2 + z - w^2$

16.8 Summary: A Strategy for Factoring a Polynomial

While any one factoring technique can be mastered with practice, the major problem encountered by many students is deciding which technique is applicable to a given problem. By isolating these techniques in separate sections, we have attempted to familiarize you with the major ones. In this section we combine these techniques to enable you to factor a polynomial without being told which technique to use.

Although it is possible to factor a polynomial in different ways and still obtain prime factors, the most efficient method is illustrated by the following example. Notice that when the instructions say **factor completely**, this means that each factor, other than a monomial factor, should be prime.

Example 1   Factor $8x^4y^2 - 40x^3y^2 - 192x^2y^2$ completely.

$$8x^4y^2 - 40x^3y^2 - 192x^2y^2 = (2x^2y - 16xy)(4x^2y + 12xy)$$
$$= 2xy(x - 8)(4xy)(x + 3)$$
$$= 8x^2y^2(x - 8)(x + 3)$$

Alternate Solution:

$$8x^4y^2 - 40x^3y^2 - 192x^2y^2 = 8x^2y^2(x^2 - 5x - 24)$$
$$= 8x^2y^2(x - 8)(x + 3)$$

In Example 1 the first solution is for most people the more difficult one since the trial and error factorization into two binomials has so many possibilities. The second solution is easier since once the GCF is removed, the remaining trinomial has fewer possible factors. For this reason we suggest that you always look for a GCF when you start to factor a polynomial. This suggestion is incorporated into the *procedure for factoring a polynomial over the integers.*

This procedure covers only the methods of factoring presented in this text. Some polynomials which cannot be factored by these techniques are factorable by other methods. However, in this text, the polynomials which are considered are either factorable by the methods considered here or are prime over the integers.

PROCEDURE FOR FACTORING A POLYNOMIAL OVER THE INTEGERS

Determine the number of terms in the polynomial, and factor using the procedure indicated in the appropriate column.

Binomial	*Trinomial*	*Polynomial Containing More Than 3 Terms*
1. Factor out the GCF.	1. Factor out the GCF.	1. Factor out the GCF.
2. Factor the binomial by looking for one of the following special products: a. difference of two squares b. sum of two cubes c. difference of two cubes	2. If the trinomial is a perfect square trinomial, factor as a special product.	2. Factor by grouping.
	3. If the trinomial is not a perfect square, factor by trial and error.	

Example 2   Factor $5x^5y - 5xy^5$ completely.

$$5x^5y - 5xy^5 = 5xy(x^4 - y^4) \qquad \text{recognized a common factor}$$
$$= 5xy[(x^2)^2 - (y^2)^2]$$
$$= 5xy(x^2 + y^2)(x^2 - y^2) \qquad \text{recognized the difference of two squares}$$
$$= 5xy(x^2 + y^2)(x + y)(x - y)$$

Thus, $5x^5y - 5xy^5 = 5xy(x^2 + y^2)(x + y)(x - y)$.

Note that each of these factors is prime. In order to obtain these prime factors, it was necessary to examine the factors obtained in each step to determine if further factoring was possible.

Example 3   Factor $7a^3 - 42a^2b + 63ab^2$.

$$7a^3 - 42a^2b + 63ab^2$$
$$= 7a(a^2 - 6ab + 9b^2) \qquad \text{recognized a common factor}$$
$$= 7a[a^2 - 2(3ab) + (3b)^2] \qquad \text{recognized a perfect square trinomial}$$
$$= 7a(a - 3b)^2$$

Example 4   Factor $3s^3 - 15s^2 - 18s$.

$$3s^3 - 15s^2 - 18s = 3s(s^2 - 5s - 6) \qquad \text{recognized a common factor}$$
$$= 3s(s + 1)(s - 6) \qquad \text{factored by trial and error}$$

Example 5   Factor $2ax^2 + 2ax + 2ay - 2ay^2$.

$$2ax^2 + 2ax + 2ay - 2ay^2$$
$$= 2a(x^2 + x + y - y^2) \qquad \text{recognized a common factor}$$
$$= 2a(x^2 - y^2 + x + y)$$
$$= 2a[(x^2 - y^2) + (x + y)] \qquad \text{regrouped terms}$$
$$= 2a[(x + y)(x - y) + (x + y)(1)] \qquad \text{factored the difference of two}$$
$$\text{squares}$$
$$= 2a(x + y)(x - y + 1) \qquad \text{factored out the common factor } (x + y)$$

Thus, $2ax^2 + 2ax + 2ay - 2ay^2 = 2a(x + y)(x - y + 1)$.

The technique for factoring second-degree trinomials can also be used to factor trinomials which have degree greater than two. This is illustrated in Examples 6 and 7.

Example 6   Factor $x^4 + 3x^2 + 2$.

$$x^4 + 3x^2 + 2 = (x^2 + 1)(x^2 + 2)$$

Example 7   Factor $3as^4 + 18as^2t^2 - 12a + 27at^4$.

$$3as^4 + 18as^2t^2 - 12a + 27at^4$$
$$= 3a(s^4 + 6s^2t^2 - 4 + 9t^4) \qquad \text{factored out the GCF}$$
$$= 3a(s^4 + 6s^2t^2 + 9t^4 - 4) \qquad \text{reordered terms}$$
$$= 3a[(s^4 + 6s^2t^2 + 9t^4) - 4] \qquad \text{grouped 3-1}$$
$$= 3a[(s^2 + 3t^2)^2 - 4]$$
$$= 3a[(s^2 + 3t^2)^2 - (2)^2]$$
$$= 3a(s^2 + 3t^2 + 2)(s^2 + 3t^2 - 2) \qquad \text{factored by grouping as the}$$
$$\text{difference of two squares}$$

Thus, $3as^4 + 18as^2t^2 - 12a + 27at^4 = 3a(s^2 + 3t^2 + 2)(s^2 + 3t^2 - 2)$.

Example 8   Factor $x^2 + 2ax + a^2 - y^2 + 2by - b^2$.

$$x^2 + 2ax + a^2 - y^2 + 2by - b^2$$
$$= (x^2 + 2ax + a^2) - (y^2 - 2by + b^2) \qquad \text{grouped 3-3}$$
$$= (x + a)^2 - (y - b)^2$$
$$= [(x + a) + (y - b)][(x + a) - (y - b)] \qquad \text{factored as the difference}$$
$$\text{of two squares}$$
$$= (x + y + a - b)(x - y + a + b)$$

Example 9 gives two polynomials which are factorable but which cannot be factored by the techniques presented in this text. The factors can be verified by multiplication.

Example 9   a. $x^4 + x^2y^2 + y^4 = (x^2 - xy + y^2)(x^2 + xy + y^2)$

b. $x^5 - y^5 = (x - y)(x^4 + x^3y + x^2y^2 + xy^3 + y^4)$

16.8   Exercises

Factor the polynomials in Exercises 1–65 completely over the integers. If the polynomial cannot be factored, indicate that it is prime.

1. $64s^2 - 9t^2$
2. $3ax - 3a$
3. $49a^2 - 28a + 4$
4. $bw^3 - b$
5. $12x^2 - 27x + 15$
6. $25a^2 - 10a + 1$
7. $4ay^2 - 4ay$
8. $49b^2 + 126bc + 81c^2$
9. $4x^{10} - 600x$
10. $25a^2 - 144b^2$
11. $4x^{10} + 12x^5y^3 + 9y^6$
12. $200x^2 + 2$
13. $cz^3 + 8c$
14. $25y^2 - 30yz + 9z^2$
15. $36s^4 - 49$
16. $10w^2 - 6w - 21$
17. $35x^2 + 37x + 6$
18. $71ax^4 - 71a$
19. $8h^3 - 125j^3$
20. $8x^6 - y^3$
21. $x^2 + 6x + 5$
22. $3ax^2 + 33ax + 72a$
23. $80x^2 + 245y^2$
24. $81x^2 + 18xy + y^2$
25. $4m^3 + 4m^2 + m$
26. $5a^2bc - 5b^3c$
27. $3ax^2 + 3ay^2$
28. $4bx^3 - 32$
29. $12x^3y - 12xy^3$
30. $9s^2 - 63$
31. $12ax^2 - 10axy - 12ay^2$
32. $x^3 + 4x^2y + 4xy^2$
33. $10x^2 + 40$
34. $3x^3 - 48x$
35. $7as^4 - 189as$
36. $63x^2 + 30x - 72$
37. $7s^5t - 7st^5$
38. $-6ax^3 + 24ax$
39. $27x^3y + 72x^2y^2 + 48xy^3$
40. $18a^3 + 63a^2 - 36a$
41. $20x^3y - 245xy^3$
42. $18x^3 - 21x^2y - 60xy^2$
43. $4a^7 + 32ab^3$
44. $11s^5 + 11st^2$
45. $18x^2 - 3xy - 10y^2$
46. $7a^2d - 28b^2c^2d$
47. $63a^3b - 175ab$
48. $100s^4 + 120s^3t + 36s^2t^2$
49. $8ax^2 - 648ay^4$
50. $x^6 + 4x^3y + 4y^2$
51. $12x^2 - 12xy + 3y^2$
52. $5x^2 - 55$
53. $6kx - 6k + 6jx - 6j$
54. $3s^2 + 3s + 3t - 3t^2$
55. $cx + cy + dx + dy$
56. $ax^2 + ax + bxy + by$
57. $x(a - b) + y(a - b)$
58. $a(x - y) - b(x - y)$
59. $3ax^2 - 3ay^2 + 6ay - 3a$
60. $7abx + 35ax + 7bx + 35x$
61. $20ax^4 + 220ax^2y + 605ay^2$
62. $x^3 - y^3 + x - y$
63. $x^2 + 2xy + y^2 - 16z^2$
64. $9x^2 - 6x + 1 - 25y^2$
65. $2x^2 + 2x + 2y - 2y^2$

Important Terms Used in this Chapter	common factor	prime over the integers
	factor completely	linear term
	factorable	perfect square trinomial
	factoring by grouping	quadratic term
	greatest common factor (GCF)	quadratic trinomial

REVIEW EXERCISES

Find the GCF for Exercises 1–4.

1. $90, 165, 300$
2. $360, 432, 936$
3. $18ax^4 - 54ax^3 + 66ax^2$
4. $98a^4b^3c^2 - 28a^3b^3c^2 - 154a^2b^2c^2 - 42ab^2c$

Multiply the binomials in Exercises 5–10 by inspection.

5. $(6x - 7y)(5x - 2y)$
6. $(4s + 9t)(10s - 3t)$
7. $(3x - 2y)(3x + 2y)$
8. $(3x - 2y)(3x - 2y)$
9. $(3x + 2y)(3x + 2y)$
10. $(11x + 12y)(11x - 12y)$

Factor the polynomials in Exercises 11–40 completely over the integers.

11. $12ax^2 - 24ax$
12. $49x^2 - 4$
13. $64t^3 - 1$
14. $12x^2 - 35x + 18$
15. $25x^2 - 90xy + 81y^2$
16. $16as^3 + 2at^3$
17. $16x^3 + y^3$
18. $7at^4 - 7a$
19. $x^4 + x^2 - 42$
20. $28ax^2 + 6ax - 72a$
21. $x^2 - x - 20$
22. $9a^2 - 16$
23. $4x^3 + 1$
24. $x^3y - 5x^2y + 4xy$
25. $12x^2 - 22x - 144$
26. $2x^5 - 32x$
27. $ax^2 - ax - 6a^2$
28. $x^4 - x^2y^2$
29. $11x^2 - 11y^2 + 33x + 33y$
30. $10x^3 - 10x^2y + 5ax^2 - 5axy$
31. $m^8 - n^8$
32. $36x^2 - 49y^2 + 24x + 4$
33. $a^2 + 2ab + b^2 + 4a + 4b + 3$ (group 3-2-1)
34. $16m^2 - 9 - n^2 + 6n$
35. $26a^2 + 2a^2b - 13bc - b^2c$
*36. $x^{2n} - 1$
*37. $6x^{2n} - 5x^n + 1$
*38. $16x^{6n} - 25$
*39. $x^{3n} + 8$
*40. $x^{3n} + 4x^{2n} - x^n$

*These exercises are optional.

17

OPERATIONS WITH RATIONAL EXPRESSIONS

OBJECTIVES

Upon completion of this chapter you should be able to:

1. Reduce a rational expression to lowest terms.	17.1
2. Produce equivalent rational expressions by multiplying the numerator and denominator by the same polynomial.	17.1
3. Multiply and divide rational expressions.	17.2
4. Add and subtract rational expressions which have the same denominator.	17.3
5. Find the least common multiple of a set of polynomials.	17.4
6. Add and subtract rational expressions which have different denominators.	17.5
7. Simplify complex fractions.	17.6
8. Simplify expressions involving addition, subtraction, multiplication, and division of rational expressions.	17.6
9. Solve equations involving rational expressions.	17.7

PRETEST

1. Reduce the following rational expressions to lowest terms (Objective 1).

   a. $\dfrac{4a - 10b}{2a - 5b}$ b. $\dfrac{a - b}{b - a}$

   c. $\dfrac{a^2 + 2ab + b^2}{a^2 - b^2}$ d. $\dfrac{x^3 - 1}{x - 1}$

2. Fill in the missing numerators and denominators so that the resulting rational expressions will be equivalent (Objective 2).

   a. $\dfrac{7}{36a^2b} = \dfrac{?}{72a^2b^3}$ b. $\dfrac{5}{24b^3} = \dfrac{15a^2}{?}$

   c. $\dfrac{7}{12 - x - x^2} = \dfrac{?}{(-1)(x + 4)(x - 3)(x - 5)}$

   d. $\dfrac{a - 2}{a^2 - 25} = \dfrac{?}{6(a - 5)(a + 5)(a + 2)}$

3. Perform the indicated operations, and simplify (Objective 3).

   a. $\dfrac{ax - bx}{x^2} \cdot \dfrac{a^2 + 2ab + b^2}{a^2 - b^2}$ b. $\dfrac{1 - x^2}{1 + x} \div \dfrac{x^2 - 3x + 2}{x - 2}$

4. Perform the indicated operations, and simplify (Objective 4).

   a. $\dfrac{7}{36a^2b} + \dfrac{5}{36a^2b}$

   b. $\dfrac{7x}{x^2 + x - 12} + \dfrac{6x - 3}{x^2 + x - 12}$

   c. $\dfrac{a}{3a - 9} - \dfrac{2a - 3}{3a - 9}$

5. Find the least common multiple of the following sets of polynomials (Objective 5).

   a. $36a^2b, \quad 24b^3$
   b. $x^2 + x - 12, \quad x^2 - 8x + 15$
   c. $3a - 9, \quad 5a - 15$
   d. $a^2 - 25, \quad 3a + 15, 2a - 10$

6. Perform the indicated operations, and simplify (Objective 6.)

   a. $\dfrac{7}{36a^2b} + \dfrac{5}{24b^3}$

b. $\dfrac{7}{x^2+x-12}+\dfrac{2}{x^2-8x+15}$

c. $\dfrac{a}{3a-9}-\dfrac{2a-3}{5a-15}$

7. Simplify each of the following complex fractions (Objective 7).

a. $\dfrac{1-\dfrac{1}{x+1}}{1+\dfrac{1}{x-1}}$

b. $\dfrac{\dfrac{x}{3}-2+\dfrac{3}{x}}{1-\dfrac{3}{x}}$

c. $\dfrac{\dfrac{3x^2y^6}{20a^2b^2}}{\dfrac{7x^3y^6}{30a^2b^4}}$

8. Perform the indicated operations, and simplify (Objective 8).

a. $\dfrac{a}{a^2-25}-\dfrac{2}{3a+15}+\dfrac{3}{2a-10}$

b. $\left(1-\dfrac{1}{x+1}\right)\left(1+\dfrac{1}{x-1}\right)$

c. $\left(\dfrac{a}{1-a}+\dfrac{1+a}{a}\right)\div\left(\dfrac{1-a}{a}+\dfrac{a}{1+a}\right)$

9. Solve the following equations (Objective 9).

a. $\dfrac{3}{x-2}=5-\dfrac{2}{x-2}$ b. $10+\dfrac{2}{x-3}=\dfrac{2}{x-3}$

c. $\dfrac{x}{x-2}=\dfrac{x}{x-10}$

A fraction which contains at least one variable in the denominator is frequently called an **algebraic fraction**. An algebraic fraction which is the ratio of two polynomials is called a **rational algebraic expression**. Consider the following four examples.

$\dfrac{4}{3}$ This fraction does not contain a variable so it is an arithmetic fraction, not an algebraic fraction. $\dfrac{4}{3}$ is a rational number since it is the ratio of two integers.

$\dfrac{4}{\sqrt{x}}$ This fraction contains the variable x in the denominator so it is an algebraic fraction. It is *not* a rational expression since $\sqrt{x}$ is not a polynomial.

$\dfrac{x^2-1}{x^3+3}$ This algebraic fraction is a rational expression since both the numerator and denominator are polynomials.

$\dfrac{x^2+3xy+2y^2}{x-y}$ This algebraic fraction is a rational expression.

In this chapter we will consider only rational expressions. The procedures and techniques which were used in Chapter 2 to simplify and combine arithmetic fractions will be used with rational expressions.

17.1 Equivalent Rational Expressions

One of the first tasks in working with fractions is to learn how to determine whether or not two fractions are equivalent. It is obvious from our work with arithmetic fractions that equivalent fractions cannot always be recognized immediately. For example, is $\dfrac{7}{9}$ equal to $\dfrac{287}{329}$ or $\dfrac{281}{329}$? We learned in Chapter 2 that equivalent arithmetic fractions can be obtained when both the numerator and the denominator are multiplied by the same nonzero number (building factor) or when the numerator and the denominator are divided by the same nonzero number (reducing factor). These properties are, of course, still true for algebraic fractions, but some additional techniques for recognizing equivalent fractions must also be considered.

In working with algebraic fractions, it is helpful to think of three signs associated with any given fraction: the sign of the numerator, the sign of the denominator, and the sign of the fraction. Two fractions are equivalent if they differ only in two of these three signs. Thus $\frac{-3}{4} = -\frac{3}{4}$ since the two fractions differ only in the sign of the numerator and the sign of the fraction. (Notice that the denominators have the same sign.)

It is important to realize that the sign of the numerator and denominator is associated with every term in the numerator or denominator. This means that two fractions having different signs for the numerators must have different signs for every term in the numerators. For example $\frac{a+b}{3} = \frac{-a-b}{-3}$ since the two fractions differ only in the signs of the numerators and the signs of the denominators. (The signs of the fractions are the same.) $\frac{a+b}{3} \neq \frac{a-b}{-3}$ (Why?)

Example 1   Write a fraction equivalent to the given fractions, but differing in the signs indicated.

a. $\frac{a-b}{3}$ Fraction and numerator

$$\frac{a-b}{3} = -\frac{-a+b}{3} \quad \text{or} \quad -\frac{b-a}{3}$$

b. $\frac{a-b}{3}$ Numerator and denominator

$$\frac{a-b}{3} = \frac{b-a}{-3}$$

c. $\frac{a-b}{3}$ Fraction and denominator

$$\frac{a-b}{3} = -\frac{a-b}{-3}$$

This relationship is also used to write final answers in the desired form. For example you may want as few minus signs as possible.

Example 2   Express the following rational expressions with as few minus signs as possible.

a. $\frac{-a-b}{-3} = \frac{a+b}{3}$

b. $-\frac{a-b}{3} = \frac{b-a}{3}$

c. $-\frac{a-b}{-3} = \frac{a-b}{3}$

d. $\frac{-a-b}{3} = \frac{a+b}{-3} = -\frac{a+b}{3}$

A rational expression is in **lowest terms** when the numerator and denominator have no common factor other than 1 or −1. When it is possible to recognize factors common to both the numerator and the denominator of a rational expression, the expression may be reduced by dividing both the numerator and denominator by the common factors. The technique we used to reduce fractions in Chapter 2 was to factor both the numerator and the denominator and examine these factorizations for common factors. The only new feature is that we will be factoring polynomials rather than integers. Thus, the techniques of factoring covered in the previous chapter are important prerequisites for this material.

PROCEDURE FOR REDUCING A RATIONAL EXPRESSION	EXAMPLE Reduce $\dfrac{x^2 - y^2}{5x - 5y}$.
	Solution:
1. Factor both the numerator and the denominator of the rational expression.	$\dfrac{x^2 - y^2}{5x - 5y} = \dfrac{(x + y)(x - y)}{5(x - y)}$
2. Divide the numerator and denominator by any common factors.	$\dfrac{(x + y)\cancel{(x - y)}^{1}}{5\cancel{(x - y)}_{1}} = \dfrac{x + y}{5}$
	Thus, $\dfrac{x^2 - y^2}{5x - 5y} = \dfrac{x + y}{5}$

You should recall from your work with subtraction that the opposite of a polynomial is found by taking the opposite of each term in the polynomial. Thus the opposite of $a - b$ is $-a + b$ or $b - a$. Also, the quotient of a number and its opposite is −1. Thus $\dfrac{-3}{3} = -1$, $\dfrac{4}{-4} = -1$, and $\dfrac{a - b}{b - a} = -1$. If you can recognize opposites when reducing fractions, it will simplify your work.

Example 3   Reduce the following rational expressions to lowest terms.

a. $\dfrac{5 - x}{x - 5} = -1$

Notice that $5 - x$ is the opposite of $x - 5$.

b. $\dfrac{2(a - b)}{b - a} = \dfrac{2\cancel{(a - b)}^{-1}}{\cancel{(b - a)}_{1}} = -2$

c. $\dfrac{w^2 - 16}{4 - w} = \dfrac{(w + 4)\cancel{(w - 4)}^{-1}}{\cancel{(4 - w)}_{1}} = -(w + 4) = -w - 4$

Example 4   Reduce each of the following rational expressions to lowest terms.

a. $\dfrac{6x^2y^3z}{8xyz^2}$

$$\frac{6x^2y^3z}{8xyz^2} = \frac{(2xyz)(3xy^2)}{(2xyz)(4z)}$$

$$= \frac{3xy^2}{4z}$$

b. $\dfrac{18x^3 + 30x^2y + 42xy^2}{6xy^2}$

$$\frac{18x^3 + 30x^2y + 42xy^2}{6xy^2} = \frac{6x(3x^2 + 5xy + 7y^2)}{6x(y^2)}$$

$$= \frac{3x^2 + 5xy + 7y^2}{y^2}$$

c. $\dfrac{77ab^3}{63a^2b^2c - 35ab^2c^2}$

$$\frac{77ab^3}{63a^2b^2c - 35ab^2c^2} = \frac{7ab^2(11b)}{7ab^2(9ac - 5c^2)}$$

$$= \frac{11b}{9ac - 5c^2}$$

d. $\dfrac{x^2 - y^2}{3x + 3y}$

$$\frac{x^2 - y^2}{3x + 3y} = \frac{(x - y)(x + y)}{3(x + y)}$$

$$= \frac{x - y}{3}$$

e. $\dfrac{x^2 - 9}{x^2 - 5x + 6}$

$$\frac{x^2 - 9}{x^2 - 5x + 6} = \frac{(x + 3)(x - 3)}{(x - 2)(x - 3)}$$

$$= \frac{x + 3}{x - 2}$$

f. $\dfrac{-3a + 6b - 9c}{4a - 8b + 12c}$

$$\frac{-3a + 6b - 9c}{4a - 8b + 12c} = \frac{-3\,(\cancel{a - 2b + 3c})^{1}}{4\,(\cancel{a - 2b + 3c})_{1}}$$

$$= \frac{-3}{4}$$

g. $\dfrac{w^2 - 16}{8 - 2w}$

$$\frac{w^2 - 16}{8 - 2w} = \frac{(w + 4)(\cancel{w - 4})^{-1}}{2\,(\cancel{4 - w})_{1}}$$

$$= \frac{-w - 4}{2} \quad \text{or} \quad \frac{w + 4}{-2} \quad \text{or} \quad -\frac{w + 4}{2}$$

Note: Recall that $w - 4$ and $4 - w$ are opposites.

h. $\dfrac{5x^3 + 5y^3}{10x^2 - 10xy + 10y^2}$

$$\frac{5x^3 + 5y^3}{10x^2 - 10xy + 10y^2} = \frac{5\,(x^3 + y^3)}{10\,(x^2 - xy + y^2)}$$

$$= \frac{\cancel{5}^{1}(x + y)(\cancel{x^2 - xy + y^2})^{1}}{\cancel{10}_{2}\,(\cancel{x^2 - xy + y^2})_{1}}$$

$$= \frac{x + y}{2}$$

i. $\dfrac{5x + 10y - ax - 2ay}{bx + 2by}$

$$\frac{5x + 10y - ax - 2ay}{bx + 2by} = \frac{5\,(x + 2y) - a(x + 2y)}{b\,(x + 2y)}$$

$$= \frac{(\cancel{x + 2y})^{1}(5 - a)}{b\,(\cancel{x + 2y})_{1}}$$

$$= \frac{5 - a}{b} \quad \text{or} \quad \frac{-1\,(a - 5)}{b}$$

$$= -\frac{a - 5}{b}$$

j. $\dfrac{7x^2 - 61x - 18}{x^2 - 5x - 36}$.

$$\frac{7x^2 - 61x - 18}{x^2 - 5x - 36} = \frac{(7x + 2)(\cancel{x - 9})^{1}}{(x + 4)(\cancel{x - 9})_{1}}$$

$$= \frac{7x + 2}{x + 4}$$

k. $\dfrac{x^2 - 5x - 6}{x^2 - 4}$

$$\frac{x^2 - 5x - 6}{x^2 - 4} = \frac{(x - 6)(x + 1)}{(x - 2)(x + 2)}$$

Note: Both the numerator and denominator are factored completely. Since they do not have any common factor other than 1 or −1,

$$\frac{x^2 - 5x - 6}{x^2 - 4}$$

is already in lowest terms.

As noted previously, equivalent rational expressions can also be produced by multiplying both the numerator and the denominator by the same value. This skill is necessary in addition problems.

Example 5   Fill in the missing numerators and denominators so that the resulting rational expressions will be equivalent to the original ones.

a. $\dfrac{3}{7} = \dfrac{?}{28}$

Since $\dfrac{3}{7} = \dfrac{3 \cdot 4}{7 \cdot 4}, \dfrac{3}{7} = \dfrac{12}{28}$. The building factor, 4, can be determined by dividing 28 by 7.

b. $\dfrac{9}{11} = \dfrac{?}{143}$

$143 \div 11 = 13$; thus, the building factor is 13.

$$\frac{9}{11} = \frac{9 \cdot 13}{11 \cdot 13} = \frac{117}{143}$$

c. $\dfrac{x^2y}{wz} = \dfrac{?}{w^3z^2}$

$w^3z^2 \div wz = w^2z$; thus, the building factor is w^2z.

$$\frac{x^2y}{wz} = \frac{x^2y \cdot w^2z}{wz \cdot w^2z} = \frac{w^2x^2yz}{w^3z^2}$$

d. $\dfrac{7}{3} = \dfrac{?}{6x - 3y}$

Since $(6x - 3y) \div 3 = 2x - y$, the building factor is $2x - y$.

$$\dfrac{7}{3} = \dfrac{7(2x - y)}{3(2x - y)} = \dfrac{14x - 7y}{6x - 3y}$$

e. $\dfrac{x + 3}{x - 2} = \dfrac{x^2 + 2x - 3}{?}$

Since $(x^2 + 2x - 3) \div (x + 3) = \dfrac{(x - 1)(x + 3)}{(x + 3)} = x - 1$, the building factor is $x - 1$.

$$\dfrac{x + 3}{x - 2} = \dfrac{(x + 3)(x - 1)}{(x - 2)(x - 1)} = \dfrac{x^2 + 2x - 3}{x^2 - 3x + 2}$$

f. $\dfrac{x + 1}{x - 1} = \dfrac{?}{x^3 - 1}$

Since $(x^3 - 1) \div (x - 1) = (x^2 + x + 1)$, the building factor is $x^2 + x + 1$.

$$\dfrac{x + 1}{x - 1} = \dfrac{(x + 1)(x^2 + x + 1)}{(x - 1)(x^2 + x + 1)} = \dfrac{x^3 + 2x^2 + 2x + 1}{x^3 - 1}$$

Before you try the exercises, we must give you one brief message of caution. Since a fraction can be interpreted as division of the numerator by the denominator, you must be careful not to divide by zero. If you replace the variables by constants, do not use any values that would make the denominator zero. For example, in the rational expression $\dfrac{x - 3}{x - 2}$, the expression is undefined when x is 2. We will assume that *unless otherwise stated, the replacement set for the variables is the set of real numbers for which the expression is defined.*

17.1 Exercises

Write a fraction equivalent to the given fraction in Exercises 1–5, but differing in the signs indicated.

1. $\dfrac{x - y}{3}$ Fraction and denominator

2. $\dfrac{x - y}{3}$ Numerator and denominator

3. $\dfrac{x - y}{3}$ Fraction and numerator

4. $\dfrac{3a + b}{a - b}$ Numerator and denominator

5. $\dfrac{3a + b}{a - b}$ Fraction and denominator

Write the expressions in Exercises 6–8 with as few minus signs as possible.

6. $\dfrac{-x - y}{3}$

7. $-\dfrac{a - b}{-6}$

8. $\dfrac{a - b}{-6}$

Write the expressions in Exercises 9 and 10 so that there are no minus signs in denominators and no minus signs for the fractions.

9. $-\dfrac{a - b}{-6}$

10. $\dfrac{a - b}{-6}$

Reduce each of the rational expressions in Exercises 11-60 to lowest terms.

11. $\dfrac{10x^7}{35x^4}$

12. $\dfrac{63s^4}{42s^3}$

13. $\dfrac{45a^2b^3c}{105a^3b^2c}$

14. $\dfrac{154x^7y^5z^8}{385x^4y^4z^2}$

15. $\dfrac{30x^2y^3 - 95xy^4}{5xy^2}$

16. $\dfrac{46x^4y^2z^2 - 69x^3y^2z^3}{23x^2yz^2}$

17. $\dfrac{121s^2t^5}{-132s^2t^4 - 99st^5}$

18. $\dfrac{119a^3b^6}{34a^4b^3 - 68a^3b^9}$

19. $\dfrac{16x - 16y}{8x + 8y}$

20. $\dfrac{14ax + 21a}{35ay + 42az}$

21. $\dfrac{15a^3x^2 + 6a^2x + 21ax}{33a^2x^2 + 36ax}$

22. $\dfrac{48a^3b^2x - 26abx^2}{-78a^2bx - 14ab^2x}$

23. $\dfrac{ax - ay}{bx - by}$

24. $\dfrac{st - 5t}{ks - 5k}$

25. $\dfrac{7t - 14}{6 - 3t}$

26. $\dfrac{28 - 42a}{15a - 10}$

27. $\dfrac{x^2 - y^2}{3x + 3y}$

28. $\dfrac{x^2 - y^2}{7x + 7y}$

29. $\dfrac{9x^2 - 16y^2}{27x + 36y}$

30. $\dfrac{25x^2 - 4}{14 - 35x}$

31. $\dfrac{60x - 24}{40x - 16}$

32. $\dfrac{28x - 52}{21x - 39}$

33. $\dfrac{33ay - 88ax}{64x^2 - 9y^2}$

34. $\dfrac{2a^2 - ab - b^2}{a^2 - b^2}$

35. $\dfrac{3x^2 - 2xy}{3x^2 - 5xy + 2y^2}$

36. $\dfrac{3s^2 - 13st - 10t^2}{s^3 - 125t^3}$

37. $\dfrac{3a^2 + 14ab - 5b^2}{3a^2 - 4ab + b^2}$

38. $\dfrac{14x^2 - 9xy + y^2}{y^2 - 7xy}$

39. $\dfrac{11a^5b - 11ab^5}{33a^2 + 33b^2}$

40. $\dfrac{10x^2 + 13xy - 3y^2}{5x^2 - 26xy + 5y^2}$

41. $\dfrac{90x^2 - 40}{3ax - 2a - 3bx + 2b}$

42. $\dfrac{4x^2 + 12xy + 9y^2}{2x + 3y}$

43. $\dfrac{x^3 + y^3}{13x^2 - 13xy + 13y^2}$

44. $\dfrac{x^2 - y^2}{x^3 - y^3}$

45. $\dfrac{8x^3 - 27}{18 - 12x}$

46. $\dfrac{9x^2 - 6xy + y^2}{9x^2 - y^2}$

47. $\dfrac{63s^2 + 30s - 72}{84s^2 + 70s - 56}$

48. $\dfrac{12x^2 + 24xy + 12y^2}{16x^2 - 16y^2}$

49. $\dfrac{15a^3 - 75a^2 - 90a}{18a^2 - 153a + 270}$

50. $\dfrac{y^2 - 169}{y^2 + 12y - 13}$

51. $\dfrac{x(a + b) - y(a + b)}{5x - 5y}$

52. $\dfrac{s(x + y) - t(x + y)}{s^2 - t^2}$

53. $\dfrac{7x - 14y + ax - 2ay}{3x - 6y}$

54. $\dfrac{x^2 + 9x + 14}{ax + 7a - 3x - 21}$

55. $\dfrac{a^2 - 4ab + 4b^2 - z^2}{5a - 10b + 5z}$

56. $\dfrac{b^2 - 2b + 1 - a^2}{5ab - 5a + 5a^2}$

57. $\dfrac{a^2 + 2a + 1 + ab + b}{9a + 9b + 9}$ (*Hint:* Group 3-2.)

58. $\dfrac{ax - y - z - 15}{15 + z + y - ax}$

59. $\dfrac{kx - k + jx - j}{ka - k + ja - j}$

60. $\dfrac{x^2 + x + y - y^2}{2x^3 + 2y^3}$ (*Hint:* Reorder terms and group 2-2.)

Fill in the missing numerators and denominators in Exercises 61-80 so that the resulting rational expressions will be equivalent.

61. $\dfrac{5}{9} = \dfrac{?}{72}$

62. $\dfrac{4}{13} = \dfrac{?}{52}$

63. $\dfrac{7}{144} = \dfrac{21}{?}$

64. $\dfrac{35}{36} = \dfrac{385}{?}$

65. $\dfrac{a}{x} = \dfrac{?}{x^2}$

66. $\dfrac{w}{y} = \dfrac{w^2}{?}$

67. $\dfrac{12xy}{9w} = \dfrac{?}{63wz}$

68. $\dfrac{22x}{15ab} = \dfrac{154x^2y}{?}$

69. $\dfrac{x - y}{x + y} = \dfrac{?}{3x + 3y}$

70. $\dfrac{2s + t}{s + 2t} = \dfrac{?}{12s + 24t}$

71. $\dfrac{a + b}{10} = \dfrac{a^2 - b^2}{?}$

72. $\dfrac{6}{x - y} = \dfrac{?}{x^2 - y^2}$

73. $\dfrac{3}{s + t} = \dfrac{?}{s^3 + t^3}$

74. $\dfrac{2x - y}{x + 3y} = \dfrac{?}{x^2 + 5xy + 6y^2}$

75. $\dfrac{2x - y}{x + 3y} = \dfrac{2x^2 + xy - y^2}{?}$

76. $\dfrac{2x + 3y}{4x - 5y} = \dfrac{?}{72x^2y - 90xy^2}$

77. $\dfrac{x + 5y}{7x - y} = \dfrac{?}{28x^3y - 4x^2y^2}$

78. $\dfrac{x + y}{x - y} = \dfrac{(x + y)^3}{?}$

79. $\dfrac{2x + y}{x + 3y} = \dfrac{(2x + y)^4}{?}$

80. $\dfrac{5}{x + y} = \dfrac{?}{(x + y)(x + 3y)(x - 2y)}$

17.2   Multiplication and Division of Rational Expressions

Since the variables in a rational expression represent real numbers, the rules and procedures for performing operations on rational expressions are the same as those for performing operations on arithmetic fractions. Recall that

$$\frac{3}{4} \cdot \frac{5}{7} = \frac{3 \cdot 5}{4 \cdot 7} = \frac{15}{28}$$

Similarly, $\dfrac{w}{x} \cdot \dfrac{y}{z} = \dfrac{wy}{xz}$.

In order to express the product in lowest terms, always examine the numerator and denominator for common factors before performing the multiplication. The product of the following fractions illustrates this.

$$\frac{6}{35} \cdot \frac{77}{39} = \frac{2 \cdot \cancel{3} \cdot \cancel{7} \cdot 11}{5 \cdot \cancel{7} \cdot \cancel{3} \cdot 13} = \frac{22}{65}$$

The *procedure for multiplying rational expressions* incorporates these same steps.

PROCEDURE FOR MULTIPLYING RATIONAL EXPRESSIONS	EXAMPLE
	$\dfrac{x^2 - y^2}{3a + 3b} \cdot \dfrac{18c}{x^2 - 2xy + y^2} = ?$ *Solution:*
1. Indicate the product of the numerators and of the denominators.	$\dfrac{x^2 - y^2}{3a + 3b} \cdot \dfrac{18c}{x^2 - 2xy + y^2}$ $= \dfrac{(x^2 - y^2)(18c)}{(3a + 3b)(x^2 - 2xy + y^2)}$
2. Reduce this fraction by factoring the numerator and denominator and then dividing the numerator and denominator by the common factors.	$= \dfrac{\overset{1}{\cancel{(x - y)}}(x + y)(2)\overset{1}{(\cancel{3})}(3c)}{\underset{1}{\cancel{3}}(a + b)\underset{1}{\cancel{(x - y)}}(x - y)}$
3. Multiply or indicate the product of any remaining factors in the numerator or denominator.	$= \dfrac{6c(x + y)}{(a + b)(x - y)}$
	Thus, $\dfrac{x^2 - y^2}{3a + 3b} \cdot \dfrac{18c}{x^2 - 2xy + y^2}$ $= \dfrac{6c(x + y)}{(a + b)(x - y)}$

Example 1   a. $\dfrac{3x - 6y}{4a - 12b} \cdot \dfrac{24a - 72b}{6x - 12y} = \dfrac{(3x - 6y)(24a - 72b)}{(4a - 12b)(6x - 12y)}$

$$= \frac{3\overset{1}{\cancel{(x - 2y)}}\overset{{}^1 4}{\cancel{(24)}}\overset{1}{\cancel{(a - 3b)}}}{\underset{1}{\cancel{4}}\underset{1}{\cancel{(a - 3b)}}\underset{1}{\cancel{(6)}}\underset{1}{\cancel{(x - 2y)}}}$$

$$= \frac{3}{1}$$

$$= 3$$

b. $\dfrac{12x^3y - 6x^2y^2}{4x^2 - 4y^2} \cdot \dfrac{x^2 + 2xy + y^2}{2x^2 + xy - y^2} = \dfrac{\overset{3}{\cancel{6}}x^2y(2x\cancel{-y})(x\cancel{+y})(x\cancel{+y})}{\underset{2}{\cancel{4}}(x-y)(x\cancel{+y})(2x\cancel{-y})(x\cancel{+y})}$

$$= \dfrac{3x^2y}{2(x-y)}$$

Note that steps 1 and 2 of the rule are combined.

c. $\dfrac{8ax^3 - 8ay^3}{5ax^2 - 5ay^2} \cdot \dfrac{10bx + 10by}{16x^2 + 16xy + 16y^2} = \dfrac{8a(x^3 - y^3)(10b)(x + y)}{5a(x^2 - y^2)(16)(x^2 + xy + y^2)}$

$$= \dfrac{\cancel{8}\cancel{a}(x\cancel{-y})(x^2\cancel{+xy+y^2})(\cancel{10}b)(x\cancel{+y})}{\cancel{8}\cancel{a}(x\cancel{-y})(x\cancel{+y})(\cancel{16})(x^2\cancel{+xy+y^2})}$$

$$= \dfrac{b}{1} = b.$$

d. $\dfrac{3a}{2x^2 - 2y^2} \cdot \dfrac{x^2 - 2xy + y^2}{6b} \cdot \dfrac{5ab - b}{x + 2y}$

$$= \dfrac{\overset{1}{\cancel{3}}a}{2(x\cancel{-y})(x + y)} \cdot \dfrac{(x\cancel{-y})(x - y)}{\underset{2}{\cancel{6}}\underset{1}{\cancel{b}}} \cdot \dfrac{\cancel{b}(5a - 1)}{x + 2y} = \dfrac{a(x - y)(5a - 1)}{4(x + y)(x + 2y)}$$

In Example 1d, the numerators and denominators are factored and common factors eliminated before the product is expressed as a single fraction. This is an acceptable alternate method of solution.

When the polynomial expressions are monomials, there usually isn't any advantage to expressing them in factored form. The factors will be obvious and the product can be simplified by dividing by the common factors. This is illustrated in Example 2.

Example 2 $\dfrac{4a^4}{5b} \cdot \dfrac{25b^3}{8a^5} = \dfrac{(\overset{1}{\cancel{4a^4}})(\overset{5b^2}{\cancel{25b^3}})}{(\underset{1}{\cancel{5b}})(\underset{2a}{\cancel{8a^5}})}$

$$= \dfrac{5b^2}{2a}$$

The technique for dividing rational expressions is the same as that for dividing arithmetic fractions.

PROCEDURE FOR DIVIDING RATIONAL EXPRESSIONS	*EXAMPLE* $\dfrac{x^2-y^2}{3x} \div \dfrac{x+y}{6x} = ?$

EXAMPLE $\dfrac{x^2-y^2}{3x} \div \dfrac{x+y}{6x} = ?$

Solution:

1. Indicate the product of the dividend and the reciprocal of the divisor.

$$\frac{x^2-y^2}{3x} \div \frac{x+y}{6x} = \frac{x^2-y^2}{3x} \cdot \frac{6x}{x+y}$$

2. Perform the multiplication using the rule for multiplying rational expressions.

$$= \frac{(x-y)\overset{1}{\cancel{(x+y)}}\overset{2}{\cancel{(6x)}}}{\underset{1}{\cancel{3x}}\underset{1}{\cancel{(x+y)}}}$$

$$= 2(x-y)$$

Thus, $\dfrac{x^2-y^2}{3x} \div \dfrac{x+y}{6x}$

$$= 2(x-y).$$

Example 3   **a.** $\dfrac{10ab}{x-3y} \div \dfrac{-5a^3}{7x-21y} = \dfrac{10ab}{x-3y} \cdot \dfrac{7x-21y}{-5a^3}$

$$= \frac{\overset{2b}{\cancel{(10ab)}}(7)\overset{1}{\cancel{(x-3y)}}}{\underset{1}{\cancel{(x-3y)}}\underset{-a^2}{\cancel{(-5a^3)}}}$$

$$= \frac{14b}{-a^2} \quad \text{or} \quad -\frac{14b}{a^2}$$

b. $\dfrac{2a-2b}{6a} \div \dfrac{a^2-2ab+b^2}{a^2-b^2} = \dfrac{2a-2b}{6a} \cdot \dfrac{a^2-b^2}{a^2-2ab+b^2}$

$$= \frac{\overset{1}{\cancel{2}}\overset{1}{\cancel{(a-b)}}\overset{1}{\cancel{(a-b)}}(a+b)}{\underset{3}{\cancel{6}}a\underset{1}{\cancel{(a-b)}}\underset{1}{\cancel{(a-b)}}}$$

$$= \frac{a+b}{3a}$$

c. $\dfrac{14x^2-21x}{42x-63} \div \dfrac{24x-16}{12x-8} = \dfrac{14x^2-21x}{42x-63} \cdot \dfrac{12x-8}{24x-16}$

$$= \frac{\overset{1}{\cancel{7}}x\overset{1}{\cancel{(2x-3)}}\overset{1}{\cancel{(4)}}\overset{1}{\cancel{(3x-2)}}}{\underset{3}{\cancel{21}}\underset{1}{\cancel{(2x-3)}}\underset{2}{\cancel{(8)}}\underset{1}{\cancel{(3x-2)}}}$$

$$= \frac{x}{6}$$

d. $\dfrac{x^2 - 16}{10x + 10y} \cdot \dfrac{30a^2b}{x^2 - 2x - 8} \div \dfrac{3x + 12}{5x + 10}$

$= \dfrac{(x^2 - 16)(30a^2b)}{(10x + 10y)(x^2 - 2x - 8)} \cdot \dfrac{5x + 10}{3x + 12}$

$= \dfrac{\overset{1}{\cancel{(x-4)}}\overset{1}{\cancel{(x+4)}}\overset{3}{\cancel{(30}}a^2b)(5)\overset{1}{\cancel{(x+2)}}}{\underset{1}{\cancel{10}}(x+y)\underset{1}{\cancel{(x-4)}}\underset{1}{\cancel{(x+2)}}\underset{1}{\cancel{(3)}}\underset{1}{\cancel{(x+4)}}}$

$= \dfrac{5a^2b}{x + y}$

e. $\dfrac{x^2 - 25}{x^2 - x - 12} \div \dfrac{x^2 - x - 20}{3x - 3} \div \dfrac{x^2 + 4x - 5}{x^2 - 16}$

$= \dfrac{x^2 - 25}{x^2 - x - 12} \cdot \dfrac{3x - 3}{x^2 - x - 20} \cdot \dfrac{x^2 - 16}{x^2 + 4x - 5}$

$= \dfrac{\overset{1}{\cancel{(x-5)}}\overset{1}{\cancel{(x+5)}}(3)\overset{1}{\cancel{(x-1)}}\overset{1}{\cancel{(x+4)}}\overset{1}{\cancel{(x-4)}}}{\underset{1}{\cancel{(x-4)}}(x+3)\underset{1}{\cancel{(x-5)}}\underset{1}{\cancel{(x+4)}}\underset{1}{\cancel{(x-1)}}\underset{1}{\cancel{(x+5)}}}$

$= \dfrac{3}{x + 3}$

f. $\dfrac{12x^3y^2}{x + y} \div \left[\dfrac{5x - 5}{x^2 - y^2} \cdot \dfrac{3x^3y^3}{xy - y} \right] = \dfrac{12x^3y^2}{x + y} \div \dfrac{(5x - 5)(3x^3y^3)}{(x^2 - y^2)(xy - y)}$

$= \dfrac{12x^3y^2}{x + y} \cdot \dfrac{(x^2 - y^2)(xy - y)}{(5x - 5)(3x^3y^3)}$

$= \dfrac{\overset{4}{\cancel{(12}}x^3y^2)(x - y)\overset{1}{\cancel{(x+y)}}\overset{1}{\cancel{(y)}}\overset{1}{\cancel{(x-1)}}}{\underset{1}{\cancel{(x+y)}}(5)\underset{1}{\cancel{(x-1)}}\underset{y_1}{\cancel{(3x^3y^3)}}}$

$= \dfrac{4(x - y)}{5}$

17.2 Exercises

Find the products and quotients in Exercises 1–50. Express all answers in reduced form.

1. $\dfrac{96}{121} \cdot \dfrac{77}{30}$

2. $\dfrac{45}{26} \cdot \dfrac{91}{63}$

3. $\dfrac{-94}{40} \div \dfrac{141}{46}$

4. $\dfrac{36}{143} \div \dfrac{-45}{22}$

5. $\dfrac{-4}{7} \cdot \dfrac{27}{55} \div \dfrac{-81}{88}$

6. $\dfrac{6}{-92} \div \dfrac{-30}{7} \div \dfrac{8}{-105}$

7. $\dfrac{-66}{169} \cdot \left[\dfrac{48}{39} \div \dfrac{-77}{26} \right]$

8. $\dfrac{-770}{-57} \div \left[\dfrac{-35}{38} \cdot \dfrac{-22}{9} \right]$

9. $\dfrac{36s^{12}t^4}{-60s^{10}t^7} \cdot \dfrac{15t^6}{81s}$

10. $\dfrac{57w^3x^2y}{38wx^2y^3} \cdot \dfrac{8y^4}{15w}$

11. $\dfrac{96a^3b^2}{57bc^3} \cdot \dfrac{76c^4d^2}{52a^2b}$

12. $\dfrac{20x^2yz^2}{-27wz} \div \dfrac{60x^3z}{9wy^2}$

13. $\dfrac{-69x^5y}{-35x^2y^3} \div \dfrac{46x^4y^4}{-84x^6y}$

14. $\left(\dfrac{3xy}{2z} \right)^2$

15. $\dfrac{(-2xy)^3}{3x^3z} \cdot \dfrac{-42x}{112x^2z^2}$

16. $\dfrac{(-3a^2b^3)^4}{a^{12}b^9} \div \dfrac{(-3ab^2)^3}{(2a^2b)^2}$

17. $\left(\dfrac{-4xy^2z^3}{3a^2bc}\right)^2 \div \left(\dfrac{2x^2yz^2}{-abc^2}\right)^3$

18. $\dfrac{2x - 6y}{4xy} \cdot \dfrac{20x^2}{5x - 10y}$

19. $\dfrac{15a - 3b}{9xy^2} \cdot \dfrac{121x^2y}{55a - 11b}$

20. $\dfrac{14x - 49y}{a^2 - b^2} \cdot \dfrac{a - b}{35y - 10x}$

21. $\dfrac{3x^2 - 12}{14x - 28} \cdot \dfrac{7x}{11x + 22}$

22. $\dfrac{x^2 - 5x + 6}{10x - 20} \cdot \dfrac{5x - 15}{x^2 - 9}$

23. $\dfrac{10y - 14x}{x - 1} \cdot \dfrac{x^2 - 2x + 1}{21x - 15y}$

24. $\dfrac{a^2 - 9b^2}{4a^2 - b^2} \cdot \dfrac{4a^2 - 4ab + b^2}{2a^2 - 7ab + 3b^2}$

25. $\dfrac{39w - 65z}{26w^2z} \cdot \dfrac{18w^2z + 30wz^2}{9w^2 - 25z^2}$

26. $\dfrac{4x^2 - 1}{18xy} \div \dfrac{6x - 3}{16x^2 + 8x}$

27. $\dfrac{x^2 - y^2}{x^2 - 2xy + y^2} \div \dfrac{3x + 3y}{7x - 21}$

28. $\dfrac{5x^2y - 15xy}{x^2 - 4} \div \dfrac{x^2 - 9}{10x^2 - 20x}$

29. $\dfrac{4x^2 + 12x + 9}{5x^3 - 2x^2} \div \dfrac{14x^2 + 21x}{10x^2y^2}$

30. $\dfrac{x^3 - y^3}{6x^2 - 6y^2} \div \dfrac{2x^2 + 2xy + 2y^2}{9xy}$

31. $\dfrac{9x^2 - 9xy + 9y^2}{5x^2y + 5xy^2} \div \dfrac{81xy}{3x^3 + 3y^3}$

32. $\dfrac{x^2 - 3x + 2}{2x^2 - 3x + 1} \div \dfrac{x^2 - 2x + 1}{3x^2 - 12}$

33. $\dfrac{15x^2 - 15y^2}{11x^3 + 11y^3} \div \dfrac{5x^3y^2 - 5x^2y^3}{4x^3 - 4x^2y + 4xy^2}$

34. $\dfrac{18x^2y - 42xy^2}{3x^2 - 6xy + 3y^2} \div \dfrac{56x^2y - 24x^3}{5x^2 - 5y^2}$

35. $\dfrac{8x^3 + 27}{x^3 - 64} \div \dfrac{10x + 15}{3x^2 + 12x + 48}$

36. $\dfrac{121ab}{33a^2 + 44a} \div \dfrac{3a^2 + a - 4}{36ab^2 + 96ab + 64a}$

37. $\dfrac{22a - 33b + 11c}{x^2y^2 - 3xy} \cdot \dfrac{6 - 2xy}{-4a + 6b - 2c}$

38. $\dfrac{42x^2 - 42y^2}{-21x + 28y - 35z} \cdot \dfrac{6x^2 - 8xy + 10xz}{4x + 4y}$

39. $\dfrac{x^4 - y^4}{-2x - 3y} \cdot \dfrac{2x + 3y}{7x^2 + 7y^2} \cdot \dfrac{12xy}{8x^2 + 8xy}$

40. $\dfrac{x^3 - y^3}{18x^2y^3} \cdot \dfrac{9x^2y + 9xy^2}{x^2 - y^2} \div \dfrac{x^2 + xy + y^2}{36y^4}$

41. $\dfrac{3x - 6y}{7x^2y} \div \dfrac{17x^3 - 17y^3}{34xy^2} \div \dfrac{3x}{2x^2 + 2xy + 2y^2}$

42. $\dfrac{5a - b}{a^2 - 5ab + 4b^2} \div \left[\dfrac{6ab}{3a - 12b} \cdot \dfrac{b^2 - 5ab}{4a - 4b}\right]$

43. $\dfrac{x^2 - y^2 + 6y - 9}{76x^2 - 19} \cdot \dfrac{19xy}{7x^2 + 7xy - 21x}$ *(Hint:* Group 1-3.)

44. $\dfrac{5(a + b) - x(a + b)}{2a^2 - 2b^2} \cdot \dfrac{6ax}{15 - 3x}$

45. $\dfrac{7(c - y) - a(c - y)}{2c^2 - 7cy + 5y^2} \cdot \dfrac{38c - 95y}{17c - 17y}$

46. $\dfrac{x^2 + x - y^2 - y}{3x^2 - 3y^2} \div \dfrac{5x + 5y + 5}{7x^2y + 7xy^2}$

(Hint: Reorder for grouping.)

47. $\dfrac{x(a - b + 2c) - 2y(a - b + 2c)}{x^2 - 4xy + 4y^2} \div \dfrac{2a^2 - 2ab + 4ac}{7xy^2 - 14y^3}$

48. $\dfrac{6ax - 2ay + 24az - 3bx + by - 12bz}{4a^2 - b^2} \div \dfrac{24x - 8y + 96z}{18a}$

(Hint: Group 3-3.)

49. $\dfrac{a(2x - y) - b(2x - y)}{a^2 - b^2} \div \dfrac{6x - 3y}{4a + 4b}$

50. $\dfrac{20a^2 + 22a - 12}{36a^2 + 19ab - 6b^2} \div \dfrac{10a^2 - 19a + 6}{18a^2 - 4ab - 27a + 6b}$

17.3 Addition and Subtraction of Rational Expressions with the Same Denominator

It is very easy to add rational expressions which have the same denominator. The common denominator is retained as the denominator of the result and the numerators are added to obtain the numerator of the result. Special care should be given to reducing the result to lowest terms. In fact, the major portion of your work in many addition problems will be the simplification of the sum to lowest terms.

Subtraction is performed similarly to addition except that numerators are subtracted rather than added.

Example   a. $\dfrac{3}{17} + \dfrac{8}{17} = \dfrac{3 + 8}{17}$

$= \dfrac{11}{17}$

b. $\dfrac{5}{24} + \dfrac{7}{24} = \dfrac{5+7}{24}$

$\hspace{2.6cm} = \dfrac{12}{24}$

$\hspace{2.6cm} = \dfrac{1}{2}$

c. $\dfrac{11}{18} - \dfrac{5}{18} = \dfrac{11-5}{18}$

$\hspace{2.6cm} = \dfrac{6}{18}$

$\hspace{2.6cm} = \dfrac{1}{3}$

d. $\dfrac{5}{y} + \dfrac{8}{y} = \dfrac{5+8}{y}$

$\hspace{2.4cm} = \dfrac{13}{y}$

e. $\dfrac{7}{z} - \dfrac{9}{z} = \dfrac{7-9}{z}$

$\hspace{2.4cm} = \dfrac{-2}{z}$ or $-\dfrac{2}{z}$

f. $\dfrac{2x-1}{x^2} + \dfrac{4x+1}{x^2} = \dfrac{(2x-1)+(4x+1)}{x^2}$

$\hspace{3.6cm} = \dfrac{6x}{x^2}$

$\hspace{3.6cm} = \dfrac{6}{x}$

Note: Similar terms in the numerator of the sum are combined and the result is then simplified.

g. $\dfrac{5x+7}{2x+2} - \dfrac{3x-1}{2x+2} = \dfrac{(5x+7)-(3x-1)*}{2x+2}$

$\hspace{3.5cm} = \dfrac{5x+7-3x+1}{2x+2}$

$\hspace{3.5cm} = \dfrac{2x+8}{2x+2}$

$\hspace{3.5cm} = \dfrac{\cancel{2}(x+4)}{\cancel{2}(x+1)}$

$\hspace{3.5cm} = \dfrac{x+4}{x+1}$

*Note: A common error is to write $5x + 7 - 3x - 1$ as the numerator. The correct numerator is $5x + 7 - 3x + 1$. To avoid this error we suggest enclosing the numerators in parentheses (as indicated above) before combining like terms.

h. $\dfrac{x^2}{x^2-1} - \dfrac{2x-1}{x^2-1} = \dfrac{x^2-(2x-1)}{x^2-1}$

$\qquad = \dfrac{x^2-2x+1}{x^2-1}$

$\qquad = \dfrac{(x-1)(x-1)}{(x-1)(x+1)} = \dfrac{x-1}{x+1}$

i. $\dfrac{3x^2+4x+3}{x^3-1} + \dfrac{4x^2+x+1}{x^3-1} - \dfrac{2x^2-1}{x^3-1}$

$\qquad = \dfrac{(3x^2+4x+3)+(4x^2+x+1)-(2x^2-1)}{x^3-1}$

$\qquad = \dfrac{3x^2+4x+3+4x^2+x+1-2x^2+1}{x^3-1}$

$\qquad = \dfrac{5x^2+5x+5}{x^3-1}$

$\qquad = \dfrac{5(x^2+x+1)}{(x-1)(x^2+x+1)} = \dfrac{5}{x-1}$

j. $\dfrac{7x+4}{x^2-3x-5} - \dfrac{x^2+2x-1}{x^2-3x-5} - \dfrac{2x}{x^2-3x-5}$

$\qquad = \dfrac{(7x+4)-(x^2+2x-1)-(2x)}{x^2-3x-5}$

$\qquad = \dfrac{7x+4-x^2-2x+1-2x}{x^2-3x-5}$

$\qquad = \dfrac{-x^2+3x+5}{x^2-3x-5}$

$\qquad = \dfrac{-(x^2-3x-5)}{x^2-3x-5} = -1$

17.3 Exercises

Perform the indicated operations in Exercises 1–40 and reduce the results to lowest terms.

1. $\dfrac{7}{30} + \dfrac{11}{30}$

2. $\dfrac{23}{135} + \dfrac{22}{135}$

3. $\dfrac{13}{56} - \dfrac{5}{56}$

4. $\dfrac{53}{209} - \dfrac{34}{209}$

5. $\dfrac{18}{w} + \dfrac{5}{w}$

6. $\dfrac{20}{a} - \dfrac{17}{a}$

7. $\dfrac{8}{x^2} - \dfrac{11}{x^2}$

8. $\dfrac{19}{x+1} + \dfrac{13}{x+1}$

9. $\dfrac{5b+13}{3b^2} + \dfrac{b-4}{3b^2}$

10. $\dfrac{3x^2+1}{8x^3} - \dfrac{1-3x^2}{8x^3}$

11. $\dfrac{3s+7}{s^2-9} + \dfrac{s+5}{s^2-9}$

12. $\dfrac{7}{x-7} - \dfrac{x}{x-7}$

13. $\dfrac{5}{a-5} - \dfrac{8}{a-5}$

14. $\dfrac{28}{t^2-16} - \dfrac{7t}{t^2-16}$

15. $\dfrac{a-1}{a+7} + \dfrac{8}{a+7}$

16. $\dfrac{2x-7y}{3x-5y} + \dfrac{x+2y}{3x-5y}$

17. $\dfrac{5s-22}{s^2-5s+6} + \dfrac{4s-5}{s^2-5s+6}$

18. $\dfrac{9y+8}{y^2-5y-6} + \dfrac{8y+9}{y^2-5y-6}$

19. $\dfrac{5a+21}{a^2+5a+6} - \dfrac{5-3a}{a^2+5a+6}$

20. $\dfrac{13b+69}{b^2+5b-6} - \dfrac{2b+3}{b^2+5b-6}$

21. $\dfrac{4}{x} + \dfrac{5}{x} + \dfrac{7}{x}$

22. $\dfrac{6}{y} + \dfrac{7}{y} + \dfrac{2}{y}$

23. $\dfrac{14}{t^2} - \dfrac{8}{t^2} - \dfrac{1}{t^2}$

24. $\dfrac{70}{c^2} - \dfrac{60}{c^2} - \dfrac{4}{c^2}$

25. $\dfrac{7}{4x} + \dfrac{5}{4x} + \dfrac{3}{4x} + \dfrac{1}{4x}$

26. $\dfrac{19}{10w} + \dfrac{23}{10w} + \dfrac{5}{10w} + \dfrac{13}{10w}$

27. $\dfrac{8}{3s} + \dfrac{14}{3s} - \dfrac{16}{3s}$

28. $\dfrac{7t^2}{5t} - \dfrac{3t^2}{5t} + \dfrac{t^2}{5t}$

29. $\dfrac{7x}{2x+3} + \dfrac{2x+10}{2x+3} - \dfrac{x-2}{2x+3}$

30. $\dfrac{x+4}{x^2-1} + \dfrac{x-2}{x^2-1}$

31. $\dfrac{a^2+2a-1}{3a^2-3a} - \dfrac{3a^2-1}{3a^2-3a}$

32. $\dfrac{5x-7y}{12x^2y-6xy} - \dfrac{3x-6y}{12x^2y-6xy}$

33. $\dfrac{z^2}{z^3+z} - \dfrac{z}{z^3+z}$

34. $\dfrac{3z^2}{z^3+z} + \dfrac{3}{z^3+z}$

35. $\dfrac{3x+2}{x^2-8x+7} + \dfrac{12-x}{x^2-8x+7}$

36. $\dfrac{x^2-2}{x^2-x-12} - \dfrac{3x+2}{x^2-x-12}$

37. $\dfrac{2x^2+1}{2x^2-5x-12} - \dfrac{4-x}{2x^2-5x-12}$

38. $\dfrac{x+a}{x(a+b)+y(a+b)} + \dfrac{b-x}{x(a+b)+y(a+b)}$

39. $\dfrac{y^2+b}{by+5b-3y-15} - \dfrac{y^2+3}{by+5b-3y-15}$

40. $\dfrac{2x-y+4z}{x^2-2xy+y^2-z^2} + \dfrac{x-2y-z}{x^2-2xy+y^2-z^2}$

17.4 Least Common Denominator

You can add and subtract rational expressions only if they have the same denominator. If two rational expressions do not have the same denominator, then you must find a common denominator before the terms can be added. Although any common denominator can be used, you can simplify your work considerably by using the least common denominator (LCD).

The LCD of rational expressions is calculated by the same procedure used in Chapter 2 to calculate the LCD of fractions. Recall that the LCD is the least common multiple (LCM) of the denominators. The following example reviews the steps taken to obtain the LCM of a set of whole numbers.

Example 1   Find the least common multiple (LCM) of 45, 24, and 20.

$$45 = 3 \cdot 3 \cdot 5$$
$$24 = 2 \cdot 2 \cdot 2 \cdot 3$$
$$20 = 2 \cdot 2 \cdot 5$$

The LCM is formed by using each of the prime factors the greatest number of times it occurs in any single factorization:

Factor 2 is used at most 3 times in any single factorization.

Factor 3 is used at most 2 times in any single factorization.

Factor 5 is used at most 1 time in any single factorization.

Thus, the LCM $2 \cdot 2 \cdot 2 \cdot 3 \cdot 3 \cdot 5 = 360$.

Since exponents were introduced in Section 6.2, you should now be able to write the repeated factors of a number in exponential form. Expressing factors in this form facilitates the process of determining the maximum number of times any factor is used. This form also prepares us to determine the LCM of a set of polynomials. Example 1 will now be reworked to illustrate this form.

$$45 = 3^2 \cdot 5$$
$$24 = 2^3 \cdot 3$$
$$20 = 2^2 \cdot 5$$

Thus, the LCM $= 2^3 \cdot 3^2 \cdot 5 = 360$.

Example 2   Find the LCM of 50, 54, and 882. Use exponential form.

$$50 = 2 \cdot 5^2$$
$$54 = 2 \cdot 3^3$$
$$882 = 2 \cdot 3^2 \cdot 7^2$$

Thus, the LCM $= 2 \cdot 3^3 \cdot 5^2 \cdot 7^2 = 66,150$.

To define the **least common multiple** (LCM) of a set of polynomials with integral coefficients, we first define a common multiple. A common multiple is a polynomial which each of the given polynomials will divide exactly. Among these common multiples, the LCM has the lowest degree and the smallest coefficient on the highest degree term.

Example 3   Among the common multiples of the polynomials x and x − 1 are

$$x^2 - x, \; x^3 - x^2, \; 5x^2 - 5x, \; 7x^3 - 7x^2$$

The one of lowest degree with the smallest positive leading coefficient is $x^2 - x$.

The procedure for finding the LCM of a set of polynomials is very similar to the procedure for finding the LCM of a set of integers.

PROCEDURE FOR FINDING THE LCM OF A SET OF POLYNOMIALS	EXAMPLE Find the LCM of $18x - 18y$, $12x^2 - 12y^2$, and $x^2 - 2xy + y^2$. *Solution:*
1. Factor each of the polynomials completely, including constant factors. Express repeated factors in exponential form.	$18x - 18y = 18(x - y)$ $= 2 \cdot 3^2(x - y)$ $12x^2 - 12y^2 = 12(x^2 - y^2)$ $= 2^2 \cdot 3(x - y)(x + y)$ $x^2 - 2xy + y^2 = (x - y)^2$
2. List each factor the greatest number of times it occurs in any single factorization.	2^2 3^2 $(x - y)^2$ $(x + y)$
3. Form the LCM by multiplying each of the factors from step 2.	LCM $= 2^2 \cdot 3^2(x - y)^2(x + y)$ $= 36(x - y)^2(x + y)$.
	(Note that the polynomial factors of the LCM are usually not multiplied. Instead, parentheses are used to indicate the multiplication. This form is much more useful and saves us the effort of multiplying these factors. The monomial factors, however, are usually multiplied.)

Example 4   Find the LCM for each of the following sets of polynomials.

a. $18x^2y^3z^2$ and $24x^3y^2z^4$

$$18x^2y^3z^2 = 2 \cdot 3^2x^2y^3z^2$$
$$24x^3y^2z^4 = 2^3 \cdot 3x^3y^2z^4$$

LCM $= 2^3 3^2 x^3 y^3 z^4 = 72x^3y^3z^4$

b. $27x^3 - 27xy^2$ and $45x^3 - 45x^2y$

$$27x^3 - 27xy^2 = 27x(x^2 - y^2)$$
$$= 3^3 x(x - y)(x + y)$$
$$45x^3 - 45x^2y = 45x^2(x - y)$$
$$= 3^2 \cdot 5x^2(x - y)$$

LCM $= 3^3 \cdot 5 \cdot x^2(x - y)(x + y) = 135x^2(x - y)(x + y)$.

c. $49x^4 - 49x$, $28x^4 - 28x^3$, and $8x^2 - 8y^2$

$$49x^4 - 49x = 49x(x^3 - 1)$$
$$= 7^2 \cdot x(x - 1)(x^2 + x + 1)$$
$$28x^4 - 28x^3 = 28x^3(x - 1)$$
$$= 2^2 \cdot 7x^3(x - 1)$$
$$8x^2 - 8y^2 = 8(x^2 - y^2)$$
$$= 2^3(x - y)(x + y)$$

LCM $= 2^3 \cdot 7^2 x^3(x - y)(x - 1)(x + y)(x^2 + x + 1)$
$$= 392x^3(x - y)(x - 1)(x + y)(x^2 + x + 1)$$

d. $6x^2 - 30x + 36$ and $9x^2 + 9x - 108$

$$6x^2 - 30x + 36 = 6(x^2 - 5x + 6)$$
$$= 2 \cdot 3(x - 2)(x - 3)$$
$$9x^2 + 9x - 108 = 9(x^2 + x - 12)$$
$$= 3^2(x + 4)(x - 3)$$

LCM $= 2 \cdot 3^2(x - 2)(x - 3)(x + 4) = 18(x - 2)(x - 3)(x + 4)$

e. $x^2 - xy - 2y^2$ and $6xy - 3x^2$

$$x^2 - xy - 2y^2 = (x + y)(x - 2y)$$
$$6xy - 3x^2 = 3x(2y - x) = -3x(x - 2y)$$

LCM $= -3x(x + y)(x - 2y)$
Note that $(2y - x) = -(x - 2y)$; thus, $3x(2y - x) = -3x(x - 2y)$.

17.4 Exercises

Find the LCM of each of the sets in Exercises 1–30.

1. 16, 40, and 50

2. 18, 45, and 75

3. 32, 48, and 15

4. 60, 54, and 36

5. $6x^2y$ and $21xy^5$

6. $15x^2y^3$ and $27x^3y$

7. $12ab$ and $15a^2c^3$

8. $35s^2t^3w^4$ and $55s^7t^3w^2$

9. $16a^2b^2$, $40ab^3$, and $75x^3y^2z^8$

10. $32x^2y$, $24xy^3z$, and $72x^3y^2z^8$

11. $12wxy^2z$, $44wy^3z$, $121wx^5yz^4$, and $33wxyz^7$

12. $15a^3b^5c^7d^{11}$, $77a^5b^7c^{11}d^3$, $35a^7b^5c^3d^{11}$, and $33a^{11}b^7c^5d^3$

13. $12(x + y)$ and $15(x + y)$

14. $21(3a - b)$ and $14(3a - b)$

15. $12x + 8y$ and $15x + 10y$

16. $78s - 12t$ and $195s - 30t$

17. $18a - 27b + 9c$ and $42a - 63b + 21c$

18. $110w - 154x - 66y$ and $165w - 231x - 99y$

19. $x^2 - y^2$ and $5x - 5y$ 20. $7x + 7y$ and $x^2 - y^2$

21. $18x + 18y$ and $6x^2 + 6xy$

22. $4x^2y - 4xy^2$ and $26x^2 - 26y^2$

23. $x^2 - 36$ and $x^2 - 5x - 6$

24. $x^2 - 5x + 6$ and $x^2 - 9$

25. $4x^2y - 8xy$ and $3x^3 + 15x^2 - 42x$

26. $x^3 - 27$, $5x^2 - 45$, and $5x^2 + 15x + 45$

27. $48y^2 - 12$, $8y^3 - 1$, and $36y^2 + 18y + 9$

28. $3y^2 - 27$, $y^3 - 27$, and $y^2 - 6y + 9$

29. $25x^2 - 70xy + 49y^2$, $25x^2 - 49y^2$, and $25x^2 + 35xy$

30. $w^2 + w - 2$, $w^2 - w - 6$, and $w^2 - 4w + 3$

17.5 Addition and Subtraction of Rational Expressions

The sum of two fractions is calculated by first expressing those fractions as equivalent fractions with a common denominator, that is,

$$\frac{5}{6} + \frac{1}{9} = \frac{15}{18} + \frac{2}{18}$$

Once the fractions are expressed with the same denominator the numerators can be added to calculate the sum. Therefore, $\frac{5}{6} + \frac{1}{9} = \frac{15}{18} + \frac{2}{18} = \frac{15 + 2}{18} = \frac{17}{18}$.

If rational expressions do not have the same denominator, we must follow a similar procedure in order to combine these expressions. You should be prepared to calculate the LCD from your work in the previous section.

PROCEDURE FOR ADDING RATIONAL EXPRESSIONS	*EXAMPLE*
	$\dfrac{xy}{x^2 - y^2} + \dfrac{y}{x - y} = ?$ *Solution:*
1. Factor the denominator of each rational expression, and calculate their LCD.	$x^2 - y^2 = (x - y)(x + y)$ $x - y = x - y$ $\text{LCD} = (x - y)(x + y).$
2. Convert each of these terms to an equivalent rational expression whose denominator is the LCD, and indicate their sum.	$\dfrac{xy}{x^2 - y^2} + \dfrac{y}{x - y}$ $= \dfrac{xy}{(x - y)(x + y)} + \dfrac{y(x + y)}{(x - y)(x + y)}$
3. Retain the LCD as the denominator, and add the numerators to form the sum.	$= \dfrac{xy + y(x + y)}{(x - y)(x + y)}$ $= \dfrac{xy + xy + y^2}{(x - y)(x + y)}$ $= \dfrac{y^2 + 2xy}{(x - y)(x + y)}$
4. Reduce the expression to lowest terms.	$= \dfrac{y(y + 2x)}{(x - y)(x + y)}$
	Thus, $\dfrac{xy}{x^2 - y^2} + \dfrac{y}{x - y} = \dfrac{y(y + 2x)}{(x - y)(x + y)}$

Example   Perform the following additions and subtractions, and simplify the results.

a. $\dfrac{3}{x} + \dfrac{4}{y} = ?$

$$\frac{3}{x} + \frac{4}{y} = \frac{3 \cdot y}{x \cdot y} + \frac{4 \cdot x}{y \cdot x}$$

$$= \frac{3y + 4x}{xy} \quad \text{or} \quad \frac{4x + 3y}{xy}$$

b. $\dfrac{3}{2x} + \dfrac{5}{6x} = ?$

$$\frac{3}{2x} + \frac{5}{6x} = \frac{9}{6x} + \frac{5}{6x}$$

$$= \frac{9 + 5}{6x}$$

$$= \frac{14}{6x}$$

$$= \frac{7}{3x}$$

c. $\dfrac{x}{yz} + \dfrac{y}{xz} + \dfrac{z}{xy} = ?$ The LCD is xyz. $\dfrac{x}{yz} = \dfrac{x^2}{xyz}$

$$\frac{y}{xz} = \frac{y^2}{xyz}$$

$$\frac{z}{xy} = \frac{z^2}{xyz}$$

Thus, $\dfrac{x}{yz} + \dfrac{y}{xz} + \dfrac{z}{xy} = \dfrac{x^2}{xyz} + \dfrac{y^2}{xyz} + \dfrac{z^2}{xyz} = \dfrac{x^2 + y^2 + z^2}{xyz}.$

d. $\dfrac{2}{x^2 - y^2} + \dfrac{3}{x^2 + 2xy + y^2} = ?$ $x^2 - y^2 = (x - y)(x + y)$

$$x^2 + 2xy + y^2 = (x + y)^2$$

$$\text{LCD} = (x - y)(x + y)^2.$$

$$\frac{2}{x^2 - y^2} + \frac{3}{x^2 + 2xy + y^2} = \frac{2}{(x - y)(x + y)} + \frac{3}{(x + y)(x + y)}$$

$$= \frac{2(x + y)}{(x - y)(x + y)(x + y)} + \frac{3(x - y)}{(x - y)(x + y)(x + y)}$$

$$= \frac{2(x + y) + 3(x - y)}{(x - y)(x + y)(x + y)}$$

$$= \frac{2x + 2y + 3x - 3y}{(x - y)(x + y)(x + y)}$$

$$= \frac{5x - y}{(x - y)(x + y)^2}$$

e. $\dfrac{7a}{18a + 36} + \dfrac{5}{4a + 8} = ?$ $18a + 36 = 18(a + 2) = 2 \cdot 3^2(a + 2)$

$4a + 8 = 4(a + 2) = 2^2(a + 2)$

$\text{LCD} = 2^2 \cdot 3^2(a + 2) = 36(a + 2).$

$$\dfrac{7a}{18a + 36} + \dfrac{5}{4a + 8} = \dfrac{7a}{18(a + 2)} + \dfrac{5}{4(a + 2)}$$

$$= \dfrac{7a(2)}{36(a + 2)} + \dfrac{5(9)}{36(a + 2)}$$

$$= \dfrac{14a + 45}{36(a + 2)}$$

f. $\dfrac{3x}{3x - y} - \dfrac{3x}{3x + y} - \dfrac{2y^2}{9x^2 - y^2} = ?$

Since $9x^2 - y^2 = (3x - y)(3x + y)$, the LCD is $(3x - y)(3x + y)$.
Therefore,

$$\dfrac{3x}{3x - y} - \dfrac{3x}{3x + y} - \dfrac{2y^2}{9x^2 - y^2} = \dfrac{3x(3x + y) - 3x(3x - y) - 2y^2}{(3x - y)(3x + y)}$$

$$= \dfrac{9x^2 + 3xy - 9x^2 + 3xy - 2y^2}{(3x - y)(3x + y)}$$

$$= \dfrac{6xy - 2y^2}{(3x - y)(3x + y)}$$

$$= \dfrac{2y\cancel{(3x - y)}}{\cancel{(3x - y)}(3x + y)}$$

$$= \dfrac{2y}{3x + y}$$

Example f illustrates an alternate format which combines steps 3 and 4 in the procedure for adding rational expressions. This and other alternate forms for working problems are acceptable.

17.5 Exercises

Perform the additions and subtractions in Exercises 1-40 and simplify the results.

1. $\dfrac{5}{a} + \dfrac{7}{b}$

2. $\dfrac{8}{s} - \dfrac{5}{t}$

3. $\dfrac{4}{9w} + \dfrac{7}{6w}$

4. $\dfrac{8}{15x} + \dfrac{11}{35x}$

5. $\dfrac{5}{x^2y} - \dfrac{6}{xy^2}$

6. $\dfrac{a}{xy^2z} - \dfrac{b}{x^2yz^3}$

7. $\dfrac{4}{ab} + \dfrac{3}{bc} - \dfrac{1}{ac}$

8. $\dfrac{7}{a^2bc} - \dfrac{8}{ab^2c} + \dfrac{18}{abc^2}$

9. $\dfrac{5x}{48} + \dfrac{7x}{40}$

10. $\dfrac{6s}{35} + \dfrac{12s}{49}$

11. $\dfrac{a - b}{30} - \dfrac{a + b}{42}$

12. $\dfrac{2x - 1}{15} + \dfrac{x + 1}{33}$

13. $\dfrac{4t - w}{65} + \dfrac{3w + t}{26}$

14. $\dfrac{13k + 6}{133} - \dfrac{k - 1}{38}$

15. $\dfrac{ab}{a^2 - b^2} - \dfrac{b}{a - b}$

16. $\dfrac{3}{x - y} + \dfrac{1}{x + y}$

17. $\dfrac{y}{2x + 3y} - \dfrac{x}{2x - 3y}$

18. $\dfrac{a}{7a - 5b} - \dfrac{b}{7a + 5b}$

19. $\dfrac{3x}{x + 2y} - \dfrac{2y}{x - 2y}$

20. $\dfrac{t}{8s - 6t} - \dfrac{s}{16s - 12t}$

21. $\dfrac{x}{77x - 121y} - \dfrac{y}{49x - 77y}$

22. $\dfrac{1}{x - 5} + \dfrac{1}{x + 5} - \dfrac{10}{x^2 - 25}$

23. $\dfrac{1}{a - 3b} + \dfrac{b}{a^2 - 7ab + 12b^2} + \dfrac{1}{a - 4b}$

24. $\dfrac{3}{s - 5t} + \dfrac{7}{s - 2t} - \dfrac{9t}{s^2 - 7st + 10t^2}$

25. $\dfrac{42}{x^2 - 49} + \dfrac{3}{x + 7} + \dfrac{3}{x - 7}$

26. $\dfrac{5x}{5x - y} + \dfrac{2y^2}{25x^2 - y^2} - \dfrac{5x}{5x + y}$

27. $\dfrac{1}{5a^2 - 7ab} + \dfrac{1}{7a^2 - 5ab}$ 28. $2 + \dfrac{8y}{6x - 4y}$

29. $3 + \dfrac{15y}{2x - 5y} - \dfrac{6x}{2x + 5y}$

30. $\dfrac{3}{4a^2 - 1} - \dfrac{3a}{2a + 1} - \dfrac{2a}{1 - 2a}$

31. $\dfrac{2x^2 - 5y^2}{x^3 - y^3} + \dfrac{x - y}{x^2 + xy + y^2} + \dfrac{1}{x - y}$

32. $\dfrac{5}{2x + 2} + \dfrac{x + 5}{2x^2 - 2} - \dfrac{3}{x - 1}$ 33. $\dfrac{1}{a + b} - \dfrac{2b^2 - ab}{a^3 + b^3}$

34. $\dfrac{x + y}{xy} + \dfrac{y}{x(x - y)} - \dfrac{1}{y}$

35. $\dfrac{2x + y}{(x - y)(x - 2y)} - \dfrac{x + 4y}{(x - 3y)(x - y)} - \dfrac{x - 7y}{(x - 3y)(x - 2y)}$

36. $\dfrac{ac}{(b - a)(b - c)} - \dfrac{ac}{(a - b)(a - c)} - \dfrac{bc}{(a - c)(b - c)}$

37. $\dfrac{2w - 7}{w^2 - 5w + 6} - \dfrac{2 - 4w}{w^2 - w - 6} + \dfrac{5w + 2}{4 - w^2}$

38. $\dfrac{9w + 2}{3w^2 - 2w - 8} - \dfrac{7}{4 - w - 3w^2}$

39. $\dfrac{2}{2x + 2y} + \dfrac{9xy}{3x^3 + 3y^3} - \dfrac{6x}{6x^2 - 6xy + 6y^2}$

40. $\dfrac{5x}{x^2 + xy - 6y^2} + \dfrac{2x - y}{x^2 - 3xy + 2y^2} + \dfrac{2x + y}{x^2 + 2xy - 3y^2}$

17.6   Complex Rational Expressions and Combined Operations with Rational Expressions

A **complex rational expression** is a rational expression whose numerator or denominator, or both, is also a rational expression.

Example 1   a. $\dfrac{\dfrac{1}{x}}{y}$

This is a complex rational expression since the numerator, $\dfrac{1}{x}$, is a rational expression.

b. $\dfrac{b}{\dfrac{a}{a - b}}$

This is a complex rational expression since the denominator, $\dfrac{a}{a - b}$, is a rational expression.

There are two methods for simplifying a complex rational expression, also called a **complex fraction.** You can use whichever of these methods you prefer. Both rules will be given, and then the same examples will be worked by both methods.

The fraction bar of a complex fraction is a grouping symbol that separates the numerator from the denominator. In the first method, the numerator and denominator are simplified separately before the division indicated by the fraction bar is performed. This is outlined in the *procedure for simplifying a complex rational expression.*

PROCEDURE FOR SIMPLIFYING A COMPLEX RATIONAL EXPRESSION

1. Express the numerator and denominator as a single fraction if they are not already in this form.

2. Indicate the quotient of the numerator and denominator.

3. Divide, using the rule for dividing rational expressions.

EXAMPLE

$$\frac{\dfrac{s}{t} - \dfrac{t}{s}}{\dfrac{s}{t}} = ?$$

Solution:

$$\frac{\dfrac{s}{t} - \dfrac{t}{s}}{\dfrac{s}{t}} = \frac{\dfrac{s^2 - t^2}{st}}{\dfrac{s}{t}}$$

$$= \frac{s^2 - t^2}{st} \div \frac{s}{t}$$

$$= \frac{s^2 - t^2}{s\not{t}} \cdot \frac{\not{t}}{s}$$

$$= \frac{s^2 - t^2}{s^2}$$

Thus, $\dfrac{\dfrac{s}{t} - \dfrac{t}{s}}{\dfrac{s}{t}} = \dfrac{s^2 - t^2}{s^2}$

Another technique which is sometimes used to simplify a complex fraction is to multiply both the numerator and the denominator by the LCD of the numerator and denominator. This method is usually much shorter, and is considered by some to be the preferred method. However, it is difficult to classify one method as superior, since the first method is better for some problems and the second is better for others. For this reason it is advisable that you become proficient in the use of both methods.

PROCEDURE FOR SIMPLIFYING A COMPLEX RATIONAL EXPRESSION

1. Calculate the LCD for all the fractions in the numerator and denominator of the complex fraction.

2. Multiply both the numerator and denominator of the complex fraction by the LCD found in step 1.

3. Simplify the result from step 2.

EXAMPLE

$$\frac{\dfrac{s}{t} - \dfrac{t}{s}}{\dfrac{s}{t}} = ?$$

Solution:

$$\text{LCD} = st$$

$$\frac{\left(\dfrac{s}{t} - \dfrac{t}{s}\right)st}{\left(\dfrac{s}{t}\right)st} = \frac{s^2 - t^2}{s^2}$$

Thus, $\dfrac{\dfrac{s}{t} - \dfrac{t}{s}}{\dfrac{s}{t}} = \dfrac{s^2 - t^2}{s^2}.$

Example 2   a. $\dfrac{\frac{2}{3}}{\frac{5}{6}} = ?$

$$\dfrac{\frac{2}{3}}{\frac{5}{6}} = \frac{2}{3} \div \frac{5}{6}$$

$$= \frac{2}{\cancel{3}} \cdot \frac{\cancel{6}^{\,2}}{5}$$

$$= \frac{4}{5}$$

Alternate Solution:

$$\dfrac{\frac{2}{3}}{\frac{5}{6}} = \dfrac{\frac{2}{\cancel{3}} \cdot \cancel{6}^{\,2}}{\frac{5}{\cancel{6}} \cdot \cancel{6}} = \frac{4}{5}$$

b. $\dfrac{\frac{3x}{4y^2}}{\frac{9x^2}{8y}} = ?$

$$\dfrac{\frac{3x}{4y^2}}{\frac{9x^2}{8y}} = \frac{3x}{4y^2} \div \frac{9x^2}{8y}$$

$$= \frac{\cancel{3x}}{\cancel{4y^2}_{\,y}} \cdot \frac{\cancel{8y}^{\,2}}{\cancel{9x^2}_{\,3x}}$$

$$= \frac{2}{3xy}$$

Alternate Solution:

$$\dfrac{\frac{3x}{4y^2}}{\frac{9y^2}{8y}} = \dfrac{\frac{3x}{\cancel{4y^2}} \cdot \cancel{8y^2}^{\,2}}{\frac{9x^2}{\cancel{8y}} \cdot \cancel{8y^2}_{\,y}}$$

$$= \frac{6x}{9x^2y}$$

$$= \frac{2}{3xy}$$

c. $\dfrac{x - y}{\frac{1}{x} - \frac{1}{y}} = ?$

$$\dfrac{x - y}{\frac{1}{x} - \frac{1}{y}} = \dfrac{x - y}{\frac{y - x}{xy}}$$

$$= (x - y) \div \frac{y - x}{xy}$$

$$= \frac{x - y}{1} \cdot \frac{xy}{y - x}$$

$$= \frac{x - y}{1} \cdot \frac{-xy}{x - y}$$

$$= -xy$$

Alternate Solution:

$$\dfrac{x - y}{\frac{1}{x} - \frac{1}{y}} = \dfrac{(x - y) \cdot xy}{\left(\frac{1}{x} - \frac{1}{y}\right) \cdot xy}$$

$$= \frac{x^2 y - xy^2}{y - x}$$

$$= \frac{xy(\cancel{x - y})^{-1}}{(\cancel{y - x})_{1}}$$

$$= -xy$$

d. $\dfrac{\dfrac{x^2 - y^2}{15xy^2}}{\dfrac{x + y}{6x^2y}} = ?$

By first method:

$$\dfrac{\dfrac{x^2 - y^2}{15xy^2}}{\dfrac{x + y}{6x^2y}} = \dfrac{x^2 - y^2}{15xy^2} \cdot \dfrac{6x^2y}{x + y}$$

$$= \dfrac{\overset{1}{\cancel{(x + y)}}(x - y)}{\underset{5y}{\cancel{15xy^2}}} \cdot \dfrac{\overset{2x}{\cancel{6x^2y}}}{\underset{1}{\cancel{x + y}}}$$

$$= \dfrac{2x(x - y)}{5y}$$

e. $\dfrac{x - \dfrac{1}{y}}{y - \dfrac{1}{x}} = ?$

By second method: LCD = xy.

$$\dfrac{x - \dfrac{1}{y}}{y - \dfrac{1}{x}} = \dfrac{\left(x - \dfrac{1}{y}\right)xy}{\left(y - \dfrac{1}{x}\right)xy}$$

$$= \dfrac{x^2y - x}{xy^2 - y}$$

$$= \dfrac{x(\cancel{xy - 1})}{y(\cancel{xy - 1})}$$

$$= \dfrac{x}{y}$$

Recall that if more than one operation is indicated in a problem, the expression inside the innermost grouping symbol is evaluated first. Within a pair of grouping symbols, or when there are no grouping symbols, all multiplications and divisions are performed from left to right, and then all additions and subtractions are performed from left to right.

Example 3   a. $\dfrac{2}{x^2} \cdot \dfrac{x^3}{4} - \dfrac{x}{3} = \dfrac{2x^3}{4x^2} - \dfrac{x}{3}$

$$= \dfrac{x}{2} - \dfrac{x}{3}$$

$$= \dfrac{3x}{6} - \dfrac{2x}{6}$$

$$= \dfrac{3x - 2x}{6}$$

$$= \dfrac{x}{6}$$

b. $\dfrac{2x+3}{x^2-y^2} + \dfrac{x^2+xy+y^2}{6xy} \cdot \dfrac{18y}{x^3-y^3}$

$$= \dfrac{2x+3}{(x-y)(x+y)} + \dfrac{\overset{3}{\cancel{18y}}(\cancel{x^2+xy+y^2})}{\cancel{6xy}(x-y)(\cancel{x^2+xy+y^2})}$$

$$= \dfrac{2x+3}{(x-y)(x+y)} + \dfrac{3}{x(x-y)}$$

$$= \dfrac{x(2x+3)}{x(x-y)(x+y)} + \dfrac{3(x+y)}{x(x-y)(x+y)}$$

$$= \dfrac{(2x^2+3x)+(3x+3y)}{x(x-y)(x+y)}$$

$$= \dfrac{2x^2+6x+3y}{x(x-y)(x+y)}$$

c. $\left(\dfrac{2}{x}-\dfrac{1}{y}\right) \div \left(\dfrac{6}{x}-\dfrac{3x}{x^2}\cdot\dfrac{x}{y}\right) = \left(\dfrac{2y-x}{xy}\right) \div \left(\dfrac{6}{x}-\dfrac{3}{y}\right)$

$$= \left(\dfrac{2y-x}{xy}\right) \div \left(\dfrac{6y-3x}{xy}\right)$$

$$= \dfrac{2y-x}{\cancel{xy}} \cdot \dfrac{\cancel{xy}}{6y-3x}$$

$$= \dfrac{2y-x}{6y-3x}$$

$$= \dfrac{\cancel{2y-x}}{3(\cancel{2y-x})}$$

$$= \dfrac{1}{3}$$

In Example 3c, note that a single fraction must be obtained in each expression within parentheses before the division is performed. Note also that you cannot invert the divisor unless it is a single fraction.

17.6   Exercises

Simplify the complex rational expressions in Exercises 1–10.

1. $\dfrac{\frac{2}{3}}{4}$

2. $\dfrac{2}{\frac{3}{4}}$

3. $\dfrac{\frac{5}{6}}{\frac{2}{3}}$

4. $\dfrac{\frac{3}{5}}{\frac{7}{10}}$

5. $\dfrac{\frac{a}{2}}{4}$

6. $\dfrac{a}{\frac{2}{4}}$

7. $\dfrac{\frac{a}{2}}{\frac{a}{3}}$

8. $\dfrac{\frac{4}{a}}{6}$

9. $\dfrac{4}{\frac{a}{6}}$

10. $\dfrac{\frac{a}{2}}{\frac{a}{2}}$

Perform the indicated operations in Exercises 11–45 and simplify.

11. $\dfrac{1+\frac{1}{5}}{1-\frac{1}{5}}$

12. $\dfrac{2-\frac{3}{7}}{6+\frac{2}{7}}$

13. $\dfrac{\dfrac{3}{5}-\dfrac{5}{3}}{\dfrac{1}{3}+\dfrac{1}{5}}$

14. $\dfrac{\dfrac{1}{2}+\dfrac{2}{3}}{\dfrac{3}{2}+2}$

29. $\dfrac{x+2-\dfrac{6}{2x+3}}{x+\dfrac{8x}{2x-1}}$

30. $\dfrac{\dfrac{121}{x^2}-\dfrac{66}{x}+9}{3-\dfrac{8}{x}-\dfrac{11}{x^2}}$

15. $\dfrac{2+\dfrac{1}{4}}{2-\dfrac{2}{2+\dfrac{1}{2}}}$

16. $\dfrac{1+\dfrac{1}{1+\dfrac{1}{6}}}{2+\dfrac{3}{5}}$

31. $\dfrac{\dfrac{1}{a}+\dfrac{1}{b}}{\dfrac{a}{b}-\dfrac{b}{a}}$

32. $\dfrac{\dfrac{s}{t}-\dfrac{t}{s}}{\dfrac{1}{s}-\dfrac{1}{t}}$

17. $\dfrac{\dfrac{w-z}{x^2y^2}}{\dfrac{w^2-z^2}{2xy}}$

18. $\dfrac{\dfrac{a^2-b^2}{6a^2b^3}}{\dfrac{a+b}{9a^3b}}$

33. $\dfrac{\dfrac{w-a}{w+a}-\dfrac{w+a}{w-a}}{\dfrac{w^2+a^2}{w^2-a^2}}$

34. $\dfrac{\dfrac{x^2+b^2}{x^2-b^2}}{\dfrac{x-b}{x+b}+\dfrac{x+b}{x-b}}$

19. $\dfrac{\dfrac{3x^2-27x+42}{34x^4y^5}}{\dfrac{5x^2-20}{51x^5y^3}}$

20. $\dfrac{\dfrac{5s^2+5st+5t^2}{8s^3-8t^3}}{\dfrac{7s^2+28st+7t^2}{14s^2-14t^2}}$

35. $\dfrac{\dfrac{b^2}{c^2}-\dfrac{c^2}{b^2}}{\dfrac{b}{c}+\dfrac{c}{b}}$

36. $\dfrac{x^4}{4}\cdot\dfrac{6}{x^2}-\dfrac{x^2}{5}$

21. $\dfrac{2-\dfrac{1}{x}}{4-\dfrac{1}{x^2}}$

22. $\dfrac{\dfrac{1}{x^2}-49}{7-\dfrac{1}{x}}$

37. $\dfrac{x^3}{45}\cdot\dfrac{15}{x^2}+\dfrac{x}{7}$

38. $\dfrac{x^2-y^2}{x^2}\cdot\dfrac{xy}{x+y}-\dfrac{y^2}{x}$

23. $\dfrac{a-\dfrac{1}{a}}{a+\dfrac{1}{a}}$

24. $\dfrac{s-\dfrac{9}{s}}{1-\dfrac{3}{s}}$

39. $\dfrac{x-y}{xy}+\dfrac{x^3+y^3}{x^3y^3}\cdot\dfrac{x^2y^2}{4x^2-4xy+4y^2}$

40. $\left(\dfrac{x}{y}-\dfrac{y}{x}\right)\left(\dfrac{x^2y^2}{x-y}\right)$ 41. $\left(\dfrac{a^2}{b}+\dfrac{b^2}{a}\right)\left(\dfrac{3a^3b^3}{a+b}\right)$

25. $\dfrac{1-\dfrac{1}{x^3}}{1+\dfrac{1}{x}+\dfrac{1}{x^2}}$

26. $\dfrac{\dfrac{3}{x^2}-\dfrac{3}{x}+3}{18+\dfrac{18}{x^3}}$

42. $\left(3-\dfrac{b^2}{3a^2}\right)\left(\dfrac{9a^2-ab}{3a-b}-a\right)$

43. $\left(\dfrac{25w^2-wz}{5w-z}-w\right)\left(5-\dfrac{z^2}{5w^2}\right)$

27. $\dfrac{\dfrac{1}{x}-\dfrac{8}{x^2}+\dfrac{15}{x^3}}{1-\dfrac{5}{x}}$

28. $\dfrac{3+\dfrac{9}{x}}{\dfrac{15}{x^3}+\dfrac{8}{x^2}+\dfrac{1}{x}}$

44. $\left(\dfrac{4}{5x+25}+\dfrac{1}{5x-25}\right)\div\left(\dfrac{x^2-x-6}{x+5}\right)$

45. $\dfrac{x+2}{x+3}-\left(\dfrac{x^3-8}{x^2-9}\div\dfrac{x^2+2x+4}{x-3}\right)$

17.7   Solving Equations Involving Rational Expressions

In Chapter 10 we discussed the solution of equations involving fractions. When rational expressions are involved, the same procedure applies. The multiplication principle is used to clear the equation of fractions. However, when the denominator contains a variable, extra caution must be exercised so that division by zero does not occur. Consider the equation

$$\frac{3}{x-2}=5+\frac{3}{x-2}$$

First clear the equations of fractions, by multiplying both members of the equation by x − 2, and then solve for x.

$$\frac{3}{x - 2} = 5 + \frac{3}{x - 2}$$

$$\frac{\overset{1}{(x-2)}}{1} \cdot \frac{3}{\underset{1}{x-2}} = (x - 2)5 + \frac{\overset{1}{(x-2)}}{1} \cdot \frac{3}{\underset{1}{x-2}}$$

$$3 = 5x - 10 + 3$$

$$3 = 5x - 7$$

$$5x = 10$$

$$x = 2$$

This would seem to be a solution, but checking the problem leads to an undefined expression, $\frac{3}{0}$.

Check:

$$\frac{3}{2 - 2} = 5 + \frac{3}{2 - 2}$$

$$\frac{3}{0} = 5 + \frac{3}{0}$$

Even though the computations are correct, 2 is not a solution, since the last statement is false. Two is called an **extraneous root** and the equation has no solution. Any value which yields zero as a denominator when it is substituted for the variable must be excluded from the set of possible solutions for the equation. Thus, when solving equations containing variables in the denominator, you must always check the solutions obtained to determine whether they are extraneous roots.

Example 1   What are the excluded values in each of the following equations?

a. $\dfrac{2}{x - 5} + 6 = \dfrac{8}{x - 5}$

x = 5 is an excluded value.

b. $\dfrac{2}{x - 1} = \dfrac{5}{2x + 3}$

x = 1 and x = $-\dfrac{3}{2}$ are excluded values.

Example 2   Solve the following equations.

a. $\dfrac{2}{x-5} + 6 = \dfrac{8}{x-5}$

$$\dfrac{2}{x-5} + 6 = \dfrac{8}{x-5}$$

$$(x-5) \cdot \dfrac{2}{x-5} + (x-5)6 = (x-5) \cdot \dfrac{8}{x-5}$$

$$2 + 6x - 30 = 8$$

$$6x - 28 = 8$$

$$6x = 36$$

$$x = 6$$

Check: $\dfrac{2}{6-5} + 6 = \dfrac{8}{6-5}$

$$2 + 6 = 8$$

$$8 = 8$$

b. $\dfrac{2}{x-1} = \dfrac{5}{x+3}$

$$\dfrac{2}{x-1} = \dfrac{5}{x+3}$$

$$(x-1)(x+3) \cdot \dfrac{2}{x-1} = (x-1)(x+3) \cdot \dfrac{5}{x+3}$$

$$2x + 6 = 5x - 5$$

$$-3x = -11$$

$$x = \dfrac{11}{3}$$

Check: $\dfrac{2}{\dfrac{11}{3} - 1} = \dfrac{5}{\dfrac{11}{3} + 3}$

$$\dfrac{6}{11 - 3} = \dfrac{15}{11 + 9}$$

$$\dfrac{6}{8} = \dfrac{15}{20}$$

$$\dfrac{3}{4} = \dfrac{3}{4}$$

c. $\dfrac{5}{x-4} = 3 + \dfrac{5}{x-4}$

$$(x-4) \cdot \dfrac{5}{x-4} = (x-4) \cdot 3 + (x-4) \cdot \dfrac{5}{x-4}$$

$$5 = 3x - 12 + 5$$

$$5 = 3x - 7$$

$$3x = 12$$

$$x = 4$$

$$Check: \quad \frac{5}{4-4} = 3 + \frac{5}{4-4}$$

$$\frac{5}{0} = 3 + \frac{5}{0}$$

Since $x = 4$ makes the denominator of the fraction 0, and $\frac{5}{0}$ is undefined, $x = 4$ is an extraneous root, and the equation has no solution.

17.7 Exercises

What are the excluded values in each of the equations in Exercises 1–10?

1. $\dfrac{3}{x-1} = 5$

2. $\dfrac{6}{x-2} = 3$

3. $\dfrac{3}{x-1} + 2 = \dfrac{5}{x-1}$

4. $\dfrac{2}{x+2} + 3 = \dfrac{2}{x+2}$

5. $\dfrac{3}{x-3} = \dfrac{5}{x-2}$

6. $\dfrac{x+2}{x-5} = 10$

7. $\dfrac{x-2}{3} = \dfrac{3}{x-2}$

8. $\dfrac{x}{x-3} = 4$

9. $\dfrac{7}{x-4} = \dfrac{2}{x+1}$

10. $\dfrac{2x}{x-1} = \dfrac{3x}{x+5}$

Solve the equations in Exercises 11–25.

11. $\dfrac{3}{x-1} = 3$

12. $\dfrac{6}{x-2} = 3$

13. $\dfrac{3}{x-1} + 2 = \dfrac{5}{x-1}$

14. $\dfrac{2}{x+2} + 3 = \dfrac{2}{x+2}$

15. $\dfrac{x}{x-1} = 4$

16. $5 = \dfrac{x}{2x-1}$

17. $\dfrac{-3}{x+2} = \dfrac{-8}{x-3}$

18. $\dfrac{7}{x-4} = \dfrac{2}{x+1}$

19. $\dfrac{6}{x-4} = 3 + \dfrac{6}{x-4}$

20. $\dfrac{1}{x+3} + 9 = \dfrac{55}{x+3}$

21. $\dfrac{-1}{x-1} = \dfrac{-9}{x-9}$

22. $\dfrac{2}{x-1} = \dfrac{3}{x+5}$

23. $\dfrac{10}{x-3} = \dfrac{34}{2x+1}$

24. $\dfrac{6}{x+3} = \dfrac{6}{2x-1}$

25. $\dfrac{7}{3x-1} = \dfrac{2}{x+2}$

Important Terms Used in this Chapter

algebraic fraction
complex rational expression
excluded value
extraneous root

least common multiple of a set of polynomials
lowest terms
rational expression

REVIEW EXERCISES

Reduce the rational expressions in Exercises 1–4 to lowest terms.

1. $\dfrac{6x-18y}{8x-24y}$

2. $\dfrac{3x^2-3y^2}{15x+15y}$

3. $\dfrac{4a^2-12ab+9b^2}{4a^2-9b^2}$

4. $\dfrac{7s^3-7t^3}{63s^2+63st+63t^2}$

Fill in the missing numerators and denominators in Exercises 5–8 so that the resulting rational expressions will be equivalent.

5. $\dfrac{5x}{6y} = \dfrac{?}{30y^2}$

6. $\dfrac{x}{x^2-y^2} = \dfrac{3x}{?}$

7. $\dfrac{x^2+xy+y^2}{4x} = \dfrac{x^3-y^3}{?}$

8. $\dfrac{2x+1}{x-3} = \dfrac{?}{2x^2+4x-30}$

Perform the indicated operations in Exercises 9–25 and simplify the results.

9. $\dfrac{5x}{x-y} - \dfrac{5y}{x-y}$

10. $\dfrac{6xy}{x-y} \cdot \dfrac{5x-5y}{42x^2}$

11. $\dfrac{5x+7y}{2x+3y} + \dfrac{x+2y}{2x+3y}$

12. $\dfrac{x^2-6x-7}{x^3-8} \div \dfrac{x^2-49}{x^2-4x-4}$

13. $\dfrac{3}{x} + \dfrac{4}{y} - \dfrac{5}{z}$

14. $\dfrac{x}{x-y} - \dfrac{y}{x+y} - \dfrac{2y^2}{x^2-y^2}$

15. $\dfrac{5x-20}{6x+9} \cdot \dfrac{4x^2+12x+9}{x^2-6x+8} \cdot \dfrac{11x}{50x+75}$

16. $\dfrac{\dfrac{x^3+y^3}{x^2y^2}}{\dfrac{x+y}{xy}}$

17. $\dfrac{1}{3x^2-3y^2} - \dfrac{1}{5x^2+10xy+5y^2} - \dfrac{1}{7x^2+14xy+7y^2}$

18. $\left(\dfrac{1}{x} - \dfrac{1}{y}\right)\left(\dfrac{3xy}{5x-5y}\right)$

19. $\left(\dfrac{5}{y} - \dfrac{15}{x}\right)\left(\dfrac{1}{x-3y} - \dfrac{1}{x+3y}\right)$

20. $\dfrac{\dfrac{28x^2y^3}{15zw^2}}{\dfrac{16x^3y}{25z^2w}}$

21. $\dfrac{2x-y}{12xy} \cdot \dfrac{18x^2y^2}{4x^2-4xy+y^2} - \dfrac{4x^3y^3}{2x^2-xy-6y^2} \cdot \dfrac{10x+15y}{40x^2y^2}$

22. $\dfrac{1+\dfrac{2}{x}}{\dfrac{6}{x}+3}$

23. $\dfrac{x - \dfrac{y}{\dfrac{x}{y}}}{\dfrac{1}{3} + \dfrac{3}{y}}$

24. $\dfrac{1}{2x+y} + \dfrac{y}{4x^2-y^2} + \dfrac{3}{4x^2-4xy+y^2} + \dfrac{2x}{y^2-4x^2}$

25. $\dfrac{5x}{y-\dfrac{x^2}{y}} \cdot \dfrac{\dfrac{x+y}{35}}{\dfrac{x^2y^2}{6}}$

Solve the equations in Exercises 26–30.

26. $\dfrac{10}{x-2} = 10$

27. $\dfrac{1}{x-3} = 2 + \dfrac{1}{x-3}$

28. $\dfrac{2}{x-4} = \dfrac{12}{x+1}$

29. $\dfrac{4}{3x-1} = \dfrac{4}{x+5}$

30. $\dfrac{8}{2x+5} = 4 + \dfrac{2}{2x+5}$

18

SQUARE ROOTS, RADICALS, AND QUADRATIC EXPRESSIONS

OBJECTIVES

Upon completion of this chapter you should be able to:

1. Add, subtract, multiply, and divide square roots.   18.1
2. Simplify expressions containing square roots.   18.1
3. Recognize a quadratic equation, and write a quadratic equation in standard form.   18.2
4. Solve quadratic equations by factoring.   18.2
5. Solve quadratic equations by using the Quadratic Formula.   18.3
6. Solve word problems using quadratic equations.   18.4

PRETEST

1. Find the following sums and differences (Objective 1).
   a. $5\sqrt{2} + 3\sqrt{2} - 10\sqrt{2}$
   b. $\sqrt{18} + \sqrt{200}$
   c. $\sqrt{9x} + 2\sqrt{x} - \sqrt{49x}$

   Find the following products and simplify the results (Objective 1).
   d. $\sqrt{2} \cdot \sqrt{10}$
   e. $\sqrt{4x} \cdot \sqrt{4x}$
   f. $\sqrt{48} \cdot \sqrt{5}$

   Find the following quotients and simplify the results (Objective 1).
   g. $\dfrac{\sqrt{50}}{\sqrt{5}}$

   h. $\dfrac{\sqrt{150}}{\sqrt{90}}$

   i. $\dfrac{\sqrt{6x^3}}{\sqrt{x}}$

2. Simplify the following square roots (Objective 2).
   a. $\sqrt{96}$
   b. $\sqrt{\dfrac{5}{7}}$

   c. $\sqrt{x^2 y^4}$
   d. $\sqrt{588}$
   e. $\sqrt{81x^3 y^{10} z}$
   f. $\dfrac{1}{\sqrt{5x}}$

3. Which of the following are quadratic equations (Objective 3)?
   a. $x^2 - 19x = 0$ b. $24x - 39 = 0$
   c. $2x^2 - 6x + 13 = 0$ d. $5x^2 = 191$
   e. $3x^2 - 5x + 21$ f. $10x^2 - 8x = 49$

   Write the following quadratic equations in standard form (Objective 3).
   g. $12x^2 = 6x - 9$ h. $5x + 3x^2 = 8$
   i. $x^2 - 6x + 17 = 0$ j. $0 = x^2 - 6x + 35$
   k. $x^2 = 12$ l. $3x^2 = 81x$

4. Identify the quadratic equations that can be solved by factoring over the integers. Solve and check the solutions for the equations which can be solved by factoring (Objective 4).
   a. $x^2 - 10x + 16 = 0$ b. $4x^2 = 49$
   c. $3x^2 + x + 1 = 0$ d. $15x^2 = 10$
   e. $5x^2 - x - 35 = 0$ f. $x^2 + 6x = 27$

5. Use the Quadratic Formula to solve the following equations. Express answers in simplest form. (In e, give the rational approximations of the roots to the nearest tenth.) (Objective 5).
   a. $x^2 + 2x - 1 = 0$ b. $2x^2 + 3x + 1 = 0$
   c. $3x^2 + 2x = 1$ d. $x^2 = 10$
   e. $x^2 - 5x - 1 = 0$

6. Use equations to solve the following problems (Objective 6).
   a. Jim is 3 years older than Jack. In 3 years the product of their ages will be 208. How old is each now?
   b. The length of a swimming pool is 10 meters more than the width. The surface area is 171 sq meters. Find the length and width of the pool.

A second-degree equation in one variable is called a **quadratic equation.** In this chapter we are primarily interested in finding the solutions to such equations. One technique for solving quadratic equations involves the use of a formula called the Quadratic Formula. To apply this formula, you need to be able to simplify square roots. For this reason, we will begin this chapter with a discussion of square roots. We will consider both simplifying square roots and operations with square roots.

18.1 Square Roots

Recall that to find the square root of a number, you find a number whose square equals the given number. Every positive number has two square roots, one positive and one negative. For example, the square roots of 25 are +5 and −5 since $(5)^2 = 25$ and $(-5)^2 = 25$. The square root of 0 is 0, while the square root of a negative number is not a real number. In this section we will restrict our attention to square roots of positive numbers.

The symbol $\sqrt{}$ is called the **radical symbol.** In a radical expression, the expression under the radical is called the **radicand.** For example, in the expression $\sqrt{5}$, 5 is the radicand. In the expression $\sqrt{4x^2y}$, $4x^2y$ is the radicand. The radical symbol $\sqrt{}$ denotes the *principal square root,* that is, the *positive* square root. Thus $\sqrt{25} = 5$. To denote the negative square root, we write $-\sqrt{25} = -5$.

Example 1   Evaluate each of the following.

   a. $\sqrt{16} = 4$ b. $\sqrt{144} = 12$
   c. $-\sqrt{81} = -9$ d. $-\sqrt{4} = -2$

Frequently the radicand will contain variables. (We will assume in this text that the variables represent positive numbers.) However, the square root is defined in the same way. The square root is an expression whose square is equal to the radicand.

Example 2   Evaluate each of the following square roots.

   a. $\sqrt{x^2} = x$ since the square of x is x^2.
      Note: x^2 is called a **perfect square.**
   b. $\sqrt{a^4} = a^2$ since $(a^2)^2 = a^4$.
      Note: a^4 is a perfect square.

c. $\sqrt{y^8} = y^4$ since $(y^4)^2 = y^8$.   d. $\sqrt{n^{10}} = n^5$

e. $-\sqrt{y^2} = -y$ f. $\sqrt{(x+y)^2} = x + y$

In general, *to find the square root of an expression raised to an even power, you divide the exponent by 2.*

To multiply square roots, use the *rule for multiplying square roots.*

RULE FOR MULTIPLYING SQUARE ROOTS OF POSITIVE NUMBERS	EXAMPLE $\sqrt{3} \cdot \sqrt{7} = ?$
$\sqrt{a} \cdot \sqrt{b} = \sqrt{ab}$	$\sqrt{3} \cdot \sqrt{7} = \sqrt{21}$

Example 3   Express each of the following products as a single square root.

a. $\sqrt{2} \cdot \sqrt{5} = \sqrt{10}$ b. $\sqrt{5} \cdot \sqrt{3} = \sqrt{15}$

c. $\sqrt{10} \cdot \sqrt{7} = \sqrt{70}$ d. $\sqrt{15} \cdot \sqrt{x} = \sqrt{15x}$

e. $\sqrt{a} \cdot \sqrt{y} = \sqrt{ay}$ f. $\sqrt{7} \cdot \sqrt{7} = \sqrt{49} = 7$

g. $\sqrt{8} \cdot \sqrt{2} = \sqrt{16} = 4$

If the members of the equation in the rule for multiplying square roots are interchanged, another form is obtained, $\sqrt{ab} = \sqrt{a} \cdot \sqrt{b}$. This formula can be used to simplify a square root.

Example 4   Simplify the following square roots.

a. $\sqrt{20} = \sqrt{4 \cdot 5} = \sqrt{4} \cdot \sqrt{5} = 2\sqrt{5}$ Perfect square factor

b. $\sqrt{27} = \sqrt{9 \cdot 3} = \sqrt{9} \cdot \sqrt{3} = 3\sqrt{3}$ Perfect square factor

c. $\sqrt{300} = \sqrt{100 \cdot 3} = \sqrt{100}\sqrt{3} = 10\sqrt{3}$ Perfect square factor

In these examples, the radicand was factored so that one factor was a perfect square and the other was not. The *largest* perfect square factor was chosen so that in the answer, the radicand would not contain a perfect square factor. If you do not use the largest perfect square factor, you must repeat the process.

For instance, in Example 4c, we could have factored 300 as $4 \cdot 75$. Then,

$$\sqrt{300} = \sqrt{4 \cdot 75} \quad \text{Perfect square factor}$$
$$= \sqrt{4} \cdot \sqrt{75}$$
$$= 2\sqrt{75}$$

This expression is not in simplest form because the radicand still contains a perfect square factor, namely 25 since $75 = 3 \cdot 25$.

$$2\sqrt{75} = 2 \cdot \sqrt{25} \cdot \sqrt{3}$$
$$= 2 \cdot 5\sqrt{3}$$
$$= 10\sqrt{3}$$

Now 3 does not contain a perfect square factor greater than 1 so this expression is in simplest form.

In general, *a square root is not in simplest form if the radicand contains a perfect square factor other than 1*. To get a radical to meet this condition, follow the procedure given in the *rule for simplifying square roots*.

RULE FOR SIMPLIFYING SQUARE ROOTS	EXAMPLE Simplify $\sqrt{18}$.
	Solution:
1. Determine the largest perfect square factor contained in the radicand.	18 has the perfect square 9 as a factor.
2. Express the square root as a product of two square roots, one of which is a perfect square.	$\sqrt{18} = \sqrt{9 \cdot 2}$
	$= \sqrt{9} \cdot \sqrt{2}$
3. Evaluate the square root of the perfect square factor.	$= 3\sqrt{2}$

Example 5 Simplify the following square roots.

a. $\sqrt{32} = \sqrt{16 \cdot 2} = \sqrt{16} \cdot \sqrt{2} = 4\sqrt{2}$

b. $\sqrt{28} = \sqrt{4 \cdot 7} = \sqrt{4} \cdot \sqrt{7} = 2\sqrt{7}$

c. $\sqrt{9x} = \sqrt{9 \cdot x} = 3\sqrt{x}$

d. $\sqrt{3x^2} = \sqrt{x^2 \cdot 3} = \sqrt{x^2} \cdot \sqrt{3} = x\sqrt{3}$

e. $\sqrt{a^3} = \sqrt{a^2 \cdot a} = a\sqrt{a}$

f. $\sqrt{4a^5} = \sqrt{4a^4 \cdot a} = \sqrt{4a^4} \cdot \sqrt{a} = 2a^2\sqrt{a}$ $(\sqrt{4a^4} = \sqrt{4}\sqrt{a^4} = 2a^2)$

If the radicand contains a large number, prime factorization can be used to simplify the square root. This is illustrated in Example 6.

Example 6 Simplify the following square roots.

a. $\sqrt{363} = \sqrt{3 \cdot 11^2}$

$\qquad = \sqrt{3} \cdot \sqrt{11^2}$

$\qquad = \sqrt{3} \cdot 11$

$\qquad = 11\sqrt{3}$

$$3\underline{|363}$$
$$11\underline{|121}$$
$$11$$

$$363 = 3 \cdot 11^2$$

b. $\sqrt{900} = \sqrt{2^2 \cdot 3^2 \cdot 5^2}$

$\qquad = \sqrt{2^2} \cdot \sqrt{3^2} \cdot \sqrt{5^2}$

$\qquad = 2 \cdot 3 \cdot 5$

$\qquad = 30$

$$2\underline{|900}$$
$$2\underline{|450}$$
$$3\underline{|225}$$
$$3\underline{|75}$$
$$5\underline{|25}$$
$$5$$

$$900 = 2^2 \cdot 3^2 \cdot 5^2$$

Simplification of square roots can also be used to find approximations.

Example 7 Find a rational approximation of $\sqrt{32}$, given that $\sqrt{2} \doteq 1.414$.

$$\sqrt{32} = \sqrt{16} \cdot \sqrt{2}$$
$$= 4\sqrt{2}$$
$$\doteq 4(1.414)$$
$$= 5.656$$

Thus $\sqrt{32} \doteq 5.656$. Note that $\sqrt{2} \doteq 1.414$ can be determined by the technique given in Section 5.2 or from a square root table.

Example 8 Multiply the following square roots and simplify the results.

    a. $\sqrt{20} \cdot \sqrt{10} = \sqrt{200} = \sqrt{100 \cdot 2} = \sqrt{100} \cdot \sqrt{2} = 10\sqrt{2}$

    b. $\sqrt{4y}\,\sqrt{4y} = \sqrt{16y^2} = 4y$

    c. $\sqrt{3x} \cdot \sqrt{6x} = \sqrt{18x^2} = \sqrt{9x^2} \cdot \sqrt{2} = 3x\sqrt{2}$

To divide square roots, use *the rule for dividing square roots of positive numbers.*

RULE FOR DIVIDING SQUARE ROOTS OF POSITIVE NUMBERS $$\frac{\sqrt{a}}{\sqrt{b}} = \sqrt{\frac{a}{b}}$$	*EXAMPLE:* $\dfrac{\sqrt{10}}{\sqrt{5}} = ?$ $$\frac{\sqrt{10}}{\sqrt{5}} = \sqrt{\frac{10}{5}} = \sqrt{2}$$

Example 9 Express each of the following as a single square root and then simplify.

    a. $\dfrac{\sqrt{10}}{\sqrt{2}} = \sqrt{\dfrac{10}{2}} = \sqrt{5}$

    b. $\dfrac{\sqrt{14x}}{\sqrt{2x}} = \sqrt{\dfrac{14x}{2x}} = \sqrt{7}$

    c. $\dfrac{\sqrt{16x^2}}{\sqrt{2x}} = \sqrt{\dfrac{16x^2}{2x}} = \sqrt{8x} = \sqrt{4 \cdot 2x} = \sqrt{4} \cdot \sqrt{2x} = 2\sqrt{2x}$

    d. $\dfrac{\sqrt{4x^3}}{\sqrt{x}} = \sqrt{\dfrac{4x^3}{x}} = \sqrt{4x^2} = 2x$

A radical expression is not in simplest form if the radicand contains a fraction. If the denominator of the fraction is a perfect square, we can use the fact that $\sqrt{\dfrac{a}{b}} = \dfrac{\sqrt{a}}{\sqrt{b}}$ to simplify the radical.

Example 10 a. $\sqrt{\dfrac{2}{9}} = \dfrac{\sqrt{2}}{\sqrt{9}} = \dfrac{\sqrt{2}}{3}$

b. $\sqrt{\dfrac{4}{25}} = \dfrac{\sqrt{4}}{\sqrt{25}} = \dfrac{2}{5}$

c. $\sqrt{\dfrac{3}{x^2}} = \dfrac{\sqrt{3}}{\sqrt{x^2}} = \dfrac{\sqrt{3}}{x}$

If the denominator of a fraction is not a perfect square, the problem is more difficult. Consider $\sqrt{\dfrac{2}{3}}$. If we proceed as in Example 10, we get $\sqrt{\dfrac{2}{3}} = \dfrac{\sqrt{2}}{\sqrt{3}}$. Although the radicand does not contain a fraction, the denominator contains a square root, and *a fraction is not in simplest form if the denominator contains a square root*. To simplify this expression, and eliminate the square root in the denominator, we use the multiplication property of fractions.

$\dfrac{\sqrt{2}}{\sqrt{3}} = \dfrac{\sqrt{2} \cdot \sqrt{3}}{\sqrt{3} \cdot \sqrt{3}}$ Multiply the numerator and denominator by $\sqrt{3}$.

$= \dfrac{\sqrt{6}}{\sqrt{9}}$ Multiply the square roots.

$= \dfrac{\sqrt{6}}{3}$ Simplify the square root(s).

In using this procedure, it is necessary to choose a multiplier that will make the denominator a perfect square. Then the radicand in the denominator can be eliminated. The procedure is given in the *rule for rationalizing denominators*.

RULE FOR RATIONALIZING MONOMIAL SQUARE ROOT DENOMINATORS	EXAMPLE $\dfrac{\sqrt{3}}{\sqrt{10}} = ?$
1. If possible, simplify the square root in the denominator.	$\sqrt{10}$ is in simplest form.
2. Multiply the numerator and denominator of the fraction by the square root which appears in the denominator.	$\dfrac{\sqrt{3}}{\sqrt{10}} = \dfrac{\sqrt{3} \cdot \sqrt{10}}{\sqrt{10} \cdot \sqrt{10}}$ $= \dfrac{\sqrt{30}}{\sqrt{100}}$
3. Simplify the square roots.	$= \dfrac{\sqrt{30}}{10}$

Example 11 Rationalize the denominators of the following fractions.

a. $\dfrac{6}{\sqrt{5}} = \dfrac{6 \cdot \sqrt{5}}{\sqrt{5} \cdot \sqrt{5}} = \dfrac{6\sqrt{5}}{\sqrt{25}} = \dfrac{6\sqrt{5}}{5}$

b. $\dfrac{\sqrt{3}}{\sqrt{7}} = \dfrac{\sqrt{3} \cdot \sqrt{7}}{\sqrt{7} \cdot \sqrt{7}} = \dfrac{\sqrt{21}}{\sqrt{49}} = \dfrac{\sqrt{21}}{7}$

c. $\dfrac{3}{\sqrt{x}} = \dfrac{3 \cdot \sqrt{x}}{\sqrt{x} \cdot \sqrt{x}} = \dfrac{3\sqrt{x}}{\sqrt{x^2}} = \dfrac{3\sqrt{x}}{x}$

d. $\dfrac{\sqrt{5}}{\sqrt{8}} = \dfrac{\sqrt{5} \cdot \sqrt{2}}{2\sqrt{2} \cdot \sqrt{2}} = \dfrac{\sqrt{10}}{2\sqrt{4}} = \dfrac{\sqrt{10}}{4}$

We have discussed three conditions that expressions containing square roots must meet in order to be in simplest form. They are:

1. The radicand cannot contain any perfect square factor other than 1.
2. The radicand cannot contain a fraction.
3. The denominator of a fraction cannot contain a square root.

Example 12   Simplify each of the following expressions so that they meet all three conditions.

a. $\sqrt{12x^3y^4} = \sqrt{12}\sqrt{x^3}\sqrt{y^4}$

$\qquad\qquad\; = \sqrt{4}\sqrt{3}\sqrt{x^2}\sqrt{x}\sqrt{y^4}$

$\qquad\qquad\; = 2xy^2\sqrt{3x}$

b. $\sqrt{\dfrac{18}{x^4}} = \dfrac{\sqrt{18}}{\sqrt{x^4}} = \dfrac{\sqrt{9}\sqrt{2}}{\sqrt{x^4}} = \dfrac{3\sqrt{2}}{x^2}$

c. $\dfrac{\sqrt{2}}{\sqrt{10}} = \dfrac{\sqrt{2} \cdot \sqrt{10}}{\sqrt{10} \cdot \sqrt{10}} = \dfrac{\sqrt{20}}{\sqrt{100}} = \dfrac{\sqrt{4}\sqrt{5}}{\sqrt{100}} = \dfrac{\overset{}{2}\sqrt{5}}{\underset{5}{\cancel{10}}} = \dfrac{\sqrt{5}}{5}$

d. $\sqrt{\dfrac{2}{3}} = \dfrac{\sqrt{2}}{\sqrt{3}} = \dfrac{\sqrt{2} \cdot \sqrt{3}}{\sqrt{3} \cdot \sqrt{3}} = \dfrac{\sqrt{6}}{\sqrt{9}} = \dfrac{\sqrt{6}}{3}$

We will next consider a procedure for adding and subtracting square roots.

RULE FOR ADDING OR SUBTRACTING SQUARE ROOTS OF POSITIVE NUMBERS	*EXAMPLE:* $\sqrt{18} + \sqrt{50} - \sqrt{8} = ?$
1. Simplify each radical.	$\sqrt{18} + \sqrt{50} - \sqrt{8} = \sqrt{9}\sqrt{2} + \sqrt{25}\sqrt{2} - \sqrt{4}\sqrt{2}$
	$\qquad\qquad\qquad\quad = 3\sqrt{2} + 5\sqrt{2} - 2\sqrt{2}$
2. Using the distributive property, add the coefficients of the like square roots.	$\qquad\qquad\qquad\quad = (3 + 5 - 2)\sqrt{2}$
	$\qquad\qquad\qquad\quad = 6\sqrt{2}$

Example 13   Add the following.

a. $2\sqrt{5} + 7\sqrt{5} = (2 + 7)\sqrt{5} = 9\sqrt{5}$

b. $13\sqrt{2} - \sqrt{2} = (13 - 1)\sqrt{2} = 12\sqrt{2}$

c. $4\sqrt{10} - 5\sqrt{10} = (4 - 5)\sqrt{10} = -1\sqrt{10} = -\sqrt{10}$

d. $3\sqrt{10} + 4\sqrt{5}$ cannot be added. The square roots are in simplest form but they are not alike. The radicands must be the same in like square roots.

e. $\sqrt{48} + \sqrt{27} = \sqrt{16}\sqrt{3} + \sqrt{9}\sqrt{3}$

$= 4\sqrt{3} + 3\sqrt{3}$

$= 7\sqrt{3}$

f. $\sqrt{80} - \sqrt{45} + \sqrt{5} = \sqrt{16}\sqrt{5} - \sqrt{9}\sqrt{5} + \sqrt{5}$

$= 4\sqrt{5} - 3\sqrt{5} + \sqrt{5}$

$= (4 - 3 + 1)\sqrt{5}$

$= 2\sqrt{5}$

g. $\sqrt{\tfrac{1}{2}} + \sqrt{2} = \tfrac{1}{2}\sqrt{2} + \sqrt{2}$ Note that $\sqrt{\tfrac{1}{2}} = \dfrac{\sqrt{1}}{\sqrt{2}}$

$= (\tfrac{1}{2} + 1)\sqrt{2}$ $= \dfrac{1}{\sqrt{2}}$

$= \dfrac{3}{2}\sqrt{2}$ $= \dfrac{1 \cdot \sqrt{2}}{\sqrt{2} \cdot \sqrt{2}}$

$= \dfrac{\sqrt{2}}{\sqrt{4}}$

$= \dfrac{\sqrt{2}}{2}$

$= \tfrac{1}{2}\sqrt{2}$

ODDS

18.1 Exercises

Evaluate each of the square roots in Exercises 1–30.

1. $\sqrt{36}$
2. $-\sqrt{4}$
3. $\sqrt{9}$
4. $-\sqrt{25}$
5. $-\sqrt{1}$
6. $\sqrt{0}$
7. $\sqrt{100}$
8. $\sqrt{81}$
9. $-\sqrt{144}$
10. $-\sqrt{169}$
11. $\sqrt{a^2}$
12. $-\sqrt{b^6}$
13. $\sqrt{y^{40}}$
14. $\sqrt{m^{70}}$
15. $\sqrt{x^{24}}$
16. $-\sqrt{c^{22}}$
17. $-\sqrt{u^{14}}$
18. $\sqrt{y^{36}}$
19. $\sqrt{n^4}$
20. $\sqrt{x^{10}}$
21. $\sqrt{4x^2}$
22. $\sqrt{x^2y^2}$
23. $\sqrt{16a^8}$
24. $\sqrt{144a^{50}}$
25. $\sqrt{25y^4}$
26. $\sqrt{49b^{18}}$
27. $\sqrt{100a^2}$
28. $-\sqrt{64x^4y^6}$
29. $\sqrt{81y^{10}}$
30. $\sqrt{36a^2}$

50. $\sqrt{60}$
51. $\sqrt{x^3}$
52. $\sqrt{a^{11}}$
53. $\sqrt{a^9}$
54. $\sqrt{c^3}$
55. $\sqrt{b^5}$
56. $\sqrt{d^{19}}$
57. $\sqrt{x^7}$
58. $\sqrt{u^5}$
59. $\sqrt{y^{15}}$
60. $\sqrt{b^{13}}$
61. $\sqrt{4x^3}$
62. $\sqrt{25x^7}$
63. $\sqrt{5a^2}$
64. $\sqrt{17y^4}$
65. $\sqrt{8x^5}$
66. $\sqrt{12n^3}$
67. $\sqrt{16x^2y^3}$
68. $\sqrt{36x^5y^6}$
69. $\sqrt{24xy^2}$
70. $\sqrt{44a^2b}$
71. $\sqrt{360}$
72. $\sqrt{135}$
73. $\sqrt{600}$
74. $\sqrt{294}$
75. $\sqrt{480}$
76. $\sqrt{405}$
77. $\sqrt{270x^2y^4}$
78. $\sqrt{192x^2y^2}$
79. $\sqrt{a^{15}b^{20}c}$
80. $\sqrt{400x^{10}y^{11}}$

Find the products in Exercises 31–40 and simplify.

31. $\sqrt{10} \cdot \sqrt{10}$
32. $\sqrt{5} \cdot \sqrt{5}$
33. $\sqrt{x} \cdot \sqrt{x}$
34. $\sqrt{a^5} \cdot \sqrt{a^5}$
35. $\sqrt{3a} \cdot \sqrt{3a}$
36. $\sqrt{x^3} \cdot \sqrt{x^5}$
37. $\sqrt{x} \cdot \sqrt{x^3}$
38. $\sqrt{a} \cdot \sqrt{a}$
39. $\sqrt{2x}\sqrt{2x^5}$
40. $\sqrt{3b^5} \cdot \sqrt{3b^5}$

Find the quotients in Exercises 81–90 and simplify.

81. $\dfrac{\sqrt{24}}{\sqrt{6}}$
82. $\dfrac{\sqrt{128}}{\sqrt{2}}$
83. $\dfrac{\sqrt{110}}{\sqrt{10}}$
84. $\dfrac{\sqrt{44}}{\sqrt{11}}$
85. $\dfrac{\sqrt{39}}{\sqrt{3}}$
86. $\dfrac{\sqrt{154}}{\sqrt{14}}$
87. $\dfrac{\sqrt{102}}{\sqrt{51}}$
88. $\dfrac{\sqrt{48}}{\sqrt{3}}$
89. $\dfrac{\sqrt{5x^3}}{\sqrt{x^3}}$
90. $\dfrac{\sqrt{6x^5y^3}}{\sqrt{x^4y}}$

Simplify each of the square roots in Exercises 41–80.

41. $\sqrt{8}$
42. $\sqrt{54}$
43. $\sqrt{50}$
44. $\sqrt{90}$
45. $\sqrt{32}$
46. $\sqrt{72}$
47. $\sqrt{27}$
48. $\sqrt{48}$
49. $\sqrt{98}$

Use the property $\sqrt{\dfrac{a}{b}} = \dfrac{\sqrt{a}}{\sqrt{b}}$ to express each of the radicals in Exercises 91–100 in simplest form.

91. $\sqrt{\dfrac{4}{9}}$ **92.** $\sqrt{\dfrac{25}{4}}$ **93.** $\sqrt{\dfrac{3}{64}}$

94. $\sqrt{\dfrac{5}{121}}$ **95.** $\sqrt{\dfrac{121}{100}}$ **96.** $\sqrt{\dfrac{a^2}{144}}$

97. $\sqrt{\dfrac{10}{81}}$ **98.** $\sqrt{\dfrac{31}{100}}$ **99.** $\sqrt{\dfrac{4}{x^2}}$

100. $\sqrt{\dfrac{4y^6}{49}}$

Rationalize the denominator in each of Exercises 101–120. Express the answer in simplest radical form.

101. $\dfrac{3}{\sqrt{2}}$ **102.** $\dfrac{1}{\sqrt{10}}$ **103.** $\dfrac{\sqrt{5}}{\sqrt{3}}$

104. $\dfrac{8}{\sqrt{6}}$ **105.** $\dfrac{1}{\sqrt{7}}$ **106.** $\dfrac{\sqrt{10}}{\sqrt{11}}$

107. $\dfrac{\sqrt{7}}{\sqrt{5}}$ **108.** $\dfrac{1}{\sqrt{21}}$ **109.** $\dfrac{1}{\sqrt{15}}$

110. $\dfrac{\sqrt{5}}{\sqrt{7}}$ **111.** $\dfrac{4}{\sqrt{x}}$ **112.** $\dfrac{1}{\sqrt{y}}$

113. $\dfrac{\sqrt{3}}{\sqrt{y}}$ **114.** $\dfrac{\sqrt{2x}}{\sqrt{3}}$ **115.** $\dfrac{1}{\sqrt{a}}$

116. $\dfrac{\sqrt{3}}{\sqrt{2x}}$ **117.** $\dfrac{\sqrt{3}}{\sqrt{6x}}$ **118.** $\dfrac{3}{\sqrt{ab}}$

119. $\dfrac{4}{\sqrt{3x^5}}$ **120.** $\dfrac{\sqrt{a}}{\sqrt{4a^3}}$

Find the sums and differences in Exercises 121–140.

121. $5\sqrt{2} + 3\sqrt{2}$ **122.** $-4\sqrt{10} + 5\sqrt{10}$

123. $6\sqrt{7} - 3\sqrt{7} + \sqrt{7}$ **124.** $5\sqrt{5} + 5\sqrt{5} - \sqrt{5}$

125. $-8\sqrt{19} + 7\sqrt{19}$

126. $8\sqrt{10} - 4\sqrt{10} - 5\sqrt{10}$

127. $\dfrac{1}{2}\sqrt{5} + \dfrac{5}{2}\sqrt{5}$ **128.** $7\sqrt{2} - 7\sqrt{2}$

129. $5\sqrt{x} + 3\sqrt{x} - 7\sqrt{x}$ **130.** $x\sqrt{3} + 2x\sqrt{3}$

131. $\sqrt{3} + \sqrt{12}$ **132.** $\sqrt{50} + \sqrt{5} - \sqrt{80}$

133. $\sqrt{18} + \sqrt{98} - 5\sqrt{2}$ **134.** $\sqrt{\dfrac{1}{3}} + \sqrt{27}$

135. $\sqrt{\dfrac{1}{2}} + \sqrt{18}$ **136.** $\sqrt{48} - \sqrt{27} - \sqrt{12}$

137. $\sqrt{50} + \sqrt{32} - \sqrt{2}$ **138.** $\sqrt{32} + \sqrt{12} - \sqrt{48}$

139. $3x\sqrt{x} - 2\sqrt{x^3}$ **140.** $\sqrt{a^3} - 2\sqrt{4a^3}$

Simplify each of the expressions in Exercises 141–150.

141. $\sqrt{486}$ **142.** $\dfrac{\sqrt{x}}{\sqrt{2x^3}}$

143. $\dfrac{\sqrt{5}}{\sqrt{x}}$ **144.** $\sqrt{\dfrac{1}{6a}}$

145. $\sqrt{\dfrac{3x}{y}}$ **146.** $\sqrt{180xy^2z}$

147. $\dfrac{1}{\sqrt{y^3}}$ **148.** $\sqrt{175}$

149. $\sqrt{175a^{10}b^7c^3}$ **150.** $\dfrac{1}{\sqrt{2a^5}}$

18.2   Solving Quadratic Equations by Factoring

A second-degree equation in one variable is called a **quadratic equation.** Quadratic equations can be written in the form $ax^2 + bx + c = 0$ where a, b, and c represent constants and $a \neq 0$. This is the **standard form** for a quadratic equation.

To solve a quadratic equation, you must find the value(s) of the variable which will make the original equation a true statement. Quadratic equations will have either one double root or two distinct roots.

Many quadratic equations can be solved using the techniques of factoring presented in Chapter 16. When a quadratic equation is in standard form, the right member of the equation is 0. If the left member will factor, we have the product of two or more polynomials equal to zero. We may use the zero factor property of the real numbers to solve the equation.

Zero Factor Property

If the product of two numbers is 0, at least one of the numbers must be 0. *In formula form:* If a and b are real numbers, ab = 0 if and only if a = 0 or b = 0.

Since polynomials represent real numbers, we can also use the zero factor property with them. Consider how the property is used in solving the quadratic equation in the following example.

Example 1   Solve $x^2 - 4x + 3 = 0$.

$$x^2 - 4x + 3 = 0$$
$$(x - 3)(x - 1) = 0 \qquad \text{since the left member factors}$$
$$x - 3 = 0 \text{ or } x - 1 = 0 \quad \text{from the zero factor property}$$
$$x = 3 \text{ or } x = 1 \qquad \text{solving each equation for x}$$

Note that both of these solutions check.

Check:

$x^2 - 4x + 3 = 0$	$x^2 - 4x + 3 = 0$
$(3)^2 - 4(3) + 3 = 0$	$(1)^2 - 4(1) + 3 = 0$
$9 - 12 + 3 = 0$	$1 - 4 + 3 = 0$
$0 = 0$	$0 = 0$

Thus, 3 and 1 are roots of $x^2 - 4x + 3 = 0$.

The method is summarized in the *procedure for solving quadratic equations by factoring*.

PROCEDURE FOR SOLVING QUADRATIC EQUATIONS BY FACTORING	EXAMPLE Solve $x^2 - 6x = -8$.
	Solution:
1. Write the equation in standard form.	$x^2 - 6x + 8 = 0$
2. Factor the left member.	$(x - 4)(x - 2) = 0$
3. Set each factor equal to zero.	$x - 4 = 0 \quad x - 2 = 0$
4. Solve the resulting first-degree equations.	$x - 4 = 0 \quad x - 2 = 0$
	$x = 4 \qquad x = 2$
5. The solution set for the quadratic equation consists of the values found in step 4.	The solutions are 4 and 2.

Example 2   Solve $2x^2 + x - 15 = 0$.

$$2x^2 + x - 15 = 0$$
$$(2x - 5)(x + 3) = 0$$
$$2x - 5 = 0 \text{ or } x + 3 = 0$$
$$2x = 5 \qquad x = -3$$
$$x = \frac{5}{2}$$

Check: $\quad 2\left(\dfrac{5}{2}\right)^2 + \left(\dfrac{5}{2}\right) - 15 = 0 \qquad 2(-3)^2 + (-3) - 15 = 0$

$$\dfrac{25}{2} + \dfrac{5}{2} - 15 = 0 \qquad\qquad 18 - 3 - 15 = 0$$

$$15 - 15 = 0 \qquad\qquad\qquad\quad 0 = 0$$

$$0 = 0$$

Thus the solutions are $\dfrac{5}{2}$ and -3.

Example 3   Solve $x^2 - 10x = 0$.

$$x^2 - 10x = 0$$
$$x(x - 10) = 0$$
$$x = 0 \text{ or } x - 10 = 0$$
$$x = 10$$

Check: $\quad (0)^2 - 10(0) = 0$
$$0 = 0$$
$$(10)^2 - 10(10) = 0$$
$$100 - 100 = 0$$
$$0 = 0$$

Thus, the solutions are 0 and 10.

18.2   Exercises

Which of the equations in Exercises 1–5 are quadratic equations?

1. $5x^2 - 7x - 19 = 0$

2. $7x = 19$

3. $13x^3 - 7x^2 + 5x = 11$

4. $12 = 9x^2 - 4x$

5. $-7x^2 = 5x + 4$

Rewrite each of the quadratic equations in Exercises 6–10 in standard form.

6. $4x^2 - 8x = 5$

7. $2x^2 = 4x - 8$

8. $x^2 - 5x = 6$

9. $0 = x^2 - 3x + 2$

10. $2x - 15 = x^2$

Solve the equations in Exercises 11–30 by factoring. Check all solutions.

11. $x^2 + 3x + 2 = 0$

12. $x^2 + 5x + 4 = 0$

13. $x^2 - 3x - 18 = 0$

14. $x^2 + 4x - 12 = 0$

15. $x^2 + 3x = 0$

16. $x^2 - 5x = 0$

17. $2x^2 - 7x - 15 = 0$

18. $3x^2 - 17x - 6 = 0$

19. $6x^2 + 19x + 10 = 0$

20. $10x^2 - 17x + 3 = 0$

21. $x^2 - 16 = 0$

22. $4x^2 - 25 = 0$

23. $3x^2 - 75 = 0$

24. $2x^2 - 72 = 0$

25. $x^2 + 13x = -42$

26. $x^2 - 15x = -54$

27. $14x^2 - 3 - x = 0$

28. $35x^2 - 18 + 9x = 0$

29. $121x^2 = 81$

30. $64x^2 = 9$

18.3 The Quadratic Formula

If a quadratic equation is written in standard form

$$ax^2 + bx + c = 0, \quad a \neq 0$$

the solution(s) can be represented by the quadratic formula:

$$x = \frac{-b \pm \sqrt{b^2 - 4ac}}{2a}$$

In the quadratic formula **a** is the coefficient of the second-degree term, **b** is the coefficient of the first-degree term, and **c** is the constant, when the equation is in standard form. The symbol ±, read "plus or minus," is used in the formula to give the following two roots:

$$x = \frac{-b + \sqrt{b^2 - 4ac}}{2a} \quad \text{and} \quad x = \frac{-b - \sqrt{b^2 - 4ac}}{2a}$$

Notice that if the radicand $b^2 - 4ac$ is equal to 0, the two roots will be equal. When this occurs, the equation is said to have a **double root**. If the radicand is positive, the equation will have two distinct real roots; if the radicand is negative, the roots will not be real numbers.

Note also that the quadratic formula requires that a ≠ 0. If **a** were equal to zero, then we would not have a quadratic equation. The resulting equation would be a first-degree equation. The formula if applied would result in an undefined expression.

PROCEDURE FOR SOLVING QUADRATIC EQUATIONS USING THE QUADRATIC FORMULA	*EXAMPLE* Solve $x^2 + 2x = 2$. *Solution:*
1. Write the equation in standard form.	$x^2 + 2x - 2 = 0$
2. Identify the values of a, b, and c for the equation given.	$a = 1 \quad b = +2 \quad c = -2$
3. Substitute the values of a, b, and c in the quadratic formula.	$x = \dfrac{-2 \pm \sqrt{(+2)^2 - 4(1)(-2)}}{2(1)}$ $x = \dfrac{-2 \pm \sqrt{12}}{2}$
4. Simplify the radicand if necessary.	$x = \dfrac{-2 \pm 2\sqrt{3}}{2}$
5. Express the root(s) in simplest form.	$x = -1 + \sqrt{3}$ or $x = -1 - \sqrt{3}$

Example 1   Use the quadratic formula to solve $x^2 - 2x - 3 = 0$.

We use the values a = 1, b = −2, c = −3.

$$x = \frac{-(-2) \pm \sqrt{(-2)^2 - 4(1)(-3)}}{2(1)}$$

$$x = \frac{2 \pm \sqrt{16}}{2}$$

$$x = \frac{2 \pm 4}{2}$$

$$x = \frac{2+4}{2} \text{ or } \frac{2-4}{2}$$

$$x = \frac{6}{2} \qquad x = \frac{-2}{2}$$

$$x = 3 \qquad \text{or } x = -1$$

Notice that the equation in Example 1 could have been solved by factoring. The quadratic formula can be used to solve any quadratic equation, but factoring is preferable when possible since this method is usually easier.

Example 2   Use the quadratic formula to solve $x^2 - 8 = 0$.

We use the values $a = 1$, $b = 0$, $c = -8$.

$$x = \frac{0 \pm \sqrt{0^2 - 4(1)(-8)}}{2(1)}$$

$$x = \frac{\pm\sqrt{32}}{2}$$

$$x = \frac{\pm 4\sqrt{2}}{2} \qquad (\text{Note that } \sqrt{32} = \sqrt{16} \cdot \sqrt{2} = 4\sqrt{2}.)$$

$$x = 2\sqrt{2} \text{ or } x = -2\sqrt{2}$$

Example 3   Use the quadratic formula to solve $x^2 - 2x + 2 = 0$.

We use the values $a = 1$, $b = -2$, $c = 2$.

$$x = \frac{-(-2) \pm \sqrt{(-2)^2 - 4(1)(2)}}{2}$$

$$x = \frac{2 \pm \sqrt{-4}}{2}$$

Since $\sqrt{-4}$ is not a real number, this equation has two roots which are not real numbers.*

Example 4   Use the quadratic formula to solve $x^2 - 6x = -9$.

Written in standard form, $x^2 - 6x = -9$ becomes $x^2 - 6x + 9 = 0$. We use the values $a = 1$, $b = -6$, $c = 9$.

$$x = \frac{-(-6) \pm \sqrt{(-6)^2 - 4(1)(9)}}{2(1)}$$

$$x = \frac{6 \pm \sqrt{0}}{2}$$

$$x = 3 + 0 \qquad x = 3 - 0$$

$$x = 3 \qquad \quad x = 3$$

This quadratic equation is said to have a double root, since the two roots are equal.

*The nonreal roots of quadratic equations are complex numbers.

Example 5 Solve $x^2 = 7$. (Use Table C.3 to approximate the root(s) correct to one decimal place.)

If $x^2 = 7$, then $x^2 - 7 = 0$. We use the values $a = 1$, $b = 0$, $c = -7$.

$$x = \frac{0 \pm \sqrt{0 + 28}}{2}$$

$$x = \frac{\pm\sqrt{28}}{2}$$

$$x = \frac{\pm 2\sqrt{7}}{2}$$

$$x = \pm\sqrt{7}$$

$$x \doteq 2.6 \text{ or } x \doteq -2.6$$

Equations like the one in Example 5 are called **pure quadratics**. They can be solved by using the quadratic formula, or they can be solved by taking the square root of both members. Since any positive real number has both a positive and a negative square root, this procedure yields the two solutions of the equation. As Example 6 illustrates, taking the square root of both members is a much shorter procedure than using the formula.

Example 6 **a.** Solve $x^2 = 16$.

Solution: $x^2 = 16$

$x = \pm 4$

b. Solve $x^2 = 10$.

Solution: $x^2 = 10$

$x = \pm\sqrt{10}$

c. Solve $x^2 = 28$.

Solution: $x^2 = 28$

$x = \pm\sqrt{28}$

$x = \pm 2\sqrt{7}$

18.3 Exercises

Write each of the equations in Exercises 1–22 in standard form, and identify the values that would be used in the quadratic formula for a, b, and c.

1. $x^2 - 5x - 6 = 0$
2. $-2x^2 + 10x + 12 = 0$
3. $7x^2 - 2x + 3 = 0$
4. $x^2 = 10$
5. $x^2 + 2x - 8 = 0$
6. $x^2 - 5x - 14 = 0$
7. $x^2 + x - 1 = 0$
8. $2x^2 - 2x - 1 = 0$
9. $x^2 - 6x + 9 = 0$
10. $4x^2 + 12x + 9 = 0$
11. $x^2 - 2x = 0$
12. $3x^2 + 5x = 0$
13. $6x^2 = 5$
14. $5x^2 = -8$
15. $10x^2 + 3x - 1 = 0$
16. $7x^2 + 5x - 2 = 0$
17. $x^2 + 3x - 2 = 0$
18. $2x^2 - x - 1 = 0$
19. $x^2 - 14x + 49 = 0$
20. $x^2 + 12x + 36 = 0$
21. $x^2 = 3x$
22. $3x^2 = -5x$

Use the quadratic formula to solve the quadratic equations in Exercises 23-44. Indicate for each equation whether the roots are real or nonreal. Simplify all real roots.

23. $x^2 - 5x - 6 = 0$

24. $-2x^2 + 10x + 12 = 0$

25. $7x^2 - 2x + 3 = 0$

26. $x^2 = 10$

27. $x^2 = 8$

28. $x^2 - 5x - 14 = 0$

29. $x^2 + x - 1 = 0$

30. $2x^2 - 2x - 1 = 0$

31. $x^2 - 6x + 9 = 0$

32. $4x^2 + 12x + 9 = 0$

33. $x^2 - 2x = 0$

34. $3x^2 + 5x = 0$

35. $6x^2 = 5$

36. $5x^2 = -8$

37. $10x^2 + 3x - 1 = 0$

38. $7x^2 + 5x - 2 = 0$

39. $x^2 + 3x - 2 = 0$

40. $2x^2 - x - 1 = 0$

41. $x^2 - 14x + 49 = 0$

42. $x^2 + 12x + 36 = 0$

43. $x^2 = 36$

44. $3x^2 = +5x$

Solve the equations in Exercises 45-50. Express your solutions in simplest form unless otherwise indicated. (Use either method.)

45. $4x^2 = 16$

46. $3x^2 - 5x = -7$

47. $x^2 + 6x + 8 = 0$

48. $x^2 + 3x - 3 = 0$ (Give the rational approximation of the roots correct to tenths.)

49. $3x^2 + 5x - 2 = 0$

50. $3x^2 + 2 = 7x$

18.4   Solving Problems Involving Quadratic Equations

Word problems often result in quadratic equations rather than linear equations like those in Chapter 11. This is especially true of problems requiring a product of two unknowns. For example, in finding the area of a rectangle you must find the product of the length and the width. To solve these problems, you may need to solve a quadratic equation. Equations involving rational expressions may also result in a quadratic equation. In either case, the original problem may not look like a quadratic, but in the course of setting up the problem and solving it, a quadratic equation results.

Example 1   Use an equation to find two consecutive positive integers whose product is 110.

Let x = first positive integer and let x + 1 = next consecutive integer.

$$x(x + 1) = 110$$
$$x^2 + x - 110 = 0$$
$$(x + 11)(x - 10) = 0$$
$$x = -11 \text{ or } x = 10$$

Note that this equation could be solved using the formula, but factoring is easier. Thus,

$$x = 10 \quad \text{(first integer)}$$
$$x + 1 = 11 \quad \text{(second integer)}$$

Since the problem asks for consecutive *positive* integers, one of the solutions, x = −11, does not meet the conditions of the problem.

Example 2   The length of a rectangle is twice its width. If the area of the rectangle is 288 square feet, find the length and the width.

Solution: Let x = the width of the rectangle and let 2x = the length of the rectangle.

$$2x \cdot x = 288$$
$$2x^2 = 288$$
$$x^2 = 144$$
$$x = \pm 12$$

Since we are asked to find lengths, only the positive value meets the conditions of the problem. So,

$$x = 12 \quad \text{(width)}$$
$$2x = 24 \quad \text{(length)}$$

The rectangle is 12 ft by 24 ft.

In Chapters 10 and 17 we solved equations containing fractions. The same procedure is used to solve the equation in Example 3. Both members are multiplied by the LCD of the fractions in the equation. However, in Example 3 the equation which results is a quadratic equation.

Example 3   Solve the equation $\dfrac{3}{x+1} + 2 = \dfrac{x}{x-1}$.

$$3(x-1) + 2(x+1)(x-1) = x(x+1)$$
$$3x - 3 + 2x^2 - 2 = x^2 + x$$
$$2x^2 + 3x - 5 = x^2 + x$$
$$x^2 + 2x - 5 = 0$$
$$x = \frac{-2 \pm \sqrt{4 + 20}}{2}$$
$$x = \frac{-2 \pm \sqrt{24}}{2}$$
$$x = \frac{-2 \pm 2\sqrt{6}}{2}$$
$$x = -1 \pm \sqrt{6}$$

Note that the excluded values are x = 1 and x = −1.

18.4 Exercises

Solve Exercises 1–20. Use equations to solve the word problems.

1. One number is three more than another number. If their product is 130, find the numbers.

2. The length of a rectangle is three times the width. If the area is 192 square centimeters, find the length and the width.

3. The perimeter of a rectangle is 94 feet, and its area is 496 square feet. What are its dimensions?

4. The product of two consecutive even integers is 168. Find the two integers.

5. A skating rink is 30 meters long and 20 meters wide. The owners plan to increase its area to 875 square meters by adding a strip of equal width to one side and one end, maintaining a rectangular shape. How wide should the strip be?

6. One number is 12 more than another. If their product is 288, find the two numbers.

7. The product of two consecutive positive integers is 182. Find the two integers.

8. An open box is constructed from a rectangular sheet of metal 8 inches longer than it is wide as follows: out of each corner, a square of side 2 inches is cut, and the sides are folded up. The volume of the resulting box is 256 cubic inches. What are the dimensions of the original sheet of metal?

9. If the sides of a square are increased by eight centimeters, the area of the square is increased by 160 square centimeters. What is the length of the original square?

10. $\dfrac{x}{x-3} = \dfrac{4}{x+1} - 4$

11. $\dfrac{3}{x-4} + 1 = \dfrac{2}{x+1} - 1$

12. $\dfrac{3}{x^2-2} = \dfrac{4}{x^2+1}$

13. $\dfrac{2x^2}{x-1} - 2 = x$

14. $\dfrac{12}{x+3} - 1 = \dfrac{2}{x-1}$

15. $\dfrac{2}{x-3} + 5 = \dfrac{-9}{x-5}$

16. $\dfrac{3}{x^2-1} = \dfrac{6}{x+3}$

17. $\dfrac{3}{2x^2-5} = \dfrac{6}{x^2}$

18. $\dfrac{5}{x+2} - 3 = \dfrac{-5}{x-6}$

*19. Find two consecutive even integers whose product is 143.

*20. Find two consecutive integers whose product is 88.

*These exercises are optional.

Important Terms Used in this Chapter	quadratic equation quadratic formula	rational approximation square root standard form

REVIEW EXERCISES

Perform the indicated operations in Exercises 1–5 and express the result in simplest form.

1. $9\sqrt{7} - \sqrt{7} + 3\sqrt{7}$

2. $\dfrac{\sqrt{30}}{\sqrt{10}}$

3. $\sqrt{10} \cdot \sqrt{30}$

4. $\dfrac{\sqrt{10} \cdot \sqrt{30}}{\sqrt{2}}$

5. $\sqrt{20} - \sqrt{45} - \sqrt{180}$

Express each of Exercises 6–10 in simplest form.

6. $\sqrt{88}$

7. $\sqrt{\dfrac{5a}{b}}$

8. $\sqrt{25a^2b^4}$

9. $\dfrac{1}{\sqrt{y^3}}$

10. $\sqrt{124x^2y^3}$

Which of the equations in Exercises 11–16 are quadratic equations?

11. $x^2 - 2x = 4$ 12. $3x + 4 = 0$

13. $x^3 + 2x^2 - x = 0$ 14. $x^2 + 3x = 0$

15. $5x = 25$ 16. $x^2 + 3x - 4$

Write the quadratic equations in Exercises 17–22 in standard form and identify the values that would be used for a, b, and c in the quadratic formula.

17. $2x^2 - 4x + 6 = 0$ 18. $3x^2 = 27$

19. $x^2 - 25 = 0$ 20. $x^2 + 3x = 0$

21. $x^2 + 6 = 3x$ 22. $x^2 = -3x + 2$

Solve Exercises 23–26 by factoring. Check the solutions.

23. $x^2 - 4x = 12$ 24. $x^2 = 81$

25. $3x^2 + 11x - 4 = 0$ 26. $5x^2 - 13x = 6$

27. Solve the following by using the quadratic formula. Express the solution(s) in simplest form.
$$x^2 + 3x - 9 = 0$$

28. Solve the following by using the quadratic formula. Give the rational approximation correct to tenths for the solution(s).
$$2x^2 - 3x - 6 = 0$$

Solve Exercises 29–35. Express the answers in simplest form.

29. $6x^2 - 2 = 3x$ 30. $x^2 + 3x - 6 = 0$

31. $x^2 = 50$ 32. $2x^2 - 3x + 1 = 0$

33. $x^2 + 10x + 16 = 0$

34. $\dfrac{x+1}{x-3} = \dfrac{45}{x+5}$

35. $\dfrac{4}{x+1} - 1 = \dfrac{-2}{x-3}$

Use equations to solve Exercises 36–41.

36. One number is 5 less than another. If the product of the two numbers is five times the larger number, find the two numbers.

37. Ten more than twice the product of two consecutive positive even integers is 250. Find the two integers.

38. The length of the recreation hall was 15 meters more than the width. If the area is 700 square meters, find the length and the width.

39. The product of two consecutive odd integers is 45 less than 300. Find the integers.

40. The sum of two numbers is 20 and their product is 96. Find the numbers.

41. The perimeter of a rectangle is 68 feet. If the area is 280 square feet, find the length and the width. (*Hint:* The length plus the width is 34.)

APPENDICES

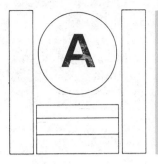

SETS AND THE REAL NUMBERS

OBJECTIVES

Upon completion of Appendix A you should be able to:

1. Correctly interpret and use set notation to denote sets and membership in a set.   A.1, A.5
2. Correctly interpret and use subset notation.   A.2
3. Write rational numbers (excluding repeating decimals) as fractions. A.3
4. Recognize that division by zero is undefined.   A.3
5. Correctly interpret and use set notation for the union of two sets. A.4
6. Correctly interpret and use set notation for the intersection of two sets.   A.4
7. Recognize the relationship between the set of real numbers and each of the following subsets: natural numbers, whole numbers, integers, rational numbers, and irrational numbers.   A.5

PRETEST

1. Write the symbolic representation of the following verbal expressions (Objective 1).
   a. The set A contains the elements 4, 17, and 35.
   b. 5 is not an element of the set S.
   c. Set A equals set C.
   d. 1 is an element of S, and 2 is an element of S.

2. Write the verbal expression associated with the following symbolic forms (Objective 2).
   a. $\{7\} \subseteq \{1, 3, 5, 7, 9, 11, \ldots\}$
   b. $\{8\} \nsubseteq \{38\}$
   c. $\{9, 3\} \subseteq \{1, 3, 9\}$

3. Rewrite the following rational numbers in fractional form (Objective 3).
   a. 0.25
   b. −18
   c. $4\frac{1}{3}$
   d. −5.07
   e. $-3\frac{1}{2}$
   f. 0

4. Answer the following true (T) or false (F) (Objective 4).
   a. $0 \div 3 = 0$
   b. $0 \div 3 = 3$
   c. $3 \div 0 = 0$
   d. $3 \div 0 = 1$
   e. $0 \div 0 = 1$

5. Complete each of the following (Objective 5).
   a. $\{2, 3, 5, 6, 8\} \cup \{2, 35, 7, 8\} = ?$
   b. $\{1, 2, 3\} \cup \{1, 2, 3, 4\} = ?$

6. Complete each of the following (Objective 6).
   a. $\{2, 3, 5, 6, 8\} \cap \{2, 35, 7, 8\} = ?$
   b. $\{1, 2, 3\} \cap \{1, 2, 3, 4\} = ?$
   c. $\{4, 7\} \cap \{5, 6\} = ?$

7. Answer the following true (T) or false (F) (Objective 7).
   Let
   N = the natural numbers
   W = the whole numbers
   I = the integers
   Q = the rationals
   T = the irrationals
   R = the reals

   a. $\frac{2}{3} \in Q$
   b. $N \subseteq W$
   c. $W \subseteq N$
   d. $T \nsubseteq Q$
   e. $0 \notin N$
   f. $0 \in W$
   g. $Q \cup T = R$
   h. $Q \cap T = 0$
   i. $\varnothing = Q \cap T$
   j. $-85 \in I$

The task of Appendix A will be to present some of the terminology and notation associated with real numbers. The history of the notation or symbolism which we use today is often very interesting. In many cases, a pioneering mathematician invented a symbol to replace several other words or symbols—a type of mathematical shorthand. Through correspondence with other mathematicians this labor-saving notation would often become widespread until the mathematician of today accepts it as customary.

Appendix A will present the names of various collections of numbers and the terminology and symbolism used to describe these numbers.

A.1 Sets

You may already be familiar with the following sets:

$$N = \{1, 2, 3, 4, 5, \ldots\}, \text{ the set of natural numbers}$$
$$W = \{0, 1, 2, 3, 4, \ldots\}, \text{ the set of whole numbers}$$
$$E = \{0, 2, 4, 6, 8, \ldots\}, \text{ the set of even whole numbers}$$
$$O = \{1, 3, 5, 7, 9, \ldots\}, \text{ the set of odd whole numbers}$$
$$P = \{2, 3, 5, 7, 11, \ldots\}, \text{ the set of prime numbers}$$
$$C = \{4, 6, 8, 9, 10, \ldots\}, \text{ the set of composite numbers}$$

The term set as used above, refers to a collection or group of numbers. The term **set**, however, can refer to any collection or group of objects. For example, the set $\{a, b, c, \ldots, z\}$ is the set of letters in the English alphabet. In this book, our main purpose is building computational skills, and we shall deal mainly with sets of numbers which are well defined. By **well defined**, we mean that the numbers in the set are described clearly enough for you to determine whether a number is in the set. For example, "the set of natural numbers less than 10" is well defined because we know what numbers are in this set, but "the set of large numbers" is not well defined since it is not clear what a large number is.

To read this material, it will be necessary to memorize the meanings of various symbols. In particular, notice in the examples that

1. Braces are used to enclose the numbers (**elements** or **members**) of a set.
2. Capital letters are usually used to name sets.
3. Commas are used to separate elements in the set.
4. Three dots (the **ellipsis notation**) indicate that elements have been omitted in the listing or that the elements continue indefinitely.
5. The order in which the elements are listed is not significant.

Example 1   a. $A = \{1, 7, 29\}$ should be read, "The set A equals the set containing the elements 1, 7, and 29" or "The set A contains the elements 1, 7, and 29."

b. $B = \{1, 7, 2, 9\}$ The set B contains four elements: 1, 2, 7, and 9.

c. $S = \{1, 2, 3, 4, \ldots, 20\}$ The set S contains twenty elements, the natural numbers 1 through 20.

d. $X = \{9, 7, 2, 1\}$ X contains the same four elements as the set B mentioned in part **b**. Therefore, $B = X$. *Note:* Sets with exactly the same elements are **equal sets**.

e. E = {2, 4, 6, 8, . . .} The three dots indicate that the even numbers E go on indefinitely. This is one example of an infinite set. Sets whose elements cannot be counted in a way so that the counting process eventually terminates are called **infinite sets**. If the elements of a set can be counted in a way so that the counting process eventually terminates, then the set is called **finite**.

f. Sets A, B, S, and X, as defined above, are finite while N, W, E, O, and P are infinite.

To designate that an object is an element of a set, the symbol ∈ (read "is an element of") is used. To designate that an object is not an element of a set, the symbol ∉ (read "is not an element of") is used.

Example 2   a. 5 ∈ H is read "five is an element of the set H."

b. 8 ∉ S is read "eight is not an element of the set S."

c. If E = {2, 4, 6, 8, . . .}, then 4 ∈ E, 12 ∈ E, 5 ∉ E, and 17 ∉ E.

d. If N = {1, 2, 3, 4, 5, . . .}, then 0 ∉ N, $1\frac{1}{2}$ ∉ N, but 10 ∈ N.

A.1 Exercises

Give the *symbolic* representation of the verbal expressions in Exercises 1–8.

1. Six is an element of the set S.

2. Eleven is not an element of the set E.

3. Four is not a member of the set of prime numbers P.

4. The set B equals the set containing the elements 2, 7, and 3.

5. The set D contains elements 5, 15, and 1.

6. Seventeen is an element of the odd numbers O.

7. The set S equals the set of elements 4, 19, and 105.

8. The set K contains the elements 7, 33, 4, and 19.

Translate into words the symbolic representations in Exercises 9–16.

9. $\frac{1}{2}$ ∉ N

10. 34 ∈ N

11. 117 ∈ W

12. 23.45 ∉ N

13. S ≠ T

14. K = {1, 5, 25}

15. A = {2, 14, 33}

16. A = B

Given A = {1, 8, 6, 5, 4}, answer true (T) or false (F) in Exercises 17–26.

17. 5 ∉ A

18. A is finite.

19. A = B when B = {1, 4, 6, 8}.

20. 7 ∈ A

21. A is well defined.

22. 8 ∈ A

23. A is infinite.

24. 54 is a member of set A.

25. A contains the element 4.

26. 11 ∉ A

27. List all of the elements in the set F = {5, 10, 15, . . . , 40}.

28. Is the set of small numbers a well-defined set?

29. Is 35 ∈ T if T = {3, 6, 9, 12, . . .}?

30. Is the set of prime numbers P infinite?

31. Does N = W?

32. What whole number is not a natural number?

33. Is the set H = {2, 4, 8, 16, . . . , 256} well defined?

34. If K = {3, 6, 9, . . .}, what are the first six elements that would be listed in this set?

35. If A = {1, 3, 2} and B = {1, 32}, does A = B?

A.2 Subsets

You have probably noticed that every natural number is also a whole number. This relationship between these two sets is the subset relation. To denote that N is a subset of W, we can write N ⊆ W. If every element of one set is also a member of a second set, the first set is said to be a **subset** of the second set. If, in addition, the second set contains at least one element which is not in the first set, then the first set is said to be a **proper subset** of the second set. (Symbolically, $N \subset W$.) However, in this text (as is customary in elementary treatments) we will only consider the subset relation.

Example a. $P \subseteq N$ is read "the set P is a subset of the set N."

b. $X \nsubseteq Y$ is read "the set X is not a subset of the set Y."

c. Let $A = \{1, 4, 7\}$ and $B = \{1, 2, 3, \ldots, 70\}$. Then $A \subseteq B$ since every element in A is also an element of B. $B \nsubseteq A$ since 2 is an element of B but is not an element of A.

d. Let $A = \{1, 4, 7\}$ and $B = \{4, 5, 6\}$. Then $A \nsubseteq B$ and $B \nsubseteq A$. Why?

e. Let $A = \{1, 2\}$. Then $A \subseteq A$. (Every set is a subset of itself.)

A.2 Exercises

Write the *symbolic* representation for the verbal expressions in Exercises 1–4.

1. The set I is a subset of the set Q.

2. The set P is not a subset of the set O.

3. N is a subset of W.

4. W is not a subset of N.

Write the *verbal expression* associated with the symbolic forms in Exercises 5–9.

5. $I \subseteq R$ 6. $C \nsubseteq E$

7. $W \neq N$ 8. $O \nsubseteq E$

9. $E \subseteq N$

Answer Exercises 10–20 as either true (T) or false (F).

10. $\{2\} \subseteq \{1, 2, 3\}$ 11. $\{2\} \subseteq \{12, 3\}$

12. $\{1, 19, 23\} \subseteq \{5, 11, 17, 23\}$

13. The set of whole numbers is a subset of the natural numbers.

14. If $A = \{0, 10, 20\}$, then $A \subseteq A$.

15. $\{0\} \subseteq N$ 16. $\{0\} \subseteq W$

17. $\{1, 25\} \nsubseteq \{1, 2, 5\}$ 18. $N \subseteq W$

19. $\{1, 2, 3\} = \{3, 2, 1\}$

20. The set of prime numbers is a subset of the natural numbers.

A.3 The Integers and the Rationals

Every point on the number line is associated with a number. These numbers are called the **real numbers**, and this line is referred to as the **real number line**.

You have already worked with several important subsets of the real numbers: the natural numbers, the whole numbers, the primes, the composite numbers, and the even and odd natural numbers. There are other subsets of the real numbers with which you should become familiar. The set of integers and the set of rational numbers are two of these sets. In Section A.5 we shall discuss briefly the set of irrational numbers, which is another subset of the real numbers.

The set of **integers** (I) consists of the whole numbers and their opposites. That is, $I = \{\ldots, -3, -2, -1, 0, 1, 2, 3, \ldots\}$. The set $\{1, 2, 3, \ldots\}$ is called the set of positive integers and the set $\{-1, -2, -3, \ldots\}$ is called the set of negative integers. The graph of the set of integers is shown below:

A **rational number** is any number which can be written as the quotient (ratio) of two integers where the denominator is not 0. Thus the rational numbers include not only the real numbers which are already expressed as quotients (fractional form) but also the integers and those decimals and mixed numbers which can be written in fractional form.

Example   **a.** $\dfrac{3}{2}$ is a rational number since it is written as the quotient of the integers 3 and 2.

       b. 3 can be written as $\dfrac{3}{1}$ so it is a rational number.

       c. $-6 = \dfrac{-6}{1}$, so -6 is rational.

       d. 0, 0.25, 1.4, and $-3\dfrac{2}{3}$ are rational because they can be written respectively as $\dfrac{0}{1}$, $\dfrac{25}{100}$, $\dfrac{14}{10}$, and $\dfrac{-11}{3}$.

Caution: The denominator of a fraction representing a rational number cannot be 0. Consider $\dfrac{3}{0}$, i.e., $0\overline{)3}$. The quotient must be some number that we can multiply by 0 to obtain 3, or $? \cdot 0 = 3$. There is no such number, so we say that division by 0 is undefined. Thus, $\dfrac{5}{0}$, $\dfrac{17}{0}$, and $\dfrac{-13}{0}$ are all undefined. $\left(\dfrac{0}{0} \text{ is also undefined.}\right)$ Do not confuse this type of problem with fractions in which the numerator is 0 and the denominator is not zero. Notice that $\dfrac{0}{3}$, or $3\overline{)0}$, is 0 since $0 \cdot 3 = 0$.

The decimals and fractions with which you have performed calculations are called the nonnegative rational numbers (positive rationals and 0). The opposites of these numbers are also rational numbers.

A.3 Exercises

Rewrite each of the rational numbers in Exercises 1–15 as the quotient of two integers.

1. 0.5

2. 17

3. $-4\dfrac{1}{3}$

4. 0

5. −33

6. −0.079

7. 605

8. 0.32

9. $-0.33\dfrac{1}{3}$

10. $5\dfrac{3}{6}$

11. $-4\dfrac{7}{100}$

12. 4.07

13. $40\dfrac{7}{10}$

14. 407

15. −0.0407

Answer Exercises 16–50 as true (T) or false (F). For these exercises, let

> N = the natural numbers
> W = the whole numbers
> I = the integers
> Q = the rationals

16. $4 \in Q$

17. $-10 \in W$

18. $-\dfrac{4}{3} \subseteq Q$

19. I is an infinite set.

20. $10 \div 0$ is undefined.

21. $I \subseteq Q$

22. $Q \not\subseteq I$

23. $N \not\subseteq W$

24. $N \subseteq I$

25. $N \subseteq Q$

26. $\left\{ -3, \dfrac{2}{3}, \dfrac{4}{3} \right\} \subseteq Q$

27. $0.245 \in I$

28. $0 \div 4$ is undefined.

29. $W \subseteq N$

30. $W \subseteq Q$

31. $4\dfrac{1}{2} \not\subseteq Q$

32. $\{ -.245 \} \not\subseteq Q$

33. $\{ 0 \} \subseteq N$

34. $\left\{ \dfrac{17}{13} \right\} \subseteq N$

35. $\{ 1, 2, 3, \ldots \} \not\subseteq W$

36. $0 \not\subseteq N$

37. $0 \div 3$ is undefined.

38. $4\dfrac{1}{5} \in Q$

39. $-13 \notin I$

40. $3.15 \in I$

41. $3 \div 0$ is undefined.

42. Q is a finite set.

43. $N \in W$

44. $0.045 \in Q$

45. $\{ 0 \} \subseteq W$

46. $5 \in I$

47. $-5 \in I$

48. $-5 \in N$

49. $-5 \in Q$

50. $0 \in Q$

A.4 Operations on Sets

In this section you will learn to perform the operations of intersection and union on sets. These operations will be used to extend your understanding of the structure of the real number system.

You may have noticed that whenever you *operate* on two numbers with addition, subtraction, multiplication, or division, the result is a number. Similarly, when you *operate* on two sets with intersection or union, the result is a set.

The first operation we shall consider is **union**. The symbol which denotes this is $\cup$. The second operation we shall consider is **intersection**. The symbol which denotes this is $\cap$.

RULE FOR FINDING THE UNION OF TWO SETS OF NUMBERS	*EXAMPLE* $\{ 1, 2 \} \cup \{ 2, 4, 6 \} = ?$ *Solution:*
Form a new set by listing all the numbers that are members of either set. (*Note:* Elements are listed only once in the union even if they occur in both sets.)	$\{ 1, 2 \} \cup \{ 2, 4, 6 \} = \{ 1, 2, 4, 6 \}$

RULE FOR FINDING THE INTERSECTION OF TWO SETS OF NUMBERS	*EXAMPLE* $\{ 1, 2 \} \cap \{ 2, 4, 6 \} = ?$ *Solution:*
Form a new set by listing only the numbers that are members of both sets.	$\{ 1, 2 \} \cap \{ 2, 4, 6 \} = \{ 2 \}$

Example a. $\{ 1, 2 \} \cup \{ 4, 5, 15 \} = \{ 1, 2, 4, 5, 15 \}$, or the union of the set containing 1 and 2 and the set containing 4, 5, and 15 is the set with elements 1, 2, 4, 5, and 15.

b. $N \cup W = \{1, 2, 3, \ldots\} \cup \{0, 1, 2, 3, \ldots\}$

$\quad = \{0, 1, 2, 3, \ldots\}$

$\quad = W$

c. $N \cap W = N$

d. The sets $\{1, 2\}$ and $\{3, 4\}$ do not have any elements in common; therefore $\{1, 2\} \cap \{3, 4\}$ is a set with no elements. This set is called the **null set** or the **empty set** and is denoted by $\emptyset$ or $\{\ \}$. Therefore, $\{1, 2\} \cap \{3, 4\} = \emptyset$. Sets such as these, which have no elements in common, are called **disjoint sets**.

e. $\emptyset \cup \{1, 2\} = \{1, 2\}$

The union of the set with no elements and the set of elements 1 and 2 is the set with elements 1 and 2.

f. $\emptyset \cap \{1, 2\} = \emptyset$

There are no elements in $\emptyset$ so there are no elements that are in both sets.

A.4   Exercises

Perform the indicated operations in Exercises 1–28.

1. $\{1, 2\} \cup \{2, 4, 6, 8\}$ **2.** $\{1, 2\} \cap \{2, 4, 6, 8\}$

3. $\{4, 8, 5, 9\} \cap \{4, 5, 6, 7\}$

4. $\{4, 8, 5, 9\} \cup \{4, 5, 6, 7\}$

5. If $A = \{6, 7, 8, 10\}$, then $A \cup \{1, 2\} = ?$

6. If $A = \{6, 7, 8, 10\}$, then $A \cap \{1, 2\} = ?$

7. If $B = \{10, 20, 30, \ldots\}$, then $B \cap \{6, 8, 10\} = ?$

8. $\{2, 4\} \cap \{1, 3\}$ **9.** $\{2, 4\} \cup \{2, 4\}$

10. $\{2, 4\} \cap \{2, 4\}$ **11.** $\{1, 3\} \cap \{2, 4\}$

12. $\{1, 3\} \cup \{2, 4\}$ **13.** $\{2, 4\} \cap \{4, 2\}$

14. $\{5, 10, 15, \ldots\} \cap \{10, 20, 30, \ldots\}$

15. $\{5, 10, 15, \ldots\} \cup \{10, 20, 30, \ldots\}$

16. $W \cap N$ **17.** $W \cap R$

18. $W \cup N$ **19.** $N \cap I$

20. $\left\{-1, \dfrac{1}{2}, \dfrac{4}{3}, 1.5, 1000, 2\right\} \cap I$

21. $\left\{-1, \dfrac{1}{2}, \dfrac{4}{3}, 1.5, 1000, 2\right\} \cap N$

22. $R \cup I$ **23.** $I \cup R$

24. $Q \cap I$ **25.** $I \cap Q$

26. $\emptyset \cup \{3, 7, 11\}$ **27.** $\emptyset \cap \{5, 13, 4\}$

28. $\{4, 8, 15\} \cup \emptyset$

29. Are the sets $\{1, 2\}$ and $\{12\}$ disjoint?

30. Are the sets $\{1, 2\}$ and $\{3, 1\}$ disjoint?

A.5   The Irrationals

As was indicated before, every point on the number line corresponds to a real number. However, the points which correspond to the rational numbers do not "fill up" the line. The real numbers which correspond to the points which remain are called irrational numbers. A brief description of these numbers will be given now but a more thorough description will be deferred to more advanced mathematics courses.

Irrational numbers are *not* rational, so they *cannot* be written in fractional form. One of the most familiar examples is $\sqrt{2}$, read "the square root of 2." To find $\sqrt{2}$, you must find a number which when used as a factor twice gives the product of 2. The required number is not obvious from inspection. In the table of square root approximations given in Appendix C, you will find $\sqrt{2} \doteq 1.414$ $\left(\text{the square root of 2 is approximately equal to } \dfrac{1414}{1000}\right)$. If you multiply

1.414 × 1.414, you get 1.999396. Consider the following approximations of $\sqrt{2}$:

$$\sqrt{2} \doteq 1.4142 \quad \text{and} \quad 1.4142 \times 1.4142 = 1.99996164$$
$$\sqrt{2} \doteq 1.41421 \quad \text{and} \quad 1.41421 \times 1.41421 = 1.9999899241$$
$$\sqrt{2} \doteq 1.414214 \quad \text{and} \quad 1.414214 \times 1.414214 = 2.000001237796$$

As you can see, when the approximations are used as a factor twice, the products approach 2 but are not exactly equal to 2. In fact, there is no rational number which when used as a factor twice will give exactly 2.* There is a formal mathematical proof of this fact which is often presented in more advanced courses.

There are an infinite number of irrationals. Some examples are

$$\sqrt{3}, \ \sqrt{24}, \ \sqrt{\frac{2}{3}}, \ \sqrt{1.6}, \ \sqrt{5}, \ \sqrt{6}, \ \sqrt{7}, \ \pi.$$

You may have used either $\frac{22}{7}$ or 3.14 as a rational approximation of π in the formula for the area of a circle ($A = \pi r^2$). Other examples of irrational numbers are provided in Chapter 12.

The chart which follows shows the relationships between the set of real numbers and its subsets. Study it carefully.

The Set of Real Numbers and Its Subsets

Real Numbers
All the numbers of the number line—fractions, decimals, integers, rationals, and irrationals.

Irrationals
Any real number which cannot be written in fractional form.
Examples : $\sqrt{2}, \pi$

Rationals
Any real number which can be written in fractional form.

Non-Integral Fractions
Examples : $\frac{2}{3}, -\frac{3}{4}, 0.66$

Integers
$\left\{ \ldots -3, -2, -1, 0, 1, 2, 3 \ldots \right\}$

Negative Integers
$\left\{ -1, -2, -3, -4 \ldots \right\}$

Whole Numbers
$\left\{ 0, 1, 2, 3, 4 \ldots \right\}$

zero
$\{0\}$

Natural Numbers
(counting numbers)
$\left\{ 1, 2, 3, 4 \ldots \right\}$

*A more accurate approximation was calculated by Dr. Dutka at Columbia University. In a two-year research project, he used a computer to calculate the square root of 2 accurate to one million decimal places. It begins 1.41421356237309504880168872. . . . You can multiply this number by itself to check his accuracy!

A.5 Exercises

Answer true (T) or false (F) in Exercises 1-20. Let

  N = the natural numbers
  W = the whole numbers
  I = the integers
  Q = the rationals
  T = the irrationals
  R = the reals

1. T and Q are disjoint.   **2.** $T \cap Q = \emptyset$

3. $T \cup Q = R$ **4.** $\sqrt{4} = 2$, thus $\sqrt{4} \in T$

5. $-17 \notin T$ **6.** $R = Q \cup T$

7. $\sqrt{2} \notin Q$ **8.** $0 \in T$

9. $\sqrt{9} \in T$ **10.** $-4\frac{2}{3} \in T$

11. $I \subseteq Q$ **12.** $W \nsubseteq T$

13. $T \subseteq Q$ **14.** $Q \subseteq T$

15. $N \cup \{0\} = W$ **16.** $N \nsubseteq Q$

17. $W \nsubseteq N$

18. $\{-1, -2, -3, \ldots\} \cup W = I$

19. $\pi \in T$ **20.** $\sqrt{3} \in T$

Fill in the chart for Exercises 21-29 by checking the sets to which each of the numbers on the left belongs.

	N	W	I	Q	T	R
21. 25						
22. −3.5						
23. π						
24. $\frac{1}{9}$						
25. 0						
26. $5\frac{3}{7}$						
27. −4						
28. $\sqrt{5}$						
29. $\sqrt{4}$						

Important Terms Used in this Appendix		
	disjoint sets	natural number
	element	null set
	ellipsis notation	number line
	empty set	rational number
	equal sets	real number
	finite	subset
	infinite	set
	integer	union
	intersection	well defined
	irrational number	whole number
	member	

Important Symbols Used in this Appendix		
	=	set equality
	≠	two sets are not equal
	∈	an object is an element of a set
	∉	an object is not an element of a set
	⊆	one set is a subset of another set
	⊄	one set is not a subset of another set
	∩	the intersection of two sets
	∪	the union of two sets
	∅	the null set

Special Symbols
Used in this Book

N   the set of natural numbers
W   the set of whole numbers
I   the set of integers
Q   the set of rational numbers
T   the set of irrational numbers
R   the set of real numbers

REVIEW EXERCISES

For Exercises 1–4, let

   N = the set of natural numbers
   W = the set of whole numbers
   I = the set of integers
   Q = the set of rational numbers
   T = the set of irrational numbers
   R = the set of real numbers
   A = $\{1, 3, 5, 7\}$
   B = $\{2, 4, 6\}$
   C = $\{5, 10, 15, \ldots\}$
   D = $\{-2, -4, -6\}$

Perform the indicated operations.

1. $A \cup B$ 2. $A \cap C$

3. $B \cap D$ 4. $\{1, 2, 3, 4, 5\} \cup C$

Write each of Exercises 5–8 in *symbols*.

5. Negative twelve

6. Five is an element of set A.

7. Set X is not a subset of Y.

8. The intersection of the set M and the set N is the empty set.

Write each of Exercises 9–12 in *words*.

9. $1.2 \in I$ 10. $T \subseteq R$

11. $A = \{1, 3, 5, 7\}$ 12. $\{4, 8\} \cap \{4\} = \{4\}$

Rewrite each of the rational numbers in Exercises 13–17 as the quotient of two integers.

13. 0.13 14. $4\frac{1}{5}$

15. $-0.2\frac{1}{4}$ 16. -0.024

17. -15

Answer true (T) or false (F) in Exercises 18–23.

18. $\{0\} = \varnothing$ 19. $\frac{4}{0} = 0$

20. $0 \in Q$ 21. $-6 \notin W$

22. $-4 \notin I$ 23. $\sqrt{2} \in R$

Answer each of Exercises 24–29 with N, W, I, Q, T, R, or $\varnothing$.

24. $Q \cap T$ 25. $\{0\} \cup N$

26. $Q \cap I$ 27. $Q \cup T$

28. $W \cap I$ 29. $T \cap R$

B

MEASUREMENT

OBJECTIVES

Upon completion of Appendix B you should be able to:

1. Identify the units in the metric system which are used to measure length, mass, and volume. **B.1**

2. Give the meaning of and symbols for the following common decimal metric prefixes: milli, centi, deci, deka, hecto, kilo. **B.1**

3. Give the meaning of and symbols for the common decimal multiples and submultiples of the liter, meter, and gram. **B.1**

4. Make conversions within the metric system. **B.1, B.3**

5. Give the approximate customary equivalents for the kilogram, liter, meter, centimeter, and kilometer. **B.2**

6. Find the product or quotient of a denominate number and a number. **B.3**

7. Find the sum, difference, product, or quotient of denominate numbers of the same unit of measure. **B.3**

8. Make conversions from one unit in the metric or customary system to another unit in the other system. **B.3**

9. Solve word problems involving measurement. **B.2**

PRETEST

1. Indicate whether each of the following units is used to measure length, mass, or volume (Objective 1).
   - **a.** centimeter
   - **b.** meter
   - **c.** kilogram
   - **d.** liter
   - **e.** kilometer
   - **f.** milliliter

2. Give the meaning of and symbol for each of the following prefixes (Objective 2).
   - **a.** centi
   - **b.** milli
   - **c.** kilo

   Give the prefix which each of the following symbols represents (Objective 2).
   - **d.** h
   - **e.** d
   - **f.** da

3. Give the symbol for each of the following metric units (Objective 3).
   - **a.** centimeter
   - **b.** meter
   - **c.** kilogram
   - **d.** kilometer
   - **e.** milliliter

Give the name of the unit which each of the following symbols represents (Objective 3).
   - **f.** cl
   - **g.** dl
   - **h.** hg
   - **i.** dam
   - **j.** kg
   - **k.** cm

4. Express in the indicated units (Objective 4).
   - **a.** 1 g = ? cg
   - **b.** 5.2 mm = ? cm
   - **c.** 0.2 kl = ? liters
   - **d.** 520 cm = ? m
   - **e.** 40.2 m = ? cm
   - **f.** 20 cl = ? dl
   - **g.** 1 m = ? km
   - **h.** 500 ml = ? liters
   - **i.** 5000.2 g = ? kg
   - **j.** 2.1 liters = ? cl
   - **k.** 1200 dag = ? kg
   - **l.** 1 liter = ? ml

5. Give the approximate customary equivalents for the kilogram, liter, meter, centimeter, and kilometer (Objective 5).

6. Evaluate each of the following expressions (Objective 6).

   a. $6(2 \text{ m}) = ?$ **b.** $\dfrac{10 \text{ kg}}{2} = ?$

   c. $4.5(3.4 \text{ liters}) = ?$ **d.** $\dfrac{5000 \text{ ml}}{100} = ?$

7. Perform the indicated operations (Objective 7).

   a. $(5m)(5m) = ?$ **b.** $4.32 \text{ m} + 15 \text{ m} = ?$

   c. $5.2 \text{ liters} - 1.4 \text{ liters} = ?$

   d. $\dfrac{28 \text{ kg}}{1.4 \text{ kg}} = ?$

8. Express in the indicated units (Objective 8).

   a. $7.5 \text{ cm} \doteq ? \text{ in.}$ **b.** $5 \text{ m} \doteq ? \text{ in.}$

   c. $2 \text{ kg} \doteq ? \text{ lbs}$ **d.** $247 \text{ km} \doteq ? \text{ mi}$

   e. $2 \text{ liters} \doteq ? \text{ qts}$ **f.** $43 \text{ in.} \doteq ? \text{ cm}$

   g. $43 \text{ in.} \doteq ? \text{ m}$ **h.** $5 \text{ lbs} \doteq ? \text{ kg}$

   i. $247 \text{ mi} \doteq ? \text{ km}$ **j.** $10 \text{ qts} \doteq ? \text{ liters}$

9. A bottle manufacturing company manufactured bottles to contain quarts, $\frac{1}{2}$ gallons, and gallons. In anticipation of a changeover to metric units, they still wanted to manufacture containers in 3 different sizes. What convenient metric measures would be closest in size to 1 quart, $\frac{1}{2}$ gallon, 1 gallon?

The notion of number and the counting process developed long before recorded history. The counting process may have been developed by man in prehistoric times as a response to a need to measure the size of difference objects in his surroundings. Such a use of numbers is a probable one for the first application of the concept of measurement.

Today, measurement is used in every facet of life—business, economics, industry, navigation, and medicine, to name a few. With the aid of special devices, we are able to measure such things as time, temperature, distance, speed, weight (mass), area, and volume. Many different units have been used to measure these quantities. The units used to measure length have included the human foot, a hand span, the length from the tip of the nose to the end of the fingers, and the length of a man's stride. These units, however, depended upon the size of the person making the measurement, so it became obvious that the units needed to be universally standardized.

Early attempts at standardization were made by some of the British monarchs who reduced some of the confusion in measurement by setting specific standards for some of the most important units. The eventual result was a system of measurement called the **customary system.** The units of this sytem include the inch, foot, yard, and mile for length; the avoirdupois ounce and the pound for weight; the fluid ounce, pint, quart, gallon, bushel, and peck for volume; the degree Fahrenheit for temperature; the ampere for electric current; the second for time; and the candela for luminous intensity. However, another system, the International System of Units (SI), is the one which is used in most of the world today. This system is a refined or modernized version of the metric system which was developed in France around 1795. In this text the term **metric system** will be used when referring to this system since this name is probably more familiar than the official name of the system. In this appendix we will consider the metric system and make conversions from one unit of measure to another.

B.1 The Metric System

The metric system is built on six base units. The unit of length is the meter; the unit of mass (commonly called weight) is the kilogram; the unit of time is the second; the unit of electric current is the ampere; the unit of temperature is the kelvin (which is usually translated into degrees Celsius, traditionally called centigrade); and the unit of luminous intensity is the candela. All other units are derived from the base units.

The metric units with which you will need to be most familiar are the units for length, volume, and mass. The **meter** is the base unit of length; the **liter** is used to measure liquid volume or capacity; and the base unit of mass is the **kilogram.**

Standard decimal prefixes are added to some of the metric units to give names for quantities that are multiples or submultiples of the unit. Table B.1 gives the meanings and symbols for the most common prefixes.

You should memorize these prefixes, their meanings, and the symbols which are used to represent them. These prefixes are used to form multiples and submultiples of the meter, liter, and gram. The resulting units are given in Table B.2. Study the symbols used to represent these units. The most commonly used units are indicated in heavier type.

Table B.1   *Common Metric Prefixes.*

Prefix	Symbol	Meaning
kilo	k	one thousand times
hecto	h	one hundred times
deka	da	ten times
deci	d	one tenth of
centi	c	one hundredth of
milli	m	one thousandth of

Table B.2   *Multiples and Submultiples of the Meter, Liter, and Gram.*

Quantity	Unit	Symbol	Meaning
Length	**kilometer**	km	1 km = 1000 m
(base unit is	hectometer	hm	1 hm = 100 m
1 meter)	dekameter	dam	1 dam = 10 m
	meter	m	1 m = 1 m
	decimeter	dm	1 dm = 0.1 m
	centimeter	cm	1 cm = 0.01 m
	millimeter	mm	1 mm = 0.001 m
Mass	**kilogram**	kg	1 kg = 1000 g
(base unit is	hectogram	hg	1 hg = 100 g
1 kilogram)	dekagram	dag	1 dag = 10 g
	gram	g	1 g = 1 g
	decigram	dg	1 dg = 0.1 g
	centigram	cg	1 cg = 0.01 g
	milligram	mg	1 mg = 0.001 g
Volume	kiloliter	kl	1 kl = 1000 liter
(base unit is	hectoliter	hl	1 hl = 100 liter
1 liter)	dekaliter	dal	1 dal = 10 liter
	liter	l	1 liter = 1 liter
	deciliter	dl	1 dl = 0.1 liter
	centiliter	cl	1 cl = 0.01 liter
	milliliter	ml	1 ml = 0.001 liter

Note that in the relationships given in Table B.2, the units on the left are arranged in order from largest to smallest, and that for any two consecutive units, either one unit is 10 times the other unit or 0.1 times the other unit. For example, consider the consecutive units millimeter and centimeter. Since 1 mm = 0.001 m and 1 cm = 0.01 m, a centimeter is 10 times a millimeter (1 cm = 10 mm) and a millimeter is 0.1 of a centimeter (1 mm = 0.1 cm). Similarly, for the meter and dekameter 1 dam = 10 m and 1 m = 0.1 dam. The following observations summarize these relationships.

Observation 1:   Each unit of a quantity is 10 times the next consecutive smallest unit of the same quantity. For example, 1 km = 10 hm, 1 hm = 10 dam, 1 dam = 10 m, 1 m = 10 dm, 1 dm = 10 cm, and 1 cm = 10 mm.

Observation 2:   Each unit of a quantity is 0.1 of the next consecutive largest unit. For example, 1 mm = 0.1 cm, 1 cm = 0.1 dm, 1 dm = 0.1 m, 1 m = 0.1 dam, 1 dam = 0.1 hm, and 1 hm = 0.1 km.

These observations form the basis for a simple procedure which can be used to convert from one unit of a quantity to another unit of the same quantity. The procedure is given in the following rules.

RULE FOR CONVERTING FROM ONE UNIT TO A LARGER UNIT WITHIN THE METRIC SYSTEM	*EXAMPLE* 235.6 dm = _____ km
Using Table B.2:	*Solution:*
1. Count the number of units from the given unit to the desired unit.	km 4 4 places hm 3 dam 2 m 1 dm
2. Move the decimal point to the left the number of places found in step 1.	0235.6
	Thus, 235.6 dm = 0.02356 km.

RULE FOR CONVERTING FROM ONE UNIT TO A SMALLER UNIT WITHIN THE METRIC SYSTEM	*EXAMPLE* 25.1 dl = _____ ml
Using Table B.2:	*Solution:*
1. Count the number of units from the given unit to the desired unit.	dl 1 cl 2 ml 2 places
2. Move the decimal point to the right the number of places found in step 1.	25.10
	Thus, 25.1 dl = 2510 ml.

Example 1   Make the following conversions from the given unit to the indicated larger unit by repositioning the decimal point.

a. 0.5 m = _____ hm

b. 1.73 liters = _____ kl

c. 2 g = _____ dag

d. 256 mg = _____ dg

e. 0.04 dag = _____ kg

f. 12300 cm = _____ km

In each case we are converting from a smaller unit to a larger unit. Thus, the decimal point will be moved to the left.

a. 0.5 m = 0.005 hm

b. 1.73 liters = 0.00173 kl

c. 2 g = 0.2 dag

d. 256 mg = 2.56 dg

e. 0.04 dag = 0.0004 kg

f. 12300 cm = 0.12300 km

Example 2   Make the following conversions from the given unit to the indicated smaller unit by repositioning the decimal point.

a. 0.5 m = _____ mm

b. 1.73 liters = _____ cl

c. 2 g = _____ dg

d. 256 hl = _____ liters

e. 0.04 dag = _____ cg

f. 12300 cm = _____ mm

In each case we are converting from a larger unit to a smaller unit. Thus, the decimal point will be moved to the right.

a. 0.5 m = 500 mm

b. 1.73 liters = 173 cl

c. 2 g = 20 dg

d. 256 hl = 25600 liters

e. 0.04 dag = 40 cg

f. 12300 cm = 123000 mm

Note that in making these conversions, when you convert a smaller unit to a larger unit you move the decimal point to the left, which results in a smaller number of units. For example, when converting from centimeters to meters, you would expect the number of meters to be less than the number of centimeters, so you would move the decimal point to the left in order to get a smaller number. Similarly, when you convert from a larger unit to a smaller unit, you would expect to obtain a larger number of units. This is accomplished by moving the decimal point to the right.

Example 3   Make the following conversions. First determine whether you are converting from a larger unit to a smaller unit or from a smaller unit to a larger unit.

a. Convert 26.52 meters to centimeters.

Since we are converting from a larger unit to a smaller unit, the number of units we obtain should be larger than 26.52. Thus, the decimal point will be moved to the right. Since there are 2 units from meters to centimeters, the decimal will be moved 2 places to the right. Thus, 26.52 m = 2652 cm.

b. Convert 111 milliliters to liters.

There are 3 units from milliliter to liter. We are converting from a smaller unit to a larger unit, so the decimal point is moved 3 places to the left. Thus, 111 ml = 0.111 liter.

c. 150 cm = _____ m

150 cm = 1.5 m

d. 10.7 liters = _____ cl

10.7 liters = 1070 cl

e. 5 kg = _____ g

5 kg = 5000 g

f. 450 g = _____ kg

450 g = 0.45 kg

g. 13.56 ml = _____ liters

13.56 ml = 0.01356 liters

h. 400 kl = _____ liters

400 kl = 400000 liters

Contrast these conversions within the metric system to conversions made within the customary system. For example, to convert from miles to yards you could multiply by 5280 and then divide by 3. Thus, it is easier to convert within the metric system since all you have to do is reposition the decimal point.

B.1 Exercises

Give the symbol which is used for each of the metric units listed in Exercises 1-20.

1. millimeter
2. kiloliter
3. meter
4. dekameter
5. decigram
6. centigram
7. kilometer
8. millimeter
9. gram
10. centiliter
11. liter
12. deciliter
13. hectogram
14. milligram
15. centimeter
16. kilogram
17. dekaliter
18. hectometer
19. decimeter
20. dekagram

21. In the metric system, what is the base unit of length? mass? volume?

22. Give the meaning of the following metric prefixes: deci, deka, hecto.

23. Give the meaning of the following metric prefixes: milli, kilo, centi.

Convert each of Exercises 24-64 to the indicated unit.

24. 1 g = ? mg
25. 1 m = ? km
26. 1 liter = ? hl
27. 1 g = ? cg
28. 1 m = ? dm
29. 1 liter = ? ml
30. 1 g = ? dag
31. 1 liter = ? dl
32. 1 liter = ? kl
33. 1 m = ? dam
34. 1 m = ? cm
35. 1 g = ? hg
36. 5261 mm = ? cm
37. 5261 mm = ? m
38. 5261 mm = ? km
39. 13.4 km = ? mm
40. 13.4 km = ? cm
41. 13.4 km = ? m
42. 52 m = ? mm
43. 52 cm = ? mm
44. 215 cm = ? m
45. 215 cm = ? km
46. 52 m = ? cm
47. 1520 m = ? km
48. 1516 ml = ? liters
49. 1.2 liters = ? ml
50. 1000 mg = ? dg
51. 1000 mg = ? kg
52. 500 g = ? mg
53. 5.04 kg = ? mg
54. 504 g = ? kg
55. 0.024 kg = ? g

56. 152.4 cm = ? m

57. 3.5 m = ? cm

58. 500 ml = ? liters

59. 0.6 liter = ? ml

60. 215,120 mm = ? m

61. 25.19 g = ? kg

62. 25.19 km = ? m

63. 428 cm = ? m

64. 2 kg = ? g

Fill in each blank with $<$, $>$, or = to make a true statement in Exercises 65-96.

65. 1 cm _____ 1 mm

66. 1 dal _____ 1 liter

67. 1 mm _____ 1 m

68. 1 hg _____ 1 dg

69. 1 km _____ 1 mm

70. 1 dl _____ 1 dal

71. 1 g _____ 1 kg

72. 1 dm _____ 1 cm

73. 1 ml _____ 1 liter

74. 1 g _____ 1 dg

75. 1 cm _____ 1 m

76. 1 ml _____ 1 dl

77. 1 km _____ 1 cm

78. 1 kl _____ 1 hl

79. 1 g _____ 1 mg

80. 1 mg _____ 1 dag

81. 1 m _____ 1 km

82. 1 m _____ 1 hm

83. 1 mg _____ 1 kg

84. 1 cl _____ 1 ml

85. 1 dag _____ 1 dg

86. 10 dg _____ 10 g

87. 1 hl _____ 1 cl

88. 5 dm _____ 5 dam

89. 2.1 kg _____ 2100 g

90. 15 cl _____ 15 hl

91. 10 liters _____ 1 kl

92. 3.75 dal _____ 5 hl

93. 5000 mm _____ 1 m

94. 5 m _____ 1 km

95. 500 cg _____ 0.5 g

96. 10 mm _____ 100 cm

The metric unit of temperature is the kelvin which is commonly translated into the degree Celsius. It is based on 0° as the freezing point of water and 100° as the boiling point of water. Normal body temperature is 37° Celsius. Thus, $0°C = 32°F$; $37°C = 98.6°F$; and $100°C = 212°F$. Use these relationships and your knowledge of temperature to answer Exercises 97-99.

97. Which of the following degrees Celsius represents the temperature on a cold day in Siberia?
    $-40°$ $15°$ $35°$ $50°$

98. Which of the following degrees Celsius represents the temperature on a hot day at the beach?
    $-40°$ $15°$ $35°$ $50°$

99. Which of the following degrees Celsius represents the temperature on a cool fall day?
    $-40°$ $15°$ $35°$ $50°$

A formula for converting degrees Fahrenheit to degrees Celsius is $°C = \dfrac{°F - 32}{1.8}$.

Use this formula to convert each of Exercises 100-105 from degrees Fahrenheit to degrees Celsius.

100. 32° F

101. 212° F

102. 98.6° F

103. 72° F (round to nearest degree)

104. 90° F (round to nearest degree)

105. 10° F (round to nearest degree)

The term cubic centimeter (cc or cm^3) is frequently used in medicine. This term refers to a quantity of liquid which is about one milliliter. Even though there is a slight difference between the two, we consider them equal for most purposes, that is 1 ml = 1 cm^3 = 1 cc. Complete each of Exercises 106-112.

106. 1 ml = ? liter

107. 1 cm^3 = ? liter

108. 1 cc = ? liter

109. 2 cm^3 = ? ml

110. 500 cc = ? ml

111. 24 cm^3 = ? ml

112. 500 cc = ? liter

B.2   Using the Metric System

Even though initial acceptance of the metric system was not overwhelming, it has become the universal language of measurement within the last 20 years. At the present time, the U.S. is the only major country in which the predominate system of measurement is the customary system. There has been resistance to changing to the metric system in the U.S. due to concern about the cost and inconvenience of a change-over. Industries will have to replace machinery and tools and maintain dual inventories during the transition period, and the public will have to be educated in the use of the system. However, even with these problems, many feel that a change-over is warranted and claim it could facilitate matters in international trade. In addition, the metric system is generally recognized to be simpler and more uniform than the customary system.

Finally, after more than 100 years of debate, a law was passed in December of 1975 known as the Metric Conversion Act of 1975. While this law does not indicate a specific transition period or make metric usage mandatory, it declares a national policy of coordinating the increased use of the metric system in the U.S. It established a board to assist in the voluntary conversion to the metric system. Thus, the U.S. is slowly changing to the metric system, and the public will soon need to be conversant with this system. Initially, you will see the use of a dual system. The familiar sizes and quantities of items will remain the same but sizes and dimensions will be printed on labels in both customary and metric units. Then, as use of the metric system increases, some of the standard sizes will be changed to the most convenient metric units and eventually only metric units will be indicated. Gasoline and milk will be sold by the liter, meat will be packaged by the kilogram, distances will be given in kilometers, temperatures will be reported in degrees Celsius, speed will be given in kilometers per hour, textiles will be sold by the meter, and recipes will call for milliliters of salt.

As the metric system becomes more commonly used in the U.S., you will need to understand the system and be able to *think metric*. Indeed, the emphasis is on teaching people to think metric rather than to make conversions between the metric and customary systems. This is certainly the ideal situation but initially a person who is familiar with the customary system will probably need to be able to make approximate conversions between the two systems. Table B.3 gives some approximate conversions.

Table B.3 *Approximate Conversions.*

	From Metric Measures		From Customary Measures
Length	1 cm $\doteq$ 0.4 in. 1 m $\doteq$ 39.37 in. 1 km $\doteq$ 0.6 mi	Length	1 in. $\doteq$ 2.5 cm 1 ft $\doteq$ 30 cm 1 yd $\doteq$ 0.9 m 1 mi $\doteq$ 1.6 km
Mass	1 g $\doteq$ 0.04 oz 1 kg $\doteq$ 2.2 lbs	Mass	1 oz $\doteq$ 28.4 g 1 lb $\doteq$ 0.45 kg
Liquid and dry Measure	1 ml $\doteq$ 0.2 tsp 1 liter $\doteq$ 1.06 qts or 4.2 c	Liquid and dry Measure	1 tsp $\doteq$ 5 ml 1 cup $\doteq$ 237 ml 1 qt $\doteq$ 0.95 liter 1 gal $\doteq$ 3.8 liters

Table B.3 is provided to help you become familiar with the metric system units. It will probably be sufficient for you to learn only the following very rough equivalents.

1 kg is a little more than 2 lbs.

1 m is a little more than a yd.

1 km is a little more than $\frac{1}{2}$ mi.

1 cm is a little less than $\frac{1}{2}$ in.

1 liter is a little more than 1 qt.

5 ml is about 1 tsp.

Example 1   Give the metric equivalents for each of the following.

   a. 1 yard $\doteq$ 0.9 m   b. 1 mile $\doteq$ 1.6 km
   c. 1 quart $\doteq$ 0.95 liter   d. 1 pound $\doteq$ 0.45 kg

Example 2   Insert $>$ or $<$ in each of the following blanks.

   a. 1 in. __$>$__ 1 cm   b. 1 yd __$<$__ 1 m
   c. 1 ft __$<$__ 1 m   d. 1 mi __$>$__ 1 km
   e. 1 qt __$<$__ 1 liter   f. 1 gal __$>$__ 1 liter
   g. 1 c __$<$__ 1 liter   h. 1 tsp __$>$__ 1 ml
   i. 1 oz __$>$__ 1 g   j. 1 lb __$<$__ 1 kg

Example 3   Insert $>$ or $<$ in each of the following blanks.

   a. 5 lbs __$>$__ 2 kg   (since 5 lbs = 5(0.45)kg = 2.25 kg)
   b. 1 gal __$<$__ 4 liters
   c. 3 ft __$<$__ 1 m
   d. 2 mi __$>$__ 1 km

B.2   Exercises

1. Give some instances in which you have observed the use of metric units.

2. If your school library has a copy of the *U.S. Metric Study,* read the part entitled "A Metric America—A Decision Whose Time Has Come." The author of the study is Daniel V. DeSimone, and it is published by the National Bureau of Standards in Washington, D.C.

3. Will an official change to the metric system affect your daily life? Your job? If yes, how?

4. Obtain an elementary school textbook, and write a report on what is included on the metric system.

Fill each blank in Exercises 5-24 with $>$, $<$, or =.

 5. 1 in. _____ 1 cm   6. 1 mile _____ 1 km
 7. 1 yard _____ 1 m   8. 1 foot _____ 0.5 m
 9. 20 km _____ 14 miles   10. 1 ounce _____ 1 g
11. 1 pound _____ 1 kg   12. 15 in. _____ 20 cm
13. 8 m _____ 12 ft   14. 9 lbs _____ 4 kg
15. 25 ft _____ 0.5 m   16. 1 quart _____ 1 liter
17. 1 teaspoon _____ 1 ml
18. 1 cup _____ 1 liter
19. 1 gallon _____ 1 liter
20. 5 lbs _____ 3 kg   21. 5 kg _____ 10 lb
22. 5 liters _____ 1 gallon

23. 1 gallon _____ 3.8 liters
24. 50 lbs _____ 20 kg

Give the approximate metric measures for each of the measurements in Exercises 25-40. Use the approximate conversions or measure with an appropriate instrument.

25. The thickness of a dime (in mm)

26. The width of a nickel (in cm)

27. The length of a new number 2 pencil (in cm)

28. The thickness of a piece of lead in a number 2 pencil (in mm)

29. The length and width of a piece of standard size $\left(8\frac{1}{2} \times 11\right)$ notebook paper (in cm)

30. The height of an 8-foot ceiling (in cm)

31. The height of a 6-foot door (in cm)

32. The width of a door which is 4 feet wide (in cm)

33. The height of a 30-inch bike (in cm)

34. The length of a football field (in m)

35. Your height (in cm)

36. Your waist measurement (in cm)

37. Your mass or weight (in kg)

38. The width of your thumb (in cm)

39. The length of your foot (in cm)

40. The length of a paper clip (in cm)

41. A textile company manufactured cloth and made bolts of 100 yards. Would a bolt contain more cloth or less cloth if they changed to 100 meters to a bolt?

42. If the speed limit is 55 mph, what speed would appear on speed limit signs when the U.S. goes metric?

43. Sugar is generally packaged in 2 lb, 5 lb, or 10 lb bags. If the sizes remain approximately the same and if sugar is sold by the kilogram, what would be the mass (to the nearest kilogram) of corresponding bags?

B.3 Conversions

When numbers are associated with a unit of measure, they are referred to as **denominate numbers**. We have worked with denominate numbers in the metric and customary systems. We now formally state the rules for operating with denominate numbers.

RULE FOR MULTIPLYING OR DIVIDING A DENOMINATE NUMBER BY A NUMBER	*EXAMPLE* Multiply 3 meters by 4.
1. Find the product or quotient of the two numbers. 2. Affix the unit of measure associated with the denominate number to the product or quotient obtained in step 1.	*Solution:* $4 \cdot 3 = 12$ 12 meters Thus, 4(3 meters) = 12 meters.

RULE FOR ADDING (OR SUBTRACTING) DENOMINATE NUMBERS OF THE SAME UNIT OF MEASURE	*EXAMPLE* 3 liters + 6 liters = ?
1. Find the sum (or difference) of the numbers. 2. Affix the common unit of measure to the number obtained in step 1.	*Solution:* $3 + 6 = 9$ 9 liters Thus, 3 liters + 6 liters = 9 liters.

RULE FOR MULTIPLYING TWO DENOMINATE NUMBERS OF THE SAME UNIT OF MEASURE	*EXAMPLE* 7 m × 9 m = ?
1. Find the product of the numbers. 2. Affix the square of the unit to the product found in step 1.	*Solution:* $7 \cdot 9 = 63$ 63 sq m Thus, 7 m · 9 m = 63 sq m (or 63 m²).

RULE FOR FINDING THE QUOTIENT OF TWO DENOMINATE NUMBERS OF THE SAME UNIT OF MEASURE	*EXAMPLE* $\dfrac{6 \text{ kg}}{2 \text{ kg}} = ?$
1. Find the quotient of the two numbers. 2. The quotient obtained in step 1 is the quotient of the denominate numbers. (Note that the quotient of two denominate numbers of the same measure has no unit of measure associated with it.)	*Solution:* $6 \div 2 = 3$ Thus, $\dfrac{6 \text{ kg}}{2 \text{ kg}} = 3$.

We have considered conversions within the metric system, which can be accomplished by repositioning the decimal, and approximate conversions between the metric and customary systems.

In this section we will consider two additional techniques for making conversions within a system or between two systems. These techniques use some of the operations with denominate numbers given at the beginning of this section and the properties of equations given in Chapter 10.

Example 1   Make the indicated conversions.

a. Express 6 yards in terms of feet.
Since

$$1 \text{ yd} = 3 \text{ ft}$$
$$6 \cdot 1 \text{ yd} = 6 \cdot 3 \text{ ft} \qquad \text{(multiplication property of equality)}$$
$$6 \text{ yds} = 18 \text{ ft} \qquad \text{(multiplication of a denominate number by a number)}$$

b. Express 1 meter in terms of millimeters.

$$1 \text{ mm} = 0.001 \text{ m} \qquad \text{(basic relationship from Table B.2)}$$
$$\frac{1 \text{ mm}}{0.001} = \frac{0.001 \text{ m}}{0.001} \qquad \text{(division property of equality)}$$
$$1000 \text{ mm} = 1 \text{ m} \qquad \text{(division of a denominate number by a number}$$

Thus, 1 m = 1000 mm.

c. 1.7 liters = ? cl

$$1 \text{ liter} = 100 \text{ cl}$$
$$1.7(1 \text{ liter}) = 1.7(100 \text{ cl})$$
$$1.7 \text{ liter} = 170 \text{ cl}$$

d. 450 g = ? kg

$$1 \text{ g} = 0.001 \text{ kg}$$
$$450(1 \text{ g}) = 450(0.001 \text{ kg})$$
$$450 \text{ g} = 0.45 \text{ kg}$$

e. 7.2 kg ≐ ? lbs

$$1 \text{ kg} \doteq 2.2 \text{ lbs}$$
$$7.2 \text{ kg} \doteq 7.2(2.2 \text{ lbs})$$
$$\doteq 15.84 \text{ lbs}$$

Thus, 7.2 kg ≐ 15.84 lbs

f. 5 ft ≐ ? m

$$1 \text{ ft} \doteq 0.3 \text{ m}$$
$$5 \text{ ft} \doteq 5(0.3 \text{ m})$$
$$\doteq 1.5 \text{ m}$$

Thus, 5 ft ≐ 1.5 m.

g. 2 dm = ? cm

$$1 \text{ dm} = 10 \text{ cm}$$
$$2 \text{ dm} = 2(10 \text{ cm})$$
$$= 20 \text{ cm}$$

h. 0.2 kg = ? cg

$$1 \text{ kg} = 100{,}000 \text{ cg}$$
$$\mathbf{0.2} \text{ kg} = \mathbf{0.2}(100{,}000 \text{ cg})$$
$$0.2 \text{ kg} = 20{,}000 \text{ cg}$$

Note that the conversions made in parts b, c, d, g, and h of Example 1 could have been made by using one of the rules for making conversions within the metric system which was given in Section B.1. The procedure given in this section provides an alternate way of making such conversions.

Another conversion technique which is very useful in areas such as chemistry and nursing involves the following two principles:

1. The ratio of 2 equivalent units is 1.

    For example,

$$\frac{12 \text{ in.}}{1 \text{ ft}} = \frac{12 \text{ in.}}{12 \text{ in.}} = 1$$

$$\frac{1 \text{ km}}{1000 \text{ m}} = \frac{1000 \text{ m}}{1000 \text{ m}} = 1$$

2. Multiplying a quantity by 1 leaves the quantity unchanged.

These principles can be used to make conversions within the customary system, within the metric system, and between the two systems.

Example 2   Express 3 miles in terms of inches.

$$3 \text{ miles} = \frac{3 \text{ mi}}{1} \cdot \frac{5280 \text{ ft}}{1 \text{ mi}} \cdot \frac{12 \text{ in.}}{1 \text{ ft}}$$
$$= 190{,}080 \text{ in.}$$

Recall that 1 mi = 5280 ft and 1 ft = 12 in.

The procedure in Example 2 reduces to repeatedly multiplying by 1 (in the form of the quotient of two denominate numbers) until the desired unit remains. Observe that the second ratio is chosen so that the unit in the denominator is the same as the unit in the numerator of the first ratio. When these ratios are multiplied, these like units will be eliminated. How is the third ratio chosen?

Example 3   Express 2 dm in terms of centimeters.

$$2 \text{ dm} = \frac{\overset{1}{\cancel{2 \text{ dm}}}}{1} \cdot \frac{1 \cancel{\text{ m}}}{\underset{\underset{1}{\$}}{10 \cancel{\text{ dm}}}} \cdot \frac{\overset{20}{\cancel{100} \text{ cm}}}{1 \cancel{\text{ m}}}$$

$$= 20 \text{ cm}$$

Recall that 1 m = 10 dm and 1 m = 100 cm.

PROCEDURE FOR CONVERTING FROM ONE UNIT OF MEASURE TO ANOTHER UNIT WHEN THE RELATIONSHIP BETWEEN THE UNITS IS NOT DIRECTLY GIVEN	EXAMPLE 400 ml = ? kl
	Solution:
1. Write the given denominate number as a ratio with a denominator of 1.	$\dfrac{400 \text{ ml}}{1}$
2. Form a second ratio of two denominate numbers which is equal to 1 and which has the same unit of measure in the denominator as appears in the numerator of the first ratio. (Practice will help you choose the correct unit for the numerator.)	$\dfrac{1 \text{ liter}}{1000 \text{ ml}}$
3. Continue to form ratios as indicated in step 2 until the desired unit appears in the numerator of the ratio.	$\dfrac{1 \text{ kl}}{1000 \text{ liters}}$
4. Multiply the ratios. (Note that the common units are eliminated.)	$\dfrac{400 \cancel{\text{ ml}}}{1} \cdot \dfrac{1 \cancel{\text{ liter}}}{1000 \cancel{\text{ ml}}} \cdot \dfrac{1 \text{ kl}}{1000 \cancel{\text{ liters}}}$
	$= \dfrac{400}{1000000} \text{ kl}$
	$= 0.0004 \text{ kl}$

Example 4   What units should be placed in the following to convert 15.2 kg to mg?

$$15.2 \text{ kg} = \frac{15.2 \text{ kg}}{1} \cdot \frac{1000 \text{ g}}{1 \, ?} \cdot \frac{1000 \, ?}{1 \, ?}$$

Second ratio: $\dfrac{1000 \text{ g}}{1 \text{ kg}}$

Third ratio: $\dfrac{1000 \text{ mg}}{1 \text{ g}}$

Note that the third ratio is chosen so that the unit in the denominator is the same as the unit in the numerator of the second ratio.

Example 5   0.2 kg = ? cg

$$0.2 \text{ kg} = \frac{0.2 \cancel{\text{ kg}}}{1} \cdot \frac{1000 \cancel{\text{ g}}}{1 \cancel{\text{ kg}}} \cdot \frac{100 \text{ cg}}{1 \cancel{\text{ g}}}$$

$$= 0.2 \cdot 1000 \cdot 100 \text{ cg}$$

$$= 20,000 \text{ cg}$$

Example 6   Find the number of feet in 5 meters.

$$5 \text{ m} \doteq \frac{5 \text{ m}}{1} \cdot \frac{39.37 \text{ in.}}{1 \text{ m}} \cdot \frac{1 \text{ ft}}{12 \text{ in.}}$$

$$= \frac{5 \cdot 39.37}{12} \text{ ft}$$

$$= 16.4 \text{ ft}$$

Thus, 5 m $\doteq$ 16.4 ft.

Example 7   A person has a 30-inch waist. How many centimeters is this?

$$30 \text{ in.} \doteq \frac{30 \text{ in.}}{1} \cdot \frac{2.54 \text{ cm}}{1 \text{ in.}}$$

$$= 30 \cdot 2.54 \text{ cm}$$

$$= 76.2 \text{ cm}$$

Example 8   1520 ml = ? gallons

We do not know the relationship between milliliters and gallons. In fact, Table B.3 only gives us the relationship between liters and quarts. Therefore, we must go from milliliters to liters to quarts and then to gallons.

$$1520 \text{ ml} \doteq \frac{1520 \text{ ml}}{1} \cdot \frac{1 \text{ liter}}{1000 \text{ ml}} \cdot \frac{1.06 \text{ qts}}{1 \text{ liter}} \cdot \frac{1 \text{ gal}}{4 \text{ qts}}$$

$$= \frac{1520(1.06)}{4000} \text{ gal}$$

$$= 0.4028 \text{ gal}$$

Example 9   The tallest stalagmite in the world is in a cave in Aren Armand, France. Its height is 27 meters, 4 decimeters, and 3 centimeters. The longest free-hanging stalactite measures 11 meters, 5 decimeters, and 8 centimeters. How much longer is the stalagmite?

$$\begin{array}{r} 27 \text{ m} \quad 4 \text{ dm} \quad 3 \text{ cm} \\ -11 \text{ m} \quad 5 \text{ dm} \quad 8 \text{ cm} \\ \hline \end{array}$$

$$\begin{array}{r} 27 \text{ m} \quad 3 \text{ dm} \quad 13 \text{ cm} \\ -11 \text{ m} \quad 5 \text{ dm} \quad 8 \text{ cm} \\ \hline 5 \text{ cm} \end{array}$$

8 cannot be subtracted from 3, so 1 was borrowed from 4 dm (1 dm = 10 cm).

$$\begin{array}{r} 26 \text{ m} \quad 13 \text{ dm} \quad 13 \text{ cm} \\ -11 \text{ m} \quad 5 \text{ dm} \quad 8 \text{ cm} \\ \hline 15 \text{ m} \quad 8 \text{ dm} \quad 5 \text{ cm} \end{array}$$

5 cannot be subtracted from 3 so 1 m was borrowed from 27 m (1 m = 10 dm).

Alternate Solution:

Since 4 dm = 0.4 m, 3 cm = 0.03 m, 5 dm = 0.5 m, and 8 cm = 0.08 m,

$$27 \text{ m } 4 \text{ dm } 3 \text{ cm} = 27.43 \text{ m}$$

and

$$11 \text{ m } 5 \text{ dm } 8 \text{ cm} = 11.58 \text{ m}$$

So 27.43 m − 11.58 m = 15.85 m.

Example 9 illustrates a technique which can be used with denominate numbers having more than one unit. In problems such as this, you could express the number in one unit. However, this is not always the best approach. This is particularly true in situations involving times. Examples 10–14 illustrate some techniques which can be used in addition, subtraction, multiplication, and division when you have more than one unit involved. In these examples, you will need to use the facts that 1 hour = 60 minutes and 1 minute = 60 seconds.

Example 10 Simplify each of the following so that the number of minutes is less than 60 and so that the number of seconds is less than 60.

    a. 1 hr 75 min = 1 hr + 75 min
                    = 1 hr + 60 min + 15 min
                    = 2 hr + 15 min
                    = 2 hr 15 min

    b. 10 hr 92 min = 10 hr + 92 min
                     = 10 hr + 60 min + 32 min
                     = 11 hr 32 min

    c. 135 min = 60 min + 60 min + 15 min
               = 2 hr 15 min

    d. 2 hr 85 min 132 sec = 2 hr + 85 min + 132 sec
                          = 2 hr + 85 min + 60 sec + 60 sec + 12 sec
                          = 2 hr + 87 min + 12 sec
                          = 2 hr + 60 min + 17 min + 12 sec
                          = 3 hr 17 min 12 sec

With practice, you should be able to work problems such as these without showing any intermediate steps.

Example 11 Add and simplify.

    a.   5 hr 10 min
           6 hr 35 min
      +   10 min
        11 hr 55 min

    b.   15 hr 24 min
      + 2 hr 45 min
      17 hr 69 min = 18 hr 9 min

    c.   2 hr 40 min   43 sec
      +10 hr 45 min   59 sec
      12 hr 85 min 102 sec = 12 hr 86 min 42 sec
                            = 13 hr 26 min 42 sec

Example 12 Subtract the lower expression from the upper expression.

    a.   15 hr 34 min
      − 4 hr 12 min
        11 hr 22 min

b. $\overset{16}{\cancel{17}}\text{ hr }\overset{68}{\cancel{8}}\text{ min}$ You must borrow (1 hr = 60 min) in order to subtract.
$\quad\underline{-14\text{ hr }42\text{ min}}$
$\qquad 2\text{ hr }26\text{ min}$

c. $\overset{7}{\cancel{8}}\text{ hr }\overset{74}{\cancel{15}}\text{ min }\overset{92}{\cancel{32}}\text{ sec}$ You must borrow (1 min = 60 sec and 1 hr = 60 min)
$\quad\underline{-1\text{ hr }26\text{ min }42\text{ sec}}$
$\qquad 6\text{ hr }48\text{ min }50\text{ sec}$

d. $\begin{array}{l}14\text{ hr}\\ \underline{-\ 2\text{ hr }10\text{ min}}\end{array}\longrightarrow \begin{array}{l}\overset{13}{\cancel{14}}\text{ hr }60\text{ min}\\ \underline{-\ 2\text{ hr }10\text{ min}}\\ \ \ 11\text{ hr }50\text{ min}\end{array}$

Example 13   Multiply and simplify.

   a. 2(4 hr 10 min) = 8 hr 20 min

   b. 8(4 hr 10 min) = 32 hr 80 min
   $\qquad\qquad\qquad\quad\ \ = 33\text{ hr }20\text{ min}$

   c. 6(10 hr 32 min 43 sec) = 60 hr 192 min 258 sec
   $\qquad\qquad\qquad\qquad\qquad\ \ = 60\text{ hr }196\text{ min }18\text{ sec}$
   $\qquad\qquad\qquad\qquad\qquad\ \ = 63\text{ hr }16\text{ min }18\text{ sec}$

Example 14   Divide.

   a. (4 hr 10 min) ÷ 2 = 2 hr 5 min

   b. $\begin{array}{r}1\text{ hr}\quad 23\text{ min}\quad 20\text{ sec}\\ \hline 3\overline{)4\text{ hr}\quad 10\text{ min}}\\ \underline{3\text{ hr}}\qquad\qquad\\ 1\text{ hr} = \underline{60\text{ min}}\qquad\\ 70\text{ min}\qquad\\ \underline{69\text{ min}}\qquad\\ 1\text{ min} = 60\text{ sec}\\ \underline{60\text{ sec}}\end{array}$

B.3   Exercises

Use the rules for operations with denominate numbers to perform the indicated operations in Exercises 1-10.

1. 3 · 7.5 meters

2. 20 yards ÷ 7

3. 6 cm · 6 cm

4. 15 qts ÷ 3 qts

5. 10.5 miles + 50 miles

6. 52 km + 130.56 km

7. 2250 mm ÷ 150 mm

8. $2\frac{1}{2}\cdot 4$ feet

9. 28 grams ÷ $\frac{1}{2}$

10. 7 ft · 7.5 ft

In Exercises 11-20, make the indicated conversions between systems using the procedures given in this section.

11. 2 in. $\doteq$ _____ cm

12. 8 ft $\doteq$ _____ m

13. 9 oz $\doteq$ _____ g

14. 1.9 liter $\doteq$ _____ qt

15. 3300 ml $\doteq$ _____ qt

16. 1 mile $\doteq$ _____ m

17. 100 yards $\doteq$ _____ m

18. 5 kg $\doteq$ _____ oz

19. 2 gallons $\doteq$ _____ ml

20. 152 g $\doteq$ _____ lb

In Exercises 21–26, make the indicated conversions within the metric system by repositioning the decimal point. Check your result by working the problem using the techniques given in this section.

21. 4.3 kl = ? cl   **22.** 5023 cl = ? kl

23. 4.3 cm = ? km   **24.** 0.5023 ml = ? cl

25. 4.3 cg = ? mg   **26.** 0.502 km = ? cm

Simplify each of Exercises 27–34 so that the number of seconds is less than 60 and so that the number of minutes is less than 60.

27. 2 hr 85 min   **28.** 5 hr 96 min

29. 17 hr 310 min   **30.** 32 hr 150 min

31. 278 min   **32.** 10 hr 62 min

33. 4 hr 62 min 85 sec   **34.** 2 hr 18 min 93 sec

Add each of Exercises 35–40 and simplify if possible.

35. 3 hr 40 min
 1 hr 15 min

36. 10 hr 35 min
 5 hr 42 min

37. 5 hr 8 min
16 hr 42 min
 7 hr 19 min
10 hr 58 min

38. 87 min
 35 min
143 min

39. 2 hr 42 min 4 sec
15 hr 13 min 59 sec

40. 85 hr 16 min 8 sec
19 hr 45 min 24 sec

Subtract the lower expression from the upper expression in Exercises 41–46.

41. 16 hr 18 min 30 sec
 7 hr 2 min 2 sec

42. 14 hr 23 min
 7 hr 34 min

43. 72 hr
10 hr 43 min

44. 18 hr
 4 hr 8 min

45. 2 hr 24 min 19 sec
1 hr 39 min 39 sec

46. 18 hr   9 sec
 4 hr 12 min 18 sec

Multiply and then simplify in Exercises 47–52.

47. 2(2 hr 15 min)   **48.** 3(2 hr 15 min)

49. 6(2 hr 15 min)   **50.** 7(2 hr 15 min)

51. 7(14 hr 32 min 30 sec)   **52.** 4(3 hr 22 min 13 sec)

Divide in Exercises 53–58.

53. (2 hr 30 min) ÷ 2   **54.** (8 hr 16 min) ÷ 4

55. $\dfrac{15 \text{ hr } 40 \text{ min } 40 \text{ sec}}{5}$ **56.** $\dfrac{21 \text{ hr } 49 \text{ min } 7 \text{ sec}}{7}$

57. (5 hr 30 min) ÷ 3   **58.** (8 hr 15 min) ÷ 5

59. An astronaut made three trips into space. The first one was for 15 hours 16 minutes 32 seconds, the secone for 25 hours 10 minutes 40 seconds, and the third one for 73 hours 38 minutes 14 seconds. Find the total amount of time he has spent in space.

60. A marathon runner ran a road race in 1 hour 42 minutes. Another runner finished in 2 hours 13 minutes. What was the difference in their times?

61. A runner finished a race in 2 hours 37 minutes. The winner finished in half this time, and the last person to cross the finish line took twice as much time. Find the time of the winner and of the last-place finisher.

62. A child watches TV for the following number of hours per day for six days: 3 hr 30 min; 2 hr 45 min; 4 hr; 30 min; 2 hr 30 min; 5 hr 15 min. How much time did she spent watching TV? What was the average time per day?

63. The following table gives the sign-in and sign-out times for a worker during one day.

Time In	Time Out
7:01 A.M.	10:15 A.M.
10:22 A.M.	12:57 P.M.
2:00 P.M.	4:04 P.M.

Determine the total time the person worked.

64. A person worked the following lengths of time during one week:

7 hr 50 min
7 hr 45 min
8 hr 2 min
7 hr 58 min
8 hr 15 min

A worker cannot work less than 20 minutes under 40 hours without being docked, and he cannot work more than 20 minutes above 40 hours unless he is paid for overtime. Was this worker docked or paid any overtime?

65. The following is the time-in, time-out record of a worker for one day:

Time In	Time Out
6:58 A.M.	10:00 A.M.
10:10 A.M.	12:55 P.M.
1:45 P.M.	?

What time must the worker sign out in order to have an 8-hour day?

Important Terms   centi   kilo
Used in this Appendix   customary system   measure
                             deci   meter
                             deka   metric system
                             denominate number   milli
                             gram   liter
                             hecto   unit

REVIEW EXERCISES

Anwer each of Exercises 1–10 as true (T) or false (F).

1. The base unit of length in the metric system is the yard.

2. A kilometer is equal to 100 meters.

3. Centi means "0.01 of."

4. A kilogram is heavier than a milligram.

5. A hectoliter has a larger capacity than a centiliter.

6. One hundred centigrams equal one gram.

7. The symbol for milliliter is mll.

8. A new pencil is about one meter in length.

9. A dime is about one centimeter thick.

10. The basic unit of capacity in the metric system is the liter.

Express Exercises 11–26 in the indicated units.

11. 100 yds $\doteq$ _____ m   12. 100 ft $\doteq$ _____ m

13. 4000 km = _____ m   14. 4000 mm = _____ cm

15. 4000 cm = _____ mm

16. 4000 cl = _____ liters

17. 4000 m = _____ km   18. 4000 liters = _____ cl

19. 50 m $\doteq$ _____ in.   20. 120.1 cg = _____ g

21. 120.1 mg = _____ g   22. 120.1 g = _____ kg

23. 120.1 kg = _____ g   24. 120.1 cl = _____ ml

25. 50 g $\doteq$ _____ oz

26. 25 gallons $\doteq$ _____ liters

27. If hamburger meat costs 99¢ per pound, how much will it cost per kilogram? (Round to the nearest cent.)

28. A roast which weighs 5.2 kilograms is to be cut into 4 equal pieces. How much will each piece weigh?

29. If a board is 4.2 m long, what is the length in centimeters of four of these boards?

30. A seamstress needed 18.52 meters of material to make some draperies. She bought two pieces on sale. One measured 10.45 meters, and the other measured 11.4 meters. How much material was left after she made the drapes?

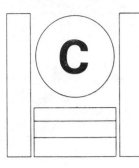

TABLES

Table C.1   Addition Facts

+	0	1	2	3	4	5	6	7	8	9	10	11	12
0	0	1	2	3	4	5	6	7	8	9	10	11	12
1	1	2	3	4	5	6	7	8	9	10	11	12	13
2	2	3	4	5	6	7	8	9	10	11	12	13	14
3	3	4	5	6	7	8	9	10	11	12	13	14	15
4	4	5	6	7	8	9	10	11	12	13	14	15	16
5	5	6	7	8	9	10	11	12	13	14	15	16	17
6	6	7	8	9	10	11	12	13	14	15	16	17	18
7	7	8	9	10	11	12	13	14	15	16	17	18	19
8	8	9	10	11	12	13	14	15	16	17	18	19	20
9	9	10	11	12	13	14	15	16	17	18	19	20	21
10	10	11	12	13	14	15	16	17	18	19	20	21	22
11	11	12	13	14	15	16	17	18	19	20	21	22	23
12	12	13	14	15	16	17	18	19	20	21	22	23	24

Table C.2   Multiplication Facts

×	0	1	2	3	4	5	6	7	8	9	10	11	12
0	0	0	0	0	0	0	0	0	0	0	0	0	0
1	0	1	2	3	4	5	6	7	8	9	10	11	12
2	0	2	4	6	8	10	12	14	16	18	20	22	24
3	0	3	6	9	12	15	18	21	24	27	30	33	36
4	0	4	8	12	16	20	24	28	32	36	40	44	48
5	0	5	10	15	20	25	30	35	40	45	50	55	60
6	0	6	12	18	24	30	36	42	48	54	60	66	72
7	0	7	14	21	28	35	42	49	56	63	70	77	84
8	0	8	16	24	32	40	48	56	64	72	80	88	96
9	0	9	18	27	36	45	54	63	72	81	90	99	108
10	0	10	20	30	40	50	60	70	80	90	100	110	120
11	0	11	22	33	44	55	66	77	88	99	110	121	132
12	0	12	24	36	48	60	72	84	96	108	120	132	144

Table C.3 Squares and Square Roots*

n	n^2	$\sqrt{n}$		n	n^2	$\sqrt{n}$
1	1	1.000		51	2,601	7.141
2	4	1.414		52	2,704	7.211
3	9	1.732		53	2,809	7.280
4	16	2.000		54	2,916	7.348
5	25	2.236		55	3,025	7.416
6	36	2.449		56	3,136	7.483
7	49	2.646		57	3,249	7.550
8	64	2.828		58	3,364	7.616
9	81	3.000		59	3,481	7.681
10	100	3.162		60	3,600	7.746
11	121	3.317		61	3,721	7.810
12	144	3.464		62	3,844	7.874
13	169	3.606		63	3,969	7.937
14	196	3.742		64	4,096	8.000
15	225	3.873		65	4,225	8.062
16	256	4.000		66	4,356	8.124
17	289	4.123		67	4,489	8.185
18	324	4.243		68	4,624	8.246
19	361	4.359		69	4,761	8.307
20	400	4.472		70	4,900	8.367
21	441	4.583		71	5,041	8.426
22	484	4.690		72	5,184	8.485
23	529	4.796		73	5,329	8.544
24	576	4.899		74	5,476	8.602
25	625	5.000		75	5,625	8.660
26	676	5.099		76	5,776	8.718
27	729	5.196		77	5,929	8.775
28	784	5.292		78	6,084	8.832
29	841	5.385		79	6,241	8.888
30	900	5.477		80	6,400	8.944
31	961	5.568		81	6,561	9.000
32	1,024	5.657		82	6,724	9.055
33	1,089	5.745		83	6,889	9.110
34	1,156	5.831		84	7,056	9.165
35	1,225	5.916		85	7,225	9.220
36	1,296	6.000		86	7,396	9.274
37	1,369	6.083		87	7,569	9.327
38	1,444	6.164		88	7,744	9.381
39	1,521	6.245		89	7,921	9.434
40	1,600	6.325		90	8,100	9.487
41	1,681	6.403		91	8,281	9.539
42	1,764	6.481		92	8,464	9.592
43	1,849	6.557		93	8,649	9.644
44	1,936	6.633		94	8,836	9.695
45	2,025	6.708		95	9,025	9.747
46	2,116	6.782		96	9,216	9.798
47	2,209	6.856		97	9,409	9.849
48	2,304	6.928		98	9,604	9.899
49	2,401	7.000		99	9,801	9.950
50	2,500	7.071		100	10,000	10.000

*Square roots of numbers which are not perfect squares have been rounded to three decimal places and are therefore approximations.

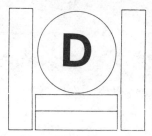

FIELD PROPERTIES

Closure Law for Addition of Real Numbers

For any two elements a and b in the set R, the sum a + b is a unique element of R.

Closure Law for Multiplication of Real Numbers

For any two elements a and b in the set R, the product a · b is a unique element of R.

Commutative Law of Addition of Real Numbers

For any two elements a and b in R, it is always true that a + b = b + a.

Commutative Law of Multiplication of Real Numbers

For any two elements a and b in R, it is always true that a · b = b · a.

Associative Law of Addition of Real Numbers

For any elements a, b, and c in R, it is always true that (a + b) + c = a + (b + c).

Associative Law of Multiplication of Real Numbers

For any elements a, b, and c in R, it is always true that (a · b) · c = a · (b · c).

The Identity for Addition of Real Numbers

There is a unique real number 0 such that for any element a in R, a + 0 = 0 + a = a.

The Identity for Multiplication of Real Numbers

There is a unique real number 1 such that for any element a in R, a · 1 = 1 · a = a.

The Additive Inverse of a Real Number

For each real number a, there exists a unique real number −a such that a + (−a) = (−a) + a = 0.

The Multiplicative Inverse of a Nonzero Real Number

For each a in R, a ≠ 0, there is a unique element, $\frac{1}{a}$, in R such that a · $\frac{1}{a}$ = $\frac{1}{a}$ · a = 1.

The Distributive Property of Multiplication Over Addition

For any elements a, b, and c in R, it is always true that a(b + c) = ab + ac.

487

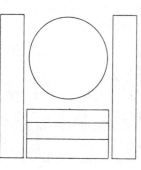

ANSWERS TO PRETESTS, ODD-NUMBERED EXERCISES, AND REVIEW EXERCISES

Answers to Pretests

Chapter 1, page 3

1. a. hundreds **b.** ones or units **c.** hundredths
d. hundred-thousandths **e.** ten-thousands **f.** thousands
2. a. 1039 **b.** 300.7 **c.** 5,300,740 **d.** 9000.000003
3. a. three hundred seven **b.** four thousand nine and
eight hundredths **c.** thirty-seven ten-thousandths
d. four million seven hundred ten and nine thousandths
4. a. 583.5 **b.** 600 **c.** 583.55 **d.** 584
5. a. 83.0818 **b.** 522.501 **6. a.** 24.38 **b.** 15.943
7. a. 323.2479 **b.** 59.5738143 **8. a.** 0.87 **b.** 14.3
9. a. $145.25 **b.** 41.04 centimeters **c.** 563.6 miles
d. $10.17 **10. a.** +, 201.209 **b.** ÷, 0.00002
c. −, 169.5 **d.** −, 1.852 **e.** −, 97.24 **f.** −, 0.1611
g. ÷, 0.11 **h.** ×, 14.5597

Chapter 2, page 32

1. a. 1, 2, 3, 6, 7, 14, 21, 42 **b.** Any four of 42, 84, 126,
168, etc. **2. a.** 3 · 37 **b.** 11 · 13 **c.** 2 · 2 · 3 · 3 · 7
d. 2 · 3 · 5 · 23 **e.** 11 · 41 **f.** 3 · 3 · 3 · 13 **3. a.** $\frac{11}{13}$
b. $\frac{8}{19}$ **c.** $\frac{7}{17}$ **d.** 35 **e.** 426 **f.** 105 **4. a.** $\frac{45}{56}$ **b.** $\frac{11}{15}$
c. $\frac{21}{8}$ **5. a.** $\frac{3}{5}$ **b.** $\frac{14}{15}$ **c.** $\frac{1}{7}$ **6. a.** 360 **b.** 180
7. a. $\frac{1}{2}$ **b.** $\frac{5}{6}$ **c.** $\frac{11}{35}$ **8. a.** $\frac{2}{3}$ **b.** $\frac{1}{6}$ **c.** $\frac{17}{30}$

9. a. 8 symbols **b.** $\frac{1}{12}$ cup **c.** 748 shares
d. $\frac{11}{12}$ of a book **e.** $\frac{1}{20}$ of the TV sets

Chapter 3, page 70

1. a. $\frac{108}{7}$ **b.** $\frac{289}{26}$ **c.** $\frac{614}{3}$ **2. a.** $3\frac{4}{5}$ **b.** $157\frac{5}{13}$
c. $22\frac{2}{23}$ **3. a.** $22\frac{1}{2}$ **b.** $9\frac{1}{6}$ **c.** $18\frac{7}{30}$ **d.** $10\frac{31}{180}$
4. a. $14\frac{4}{17}$ **b.** $6\frac{3}{4}$ **c.** $3\frac{1}{30}$ **d.** $3\frac{5}{78}$ **5. a.** 18 **b.** $43\frac{2}{7}$
c. $3\frac{5}{7}$ **d.** 14 **6. a.** 2 **b.** $2\frac{2}{3}$ **c.** $\frac{7}{10}$ **d.** $\frac{3}{8}$
7. a. 24 patients **b.** 380 horsepower **c.** $2\frac{3}{8}$ carats
d. $1\frac{7}{8}$ hours

Chapter 4, page 88

1. a. 7.25 **b.** 0.375 **c.** 2.9375 **2. a.** $\frac{1}{16}$ **b.** $\frac{9}{25}$
c. $\frac{41}{20}$ **3. a.** 0.418 **b.** 0.00072 **c.** 0.0025
4. a. $\frac{102}{25}$ **b.** $\frac{43}{50}$ **c.** $\frac{3}{500}$ **5. a.** 43.2 **b.** 132.16

c. 0.059 **6. a.** 43% **b.** 1.5% **c.** 4180%
7. a. 18.75% **b.** 62.5% **c.** 325% **8. a.** 77.8%

b. 31.5 points **c.** $411.25 **9. a.** 12.45 **b.** $\dfrac{239}{20}$

c. $\dfrac{3}{5}$ **d.** 0.1275

2. a. $4x^2 - x + 1$ **b.** $x^3 + 9x^2 + 5x + 2$
c. $5x^2 - 11x + 10$ **d.** $-x^3 + 8x^2 + 5x + 9$
e. $7h^2 - 4hk + 8k^2$ **3. a.** $-3x^2 - 4$ **b.** $-2h^2 + 3k^2$
c. $-2x^3 - 3x^2 + 2x + 4$ **d.** $-5h^2 + 6hk - 6k^2$
e. $3x^3 + 8x^2 - 15x - 1$ **4. a.** $2x + 4$ **b.** $-10x + 2y$
c. $3y + 3$ **d.** $4x + y$ **e.** x

Chapter 5, page 114

1.

$$-5 \quad -2\tfrac{3}{4}\ -1.9 \qquad\qquad 3\tfrac{1}{2}$$

$$\xleftarrow{\hspace{1em}} \begin{array}{ccccccccccccc} & \bullet & & \bullet & & \bullet & & & & \bullet & & & \\ \hline -6 & -5 & -4 & -3 & -2 & -1 & 0 & 1 & 2 & 3 & 4 & 5 & 6 \end{array} \xrightarrow{\hspace{1em}}$$

2. a. 17, 17 **b.** 0, 0 **c.** -17, 17 **d.** $-1\tfrac{3}{4}, 1\tfrac{3}{4}$

e. 403, 403 **f.** $\dfrac{2}{9}, \dfrac{2}{9}$ **3. a.** T **b.** F **c.** F **d.** F **e.** T

4. a. 49 **b.** 9 **c.** 6.5 **d.** 5.7 **e.** 400 **f.** 24

5. a. -2 **b.** 25.85 **c.** 0.99 **d.** -419 **e.** 2.9 **f.** $-\dfrac{1}{7}$

6. a. 10.7 **b.** -205.9 **c.** -10.7 **d.** -10.7 **e.** 205.9

f. $\dfrac{1}{19}$ **7. a.** -12 **b.** -58 **c.** -58 **d.** -12 **e.** -20

8. a. $-11 - 8$ **b.** $+11 - 8$ **c.** $-11 + 8$ **d.** $-11 + 8$

9. a. 51.24 **b.** $\dfrac{2}{7}$ **c.** -39.56 **d.** -0.000432 **e.** -120

10. a. -61 **b.** -0.023 **c.** 91 **d.** $-\dfrac{5}{7}$ **e.** $\dfrac{11}{10}$ or $1\tfrac{1}{10}$

11. a. -15 **b.** -14 **c.** 11 **d.** -11 **12. a.** -28 **b.** 10
c. 26 **d.** -5 **e.** 49 **f.** 292 **13.** $5505

Chapter 6, page 149

1. a. $21, 2.41, \pi, \sqrt{3}$ **b.** a, x, r, m **c.** constant factor: 3;
variable factors: a, b, c **2. a.** 0 **b.** -6 **c.** -9 **d.** 1
e. undefined **f.** 8 **g.** 12 **h.** -26 **i.** 0 **3. a.** $6(x + 5)$

b. $\tfrac{1}{3}(r + 9)$ **c.** $3x + \dfrac{8}{x}$ **d.** $(a + b)^3$ **4. a.** Four times m

plus three **b.** Four times the quantity m plus three
c. Divide the quantity a plus b by 2 **d.** a plus b divided
by 2 **5. a.** $3 \cdot a \cdot a$ **b.** $(3a)(3a)$ **c.** $x \cdot x \cdot x \cdot y$
d. $(x + y)(x + y)(x + y)$ **e.** $-2ab \cdot b \cdot b$
f. $3a \cdot a \cdot (bc)(bc)(bc)(bc)$ **g.** $-2a^2b^3$ **h.** $(3a)^3$
i. $(a + b)^2$ **j.** x^4y^3 **k.** $(-m)^4$ **l.** 3^2a^2b **m.** 8 **n.** -81
o. 108 **p.** 25 **q.** -25 **r.** -1 **6. a.** 2.4×10^3
b. 2.562×10^2 **c.** 2.9234×10 **d.** 3.45×10^4 **e.** 5×10^5
f. 3.2×10^4 **7. a.** binomial **b.** monomial **c.** trinomial
d. none of these **e.** monomial **f.** binomial **g.** $3a^2$,
$ab, -b^4$ **h.** 3, 1, -1 **i.** terms: 2, 2, 4; polynomial: 4

Chapter 7, page 163

1. a. like **b.** unlike **c.** like **d.** unlike **e.** unlike

---

Chapter 8, page 176

1. a. a^9 **b.** a^5 **c.** a^7 **d.** a^6 **e.** b^8 **2. a.** x^6 **b.** $-a^{12}$
c. $-8a^6$ **d.** $a^{12}b^8c^{18}$ **e.** $3ab^6c^2$ **3. a.** a^3b^3 **b.** $9a^2b^2$
c. $-9a^2b^2$ **d.** $a^4b^4c^4$ **e.** $a^8b^{12}c^4$ **4. a.** $6a^2b$ **b.** $30a^4b^3$
c. $-10a^2b$ **d.** $18a^4b$ **e.** $-135a^5b$ **5. a.** $3a + 3b$
b. $-4a + 4b$ **c.** $3a^4b - 12a^3b^2$ **d.** $18a^6b - 27a^5b^3$
e. $-192a^5 + 384a^4b^3$ **6. a.** $x^2 + x - 12$
b. $2x^2 + 11x - 6$ **c.** $9a^2 + 24a + 16$ **d.** $x^3 - 6x^2 - x + 30$
e. $50x^3 + 55x^2 - 28x + 3$ **7. a.** $2a^2 + 12a - 2ab + 8$
b. $-9x - 8y$ **c.** $-12a$ **d.** $-3a - 4c + 18$ **e.** $-a^2 + 10a - 25$

Chapter 9, page 189

1. a. a^3 **b.** a **c.** $\dfrac{1}{a^4}$ **d.** $\dfrac{1}{a^3}$ **e.** a^4 **2. a.** $\dfrac{a^3}{b^3}$ **b.** $\dfrac{16a^4}{b^8}$

c. $\dfrac{-27a^6}{b^9}$ **d.** $-\dfrac{9a^2}{b^2}$ **3. a.** $3a^2$ **b.** $-5bc^3$ **c.** $\dfrac{-3a^3b^3d^2}{c^3e}$

d. $\dfrac{y^9}{7z^4}$ **e.** $\dfrac{16x^9}{-27y^7}$ **4. a.** $2a + 3b$ **b.** $b^2 + b - 1$

c. $-5ab - a$ **d.** $-6a + 2a^2$ **e.** $7x^2 - 5x - 2$

5. a. $3a + 1$ **b.** $9a^2 + 3a + 1$ **c.** $5a + 1 + \dfrac{-5}{4a - 5}$

d. $3x^3 + 3x^2 + 3x - 3 + \dfrac{2}{x - 1}$

Chapter 10, page 205

1. a. F **b.** T **c.** T **d.** T **e.** F **f.** T
2. a. conditional **b.** contradiction **c.** identity
d. conditional **e.** identity **f.** contradiction
3. a. $x = 6$ **b.** $x = -2$ **c.** $x = 6$ **d.** $x = 6$ **e.** $x = -2$
f. $x = -6$ a, c, and d are equivalent; b and e are equivalent
4. a. linear equation in one variable **b.** not a linear
equation in one variable **c.** linear equation in one
variable **d.** not a linear equation; it is a second-degree
equation. **e.** linear equation in one variable
f. not a linear equation in one variable
5. a. $w = -2$ **b.** $x = -72$ **c.** $y = -45$ **d.** $z = -36$

e. $t = 2$ **f.** $s = \dfrac{5}{2}$ or $2\tfrac{1}{2}$ **6. a.** $x = \dfrac{20}{3}$ **b.** $y = -10$

c. $z = -15$ **d.** $s = -1$ **e.** $r = 0$ **f.** $v = 48$

7. a. $x = -2.5$ **b.** $n = 2$ **c.** $h = \dfrac{1}{2}$ **d.** contradiction (no solution) **8. a.** $v = -\frac{5}{2}$ **b.** $m = 4$ **c.** $s = -\frac{13}{3}$

d. $n = -4.15$ **9. a.** $t = \frac{7}{3}$ **b.** $r = -0.2$ **c.** identity

(Every real number is a solution.) **d.** $s = -2$

10. a. $x = 9$ **b.** $z = 6$ **c.** $y = -\frac{16}{5}$ **d.** $w = -12$

11. a. $4x + 3 = 13; x = 2\frac{1}{2}$ **b.** $7x + 18 = 74; x = 8$

c. $2x - 3 = 5x; x = -1$ **d.** $5 = 10x + 3; x = \frac{1}{5}$

12. a. $w = \frac{A}{l}$ **b.** $t = \frac{I}{pr}$ **c.** $y = \frac{x+5}{2}$ **d.** $b = 5c - a$

Chapter 11, page 234

1. a. $x + 9$ **b.** $x - 9 = w$ **c.** $9 - x = x + 4$ **d.** $\frac{x}{9}$

e. $\frac{2}{3}x = 7$ **f.** $2x - 4$ **g.** $6y$ **h.** $2(y + 3)$ **i.** $y^2 = 16$

j. $5(y - 2)$ **2. a.** $85 - x$ **b.** $\frac{58}{y}$ **c.** $w, 2w - 5$

d. $n + (n + 1) + (n + 2)$ or $3n + 3$ **e.** $\frac{130}{w}$

3. a. $101, 103, 105$ **b.** $-3, 2, 9$ **4. a.** $1,500$ quarters
b. 30 milliliters of 20% and 10 milliliters of 40% solution
5. a. 5.75% **b.** \$4,000 in savings account, \$26,000 in
certificate **6. a.** 11,000 feet, approximately 2 miles
b. 7 kilometers per hour and 8 kilometers per hour

Chapter 12, page 265

1. a. $\frac{4}{5}$ **b.** $\frac{1}{3}$ **c.** $\frac{2}{1}$ **d.** $\frac{2}{3}$ **e.** $\frac{2}{3}$ **f.** $\frac{2}{3}$ **g.** $\frac{2}{3}$ **h.** $\frac{2}{3}$ **i.** $\frac{5}{3}$

j. $\frac{4}{3}$ **2. a.** the 8-ounce box at a unit cost of 7.875¢ per

ounce **b.** Chick Hafey won with an average of 0.3489.
3. a. $x = 6$ **b.** $x = 57$ **c.** $x = 7$ **d.** $x = 18$ **e.** 11.2 inches
4. a. 12% **b.** 1680 defectives **c.** \$23,000
5. a. 24 inches **b.** 12 inches **c.** 28 inches
d. 18π in. $\doteq 56.52$ in. **6. a.** 14 sq in. **b.** 16 sq in.
c. 15 sq in. **d.** 36π sq in. $\doteq 113.04$ sq in. **7. a.** 24 in.³
b. 36π in.³ $\doteq 113.04$ in.³ **c.** 44π in.³ $\doteq 138.16$ in.³
d. 120π in.³ $\doteq 376.8$ in.³ **8. a.** 200' **b.** Yes, the walls
are perpendicular since $8^2 + 15^2 = 17^2$. **9. a.** rational
b. rational **c.** irrational **d.** rational **e.** irrational
10. a. $\frac{3}{8}$ **b.** $\frac{22}{9}$ **c.** $\frac{58}{11}$

Chapter 13, page 313

1. a. Two is less than ten. **b.** Five is greater than
negative ten. **c.** Six is greater than or equal to six.

d. Seven is less than or equal to twenty. **2. a.** $9 \leqslant 20$
b. $7 > 5$ **c.** $11 \geqslant 11$ **d.** $-9 < 0$ **3. a.** T **b.** F **c.** T
d. F **e.** F **f.** T **4. a.** contradiction; no solution
b. conditional inequality; $x > -6$ **c.** contradiction;
no solution **d.** conditional inequality; $x < 0$
e. absolute inequality; x can be any real number
f. conditional inequality; $x > 0$

5. a.

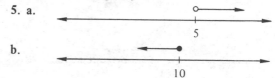

b.

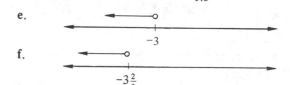

c.

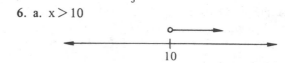

d.

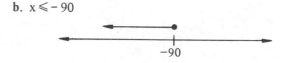

e.

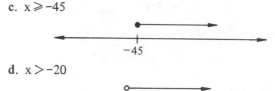

f.

6. a. $x > 10$

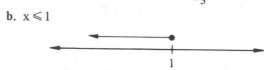

b. $x \leqslant -90$

c. $x \geqslant -45$

d. $x > -20$

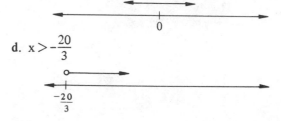

7. a. $x < -3$

b. $x \leqslant 1$

c. absolute inequality; every real number is a solution

d. $x > -\frac{20}{3}$

8. a. $x < 3$

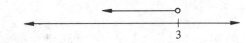

b. $b > 1$

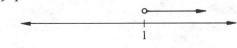

c. $x \geq \dfrac{-3}{11}$

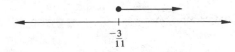

d. $x \geq \dfrac{147}{32}$ or $4\dfrac{19}{32}$

Chapter 14, page 335

1. a. 5 cents **b.** Personal income derived from government payments is increasing. **c.** \$140 **d.** D, H **e.** 14 milligrams **f.** G **g.** Food **h.** \$120 **i.** \$1440 **j.** \$20 **k.** March **l.** 5 months

2. a.

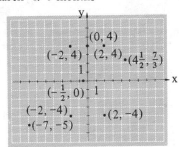

b. $(2, 4)$, I; $\left(-\dfrac{1}{2}, 0\right)$, x-axis; $(-7, -5)$, III; $(0, 4)$, y-axis;

$(-2, 4)$, II; $(2, -4)$, IV; $(-2, -4)$, III; $\left(4\dfrac{1}{2}, \dfrac{7}{3}\right)$, I

c. A, $\left(-6, -1\dfrac{1}{2}\right)$; B, $(5, 3)$; C, $(-1, 3)$; D, $(2, 0)$; E, $(0, 0)$;

F, $\left(1\dfrac{1}{2}, -1\dfrac{1}{2}\right)$; G, $\left(0, 2\dfrac{1}{2}\right)$; H, $\left(0, -3\dfrac{1}{2}\right)$; I, $(-3, 0)$;

J, $\left(-\dfrac{1}{2}, -\dfrac{1}{2}\right)$

3. a.

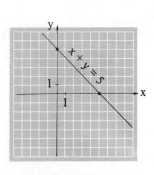

b.

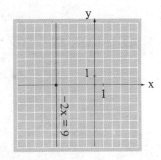

c.

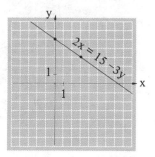

d.

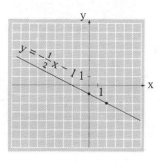

4. a. $y = -2x + 5$ **b.** $y = \dfrac{3}{4}x - \dfrac{9}{4}$ **c.** $y = \dfrac{1}{3}x - \dfrac{8}{3}$

d. $y = -\dfrac{2}{3}x + 5$ **5. a.** $-\dfrac{2}{1}$ **b.** $\dfrac{3}{4}$ **c.** $\dfrac{1}{3}$ **d.** $-\dfrac{2}{3}$

6. a.

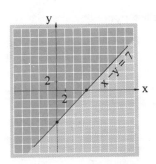

b.

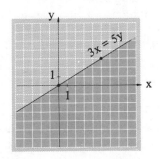

c.

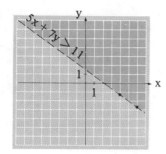

d.

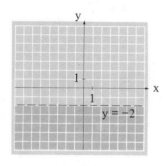

Chapter 15, page 358

1. a. $(2, 4)$ **b.** $(3, 3)$ **c.** equivalent equations; any point on the line is a solution **d.** $(6, 9)$ **2. a.** $(2, 4)$
b. $(3, 3)$ **c.** equivalent equations; an infinite number of solutions **d.** $\left(6, \frac{9}{4}\right)$ **3. a.** $(2, 4)$ **b.** $(3, 3)$
c. equivalent equations; an infinite number of solutions
d. $\left(6, \frac{9}{4}\right)$ **4. a.** unique solution **b.** no solution
c. infinite number of solutions **d.** unique solution
e. infinite number of solutions **f.** no solution
5. a. $(-2, 3)$ **b.** $(6, 2)$ **c.** $(0, 0)$ **d.** infinite number of solutions **e.** $(6, 10)$ **f.** no solution
6. 7 first number, 9 second number

Chapter 16, page 375

1. a. 15 **b.** 1 **c.** 3y **d.** $15ab^2$ **e.** $3a - 4$
2. a. $3y(6xz + 3x + z)$ **b.** $15ab^2(2a^3b^4 - 3ab^2 + 5)$
3. a. $4x^2 + 15x - 25$ **b.** $4x^2 - 15xy - 25y^2$
c. $4x^2 - 25x + 25$ **d.** $x^2 - x - 12$ **4. a.** $16x^2 - 25y^2$
b. $16x^2 - 40xy + 25y^2$ **c.** $16x^2 + 40xy + 25y^2$
5. a. $(x - 8)(x + 2)$ **b.** $(a + b)(a + 2b)$
c. $(x + 7)(x + 5)$ **d.** $(6n - 7)(n + 1)$
6. a. $(n + m)(n - m)$ **b.** $(3x + 5)^2$
c. $(x - 3)(x^2 + 3x + 9)$ **d.** $(2t + 4s)(4t^2 - 8st + 16s^2)$
7. a. $(2y + 3)(4x + 7)$ **b.** $(x - 2)(y + x + 2)$
c. $(a + b + 1)(a - b + 1)$ **8. a.** $4(5x + y)(5x - y)$
b. $12(a + 6)(a - 1)$ **c.** $2(4 - 2x + x^2)$
d. $3x(x - 3)(x^2 + 3x + 9)$ **e.** $(y - 5)(y - 5)$ **f.** prime

---

g. $(3y - 2)(4x + 9)$ **h.** $(x + 4)(x^2 - 4x + 16)$
i. $(x - 2)(x + 1)$ **j.** $(x^2 + 6)(2x - 7)$

Chapter 17, page 405

1. a. 2 **b.** -1 **c.** $\dfrac{a + b}{a - b}$ **d.** $x^2 + x + 1$ **2. a.** $\dfrac{14b^2}{72a^2b^3}$
b. $\dfrac{15a^2}{72a^2b^3}$ **c.** $\dfrac{7(x - 5)}{-1(x + 4)(x - 3)(x - 5)}$
d. $\dfrac{6(a + 2)(a - 2)}{6(a - 5)(a + 5)(a + 2)}$ **3. a.** $\dfrac{a + b}{x}$ **b.** -1
4. a. $\dfrac{1}{3a^2b}$ **b.** $\dfrac{13x - 3}{x^2 + x - 12}$ **c.** $\dfrac{-1}{3}$ **5. a.** $72a^2b^3$
b. $(x + 4)(x - 3)(x - 5)$ **c.** $15(a - 3)$ **d.** $6(a + 5)(a - 5)$
6. a. $\dfrac{15a^2 + 14b^2}{72a^2b^3}$ **b.** $\dfrac{9}{(x + 4)(x - 5)}$ **c.** $\dfrac{-a + 9}{15(a - 3)}$
7. a. $\dfrac{x - 1}{x + 1}$ **b.** $\dfrac{x - 3}{3}$ **c.** $\dfrac{9b^2}{14x}$ **8. a.** $\dfrac{11a + 65}{6(a + 5)(a - 5)}$
b. $\dfrac{x^2}{x^2 - 1}$ **c.** $\dfrac{1 + a}{1 - a}$ **9. a.** $x = 3$ **b.** no solution
c. $x = 0$

Chapter 18, page 437

1. a. $-2\sqrt{2}$ **b.** $13\sqrt{2}$ **c.** $-2\sqrt{x}$ **d.** $2\sqrt{5}$ **e.** $4x$
f. $4\sqrt{15}$ **g.** $\sqrt{10}$ **h.** $\dfrac{\sqrt{15}}{3}$ **i.** $x\sqrt{6}$ **2. a.** $4\sqrt{6}$
b. $\dfrac{\sqrt{35}}{7}$ **c.** xy^2 **d.** $14\sqrt{3}$ **e.** $9xy^5\sqrt{xz}$ **f.** $\dfrac{\sqrt{5x}}{5x}$
3. a. quadratic **b.** not quadratic **c.** quadratic
d. quadratic **e.** not quadratic **f.** quadratic
g. $12x^2 - 6x + 9 = 0$ **h.** $3x^2 + 5x - 8 = 0$
i. $x^2 - 6x + 17 = 0$ **j.** $x^2 - 6x + 35 = 0$ **k.** $x^2 - 12 = 0$
l. $3x^2 - 81x = 0$ **4. a.** $x = 8, x = 2$ **b.** $x = \dfrac{-7}{2}, x = \dfrac{7}{2}$
c. can't solve by factoring **d.** can't solve by factoring
e. can't solve by factoring **f.** $x = -9, x = 3$
5. a. $-1 + \sqrt{2}, -1 - \sqrt{2}$ **b.** $-1, -\dfrac{1}{2}$ **c.** $-1, \dfrac{1}{3}$
d. $+\sqrt{10}, -\sqrt{10}$ **e.** $5.2, -0.2$
6. a. 10 is Jack's age now, 13 is Jim's age now.
b. 9 meters is the width, 19 meters is the length.

Appendix A, page 457

1. a. $A = \{4, 17, 35\}$ **b.** $5 \notin S$ **c.** $A = C$
d. $1 \in S$ and $2 \in S$ **2. a.** The set containing seven is a subset of the set of odd natural numbers. **b.** The set containing eight is not a subset of the set with element thirty-eight. **c.** The set with elements nine and three is a subset of the set with elements one, three, and nine.

3. a. $\dfrac{25}{100}$ or $\dfrac{1}{4}$ b. $-\dfrac{18}{1}$ c. $\dfrac{13}{3}$ d. $-\dfrac{507}{100}$ e. $-\dfrac{7}{2}$ f. $\dfrac{0}{1}$

4. a. T   b. F   c. F   d. F   e. F

5. a. $\{2, 3, 5, 6, 7, 8, 35\}$ b. $\{1, 2, 3, 4\}$ 6. a. $\{2, 8\}$

b. $\{1, 2, 3\}$ c. ϕ 7. a. T   b. T   c. F   d. T   e. T

f. T   g. T   h. F   i. T   j. T

Appendix B, page 467

1. a. length   b. length   c. mass   d. volume   e. length

f. volume   2. a. 0.01 of, c   b. 0.001 of, m

c. 1000 times, k   d. hecto   e. deci   f. deka   3. a. cm

b. m   c. kg   d. km   e. ml   f. centiliter   g. deciliter

h. hectogram   i. dekameter   j. kilogram   k. centimeter

4. a. 100 cg   b. 0.52 cm   c. 200 liters   d. 5.2 m

e. 4020 cm   f. 2 dl   g. 0.001 km   h. 0.5 liters

i. 5.0002 kg   j. 210 cl   k. 12 kg   l. 1000 ml

5. a. 1 kg $\doteq$ 2.2 lb   b. 1 liter $\doteq$ 1.06 qt   c. 1 m $\doteq$ 39.37 in.

d. 1 cm $\doteq$ 0.4 in.   e. 1 km $\doteq$ 0.6 mi   6. a. 12 m

b. 5 kg   c. 15.3 liters   d. 50 ml   7. a. 25 m^2

b. 19.32 m   c. 3.8 liters   d. 20   8. a. 3 in.

b. 196.85 in.   c. 4.4 lb   d. 148.2 mi   e. 2.12 qt

f. 107.5 cm   g. 1.075 m   h. 2.25 kg   i. 395.2 km

j. 9.5 liters   9. 1 liter; 2 liters; 4 liters

Answers to Odd-Numbered Exercises and Review Exercises

Chapter 1

1.1 Exercises, page 8

1. tens   3. thousands   5. tenths   7. hundred-thousandths   9. ones or units   11. hundredths   13. millions   15. thousandths   17. 2 ten thousands; 7 thousands; 3 hundreds; 9 tens; 8 ones or units; 6 tenths; 1 hundredths; 4 thousandths   19. 5740.   21. 23.   23. 042, 42., 42.000   25. 561   27. 0.0429   29. 1800.018   31. 5602   33. 500.0072   35. 5000.05   37. five hundred eighty-two   39. seventy-two and three thousand four hundred ninety-two ten-thousandths   41. one hundred five hundred-thousandths   43. three hundred fifty-one   45. fourteen and two thousand four hundred fifty-six ten-thousandths   47. fifty-six hundred-thousandths   49. five billion seven hundred thirteen thousand four hundred

1.2 Exercises, page 10

1. 37.8   3. 0.1   5. 10.0   7. 4814.6   9. 135   11. 10   13. 754   15. 1   17. 19.96   19. 9.99   21. 122.93   23. 1,784,320.50   25. 2000   27. 1000   29. 1,784,000   31. 53,000   33. 2300   35. 0.1235   37. 19.960   39. 26,000,000

1.3 Exercises, page 13

1. 1393   3. 350   5. 4498   7. 252.5   9. 101.53   11. 41.37   13. $91.30   15. 7024.374   17. $2230.84   19. 2134.9444   21. 182.4100   23. 5472.4194   25. +, 1265.063   27. total, $1066.10   29. increased by, 2453.22   31. more than, 31.48   33. +, 5237.3575   35. total, 35.4   37. plus, 130.6808   39. +, 317.7456   41. add, 76,535.122   43. sum, 673.865   45. sum, 1004.102

1.4 Exercises, page 16

1. 33   3. 939   5. 5680   7. $18.09   9. 340.5   11. 80.366   13. 30.47   15. 226.9   17. 44.0   19. 0.2556   21. 91.8   23. 554.496   25. 751.001   27. −, 146.19   29. minus, 14.7461   31. subtract, 7678.93   33. −, 109.933   35. subtract, 340.322   37. less, 45.109   39. diminished by, 30   41. −, 0.689   43. difference or subtracted from, 5810.43   45. subtract, 49,862.638

1.5 Exercises, page 18

1. 450   3. 1932   5. 3460   7. 99.33   9. 370.666   11. 0.1458   13. 0.008472   15. 6.298

17. 4948.743   19. 356.3509   21. 431.067   23. 43106.7   25. 1238.2   27. 123000   29. 1230   31. 4.31067   33. 0.0431067   35. 12.382   37. 0.123   39. 0.00123   41. triple, 52.6293   43. ()(), 655.5   45. ·, 5600   47. multiply, 403,250.4   49. ()(), 16.7616   51. X, 64.169   53. double, 6.5   55. ·, 6,840,576   57. multiply, 78240   59. multiply, 0.000256

1.6 Exercises, page 22

1. 309   3. 500   5. 301   7. 4.89   9. 50.03   11. 1.406   13. 3.68   15. 700   17. 0.061   19. 56.5   21. 11.8728   23. 0.118728   25. 0.57841   27. 11.8   29. 0.00892   31. 1187.28   33. 118728   35. 57841   37. 1180   39. 6660   41. divide, 3.8   43. quotient or divided by, 514.9   45. quotient, 2.08   47. quotient, 131   49. ÷, 2.04   51. divide, 100.5   53. divide, 5.684   55. divide, 0.493   57. divide, 1.91   59. divide, 0.0706

1.7 Exercises, page 27

1. $275.82   3. 56.1 gal   5. $565.77   7. $57.525 ≐ $57.53   9. $73   11. $229.41   13. 15.7 miles per gal   15. 1728.3 pounds, less   17. 32 oz for $.48   19. 344 sq ft   21. 75 ft   23. $3116.25   25. $171.40   27. $2177.28   29. $.01 per mile   31. 2.05 in.   33. $734.14   35. 45.5 hr   37. 62.5   39. 8.32 centimeters, 8.323 centimeters   41. 269.5 mi   43. $196   45. $5550

Chapter 1 Review Exercises, page 30

1. F   2. T   3. T   4. T   5. T   6. F   7. F   8. T   9. F   10. F   11. F   12. F   13. T   14. F   15. F   16. 15400   17. 189.41   18. 32456   19. 0.032456   20. 0.032456   21. 324560   22. 18.59   23. 0.0052932   24. 12.03   25. 18 − 9   26. 6$\overline{)18}$ 27. 7 · 10 · 40   28. 2 · 32   29. 42$\overline{)21}$ 30. 17 − 13   31. 10 + 4   32. 22 − 17   33. 15 − 4   34. 100 − 52   35. 14 − 3   36. 10 − 2   37. 3 · 17   38. 7$\overline{)42}$ 39. 4 + 5 + 10   40. 2 · 84   41. 15 + 4   42. 45 − 15   43. 15 · 100   44. 7 + 18   45. three hundred forty-five and thirty-five thousand sixty-three hundred-thousandths   46. 8090   47. 22.1   48. 82.571   49. 0.015   50. 54.70 or 54.7   51. 28.651   52. 0.306   53. forty-two and three tenths   54. 3592.31   55. 154.591   56. 60.54344   57. 143.23   58. 1.9   59. 75.938   60. 50.558   61. $1168.61   62. 50,000 people   63. three hundred thirty-two thousand ninety-four dollars   64. 26.85 units   65. $28.50   66. $358.17   67. $58.73   68. 391 feet, 9550 square feet   69. 20.5 kilowatt hours, $1.03   70. $0.07 per mile   71. 3 hr   72. 2¢; $1.50   73. 81

Chapter 2

2.1 Exercises, page 37

1. 3 is a factor.   **3.** 3 is a factor.   **5.** 2 and 3 are factors.
7. 153   **9.** 1, 2, 3, 4, 6, 8, 12, 24   **11.** Any six of 10, 20, 30, 40, 50, 60, . . .   **13.** Any four of 1, 2, 3, 4, 6, 8, 9, 12, 18, 24, 36, 72   **15.** Any three of 1, 3, 5, 9, 15, 45
17. F   **19.** T   **21.** F   **23.** T   **25.** T   **27.** F
29. T   **31.** F   **33.** 13, 97   **35.** 11, 13, 17, 19, 23, 29, 31, 37, 41, 43, 47, 53, 59, 61, 67, 71, 73, 79, 83, 89, 97

2.2 Exercises, page 41

1. 2   **3.** 5   **5.** 79   **7.** 3   **9.** 7
11. $2 \cdot 2 \cdot 2 \cdot 3 \cdot 3$ **13.** $11 \cdot 13$ **15.** $7 \cdot 17$
17. $2 \cdot 5 \cdot 17$ **19.** $11 \cdot 17$ **21.** $2 \cdot 3 \cdot 7 \cdot 11$
23. $3 \cdot 3 \cdot 97$ **25.** $5 \cdot 7 \cdot 23$ **27.** $2 \cdot 2 \cdot 3 \cdot 37$
29. $2 \cdot 2 \cdot 2 \cdot 2 \cdot 3 \cdot 5 \cdot 5$ **31.** $2 \cdot 17 \cdot 47$
33. $7 \cdot 7 \cdot 43$ **35.** $3 \cdot 3 \cdot 5 \cdot 7 \cdot 17$
37. $3 \cdot 7 \cdot 13 \cdot 17$ **39.** $2 \cdot 2 \cdot 2 \cdot 3 \cdot 3 \cdot 3 \cdot 5 \cdot 5$

2.3 Exercises, page 47

1. $\frac{2}{3}$ **3.** $\frac{9}{11}$ **5.** $\frac{7}{11}$ **7.** $\frac{1}{41}$ **9.** 3   **11.** $\frac{13}{11}$
13. $\frac{2}{17}$ **15.** $\frac{1}{3}$ **17.** $\frac{2}{3}$ **19.** $\frac{7}{11}$ **21.** $\frac{15}{7}$ **23.** $\frac{35}{6}$
25. $\frac{143}{20}$ **27.** $\frac{23}{97}$ **29.** $\frac{5}{79}$ **31.** $\frac{1}{5}$ **33.** $\frac{77}{78}$
35. $\frac{1}{5}$ **37.** 5   **39.** $\frac{255}{97}$ **41.** $\frac{7}{3}$ **43.** $\frac{1}{15}$ **45.** $\frac{41}{63}$
47. $\frac{35}{6}$ **49.** $\frac{3}{2}$ **51.** $\frac{5}{7}$ **53.** $\frac{3}{2}$ **55.** $\frac{17}{80}$
57. $\frac{1}{85}; \frac{84}{85}$ **59.** $\frac{1}{4}$ **61.** 1; 8   **63.** 8; 8   **65.** 1; 61
67. 264; 3   **69.** 5; 42   **71.** $\frac{9}{21}$ **73.** $\frac{55}{99}$ **75.** $\frac{253}{299}$
77. $\frac{63}{144}$ **79.** $\frac{104}{144}$ **81.** 8; 4   **83.** 200; 100
85. 10; 2   **87.** 7; 1   **89.** 22; 11   **91.** 22; 2
93. 391; 23   **95.** 140; 4   **97.** 1353; 11   **99.** 1488; 3

2.4 Exercises, page 52

1. $\frac{15}{56}$ **3.** $\frac{14}{15}$ **5.** $\frac{5}{3}$ **7.** $\frac{12}{35}$ **9.** $\frac{8}{125}$ **11.** $\frac{40}{21}$
13. $\frac{3}{14}$ **15.** 9   **17.** $\frac{8}{45}$ **19.** $\frac{30}{77}$ **21.** $\frac{3}{2}$ **23.** $\frac{4}{9}$
25. $\frac{39}{44}$ **27.** 1   **29.** $\frac{1}{7}$ **31.** $\frac{8}{11}$ **33.** 50   **35.** $\frac{1}{3}$
37. 1064 yd   **39.** $\frac{21}{22}$ **41.** 2 cups   **43.** 3 tsp
45. $\frac{1}{65}$ **47.** 1   **49.** 62 lb   **51.** $64,000,000
53. $\frac{253}{4}$ **55.** $\frac{3}{25}$ **57.** $\frac{144}{5}$ **59.** 60 lb

2.5 Exercises, page 56

1. $\frac{44}{35}$ **3.** $\frac{35}{32}$ **5.** 1   **7.** $\frac{4}{15}$ **9.** 1   **11.** $\frac{1}{10}$
13. 25   **15.** 18   **17.** $\frac{64}{81}$ **19.** $\frac{1}{45}$ **21.** $\frac{4}{5}$
23. $\frac{17}{35}$ **25.** $\frac{1}{2}$ **27.** $\frac{22}{27}$ **29.** $\frac{1}{625}$ **31.** $\frac{21}{4}$
33. 30 times   **35.** $\frac{42}{715}$ **37.** $\frac{86}{85}$ **39.** $\frac{3}{20}$
41. 16 plants   **43.** $\frac{1}{28}$ **45.** $\frac{3}{20}$ lb   **47.** 104
49. $\frac{275}{102}$ **51.** 25   **53.** 90 sites   **55.** 32 slices

2.6 Exercises, page 59

1. 12   **3.** 18   **5.** 30   **7.** 27   **9.** 30   **11.** 90
13. 8   **15.** 150   **17.** 42   **19.** 1800   **21.** 18
23. 12   **25.** 231   **27.** 108   **29.** 200

2.7 Exercises, page 64

1. $\frac{3}{4}$ **3.** $\frac{6}{5}$ **5.** $\frac{7}{5}$ **7.** 2   **9.** 1   **11.** $\frac{5}{6}$ **13.** $\frac{44}{35}$
15. $\frac{31}{22}$ **17.** $\frac{19}{14}$ **19.** $\frac{53}{30}$ **21.** $\frac{65}{24}$ **23.** $\frac{53}{30}$
25. $\frac{23}{9}$ **27.** $\frac{221}{110}$ **29.** $\frac{59}{60}$ **31.** $\frac{13}{12}$ **33.** $\frac{29}{18}$
35. $\frac{47}{30}$ **37.** $\frac{10}{27}$ **39.** $\frac{31}{30}$ **41.** $\frac{31}{18}$ **43.** $\frac{17}{8}$
45. $\frac{7}{150}$ **47.** $\frac{3}{14}$ **49.** $\frac{5111}{1800}$ **51.** $\frac{9}{7}$ **53.** $\frac{9}{25}$
55. $\frac{3}{4}$ yd; $12.00   **57.** $\frac{23}{24}$ barrel   **59.** $\frac{11}{12}$

2.8 Exercises, page 67

1. $\frac{3}{7}$ **3.** $\frac{1}{9}$ **5.** $\frac{1}{5}$ **7.** 1   **9.** $\frac{1}{100}$ **11.** $\frac{1}{12}$
13. $\frac{16}{15}$ **15.** $\frac{23}{35}$ **17.** $\frac{22}{3}$ **19.** $\frac{13}{56}$ **21.** $\frac{5}{4}$
23. $\frac{1}{90}$ **25.** $\frac{27}{77}$ **27.** $\frac{7}{36}$ **29.** $\frac{1}{6}$ **31.** $\frac{7}{12}$
33. $\frac{8}{15}$ **35.** $\frac{23}{12}$ in.; no   **37.** $\frac{11}{63}$ **39.** $\frac{31}{48}$ **41.** $\frac{9}{14}$
43. $\frac{437}{500}; \frac{219}{250}$ **45.** $\frac{1}{287}$ **47.** $\frac{5}{6}$ **49.** $\frac{2}{5}$ **51.** $\frac{36}{165}$
53. $\frac{23}{360}$ **55.** $\frac{25}{72}$ **57.** $\frac{1}{4}$ mile   **59.** $\frac{13}{7}$

Chapter 2 Review Exercises, page 68

1. 138   **2.** $7 \cdot 37$ **3.** $\frac{85}{119}$ **4.** 420   **5.** $3 \cdot 5 \cdot 23$
6. $\frac{11}{13}$ **7.** 77   **8.** 28   **9.** 99   **10.** 7   **11.** 36

12. 143   **13.** 2   **14.** 697   **15.** Any three of 1, 2, 4, 8, 11, 16, 22, 44, 88, 176   **16.** Any two of 176, 352, . . .   **17.** Any four of 1, 3, 5, 15, 25, 75   **18.** Any three of 75, 150, 225, . . .   **19.** $\frac{1}{3}$ **20.** $\frac{7}{6}$ **21.** $\frac{4}{3}$ **22.** $\frac{1}{6}$ **23.** $\frac{14}{13}$ **24.** $\frac{22}{41}$ **25.** $\frac{184}{195}$ **26.** $\frac{11}{16}$ in.   **27.** $\frac{3}{40}$ **28.** $\frac{1}{15}$ **29.** $\frac{14}{15}$ **30.** $\frac{5}{2}$ **31.** $\frac{9}{11}$ **32.** $\frac{2}{5}$ **33.** $\frac{1}{10}$ **34.** $\frac{3}{5}$ **35.** $\frac{4}{5}$ **36.** $\frac{8}{21}$ **37.** 9 cubic centimeters   **38.** $\frac{2}{11}$ **39.** $\frac{32}{63}$ **40.** $\frac{1}{2}$ tsp   **41.** $6000   **42.** 40; $30   **43.** 16¢   **44.** $\frac{12}{11}$ **45.** 360 miles per day   **46.** $\frac{73}{78}$ **47.** $\frac{19}{66}$ **48.** $\frac{2}{3}$ **49.** $\frac{4}{21}$ **50.** $\frac{4}{13}$ **51.** 1   **52.** $\frac{51}{56}$ **53.** $\frac{48}{71}$ **54.** $\frac{26}{7}$ **55.** 1

Chapter 3

3.1 Exercises, page 74

1. $\frac{5}{2}$ **3.** $\frac{19}{4}$ **5.** $\frac{38}{7}$ **7.** $\frac{43}{4}$ **9.** $\frac{71}{9}$ **11.** $\frac{119}{9}$ **13.** $\frac{1615}{16}$ **15.** $\frac{127}{10}$ **17.** $\frac{748}{3}$ **19.** $\frac{77}{5}$ **21.** $4\frac{2}{3}$ **23.** 8   **25.** $25\frac{1}{2}$ **27.** $5\frac{1}{4}$ **29.** $1\frac{1}{7}$ **31.** $7\frac{16}{17}$ **33.** 14   **35.** $4\frac{18}{31}$ **37.** $3\frac{3}{4}$ **39.** 4

3.2 Exercises, page 80

1. $8\frac{1}{3}$ **3.** $22\frac{2}{3}$ **5.** 23   **7.** $18\frac{3}{8}$ **9.** $32\frac{2}{3}$ **11.** $19\frac{4}{11}$ **13.** $5\frac{3}{7}$ **15.** $22\frac{4}{25}$ **17.** $\frac{1}{3}$ **19.** $5\frac{2}{13}$ **21.** 21   **23.** 5   **25.** 14   **27.** 37   **29.** 23   **31.** 95   **33.** $2\frac{5}{7}$ **35.** $13\frac{7}{8}$ **37.** $21\frac{1}{2}$ **39.** $\frac{12}{13}$ **41.** $3\frac{4}{5}$ **43.** $8\frac{1}{4}$ **45.** $21\frac{5}{7}$ **47.** $6\frac{1}{11}$ **49.** $8\frac{2}{5}$ **51.** $13\frac{6}{7}$ **53.** $2\frac{5}{12}$ **55.** yes   **57.** $123\frac{41}{72}$ **59.** $4\frac{29}{200}$ in.   **61.** $5\frac{45}{56}$ **63.** $217\frac{1}{4}$ lb   **65.** $32\frac{7}{12}$ in.   **67.** $3\frac{1}{16}$ **69.** $91\frac{9}{16}$ in.

3.3 Exercises, page 85

1. $4\frac{3}{8}$ **3.** $14\frac{2}{3}$ **5.** 4   **7.** $9\frac{1}{3}$ **9.** $1\frac{17}{32}$ **11.** $\frac{16}{33}$

13. $4\frac{2}{7}$ **15.** $\frac{25}{49}$ **17.** $1\frac{1}{2}$ **19.** 1   **21.** $1\frac{1}{2}$ **23.** $4\frac{33}{64}$ **25.** $9\frac{1}{5}$ **27.** $270\frac{2}{3}$ **29.** $\frac{5}{46}$ **31.** $\frac{2}{3}$ **33.** $1\frac{2}{3}$ **35.** $\frac{5}{7}$ **37.** $\frac{1}{3}$ **39.** $5\frac{5}{7}$ **41.** $7\frac{1}{7}$ **43.** $\frac{5}{6}$ ft   **45.** $2\frac{23}{68}$ **47.** 44   **49.** $2\frac{7}{8}$ lb   **51.** $\frac{16}{135}$ **53.** $15\frac{1}{3}$ sq yd   **55.** 39   **57.** $1\frac{43}{56}$ **59.** $5\frac{11}{14}$ **61.** 12   **63.** $3\frac{11}{23}$ **65.** 1326   **67.** $1152   **69.** $33\frac{1}{12}$

Chapter 3 Review Exercises, page 87

1. $\frac{23}{3}$ **2.** $\frac{49}{4}$ **3.** $5\frac{34}{41}$ **4.** $8\frac{7}{8}$ **5.** $60\frac{2}{3}$ **6.** $\frac{15}{16}$ cup   **7.** $4\frac{5}{7}$ **8.** $25\frac{1}{2}$ **9.** $17\frac{4}{7}$ **10.** 20,254 million pounds   **11.** $19\frac{1}{2}$ **12.** $1\frac{16}{35}$ **13.** 113 pieces   **14.** $1\frac{7}{9}$ **15.** $38\frac{1}{5}$ **16.** $9\frac{7}{18}$ **17.** $3\frac{2}{3}$ **18.** 5 "blows"   **19.** 392 sq ft   **20.** 4 bags   **21.** $10\frac{3}{20}$ **22.** $4\frac{2}{3}$ **23.** $10\frac{5}{12}$ miles   **24.** $1\frac{2}{13}$ **25.** $13\frac{4}{5}$ **26.** 9   **27.** $62\frac{3}{4}$ acres   **28.** $7\frac{4}{7}$ **29.** $\frac{2}{3}$ **30.** 4   **31.** 6   **32.** $5\frac{4}{7}$ **33.** $18\frac{1}{2}$ **34.** $2\frac{3}{10}$ **35.** $\frac{2}{5}$

Chapter 4

4.1 Exercises, page 94

1. 0.5   **3.** 1.4   **5.** 0.9   **7.** 0.625   **9.** 1.0625   **11.** 0.6   **13.** 2.75   **15.** 0.667   **17.** 0.75   **19.** 0.67   **21.** 0.875   **23.** 3.25   **25.** 0.8   **27.** 0.346   **29.** 0.0075   **31.** 4.5   **33.** 19.   **35.** 0.19   **37.** 0.0019   **39.** 31.   **41.** 0.31   **43.** 0.000031   **45.** 0.915   **47.** 0.803   **49.** 4.1325   **51.** 0.375   **53.** 7.1   **55.** $0.\overline{3}$ or $0.33\frac{1}{3}$ **57.** $0.\overline{6}$ or $0.66\frac{2}{3}$ **59.** $0.\overline{2}$ or $0.22\frac{2}{9}$ **61.** $0.\overline{45}$ or $0.45\frac{5}{11}$ **63.** $0.\overline{714285}$ or $0.71\frac{3}{7}$ **65.** $1.\overline{6}$ or $1.66\frac{2}{3}$

4.2 Exercises, page 98

1. $\frac{1}{5}$ **3.** $\frac{5}{2}$ **5.** $\frac{29}{20}$ **7.** $\frac{37}{100}$ **9.** $\frac{9}{1}$ **11.** $\frac{626}{5}$ **13.** $\frac{7}{8}$ **15.** $\frac{303}{20}$ **17.** $\frac{53}{1}$ **19.** $\frac{1001}{100}$ **21.** $\frac{4}{5}$

23. $\frac{10001}{25}$ 25. $\frac{759}{1}$ 27. $\frac{25001}{5000}$ 29. $\frac{1}{32}$ 31. $\frac{12}{25}$

33. $\frac{23}{34}$ 35. $\frac{1}{12}$ 37. $\frac{1}{8}$ 39. $\frac{512}{625}$ 41. $\frac{169}{500}$

43. $\frac{72}{35}$ 45. $\frac{38}{15}$

4.3 Exercises, page 102

1. $\frac{7}{100}$ 3. $\frac{1}{4}$ 5. $\frac{23}{50}$ 7. $\frac{1}{100}$ 9. $\frac{1}{80}$ 11. $\frac{51}{50}$

13. $\frac{2}{225}$ 15. 1 17. $\frac{513}{100}$ 19. $\frac{237}{10000}$ 21. 0.07

23. 0.25 25. 0.46 27. 0.01 29. 0.007
31. 1.0273 33. 0.00004 35. 1 37. 5.13
39. 0.006

4.4 Exercises, page 106

1. 0.28 3. 524.1 5. 0.0725 7. 5.85 9. 37.25

11. $\frac{1}{32}$ 13. $\frac{31}{40}$ 15. \$26 17. $\frac{1}{12}$ 19. $\frac{31}{7}$

21. \$95 23. 1.82 25. \$112.50 27. $\frac{8}{3}$

29. 61.6875 31. $\frac{59}{450}$ 33. $1867\frac{1}{3}$ 35. 11.66

million tons 37. \$25.52; \$344.52 39. 0.52507

41. \$28.75 43. $\frac{29}{3000}$ 45. 8165 goals

47. 127,036 residents 49. $\frac{3}{500}$ 51. 2668 females

53. \$351.36 55. Car costing \$5900

4.5 Exercises, page 108

1. 55% 3. 150% 5. 3500% 7. 0.5% 9. 1100%
11. 100% 13. 124700% 15. 5698% 17. 420%
19. 37.5% 21. 14.2% 23. 0.14% 25. 0.07%

27. 37.5% 29. $33\frac{1}{3}$%

4.6 Exercises, page 111

1. 50% 3. 62.5%° 5. 450% 7. $33\frac{1}{3}$% 9. 9%

11. 100% 13. 24% 15. $66\frac{2}{3}$% 17. 26%

19. 56.25% 21. 25% 23. 12.5% 25. 600%
27. 180% 29. 71.4% 31. 28% 33. 15.38%

35. 16% 37. 85% 39. 320% 41. $66\frac{2}{3}$%

43. 150% 45. $66\frac{2}{3}$%

Chapter 4 Review Exercises, page 112

1. 0.5, 50% 2. $\frac{1}{3}$, $0.33\frac{1}{3}$ 3. $\frac{1}{4}$, 0.25 4. $0.66\frac{2}{3}$,

$66\frac{2}{3}$% 5. $\frac{2}{5}$, 40% 6. $\frac{1}{5}$, 0.20 7. 1, 100%

8. 0.8, 80% 9. $\frac{3}{4}$, 75% 10. $\frac{3}{5}$, 60% 11. $\frac{1}{400}$,

0.25% 12. $\frac{1}{32}$, 0.03125 13. $\frac{1}{100}$, 0.01 14. $\frac{17}{50}$,

34% 15. $\frac{49}{50}$, 0.98 16. $\frac{4}{1}$, 400% 17. $\frac{17}{2000}$, 0.0085

18. 0.68, 68% 19. $\frac{25}{4}$, 625% 20. $\frac{98}{25}$, 3.92

21. 16.25, 1625% 22. $\frac{13}{2000}$, 0.65% 23. 0.007, 0.7%

24. 0.455, 45.5% 25. $\frac{1}{20}$, 0.05 26. $\frac{1}{8}$, 12.5%

27. 0.3125, 31.25% 28. $\frac{1}{200}$, 0.005 29. 0.625,

62.5% 30. $\frac{3}{8}$, 37.5% 31. \$1,875 32. 120 points

33. 25 parts 34. \$1251 35. $\frac{1}{5}$ 36. \$2625

37. \$16,845 38. 102,000 barrels 39. 570
40. \$2.61 41. 20% 42. 5 43. \$168.75

44. $7\frac{11}{24}$ 45. 80.6 46. $\frac{13}{20}$ 47. 18%

48. 0.37344 49. 0.00924 50. 40%

Chapter 5

5.1 Exercises, page 119

1.

3.

5.

7.

9. 6, 16, 30, 48, 58, 62 11. −6, −3, 0, 6, 7.5, 10

13. $4\frac{1}{2}$; $4\frac{1}{2}$ 15. 2; 2 17. −8; 8 19. −0.07; 0.07

21. $-\frac{2}{3}$ 23. −113,405 25. −36 27. 36

29. −36 31. 36 33. 35 35. −14 37. $5\frac{1}{2}$

39. 2.7 41. is greater than 43. is less than
45. is less than 47. is greater than 49. is greater than
51. is greater than 53. is less than 55. is greater than

5.2 Exercises, page 126

1. 64 3. 441 5. 16 7. 42.3801 9. 1600
11. 7 13. 20 15. 15 17. 12 19. 14
21. 5.3 23. 3.2 25. 7.5 27. 9.4 29. 4.9
31. 2.6 33. 4.6 35. 8.7 37. 3.9 39. 48

5.3 Exercises, page 129

1. 10   **3.** −60   **5.** 28   **7.** 824   **9.** −525   **11.** 4
13. −26   **15.** 0   **17.** −375   **19.** −64   **21.** −4
23. 33   **25.** −82   **27.** 38   **29.** −4   **31.** −35
33. −93   **35.** −65   **37.** 59   **39.** 26   **41.** −0.016

43. 10   **45.** 3.9   **47.** $-2\frac{2}{7}$ **49.** 5   **51.** −2

53. −6.75   **55.** $-\frac{13}{9}$ or $-1\frac{4}{9}$ **57.** 9.6

59. $-\frac{12}{11}$ or $-1\frac{1}{11}$ **61.** $-2\frac{7}{15}$ **63.** 5   **65.** 0   **67.** 1
69. 1.8

5.4 Exercises, page 130

1. −20   **3.** 6   **5.** 78   **7.** 18   **9.** −38   **11.** −1
13. 128   **15.** 27   **17.** 3   **19.** 20   **21.** 5
23. −10   **25.** −29   **27.** −57   **29.** −46   **31.** 22
33. 62   **35.** 13   **37.** −690   **39.** 0

5.5 Exercises, page 133

1. −2   **3.** −50   **5.** −18   **7.** 4   **9.** −25
11. −104   **13.** 40   **15.** 20   **17.** −1   **19.** 104
21. −22   **23.** 185   **25.** 29   **27.** −16   **29.** 0
31. −58   **33.** −48   **35.** −12   **37.** 8   **39.** −13

41. $\frac{1}{18}$ **43.** 4   **45.** 26   **47.** 14,777 ft

49. −11.23   **51.** −11   **53.** 178 lb   **55.** gained $5\frac{1}{2}$

57. −16   **59.** 6.6   **61.** 0.48   **63.** no; −$2.82

65. $2\frac{1}{3}$ **67.** $-10\frac{5}{8}$ **69.** 1.2

5.6 Exercises, page 138

1. 15   **3.** 9   **5.** 355   **7.** −15,000   **9.** −1,500
11. 2   **13.** −5   **15.** 2   **17.** −6   **19.** −1

21. −70   **23.** −17   **25.** undefined **27.** −1   **29.** −3

31. 30   **33.** 0   **35.** 0   **37.** 91   **39.** −3

41. −15272   **43.** −25   **45.** $-\frac{1}{2}$ **47.** 13   **49.** −160

51. −30   **53.** $8\frac{2}{5}$ **55.** −10   **57.** −25   **59.** 50

61. 0   **63.** $-\frac{5}{33}$ **65.** $-1\frac{3}{7}$ **67.** −27   **69.** $\frac{1}{2}$

5.7 Exercises, page 143

1. 32   **3.** −78   **5.** −8   **7.** 13   **9.** −2   **11.** ✕
13. ÷   **15.** +   **17.** +   **19.** −   **21.** 14   **23.** 82
25. −13   **27.** 3   **29.** 24   **31.** −34   **33.** −7
35. −4   **37.** 31   **39.** −25   **41.** 13   **43.** 0
45. −19   **47.** 24   **49.** −3   **51.** −19   **53.** 1
55. −2   **57.** 3   **59.** 7   **61.** −36   **63.** 3   **65.** 44
67. 6840   **69.** 6   **71.** −210   **73.** 90   **75.** 30

77. −32   **79.** 3   **81.** $5\frac{4}{9}$ **83.** $\frac{4}{5}$ **85.** 40.1

87. −1.74   **89.** $\frac{19}{40}$ **91.** 0   **93.** 6.12233   **95.** 4

97. $-\frac{1}{7}$ **99.** $-\frac{11}{42}$

Chapter 5 Review Exercises, page 144

1.

2. −11, −8, $-2\frac{1}{2}$, 0, 1, 11   **3.** −251   **4.** 29

5. 2.365   **6.** is less than   **7.** is greater than
8. is greater than   **9.** is greater than

10. is greater than   **11.** $2\frac{3}{4}$ **12.** 34   **13.** 5.56

14. 0   **15.** undefined   **16.** 0   **17.** 225   **18.** 13
19. 6.5   **20.** 10404   **21.** 0.0004   **22.** 10.7
23. 5.11   **24.** −16   **25.** 0.8   **26.** −6   **27.** 48

28. −0.5 or $-\frac{1}{2}$ **29.** 14   **30.** $-\frac{2}{5}$ **31.** 30,318 ft

32. 0   **33.** 100   **34.** 0.023   **35.** −22   **36.** −8

37. $5.10 deposit   **38.** 20   **39.** $-\frac{9}{2}$ or $-4\frac{1}{2}$ **40.** −20

41. −17   **42.** −0.72   **43.** 20   **44.** 36   **45.** −48

46. 15.1   **47.** 226.3 degrees   **48.** $-4\frac{3}{20}$ **49.** 11

50. 5   **51.** 16   **52.** −102   **53.** −8   **54.** −6
55. −4   **56.** −2   **57.** 0   **58.** 1   **59.** 0   **60.** 10

61. −30   **62.** −24   **63.** $-\frac{1}{10}$ or −0.1   **64.** 0

65. 7   **66.** 14   **67.** 5   **68.** 15   **69.** $-\frac{11}{9}$

70. −7   **71.** −10   **72.** 4.07, 4.1, 4.12, 4.1205

73. $\sqrt{2}$, 1.5, $\sqrt{3}$, π   **74.** −5, $-4\frac{7}{8}$, $-4\frac{3}{4}$, $-4\frac{5}{8}$, $-4\frac{3}{8}$, $-4\frac{1}{8}$, −3

75. $1\frac{1}{7}$

Chapter 6

6.1 Exercises, page 152

1. $-3, \sqrt{3}, 25$ **3.** 6   **5.** −29   **7.** $\frac{-21}{4}$

9. −4,769,124   **11.** −2,384.562   **13.** 1

15. undefined   **17.** −2   **19.** $\frac{33}{2}$ **21.** undefined

23. 2(w + 3)   **25.** 18 − (x + 7)   **27.** 3x   **29.** an
31. m + 6   **33.** 6 ÷ n   **35.** 5 − t   **37.** 2x + y
39. 3y − 9   **41.** m/6 or 6 ÷ m   **43.** 2(x + 3)

45. $\frac{t+2}{4}$ **47.** 6 plus t   **49.** 6 divided by t   **51.** five
times the quantity b minus three   **53.** five b minus three
55. the quotient of y minus two divided by three
57. w + 5   **59.** The sum of x and y and 7, or x + y + 7
61. *l* − 4   **63.** 0   **65.** 1

6.2 Exercises, page 156

1. $5 \cdot 5 \cdot 5$ **3.** $x \cdot x \cdot x \cdot x$ **5.** $(-2)(-2)(-2)(-2)$
7. $\frac{3}{7} \cdot \frac{3}{7} \cdot \frac{3}{7}$ **9.** $10 \cdot 10 \cdot 10$
11. $y \cdot y \cdot y \cdot y \cdot y \cdot y \cdot y \cdot y \cdot y$ **13.** $5 \cdot t \cdot t \cdot t$
15. $(5t)(5t)(5t)$ **17.** $(m+s)(m+s)(m+s)(m+s)$
19. $a \cdot a \cdot b \cdot b \cdot b$ **21.** x^5 **23.** xt^3z **25.** $\frac{1}{4^6}$
27. $(a+b)^3$ **29.** -5^6 **31.** $3a^2$ **33.** a^2b^3c **35.** 6^4
37. $(3a)^2$ **39.** $(-4)^2$ **41.** 9   **43.** 8   **45.** 36
47. -36 **49.** 36   **51.** -2048 **53.** $\frac{16}{81}$
55. $-.000001$ **57.** -1 **59.** $\frac{49}{9}$ **61.** 13   **63.** 1
65. 16   **67.** 216   **69.** $(-4)^3 = -64$ **71.** $6^2 = 36$
73. $8^2 = 64$ **75.** $(2+3)^3 = 125$ **77.** 3100
79. 20000

6.3 Exercises, page 157

1. 2.56×10^3 **3.** 3.8961×10^3 **5.** 1.2756×10^2
7. 2.53×10 **9.** 1.7689×10^2 **11.** 3.8564×10
13. 5.678×10^4 **15.** 2.9×10^3 **17.** 3.4×10^4
19. 3.8571×10

6.4 Exercises, page 160

1. 2, 3, 4   **3.** 5   **5.** 6   **7.** 4, 4, 3   **9.** 4   **11.** b, f
13. c, d, e   **15.** b   **17.** $4, -3x,$ and $7x^2$ **19.** −
21. 26   **23.** $2x^5$ **25.** $\frac{12}{13}$ **27.** e   **29.** b, f
31. 25   **33.** 5   **35.** -5 **37.** -9 **39.** -149
41. 0   **43.** -1 **45.** 15   **47.** 0   **49.** 4

Chapter 6 Review Exercises, page 162

1. trinomial; $x^2, -7x,$ and 6; 1, −7, 6; second

2. none of these; $4t^5, -2t^3, \frac{1}{3}t^2,$ and −7; 4, −2, $\frac{1}{3}$, −7; fifth

3. trinomial; $3x^5, 4x^3,$ and $-75x$; 3, 4, −75; fifth

4. binomial; $-0.3u^3$, and $-u$; −0.3, −1; third

5. binomial; $3a^6$, and $\frac{3}{4}$; 3, $\frac{3}{4}$; sixth

6. monomial; $-16b$; −16; first
7. binomial; −16, and b; −16, 1; first   **8.** none of these; $x^4, 4x^3y, 6x^2y^2, 4xy^3$ and y^4; 1, 4, 6, 4, 1; fourth
9. monomial; −2; −2; zero   **10.** monomial; $8n^2$; 8; second
11. $15tttrss$ **12.** $(2x+3)(2x+3)(2x+3)(2x+3)$
13. $(x-3y)(x-3y)(x-3y)$ **14.** $7xxxxxyyz$
15. $5mn^6$ **16.** $(us)^6$ **17.** $-x^3y^2$ **18.** $(-x)^4$
19. 9   **20.** 9   **21.** -7 **22.** 36   **23.** -16
24. 16   **25.** 3   **26.** -1 **27.** 103   **28.** $\frac{-11}{12}$
29. $2x + 8$ **30.** xy **31.** $n - m$ **32.** $\frac{1}{2}(x+1)$
33. $3x - 2$ **34.** $m \div n$ **35.** $x^3 - 3y^2$

36. the quantity b minus five   **37.** the quantity b minus five divided by four   **38.** the sum of m and 3
39. x squared minus five times y squared
40. x to the seventeenth power   **41.** three times m
42. the cube of the quantity x plus two   **43.** 1
44. -1 **45.** 0   **46.** 37400d   **47.** $x + y$
48. $t + b + 2.5$ **49.** 32, 0, −32
50. $-17, \frac{1,000,007}{43}, \pi, \sqrt{2}$ **51.** m, n   **52.** 2.563×10^2
53. 5×10^3 **54.** 3.6×10^6 **55.** 5.46×10^2
56. 5.8961×10^2 **57.** 5.2×10^6 **58.** 6×10^4
59. 6.8×10^1 **60.** 7.291×10^3

Chapter 7

7.1 Exercises, page 166

1. 2a and 3a   **3.** The terms are unlike.   **5.** The terms are similar since $-4ba = -4ab$.   **7.** $4x, -7x,$ and $-0.7x$; $3y$ and $\frac{1}{2}y$; $11z$ and $9z$ **9.** The terms are unlike.
11. $5a$ **13.** $3a$ **15.** $9x^2$ **17.** $-4m$ **19.** $8a$
21. 0   **23.** $10y^2$ **25.** $3n$ **27.** $2a$ **29.** $7x^2$
31. $-2ab$ **33.** $\frac{-11}{4}x$ **35.** $-x^2y$ **37.** $-2.08t$
39. $8a + 8$ **41.** $35n$ **43.** 3.29   **45.** $5xyz - 2xy$
47. $\frac{13x}{30}$ **49.** 0

7.2 Exercises, page 169

1. $5m + 2n$ **3.** $-10x^2$ **5.** $y^3 + y^2$ **7.** $-2a + 8b$
9. $-11x^2 + x$ **11.** $7a + 5$ **13.** $5a - 8$ **15.** $9a - 2$
17. $-5x - 9$ **19.** $3x - 2$ **21.** $3m + 8$ **23.** 0
25. $9x - y + 1$ **27.** $2.2x^3 + 1.2x^2 - 4.4$
29. $6t^3 + 4t^2 - 4t$ **31.** $14x^3 - x^2y - 12xy^2$
33. $7z^5 - 3z^4 - 9z^3 + 14z^2 - 4z - 11$ **35.** $\frac{5}{4}x + \frac{1}{3}y$
37. $x + \frac{4}{7}$ **39.** $3a^2 - 4a - 6$

7.3 Exercises, page 171

1. $3a + 1$ **3.** $a - 4$ **5.** $15a - 8$ **7.** $7x + 1$
9. $9x + 6$ **11.** $a + 4$ **13.** $-9a + 3$ **15.** $4m - 13$
17. 0   **19.** $7a - 15$ **21.** $2y + 1$ **23.** $3g + h + 4$
25. $2x^2 + 2x - 3$ **27.** $12r - 20s + 88$
29. $2.76x^2 + x - .27$ **31.** $M - 2$ **33.** $5ab + 7bc - 3cd$
35. $x^3 + 4.8$ **37.** $-9w^5 - 14w^4 + 4w^3 + 6w^2 + 2w - 8$
39. $-8a^4 + 11a^3 - 7a^2 + 4a - 30$

7.4 Exercises, page 174

1. $-3a + 8b$ **3.** $-11k + 1.03m$ **5.** $3y^3 - 4y^2 - y + 7$
7. $+w^4 + 3w^2 - 2$ **9.** $3g - 2$ **11.** $2a + b$
13. $3a - 2b$ **15.** $2a - 8$ **17.** $-5a + 4$ **19.** $-a - 2$
21. $5m - 2$ **23.** $2y - 16$ **25.** $-a - 4b$ **27.** $3w - 2z$

29. $-8x + y$ 31. $15ab + 7bc - 2cd$ 33. $n + 4y^2 + 4$
35. -1 37. $21x - 9$ 39. 0 41. 1 43. 0
45. -6 47. undefined   49. -24

Chapter 7 Review Exercises, page 175

1. $3a^3 + a^2 - 6$ 2. $m^4 + m^2 - m$ 3. $a^2b + ab^2$
4. $-a^2 + 3a + 7$ 5. $m^5 + m^4 + m^2 - m$
6. $-y^3 + y^2 + 3$ 7. $-a^2 + 5a + 6b$ 8. $a - b + c$
9. $m^3 + m - n^2$ 10. $h^2 - h - 6$ 11. F   12. T
13. T   14. F   15. T   16. F   17. $7x^2 + 2$
18. $a - 5b + 3c$ 19. $-2x^2 + 7x - 13$ 20. $5a - 7b + 4c$
21. $7a^2bc + 4ab^2c$ 22. xy^2z 23. 0
24. $-2x^4 - 3x^2 + 2x - 6$ 25. $2a^3$ 26. a 27. $2x^5$
28. 0 29. $3x^3 + x^2$ 30. $4a + 5b - 10$ 31. $4x^2 + 7y$
32. $a^2 - 2a - 24$ 33. $a - 2$ 34. $x - y - 1$
35. $-4ab + 9ac - 8bc$ 36. $3x^2y - 2xy^2 + 5y^3$
37. $3a^2 - 6ab + 3$ 38. $-a^2 - 6ab + 17$
39. $-3a^2 + 6ab - 3$ 40. $6a - 12$ 41. $a + 17x - 2$
42. $4x^3 + 5x^2 + 4x - 3$ 43. $3x - 6$
44. $x^3 + \frac{1}{2}x^2 + 2x - \frac{17}{3}$ 45. $-4bc + 4cd$
46. $4x^2 + 10x - 4$ 47. $2x^2 + 3x + 9$
48. $-3x^3 + 5x - 23$ 49. $14x^3 + x^2 - x$
50. $-22x + 7y + 5$ 51. $-2a - b - 4$ 52. $-7m - 1$
53. $18m + 6$ 54. $3b - 5c$ 55. $a - b - c + d - e + f$
56. $-4x^a$ 57. $8m^{3a}$ 58. $11a^m$
59. $15m^{2n} - 8n^{3m} + n^m$ 60. $10x^a - 7y^a$

Chapter 8

8.1 Exercises, page 180

1. x^7 3. m^6 5. a^8 7. y^{13} 9. x^4 11. m^5
13. a^2 15. x^{15} 17. a^7 19. m^2n^5 21. x^{10}
23. m^{15} 25. a^{15} 27. y^{30} 29. x^3 31. a^{18}
33. y^2 35. b^6 37. b^{15} 39. y^6 41. a^3b^3
43. $32n^5$ 45. $m^6n^6y^6$ 47. $16y^4z^4$ 49. $w^3x^3y^3z^3$
51. $\frac{1}{32}a^5b^5$ 53. $\frac{8}{27}a^3b^3$ 55. $-32a^5b^5$ 57. m^8n^{12}
59. $a^{15}b^5$ 61. $-27a^{18}b^6$ 63. $16a^2b^4$ 65. $\frac{1}{32}m^5n^{10}$
67. $\frac{9}{25}m^4$ 69. $27a^6b^9$ 71. $n^8; x^a \cdot x^b = x^{a+b}$
73. $m^3n^3; (xy)^a = x^ay^a$ 75. $h^6; (x^a)^b = x^{ab}$
77. 2^4 or 16 79. a^2b^2 81. a^{50} 83. $m^{10}s^{10}$
85. $2^2 \cdot 3^2$ or 36 87. a^{15} 89. m^{49} 91. $-32x^{10}y^{20}$
93. $(x + y)^{13}$ 95. $-x^{10}$ 97. $9a^2b^2$ 99. 128
101. To raise a power to a power, retain the base and multiply the exponents.   103. $8a^3$ 105. $-n^{15}$
107. $(a + 2)^7$ 109. $x^{20}y^{20}$

8.2 Exercises, page 182

1. xy 3. $-xy$ 5. $-2xy$ 7. $15x^2$ 9. $15m^5$
11. $4a^3$ 13. $30ab$ 15. $12a^6$ 17. $3m^{13}$
19. $4cxy$ 21. $-3u^5v$ 23. $12ab$ 25. $-3a^3b$

27. $4m^3n$ 29. $6m^5$ 31. $-x^3y^4$ 33. $-3a^4b$
35. $-8c^3d^4$ 37. $54x^3y^3$ 39. $-2835a^5b^5$
41. $-20x^4y^2$ 43. $-36xy$ 45. $-4a^4b$ 47. $-10x^2y$
49. $-2x^6y^{10}$ 51. $9a^3b^2$ 53. a^4b^6 55. $-9a^3b$
57. $25x^6y^2$ 59. $-16a^4b^4$

8.3 Exercises, page 184

1. $3a + 3b$ 3. $-3a + 3b$ 5. $3a - 6b$ 7. $a^2 + ab$
9. $ab - 6b$ 11. $3ax - 3ay$ 13. $2x + 2y$
15. $2ax - ay$ 17. $4a - 20$ 19. $4a - 24$
21. $12x + 36y$ 23. $3a^2 + 6a - 15$
25. $2x^3 + 2x^2y + 2x^2z$ 27. $8n^2p^3 - 16n^2p^4 + 12np^5$
29. $-4x^5y^2 - 16x^4y^2 - 4x^3y^4 + 12x^3y^3$ 31. $3z^2 + 2z$
33. $-5a^7b^4 - 10a^6b^5 - 5a^5b^6$ 35. $-20t^2 + 224t$
37. $a^3bc - ab^2c + abc^3$ 39. $-3x^4 - 6x^3 + 9x^2$
41. $2a^4b - a^3c$ 43. $-2a^3b - 3ab^2$ 45. $a^2b - 3ab^2$
47. $4a + 5b$ 49. $-2a^2b - 2ab^2$ 51. $5x^9 - 3x^4 - 0.23x^2$
53. $-x + 22$ 55. $4x^4 - 16x^2$ 57. $8a^6b + 2a^3b^2$
59. $15x^5y^3 - 25x^2y^5$

8.4 Exercises, page 187

1. $x^2 + 5x + 6$ 3. $x^2 + x - 6$ 5. $a^2 - b^2$
7. $x^2 + 4x + 4$ 9. $2t^2 - 11t + 12$ 11. $3a^2 + 20a - 7$
13. $3a^2 - 13a - 10$ 15. $2x^2 - 7x + 6$
17. $2x^2 + 7x - 4$ 19. $2a^2 + 17a + 30$ 21. $4x^2 - 9y^2$
23. $x^2 - 2xy + y^2$ 25. $-4x^2 - 12xy - 9y^2$
27. $x^6 - x^4y^2 - x^2y^4 + y^6$ 29. $t^5 - 1$
31. $3x^2 - 3x - 18$ 33. $4x^2 - 16x + 16$
35. $x^3 - 4x^2 - 6x + 5$ 37. $6a^2 - 8a - 30$
39. $a^4 - a^2 - 6$ 41. $3m^3 - 6m^2n + 3mn^2$
43. $6a^3b + 3a^2b^2 - 18ab^3$
45. $3a^3 - 6a^2b + 3ab^2 + 18a + 9$ 47. $6x^2 - 16x - 32$
49. -10 51. 0 53. 12 55. -40 57. -160
59. -4

Chapter 8 Review Exercises, page 188

1. a^7 multiplying powers   2. $16x^4y^4$ raising a product to a power   3. a^6 raising a power to a power   4. $25x^6y^2$ raising a product to a power, raising a power to a power
5. x^{19} multiplying powers   6. x^{18} raising a power to a power   7. x^{11} multiplying powers   8. x^{10} raising a power to a power, multiplying powers   9. $-8a^3$ raising a product to a power   10. $a^4b^6c^8$ raising a product to a power, raising a power to a power   11. product of polynomials $x^2 + 4xy + 4y^2$ 12. product of polynomials $a^2 - 2a - 24$ 13. product of a monomial and a polynomial $16y^3 - 80y^2$ 14. product of a monomial and a polynomial $-12a^4b^2 + 12ab^2 - 30ab$ 15. product of monomials $-512n^5t^5$ 16. product of polynomials $9x^3 - 6x^2 + 4x - 16$ 17. product of polynomials $2a^2 + a - 3$ 18. product of polynomials $2a^3 - 5a^2 + 7a + 5$ 19. product of monomials $12xy$ 20. product of a monomial and a polynomial $18a^4 + 27a^3 - 54a^2$

21. $12a^3x^2y^4$ **22.** $-70a^6b^5$ **23.** $-8a^6b$
24. $25x^4y^4z^8$ **25.** $-20x^4y^2 + 8xyz - 36xy$
26. $x^2 - 6x + 9$ **27.** $x^3 - 9x^2 + 27x - 27$
28. $-3a^4 + 10a^2$ **29.** $-a^2 - a + 6$ **30.** $-9a - 6$
31. $-6x - 3y + 18$ **32.** $2a^2 + 10a$ **33.** $4a^3 - 15ab$
34. $5a^2b - 27b^2$ **35.** $4ab$ **36.** $1600n^{14}t^{13}$
37. $2a^3 - 2ab^2 + a^2b - b^3$ **38.** $x^4 - 4x^3 + 6x^2 - 4x + 1$
39. $-18a^2 - 12ab + 36a + 12b - 18$ **40.** $6b^2 + 36b + 54$
41. $x^{2n} - 4x^n$ **42.** $x^{2n} - 4x^n - 5$
43. $3x^{5n} - 9x^{4n} + 18x^{3n} - 36x^{2n}$ **44.** $6x^{2n} - 19x^n - 7$
45. $x^{2n} - 1$

Chapter 9

9.1 Exercises, page 192

1. m^3 **3.** 1 **5.** $\dfrac{1}{a^4}$ **7.** a^6 **9.** $\dfrac{1}{a^5}$ **11.** 3

13. $-a^5$ **15.** 1 **17.** $\dfrac{1}{a^8}$ **19.** $\dfrac{1}{x^3}$ **21.** $-\dfrac{1}{5^3}$ or $-\dfrac{1}{125}$

23. $\dfrac{27}{b^3}$ **25.** $\dfrac{-a^3}{64}$ **27.** $\dfrac{4}{9}$ **29.** $\dfrac{a^2}{9}$ **31.** $\dfrac{x^4}{y^4}$

33. $\dfrac{-27}{x^3}$ **35.** $\dfrac{x^4}{81}$ **37.** $\dfrac{x^2}{16}$ **39.** $\dfrac{27}{125}$ **41.** $\dfrac{1}{216}$

43. $\dfrac{1}{a^9}$ **45.** $\dfrac{a^4}{81}$ **47.** 1 **49.** $\dfrac{a^2}{16b^2}$ **51.** a^9

53. $\dfrac{1}{x^2}$ **55.** x^7 **57.** 1 **59.** $\dfrac{1}{y^4}$ **61.** $\dfrac{c^2}{x^2}$ **63.** m^9

65. a^3

9.2 Exercises, page 195

1. $3a$ **3.** $3a$ **5.** $\dfrac{x^2y^2}{4}$ **7.** $\dfrac{3}{a^3}$ **9.** $\dfrac{2}{5a^4}$ **11.** $-2m^9$

13. -1 **15.** $\dfrac{-ab}{2}$ **17.** $\dfrac{1}{x^4y^2}$ **19.** $2b^5$ **21.** $-\dfrac{3a^3b^4}{2c^3}$

23. $\dfrac{7x^2}{y^5z^2}$ **25.** $\dfrac{y^5z^2}{7x^2}$ **27.** $3a^5$ **29.** $x^4y^7z^7$

31. $\dfrac{x^6}{2m^4}$ **33.** $\dfrac{a}{2b^2c^4}$ **35.** $-5b$ **37.** $-\dfrac{27a^{13}b^{13}c^3}{16}$

39. $\dfrac{81a^4}{b^8}$

9.3 Exercises, page 197

1. $a + 3b$ **3.** $a + 2b$ **5.** $a + b$ **7.** $a - 4b + 2c$
9. $-x + y$ or $y - x$ **11.** $x^2 - 2x + 3$ **13.** $3a + 1$
15. $-n + m$ or $m - n$ **17.** $x + 2$ **19.** $a - 2$
21. $10n - 1$ **23.** $-x - 5$ **25.** $2a^3 - 6a^2$
27. $ax - 5xy$ **29.** $-x^3y^2 + xy - 1$ **31.** $a^2 + a + 1$
33. $3a - b + 2c$ **35.** $ab + c^2$ **37.** $\dfrac{-x^2}{5} + x^3y^3$
39. $3x^2 - 6xy - 5yz$

9.4 Exercises, page 199

1. $a^2, 6a$ **3.** $a - 3, a^2, -3a, -3a, -3$ **5.** $a - 2$ **7.** $3x - 1$

9. $2m + 3$ **11.** $2m - 3$ **13.** $3a + 1$ **15.** $b - 2$

17. $x - 1 + \dfrac{2}{x + 5}$ **19.** $2a + 5$

21. $a^4 + 2a^3 + a - 2 + \dfrac{-1}{a - 1}$ **23.** $a^2 - 3a + 1 + \dfrac{-10}{2a - 5}$

25. $a^2 + ab + b^2$ **27.** $a - 3b$

29. $x^3 - x^2 + 2 + \dfrac{-2x - 2}{2x^2 - x - 1}$ **31.** $p^2 + 5p + 25$

33. $a + 6$ **35.** $a^4 + a^3 + a^2 + a + 1$

Chapter 9 Review Exercises, page 200

1. x^2 **2.** $\dfrac{1}{x^2}$ **3.** $\dfrac{1}{24a^4}$ **4.** $8a^3$ **5.** $\dfrac{4y^2z}{t}$

6. $2x + 1$ **7.** $-2rs + 2s - 4$ **8.** $4x + 4$ **9.** $-10x^2y^2$

10. $3x + y - 1$ **11.** $x + y$ **12.** $\dfrac{x^{15}y^{10}}{z^{20}}$ **13.** $2x^2y^2$

14. $\dfrac{5a^2b}{4} - 5a$ **15.** $3t^2 - 6t + 18$ **16.** $\dfrac{9a^2}{b^4}$

17. $4x + 2y$ **18.** $a^2 - a + 7 + \dfrac{-1}{a + 1}$ **19.** $2a - 6$

20. $\dfrac{27x^9y^{11}}{16}$ **21.** $2x$ **22.** $4a^2b - 4a^2$ **23.** $\dfrac{-5}{9ab^5}$

24. $x + \dfrac{1}{2}y$ **25.** $-x^2 - y^2 + z^2$

26. $w^4 + 2w^3 + 4w^2 + 8w + 16$ **27.** $\dfrac{32}{a^5b^{15}}$

28. $\dfrac{1}{343a^9c^{12}}$ **29.** $m^5 + m^4 + m^3 + m^2 + m + 1$

30. $\dfrac{4}{3a^5b^{10}}$ **31.** x^{3n} **32.** $5x^{2a}$ **33.** $\dfrac{1}{m^a}$

34. $x^n - 3$ **35.** $x^n + 1$

Chapter 10

10.1 Exercises, page 209

1. $2x = 18$ **3.** $\dfrac{1}{2}h = h + 9$ **5.** $5 - x = \dfrac{y}{4}$

7. $2x + 3 = 12$ **9.** $0 = (0)x$ **11.** Three times y
equals 11.   **13.** x minus 7 equals 5 times x.
15. a minus 3 equals b divided by 3.   **17.** contradiction
19. identity   **21.** conditional equation   **23.** identity
25. contradiction   **27.** conditional equation
29. identity   **31.** T   **33.** F   **35.** T   **37.** F
39. T   **41.** 4 **43.** 12 **45.** 0 **47.** $7, 1, 5$

49. $-\dfrac{1}{2}, \dfrac{1}{2}$

10.2 Exercises, page 211

1. not equivalent   **3.** not equivalent   **5.** not equivalent
7. not equivalent   **9.** equivalent   **11.** equivalent
13. not equivalent   **15.** not equivalent
17. not equivalent   **19.** equivalent

10.3 Exercises, page 214

1. a linear equation in one variable   3. Is a linear equation, but there are 2 variables.   5. not a linear equation; the equation is second degree.   7. a linear equation in one variable   9. a linear equation in one variable

11. $x = 6$ 13. $y = \frac{1}{9}$ 15. $v = 0.1$ 17. $a = 0$

19. $n = -\frac{2}{3}$ 21. $w = 49$ 23. $h = -15$ 25. $x = 702$

27. $v = \frac{2}{3}$ 29. $w = 27$ 31. $m = -18$ 33. $x = 12$

35. $s = 20$ 37. $t = -4$ 39. $v = 100$ 41. $z = 6$
43. $b = 110$ 45. $m = 1.1$ 47. $h = 25$ 49. $x = 0$

51. $x = 21$ 53. $x = -9$ 55. $f = -\frac{8}{3}$ 57. $x = \frac{1}{25}$

59. $k = -\frac{3}{8}$ 61. $t = 7$ 63. $x = \frac{177}{10}$ or 17.7

65. $x = -36$ 67. $t = -7$ 69. $x = 1124$

10.4 Exercises, page 217

1. $x = 24$ 3. $b = 2$ 5. $y = 12$ 7. $v = -1$
9. $t = 1$ 11. $n = 4$ 13. $z = 36$ 15. $a = -16$
17. $v = -6$ 19. $y = 6$ 21. $x = 9$ 23. $m = -6$

25. $a = 1$ 27. $x = \frac{8}{3}$ or $2\frac{2}{3}$ 29. $h = -18$

31. $f = -\frac{7}{2}$ or -3.5 33. $x = -\frac{7}{2}$ or -3.5

35. $x = \frac{7}{2}$ or 3.5   37. $t = 0$ 39. $w = 3.2$

41. $x = 7.1$ 43. $x = -0.02$ 45. $x = -1$ 47. $x = 21$

49. $x = \frac{15}{2}$ or $7\frac{1}{2}$ 51. $x = 1$ 53. $u = -6$

55. $x = \frac{25}{6}$ or $4\frac{1}{6}$ 57. $x = -1.5$ 59. $x = \frac{3}{2}$ or $1\frac{1}{2}$

10.5 Exercises, page 221

1. $b = -\frac{1}{3}$ 3. $z = 0$ 5. $w = 4$ 7. $t = 1$ 9. $v = 1$

11. $a = \frac{1}{2}$ 13. $k = 3$ 15. $x = 9$

17. Every real number is a solution. Identity
19. no solution; contradiction   21. $x = 0$ 23. $x = -1$
25. $y = -5$ 27. $x = -3$ 29. no solution; contradiction

31. $w = 0$ 33. $x = 2.35$ 35. $x = -\frac{1}{2}$ 37. $y = 6$

39. Every real number is a solution. Identity
41. no solution; contradiction   43. $x = 5$ 45. $x = 0.1$
47. no solution; contradiction
49. Every real number is a solution. Identity

10.6 Exercises, page 224

$r = 20$ 3. $n = -2$ 5. $m = -1$ 7. $p = -1$

9. $v = 5$ 11. $y = \frac{5}{3}$ 13. $w = 8$ 15. $s = -\frac{3}{4}$ or -0.75

17. $t = 5$ 19. $y = 1$ 21. $m = 0$ 23. $x = 4$
25. $x = 7$ 27. $x = -18$ 29. no solution; contradiction

31. $t = -26$ 33. $x = \frac{5}{2}$ or 2.5   35. $x = \frac{25}{6}$

37. $m = 0.6$ 39. no solution; contradiction
41. Every real number is a solution. Identity   43. $x = -4$

45. $z = -13$ 47. $x = 10$ 49. $x = 6$ 51. $x = \frac{1}{3}$

53. $z = -38.75$ 55. $n = -\frac{3}{2}$ or -1.5

57. Every real number is a solution. Identity

59. $x = -\frac{35}{2}$ or -17.5

10.7 Exercises, page 228

1. $w = 245$ 3. $z = 32$ 5. $t = -18$ 7. $v = 2$
9. $b = -20$ 11. $x = 36$ 13. $z = 21$ 15. $v = 165$
17. $t = -18$ 19. $x = 5$ 21. $x = 30$ 23. $w = 15$
25. $z = 14$ 27. $x = 100$ 29. Every real number is a solution. Identity   31. $x = 8$ 33. $z = -36$

35. $y = -5$ 37. $x = \frac{1}{3}$ 39. $c = 0$ 41. $n = 18$

43. $x = 30$ 45. $x = \frac{5}{4}$ or 1.25   47. $x = -15$

49. $z = 7$ 51. $w = 3$ 53. $h = -7$ 55. $x = \frac{13}{61}$

57. $h = 3$ 59. $x = \frac{25}{6}$

10.8 Exercises, page 231

1. $t = \frac{d}{r}$ 3. $y = \frac{5}{2}$ 5. $x = n + m$ 7. $a = \frac{2s - nl}{n}$

9. $F = \frac{9}{5}C + 32$ 11. $y = \frac{rst}{xz}$ 13. $w = -dr$

15. $u = \frac{m - n}{2}$ 17. $n = b + c + a$

19. $a = \frac{2A - bh}{h}$ or $a = \frac{2A}{h} - b$ 21. $b = \frac{2A}{h}$

23. $h = \frac{V}{\pi r^2}$ 25. $n = \frac{ab}{2}$ 27. $a = 5$ 29. $a = 43$

31. $g = -32$ 33. $V = 7234.56$ 35. $h = 10$

Chapter 10 Review Exercises, page 232

1. T   2. F   3. F   4. F   5. T   6. T   7. F
8. T   9. T   10. T   11. T   12. T   13. F
14. F   15. T   16. T   17. T   18. F   19. T
20. F   21. $y = 9$ 22. $w = 3$ 23. $x = -3$

24. $x = 15$ 25. $x = 2$ 26. $x = 1$ 27. $x = \frac{23}{3}$

28. Every real number is a solution. Identity   29. $x = 7$
30. $x = -77$ 31. $y = 18$ 32. $w = 6$ 33. $x = 135$
34. $x = 9$ 35. $z = 63$ 36. $x = 3$ 37. $N = 4$

38. $w = 3$ **39.** $t = 25$ **40.** $v = \dfrac{7}{3}$

41. no solution; contradiction   **42.** $w = 520$

43. Every real number is a solution. Identity   **44.** $x = 1$

45. $c = -18$ **46.** $x = \dfrac{b}{a}$ **47.** $x = \dfrac{c - b}{a}$

48. $x = -\dfrac{y^2 - 4}{3}$ or $x = \dfrac{4 - y^2}{3}$ **49.** $x = b - a$

50. $x = 3c + 2b$

Chapter 11

11.1 Exercises, page 240

1. $23x$ **3.** $7z + 4$ **5.** $\dfrac{s}{7}$ **7.** $\dfrac{7}{a}$ **9.** $w - 11$

11. $6 - s$ **13.** $3y + 11$ **15.** $4x + 7$ **17.** $w - 3 = 11$

19. $2x = y + 4$ **21.** $n - 5$ **23.** $5n$ **25.** $n - 7$

27. $\dfrac{n}{7}$ **29.** $n - 11$ **31.** $n + 9 = 0$ **33.** $3n = n + 4$

35. $2n$ **37.** $4n$ **39.** n^3 **41.** $7 - s$ **43.** $\dfrac{n}{2}$

45. $\dfrac{1}{4}(x + 19)$ **47.** $\dfrac{n}{2} = 17$ **49.** $\dfrac{n}{7} = n - 6$

51. $2(x + 5)$ **53.** $8 - x$ **55.** $5(2x + 4)$

57. $4(2x + 5)$ **59.** $2(w + 5) = 3(w - 7)$ **61.** 4

63. 8   **65.** 22   **67.** \$1100   **69.** 9 moles   **71.** −9

73. \$50 per share   **75.** 25 words per minute   **77.** 30

79. 22   **81.** −1   **83.** +1   **85.** −60

11.2 Exercises, page 249

1. $56 - x$ **3.** $c + 25$ **5.** $d - 35$ **7.** $b + 879.15$

9. $\dfrac{144}{y}$ **11.** $s - 11$ **13.** $\dfrac{1000}{c}$ gal   **15.** $85 - w$

17. $s + 11$ **19.** $l - 11$ **21.** as either $5s$ or $24 - s$

23. $145 - m$ **25.** $\dfrac{3}{4}n, 3n, n, \dfrac{n}{2}$ **27.** $n + 2, n + 4, n + 6$

29. $240 - d$ **31.** $\dfrac{f}{3}$ **33.** $d + \dfrac{1}{2}$ **35.** $\dfrac{5280}{f}$

37. $0.80p$ **39.** $m + (m + 1)$ or $2m + 1$

41. He sold $\dfrac{3}{5}$S. He still owns $\dfrac{2}{5}$S.

43. $n + (n + 2) + (n + 4)$ or $3n + 6$ **45.** $60 - l$

47. $n, n + 7, n + 14$ **49.** $100,000 - b$

51. a. 45, 46 **b.** 37, 38 **c.** −62, −63 **d.** −20, −21

53. a. 55, 57 **b.** −111, −113 **c.** −41, −43 **d.** 313, 315

55. a. 74, 76, 78 **b.** 112, 114, 116 **c.** −442, −444, −446

d. −654, −656, −658   **57.** −2, 0, 2, 4   **59.** −1, 1, 3, 5

61. 13, 31   **63.** 7, 15   **65.** 13, 20   **67.** 18, −7

69. 21, 7, 12   **71.** 19, 11, 41   **73.** width 35 ft,

length 50 ft   **75.** 12 milliliters of A, 6 milliliters of B,

18 milliliters of C   **77.** 18 A's, 12 B's   **79.** 37°

11.3 Exercises, page 254

1. 200 \$10 bills; 600 \$5 bills   **3.** 150 dimes; 600 nickels

5. 3 half dollars   **7.** 2 wooden bats; 4 aluminum bats

9. 41 Sunday subscribers

11. 0.5 oz of 50% solution; 3.5 oz of 10% solution

13. 9.17 liters of 60% solution; 73.33 liters of 15%

solution   **15.** 220 grains of 80%; 80 grains of 50%

17. 5 gal of insecticide; 495 gal of water   **19.** \$6250

11.4 Exercises, page 257

1. \$299.60   **3.** \$8.10   **5.** \$22,500   **7.** \$47,058.82

9. 7.25%   **11.** 9%   **13.** 180%   **15.** 8 months

17. 3 months   **19.** \$2500 in 8% account; \$2000 in 10%

account

11.5 Exercises, page 262

1. 1374 mi   **3.** 5500 ft; more than a mile

5. 1400 mph   **7.** 13 mph   **9.** 360 sec or 6 min

11. 3 hr 12 min   **13.** $\dfrac{1}{2}$ hr or 30 min

15. 4 mph; $4\dfrac{2}{3}$ mph   **17.** 85 mph; 115 mph

19. 15 min

Chapter 11 Review Exercises, page 263

1. $9w = 11$ **2.** $\dfrac{3}{4}y$ **3.** $2(3x + 4)$ **4.** $z - 5 = 11$

5. $x + y = 3$ **6.** $3n - 5$ **7.** $z - 7$ **8.** $\dfrac{w}{9}$ **9.** $\dfrac{9}{w}$

10. $4y = y^3$ **11.** $80 - x$ **12.** $\dfrac{168}{w}$ **13.** $s + 15$

14. $n + (n + 2)$ or $2n + 2$ **15.** $100 - w$ **16.** 53, 54

17. 36, 38   **18.** 89, 91   **19.** 59, 61, 63

20. 13, 26, 18   **21.** 10 lb   **22.** 25 @ \$1.15, 15 @ \$1.85

23. 9 months   **24.** \$42,000 @ 8.75%, \$8,000 @ 7%

25. 2,000 m   **26.** 425 mph   **27.** 12 sec   **28.** 0

29. −3   **30.** 25 mph and 29 mph

Chapter 12

12.1 Exercises, page 271

1. $\dfrac{2}{5}$ **3.** $\dfrac{6}{11}$ **5.** $\dfrac{12}{17}$ **7.** 5:7   **9.** 2:3

11. 4:9   **13.** ratio of 5 to 8   **15.** ratio of 7 to 3

17. ratio of 6 to 13   **19.** 1:4   **21.** 1:2   **23.** 17:100

25. 25%   **27.** 75%   **29.** 200%   **31.** $\dfrac{4}{3}$ **33.** $\dfrac{1}{3}$

35. $\dfrac{3}{4}$ **37.** $\dfrac{3}{5}$ **39.** $\dfrac{1}{4}$ **41.** $\dfrac{2}{1}$ **43.** $\dfrac{0.5}{1}$ **45.** $\dfrac{3}{1}$

47. $\dfrac{6}{1}$ **49.** $\dfrac{0.5}{1}$ **51.** \$0.2133   **53.** \$0.46

55. \$0.035   **57.** 0.255   **59.** $\dfrac{1}{13}$ **61.** 32 oz for 37¢

63. $\dfrac{16}{3}$ **65.** 3.8 GPA

12.2 Exercises, page 276

1. extremes: 12, 65; means: 13, 60   **3.** extremes: m, a;
means: s, t   **5.** extremes: 2x, 32; means: 13, y
7. not a proportion   **9.** proportion   **11.** proportion
13. not in proportion   **15.** not in proportion
17. in proportion   **19.** x = 12   **21.** x = 10

23. a = 2   **25.** y = 8.25   **27.** a = 5   **29.** $t = -\frac{2}{3}$

31. x = 9   **33.** equivalent   **35.** equivalent
37. not equivalent   **39.** equivalent   **41.** not equivalent
43. 4.5 cups   **45.** 2805 students   **47.** 23.5 gal
49. 50 km   **51.** 250 bulbs   **53.** 5 days
55. 17.5 cm   **57.** $0.25   **59.** 53.33 ft
61. 1416 bricks   **63.** 540 cubic ft of concrete

65. $ad = bc, \frac{a\cancel{d}}{b\cancel{d}} = \frac{\cancel{b}c}{\cancel{b}d}, \frac{a}{b} = \frac{c}{d}$

12.3 Exercises, page 282

1. 30   **3.** 500   **5.** 5%   **7.** 278.8   **9.** 3
11. 124   **13.** 3000   **15.** 0.5%   **17.** 8.5%
19. 950   **21.** 7   **23.** 20%   **25.** 200%   **27.** 90
29. 250   **31.** 75%   **33.** $1.60   **35.** $7680
37. 4%   **39.** 25 mpg   **41.** $0.45   **43.** 0.353
45. $1200   **47.** 23%   **49.** 660 tickets

12.4 Exercises, page 288

1. 18 m   **3.** 17 m   **5.** 15 m   **7.** 10 in.²

9. 7.5 in.²   **11.** $14\frac{2}{3}$ in.²   **13.** 15 m²   **15.** 6 m²

17. 12.56 m²   **19.** 18 m²

12.5 Exercises, page 291

1. 2 ft³   **3.** 4 ft³   **5.** 12 ft³   **7.** 16 ft³
9. 197.82 ft³   **11.** 20.9 ft³   **13.** 267.95 ft³
15. 108 ft³   **17.** 18.84 ft³   **19.** 46.053 ft³

12.6 Exercises, page 295

1. c = 10   **3.** a = 16   **5.** b = 9   **7.** a = 20
9. c = 25   **11.** right triangle   **13.** not a right triangle
15. right triangle   **17.** not a right triangle   **19.** 17 mi
21. 5 ft   **23.** 5 ft   **25.** 20 ft   **27.** 2.2   **29.** 4.9

12.7 Exercises, page 309

1. terminating   **3.** repeating   **5.** terminating
7. terminating   **9.** nonrepeating   **11.** repeating
13. repeating   **15.** terminating   **17.** terminating
19. nonrepeating   **21.** rational   **23.** rational
25. rational   **27.** irrational   **29.** rational
31. .875   **33.** $.2\overline{7}$ **35.** $.\overline{571428}$ **37.** $.1\overline{714285}$
39. $.\overline{153846}$ **41.** $1.\overline{5}$ **43.** .4375   **45.** $.\overline{15}$

47. $.\overline{1764705882352941}$ **49.** $2.2\overline{714285}$ **51.** $\frac{5}{16}$

53. $\frac{13}{99}$ **55.** $\frac{11719}{900}$ **57.** $\frac{5}{7}$ **59.** $\frac{5}{13}$ **61.** $\frac{20}{3}$

63. $\frac{4}{1}$ or 4   **65.** $\frac{247}{20}$ **67.** $\frac{5}{1}$ or 5   **69.** $\frac{1126}{75}$

Chapter 12 Review Exercises, page 311

1. $\frac{3}{4}$ **2.** $\frac{3}{10}$ **3.** $\frac{18}{35}$ **4.** $\frac{3}{5}$ **5.** $\frac{7}{8}$ **6.** $\frac{5}{7}$ **7.** $\frac{17}{19}$

8. $\frac{2}{1}$ **9.** $\frac{5}{1}$ **10.** $\frac{6}{1}$ **11.** a box of 125 tissues for 55¢
12. Rod Carew 0.350 average   **13.** extremes: 3, 55;
means: 11, 15   **14.** extremes: x, 35; means: 7, 13
15. extremes: 8, 21; means: 3, 56   **16.** extremes: w, b;
means: z, a   **17.** $x = 13.3\overline{3}$ **18.** y = 5   **19.** x = 55
20. a = 7   **21.** equivalent   **22.** equivalent
23. not equivalent   **24.** not equivalent
25. 1364 bricks   **26.** $333.3\overline{3}$ mi   **27.** $70.50
28. $2000   **29.** 12%   **30.** 32 m   **31.** 14 m
32. 56.52 m   **33.** 22 m   **34.** 252 ft²   **35.** 59.5 ft²
36. 42 ft²   **37.** 50.24 ft²   **38.** 216 m³
39. 133.97 m³   **40.** 4186.67 m³   **41.** a = 20 cm
42. 40 ft   **43.** 45 mi   **44.** rational   **45.** rational
46. irrational   **47.** rational   **48.** $\frac{7}{8}$ **49.** $\frac{23}{9}$
50. $\frac{17}{11}$

Chapter 13

13.1 Exercises, page 316

1. $3 \neq 7$ **3.** $3 \leqslant 7$ **5.** $18.3 \geqslant 18.3$ **7.** $12 < 20$
9. $x + 10 > x$ **11.** $16 = 4^2$ **13.** Seven is greater than
or equal to seven.   **15.** One half is greater than one third.
17. Minus seventeen is less than or equal to minus
seventeen.   **19.** Minus twelve is less than minus five.
21. y is not equal to z.   **23.** Three times the quantity x
plus one is equal to 3 times x plus 3.   **25.** $2 < 6$
27. $-2 > -6$ **29.** $2 > -6$ **31.** $0 > -2$ **33.** $0 > -6$
35. $0 > -\frac{1}{2}$ **37.** $\frac{2}{7} > \frac{3}{11}$ **39.** $-\frac{11}{23} > -\frac{13}{27}$
41. $-1 < 1$ **43.** $-8 < -5$ **45.** $-0.63 < -0.61$
47. $13.5 > 13.05$ **49.** $2 > -4$

13.2 Exercises, page 319

1. $x \geqslant 3$

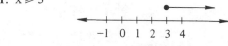

3. $-3 < x$

5. $x < 4.1$

7. $x < 4\frac{1}{2}$

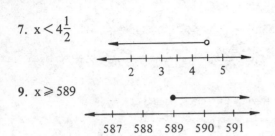

9. $x \geqslant 589$

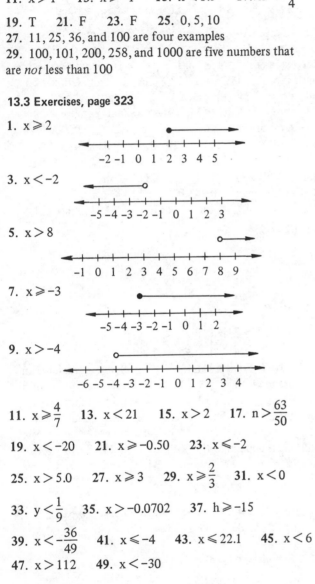

11. $x > 1$ **13.** $x \geqslant -1$ **15.** $x < 1.5$ **17.** $x \leqslant -\frac{1}{4}$

19. T   **21.** F   **23.** F   **25.** 0, 5, 10

27. 11, 25, 36, and 100 are four examples

29. 100, 101, 200, 258, and 1000 are five numbers that are *not* less than 100

13.3 Exercises, page 323

1. $x \geqslant 2$

3. $x < -2$

5. $x > 8$

7. $x \geqslant -3$

9. $x > -4$

11. $x \geqslant \frac{4}{7}$ **13.** $x < 21$ **15.** $x > 2$ **17.** $n > \frac{63}{50}$

19. $x < -20$ **21.** $x \geqslant -0.50$ **23.** $x \leqslant -2$

25. $x > 5.0$ **27.** $x \geqslant 3$ **29.** $x \geqslant \frac{2}{3}$ **31.** $x < 0$

33. $y < \frac{1}{9}$ **35.** $x > -0.0702$ **37.** $h \geqslant -15$

39. $x < -\frac{36}{49}$ **41.** $x \leqslant -4$ **43.** $x \leqslant 22.1$ **45.** $x < 6$

47. $x > 112$ **49.** $x < -30$

13.4 Exercises, page 327

1. $x > 4$

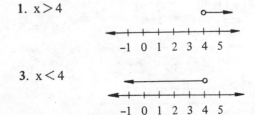

3. $x < 4$

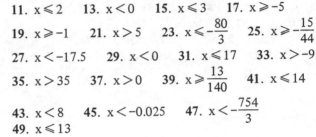

5. $x \geqslant -8$

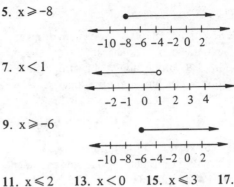

7. $x < 1$

9. $x \geqslant -6$

11. $x \leqslant 2$ **13.** $x < 0$ **15.** $x \leqslant 3$ **17.** $x \geqslant -5$

19. $x \geqslant -1$ **21.** $x > 5$ **23.** $x \leqslant -\frac{80}{3}$ **25.** $x \geqslant -\frac{15}{44}$

27. $x < -17.5$ **29.** $x < 0$ **31.** $x \leqslant 17$ **33.** $x > -9$

35. $x > 35$ **37.** $x > 0$ **39.** $x \geqslant \frac{13}{140}$ **41.** $x \leqslant 14$

43. $x < 8$ **45.** $x < -0.025$ **47.** $x < -\frac{754}{3}$

49. $x \leqslant 13$

13.5 Exercises, page 329

1. $x < -3$

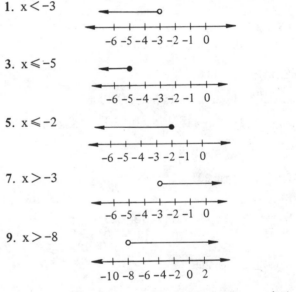

3. $x \leqslant -5$

5. $x \leqslant -2$

7. $x > -3$

9. $x > -8$

11. $x \geqslant 3$ **13.** $x > 6$ **15.** $x < 3$ **17.** no solution

19. no solution   **21.** $n \geqslant -24$ **23.** $x > -\frac{1}{2}$

25. $x \leqslant 6$ **27.** $x \leqslant \frac{11}{7}$ **29.** $x \leqslant 32$ **31.** $n < 0$

33. Every real number is a solution.   **35.** $x \leqslant 25$

37. $x > -5$ **39.** $x \geqslant -30$ **41.** $x > -5$ **43.** $x \geqslant -1$

45. Every real number is a solution.   **47.** no solution

49. $x < \frac{5}{4}$

Chapter 13 Review Exercises, page 331

1. F   **2.** T   **3.** T   **4.** T   **5.** F   **6.** T   **7.** F

8. T   **9.** T   **10.** F   **11.** F   **12.** F   **13.** T

14. T   **15.** T   **16.** T   **17.** F   **18.** T   **19.** T

20. F

21. x < 1

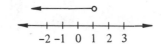

22. x < 3

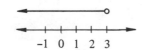

23. x < 17

24. n > 80

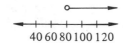

25. x ≤ 1

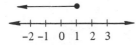

26. x < 2

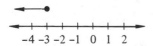

27. x ≤ -3

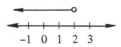

28. x > -3

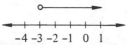

29. The solution set is the set of real numbers.

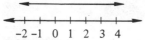

30. No solution. No points to graph.

31. x > 3

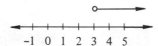

32. x ≤ $\frac{7}{3}$

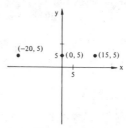

33. The solution set is the set of real numbers.

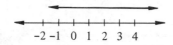

34. The solution set is the set of real numbers.

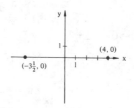

35. x ≤ 46

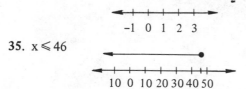

Chapter 14

14.1 Exercises, page 340

3. Social Science 600, Humanities 500, Science & Mathematics 350 **5.** 150 more **7.** C and D

9. 500 **11.** 35% **13.** $6000 **15.** $33\frac{1}{3}$% decrease

17. February **19.** yes **21.** 60% **23.** 18 boys
25. C, 20 cars

14.2 Exercises, page 343

1. A-(3, 0), B-(4, 2), C-(2, 3), D-(-2, 3), E-(-3, 0),
F(1, -2), G-$\left(-1\frac{1}{2}, -1\right)$

3.

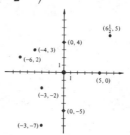

5. Any points having positive numbers as coordinates would be in the first quadrant.

7. A-I, B-II, C-I, D-IV, E-III, F-III, G-IV
9. A-(3, 3), B-(0, 0), C-(-2, -2), D-(2, 2),
E-(-1, -1), F-(-3, -3)
11. A-(20, 30), B-(-30, -10), C-(30, -10)
13. There are other possible points.

15. There are other possible points.

14.3 Exercises, page 348

1.

x	y
0	$-2\frac{3}{5}$
$6\frac{1}{2}$	0
1	$-2\frac{1}{5}$

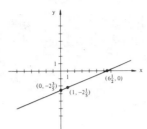

3.

x	y
0	-3
-5	-3
5	-3

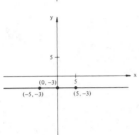

5. 0   **7.** $(0, 5), \left(\frac{1}{3}, 0\right)$ **9.** $(0, 0)$

11. $(0, -5), (10, 0)$ **13.** $(0, -4), \left(5\frac{1}{3}, 0\right)$

15. $\left(0, 2\frac{2}{5}\right), (4, 0)$

17.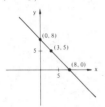

19.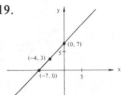

21.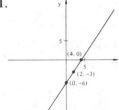

23.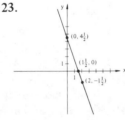

25.

27. $x = 2$ **29.** $x = -3$

14.4 Exercises, page 353

1. $9x - 3y = -6$ **3.** $2x + 6y = 12$ **5.** $5x - 2y = 10$
7. $2x - 3y + 6 = 0$ **9.** $2x + 6y - 12 = 0$
11. $5x - 2y + 10 = 0$ **13.** $y = -3x + 12$
15. $y = -3x + 5$ **17.** $y = \frac{5}{2}x - 15$ **19.** $y = \frac{9}{2}x - \frac{15}{2}$
21. $y = \frac{3}{2}x - \frac{13}{4}$ **23.** slope $= \frac{9}{1}$; intercept $(0, -15)$

25. slope $= \frac{3}{2}$; intercept $\left(0, -\frac{13}{4}\right)$

27. $y = -3x + 12$

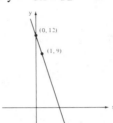

29. $y = -\frac{3}{2}x + 3$

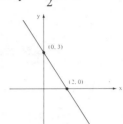

31. $y = \frac{5}{2}x - 5$

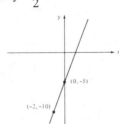

33. slope 3; point $(5, 2)$
35. slope 5; point $(3, -3)$

14.5 Exercises, page 356

1. yes   **3.** yes   **5.** no

7.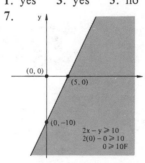

9.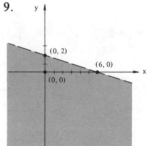

11.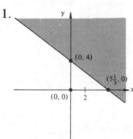

13.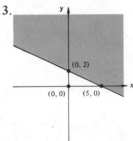

15.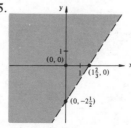

17.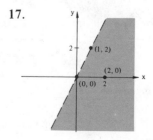

19.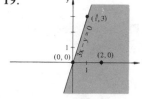

21.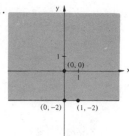

20.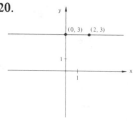

21.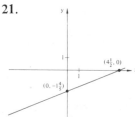

22. $y = -2x + 3$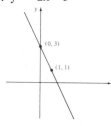

23. $y = \frac{2}{3}x - 2$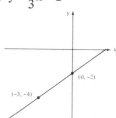

23.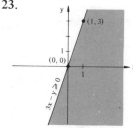

25.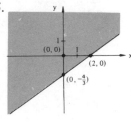

24. $y = -\frac{1}{3}x + 2$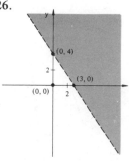

25. $y = 4x - 16$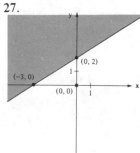

Chapter 14 Review Exercises, page 357

1. D, A, and E, B, C **2.** $10,000 **3.** C
4. $205,000 **5.** $60,000 **6.** 29% **7.** 20%
8. $\frac{1}{4}$ **9.** 9000 **10.** vocational

11.

12. (−1, −1)−III,
(−2, −2)−III,
(3, 3)−I, (4, 4)−I

26.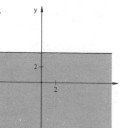

27.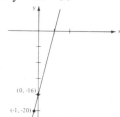

13.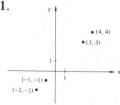

14. A−(0, 2), B−(2, 1),
C−(4, 2), D−(1, −1),
E−(−2, −1), F−(−4, 0),
G−(4, 0)
15. A−on y-axis, B−I, C−I,
D−IV, E−III,
F−on x-axis, G−x-axis

28.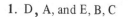

29.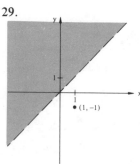

16.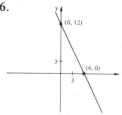

17.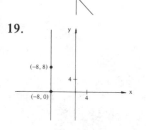

18.

19.

30.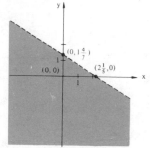

Chapter 15

15.1 Exercises, page 362

1. (6, 0)

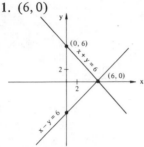

3. (−6, 4)

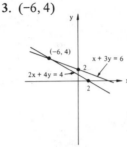

5. no solution

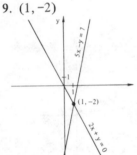

7. (6, −8)

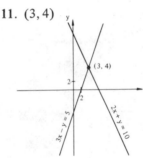

9. (1, −2)

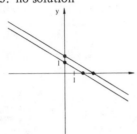

11. (3, 4)

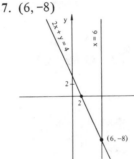

13. (2, 1)

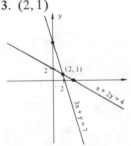

15. (5, 4)

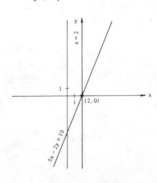

17. infinite number of solutions

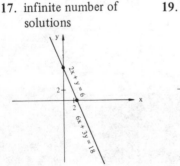

19. (4, 2)

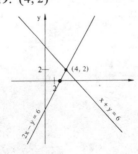

15.2 Exercises, page 367

1. (6, 0)   **3.** (2, 0)   **5.** infinite number of solutions
7. (3, 3)   **9.** (−3, 5)   **11.** $\left(\dfrac{92}{33}, \dfrac{-16}{33}\right)$

13. no solution   **15.** (0, 0)   **17.** $\left(\dfrac{17}{6}, \dfrac{25}{6}\right)$

19. (−5, 0)   **21.** 15—Paul's age; 9—Anthony's age
23. 27 dimes   **25.** 4—first integer; 14—second integer

15.3 Exercises, page 370

1. (−2, −7)   **3.** infinite number of solutions
5. $\left(\dfrac{4}{7}, \dfrac{5}{7}\right)$ **7.** (12, 9)   **9.** $\left(\dfrac{19}{5}, 4\right)$ **11.** (1, −2)
13. (2, 2)   **15.** (2, 3)   **17.** no solution   **19.** (3, 5)
21. 8 ft, 24 ft   **23.** 3, 15   **25.** 18.75 lb (45¢);
31.25 lb (37¢)

Chapter 15 Review Exercises, page 371

1. (2, 0)

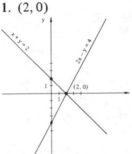

2. (2, −1)

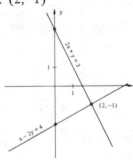

3. (−36, −24)   **4.** $\left(-\dfrac{1}{5}, \dfrac{8}{5}\right)$ **5.** $\left(\dfrac{3}{2}, -\dfrac{2}{3}\right)$
6. (1, 4)   **7.** (2, −2)   **8.** $\left(-\dfrac{1}{5}, \dfrac{8}{5}\right)$ **9.** no solution
10. (3, 3)   **11.** (1, 3)   **12.** $\left(\dfrac{3}{7}, \dfrac{5}{7}\right)$
13. infinite number of solutions   **14.** (4, −1)
15. (−2, 5)   **16.** (0, 1)
17. (2, 0)

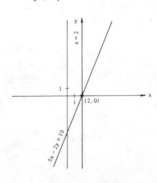

18. 83 acres of corn; 77 acres of beans
19. Tom's was $1.50; Andrew's was $1.25
20. $9000 for first year; $18000 for second year

Chapter 16

16.1 Exercises, page 379

1. 2   3. 4   5. 5   7. $5a^2$ 9. $11a^2b^2c^2$ 11. 11
13. 32   15. 70   17. 1   19. 1   21. 11   23. 11
25. 11   27. 1   29. 17   31. x^2 33. $5a^3$
35. $2b^2$ 37. a^5 39. 3   41. 3   43. $4ab^2c$
45. $70x^3y^2z^2$ 47. $12ab^2c$ 49. $6xyz$ 51. $3k^2$
53. $2ab^2$ 55. 1   57. $5x$ 59. $11h^2k$ 61. $a - b$
63. $x + y$ 65. $a + 3b$ 67. $2a - 5b$ 69. $6xy(3a - b)$

16.2 Exercises, page 382

1. $7(x - y)$ 3. $x^2(x - 1)$ 5. $2x(2 + 3x)$
7. $3m^2(m^2 + 1)$ 9. $11x^3(2x^3 - 3)$ 11. $13(x - 2y)$
13. $x^2y^3(x - y)$ 15. $7(7x^2 - 9xy + 2y)$
17. $xy(5x - 7y)$ 19. $3m^2(m^2 - 3m + 2)$
21. $4ab^2c(3ab - 5)$ 23. $70x^2y^5(2 - 3x - 5x^8)$
25. $2xy(2x^2 - 6xy - 3y^2)$ 27. $3x(x^2 - 2xy + y^2)$
29. $3k^2(-4k^2 - h^2 + 11k)$ or $-3k^2(4k^2 + h^2 - 11k)$
31. $(a + 3b)(2x - y)$ 33. $(19x - y)(3a + 5b - 7c)$
35. $(2a - 3b + c)(3x - 2y)$
37. $(5a - 4b - 3c - d)(x - 2y)$
39. $3x^2y(2a - 7c)(5x - 7y)$

16.3 Exercises, page 385

1. $2a$ 3. $-7a$ 5. $5a$ 7. $-23a$ 9. $-5x$
11. $2x^2 - xy - y^2$ 13. $9w^2 - 6wz + z^2$
15. $33x^2 - xy - 4y^2$ 17. $6s^2 - 33st + 42t^2$
19. $4a^2 - 9b^2$ 21. $21w^2 + 23wz - 20z^2$
23. $25a^2 - 49b^2$ 25. $12x^2 - 41x + 35$
27. $6a^2 - 37ab + 6b^2$ 29. $4s^2 + 44st + 121t^2$
31. $56x^2 - 94x + 30$ 33. $a^2 - a - 72$
35. $4x^2 - 4xy + y^2$ 37. $x^2y^2 + xy - 12$
39. $a^2b^2 + 2abc - 3c^2$ 41. $w^2x^2 - 2wxyz + y^2z^2$
43. $4a^2b^2c^2 - d^2$ 45. $10a^4 + 3a^2b^2 - b^4$
47. $6a^2x^4 - ax^2y - y^2$ 49. $x^6 - 9y^4$

16.4 Exercises, page 388

1. $x^2 + 2xy + y^2$ 3. $4a^2 + 4ab + b^2$ 5. $x^2 + 22x + 121$
7. $49x^2 + 70xy + 25y^2$ 9. $9a^2b^2 + 12abc + 4c^2$
11. $w^2 - 2wz + z^2$ 13. $x^2 - 6xy + 9y^2$
15. $16x^2 - 24xy + 9y^2$ 17. $16w^2 - 88w + 121$
19. $4x^2 - 36x + 81$ 21. $s^2 - t^2$ 23. $x^2 - y^2$
25. $49x^2 - 25$ 27. $25x^2y^2 - z^2$ 29. $s^6 - t^6$
31. $x^2 - 49$ 33. $y^2 - 2yz + z^2$ 35. $x^2 + 4xy + 4y^2$
37. $4x^2 - 9$ 39. $64y^2 - 48yz + 9z^2$
41. $16w^2 + 24w + 9$ 43. $81k^2 - 90nk + 25n^2$
45. $49a^2 - 64b^2$ 47. $10a^2 - 19a - 15$
49. $400x^2 + 120xy + 9y^2$ 51. $16w^2 - 50w + 6$
53. $25w^2z^4 - 10wz^2 + 1$ 55. $x^4 - 25$
57. $m^2 + 3mn - 18n^2$ 59. $64y^2z^4 + 16w^3yz^2 + w^6$

16.5 Exercises, page 394

1. $(x - 6)(x + 1)$ 3. $(x + 3)(x - 2)$
5. $(x + 6)(x + 1)$ 7. $(x - 3)(x - 2)$
9. $(w + 5)(w - 6)$ 11. $(s + 7)(s + 11)$
13. $(b + 12)(b - 3)$ 15. $(t + 8)(t - 6)$
17. $(c - 5)(c - 20)$ 19. $(r + 7)(r - 5)$
21. $(x + 2)(x + 1)$ 23. $(p - 7)(p + 1)$
25. $(x - 12)(x + 3)$ 27. $(5x + 7)(x + 1)$
29. $(7y - 13)(y + 1)$ 31. $(9c - 8)(c + 1)$
33. $(3a - 5)(a - 1)$ 35. $(2y - 1)(y - 7)$
37. $(7b - 1)(b + 5)$ 39. $(11t + 1)(t + 9)$
41. $(10a + 1)(a - 7)$ 43. $(a - 3b)(a + 2b)$
45. $(x + 4y)(x - 3y)$ 47. $(p + 7q)(p + 2q)$
49. $(3c + d)(2c + 11d)$ 51. $(5x - 2y)(3x - 4y)$
53. $(6s + 7t)(3s - t)$ 55. $(6j - 5k)(7j + 5k)$
57. $(25x - 21y)(x + y)$ 59. $(5f - g)(9f - 14g)$
61. $(21p - 5q)(3p + 2q)$ 63. $(25a + 7b)(a - 3b)$
65. $(9m + 2n)(5m + 7n)$ 67. $(12w - 7x)(2w - 5x)$
69. $(7m + 3n)(4m - 5n)$ 71. $(11xy - 2z)(9xy + 7z)$
73. $(5s + 7t)(3s + 2t)$ 75. $(5b + 4)(3b + 4)$
77. $(x + 5)(x - 5)$ 79. prime over the integers
81. prime over the integers   83. prime over the integers
85. $(x + 11)^2$ 87. $(x + 1)(x + 7)$ 89. $(x - 7)(x + 1)$
91. prime over the integers   93. $(3x + 1)(x - 5)$
95. $(3x - 5)(x + 1)$ 97. $(a - 7)(a + 4)$
99. prime over the integers

16.6 Exercises, page 398

1. $(x - 1)^2$ 3. $(a + 3)^2$ 5. $(w - 7)^2$ 7. $(x + 8y)^2$
9. $(w + z)(w - z)$ 11. $(3a + b)(3a - b)$
13. $(s + 9)(s - 9)$ 15. $(12 + x)(12 - x)$
17. $(x - y)(x^2 + xy + y^2)$ 19. $(w + z)(w^2 - wz + z^2)$
21. $(3a + 1)(9a^2 - 3a + 1)$ 23. $(x - 2)(x^2 + 2x + 4)$
25. $(x + 3)(x - 3)$ 27. $(a - 1)(a^2 + a + 1)$
29. $(a - 3)^2$ 31. $(c - d)(c^2 + cd + d^2)$ 33. $(3x - 2y)^2$
35. $(x + 6)(x - 3)$ 37. $(x + 5y)(x^2 - 5xy + 25y^2)$
39. $2(2x^2 + 9x - 4)$ 41. $(w^3 - 11)(w^3 - 11)$
43. $(4s + 11t)(4s - 11t)$ 45. $(3xy^3 + 1)(3xy^3 - 1)$
47. $(7s + 2)(7s + 2)$ 49. $5(x^2 + y^2)$
51. $(6x + y)(36x^2 - 6xy + y^2)$ 53. $(y^2 - 2)(y^2 - 2)$
55. $(x + 4)(x^2 - 4x + 16)$ 57. prime over the integers
59. $(k + j)(k^2 - kj + j^2)$ 61. $(3x - 5y)(3x - 5y)$
63.
$$
\begin{array}{r}
a^2 + ab\ + b^2 \\
a\ - b \\
\hline
a^3 + a^2b + ab^2 \\
-a^2b - ab^2 - b^3 \\
\hline
a^3\qquad\quad - b^3
\end{array}
$$
65.
$$
\begin{array}{r}
25s^2 +\ 20st\ + 16t^2 \\
5s\ -\ 4t \\
\hline
125s^3 + 100s^2t + 80st^2 \\
-100s^2t - 80st^2 - 64t^3 \\
\hline
125s^3\qquad\qquad\quad - 64t^3
\end{array}
$$

16.7 Exercises, page 401

1. $(a + b)(c + d)$ 3. $(a - 2b)(3 + 5c)$
5. $(2a + b)(3a + 1)$ 7. $(a + c)(b - d)$
9. $(a - 2b)(3 - 5c)$ 11. $(x - 1)(a + b)$

13. $(x-7)(s+1)$ 15. $(x-1)(k+j)$

17. $(3b-4+a)(3b-4-a)$ 19. $(s+k)(x^2+1)$

21. $(x+y-1)(x-y+1)$ 23. $(x+y)(3+x^2-xy+y^2)$

25. $(s+11)(s^2+1)$ 27. $(a+1)(a+1+b)$

29. $(x-y-z)(a+b)$

16.8 Exercises, page 404

1. $(8s+3t)(8s-3t)$ 3. $(7a-2)^2$

5. $3(4x-5)(x-1)$ 7. $4ay(y-1)$ 9. $4x(x^9-150)$

11. $(2x^5+3y^3)^2$ 13. $c(z+2)(z^2-2z+4)$

15. $(6s^2+7)(6s^2-7)$ 17. $(5x+1)(7x+6)$

19. $(2h-5j)(4h^2+10hj+25j^2)$ 21. $(x+5)(x+1)$

23. $5(16x^2+49y^2)$ 25. $m(2m+1)^2$ 27. $3a(x^2+y^2)$

29. $12xy(x+y)(x-y)$ 31. $2a(3x+2y)(2x-3y)$

33. $10(x^2+4)$ 35. $7as(s-3)(s^2+3s+9)$

37. $7st(s^2+t^2)(s+t)(s-t)$ 39. $3xy(3x+4y)^2$

41. $5xy(2x+7y)(2x-7y)$

43. $4a(a^2+2b)(a^4-2a^2b+4b^2)$

45. $(6x-5y)(3x+2y)$ 47. $7ab(3a+5)(3a-5)$

49. $8a(x+9y^2)(x-9y^2)$ 51. $3(2x-y)^2$

53. $6(x-1)(k+j)$ 55. $(x+y)(c+d)$

57. $(a-b)(x+y)$ 59. $3a(x+y-1)(x-y+1)$

61. $5a(2x^2+11y)^2$ 63. $(x+y+4z)(x+y-4z)$

65. $2(x+y)(x-y+1)$

Chapter 16 Review Exercises, page 404

1. 15 2. 72 3. $6ax^2$ 4. $14ab^2c$

5. $30x^2-47xy+14y^2$ 6. $40s^2+78st-27t^2$

7. $9x^2-4y^2$ 8. $9x^2-12xy+4y^2$

9. $9x^2+12xy+4y^2$ 10. $121x^2-144y^2$

11. $12ax(x-2)$ 12. $(7x+2)(7x-2)$

13. $(4t-1)(16t^2+4t+1)$ 14. $(3x-2)(4x-9)$

15. $(5x-9y)^2$ 16. $2a(2s+t)(4s^2-2st+t^2)$

17. prime over the integers   18. $7a(t^2+1)(t+1)(t-1)$

19. $(x^2+7)(x^2-6)$ 20. $2a(7x+12)(2x-3)$

21. $(x-5)(x+4)$ 22. $(3a+4)(3a-4)$

23. prime over the integers   24. $xy(x-4)(x-1)$

25. $2(3x+8)(2x-9)$ 26. $2x(x^2+4)(x+2)(x-2)$

27. $a(x^2-x-6a)$ 28. $x^2(x+y)(x-y)$

29. $11(x+y)(x-y+3)$ 30. $5x(x-y)(2x+a)$

31. $(m^4+n^4)(m^2+n^2)(m+n)(m-n)$

32. $(6x+7y+2)(6x-7y+2)$

33. $(a+b+3)(a+b+1)$

34. $(4m+n-3)(4m-n+3)$ 35. $(2a^2-bc)(b+13)$

36. $(x^n+1)(x^n-1)$ 37. $(3x^n-1)(2x^n-1)$

38. $(4x^{3n}+5)(4x^{3n}-5)$ 39. $(x^n+2)(x^{2n}-2x^n+4)$

40. $x^n(x^{2n}+4x^n-1)$

Chapter 17

17.1 Exercises, page 412

1. $-\dfrac{x-y}{-3}$ 3. $-\dfrac{y-x}{3}$ 5. $-\dfrac{3a+b}{b-a}$ 7. $\dfrac{a-b}{6}$

9. $\dfrac{a-b}{6}$ 11. $\dfrac{2x^3}{7}$ 13. $\dfrac{3b}{7a}$ 15. $y(6x-19y)$

17. $\dfrac{-11st}{12s+9t}$ 19. $\dfrac{2(x-y)}{x+y}$ 21. $\dfrac{5a^2x+2a+7}{11ax+12}$

23. $\dfrac{a}{b}$ 25. $-\dfrac{7}{3}$ 27. $\dfrac{x-y}{3}$ 29. $\dfrac{3x-4y}{9}$ 31. $\dfrac{3}{2}$

33. $\dfrac{-11a}{8x+3y}$ 35. $\dfrac{x}{x-y}$ 37. $\dfrac{a+5b}{a-b}$ 39. $\dfrac{ab(a^2-b^2)}{3}$

41. $\dfrac{10(3x+2)}{a-b}$ 43. $\dfrac{x+y}{13}$ 45. $-\dfrac{4x^2+6x+9}{6}$

47. $\dfrac{3(7s-6)}{14(2s-1)}$ 49. $\dfrac{5a(a+1)}{3(2a-5)}$ 51. $\dfrac{a+b}{5}$ 53. $\dfrac{7+a}{3}$

55. $\dfrac{a-2b-z}{5}$ 57. $\dfrac{a+1}{9}$ 59. $\dfrac{x-1}{a-1}$ 61. 40

63. 432 65. ax 67. $84xyz$ 69. $3(x-y)$

71. $10(a-b)$ 73. $3(s^2-st+t^2)$ 75. $(x+3y)(x+y)$

77. $4x^2y(x+5y)$ 79. $(x+3y)(2x+y)^3$

17.2 Exercises, page 417

1. $\dfrac{112}{55}$ 3. $-\dfrac{23}{30}$ 5. $\dfrac{32}{105}$ 7. $\dfrac{192}{1183}$ 9. $-\dfrac{st^3}{9}$

11. $\dfrac{32acd^2}{13}$ 13. $\dfrac{-18x^5}{5y^5}$ 15. $\dfrac{y^3}{xz^3}$ 17. $\dfrac{-2bc^4y}{9ax^4}$

19. $\dfrac{11x}{3y}$ 21. $\dfrac{3x}{22}$ 23. $\dfrac{-2(x-1)}{3}$ 25. $\dfrac{3}{w}$

27. $\dfrac{7(x-3)}{3(x-y)}$ 29. $\dfrac{10y^2(2x+3)}{7x(5x-2)}$ 31. $\dfrac{(x^2-xy+y^2)^2}{15x^2y^2}$

33. $\dfrac{12}{11xy^2}$ 35. $\dfrac{3(4x^2-6x+9)}{5(x-4)}$ 37. $\dfrac{11}{xy}$

39. $\dfrac{3y(x-y)}{-14}$ 41. $\dfrac{4y(x-2)}{7x^2(x-y)}$ 43. $\dfrac{y(x-y+3)}{7(2x+1)(2x-1)}$

45. $\dfrac{19(7-a)}{17(c-y)}$ 47. $\dfrac{7y^2}{2a}$ 49. $\dfrac{4}{3}$

17.3 Exercises, page 420

1. $\dfrac{3}{5}$ 3. $\dfrac{1}{7}$ 5. $\dfrac{23}{w}$ 7. $\dfrac{-3}{x^2}$ 9. $\dfrac{2b+3}{b^2}$ 11. $\dfrac{4}{s-3}$

13. $\dfrac{-3}{a-5}$ 15. 1 17. $\dfrac{9}{s-2}$ 19. $\dfrac{8}{a+3}$ 21. $\dfrac{16}{x}$

23. $\dfrac{5}{t^2}$ 25. $\dfrac{4}{x}$ 27. $\dfrac{2}{s}$ 29. 4 31. $\dfrac{-2}{3}$

33. $\dfrac{z-1}{z^2+1}$ 35. $\dfrac{2(x+7)}{(x-7)(x-1)}$ 37. $\dfrac{x-1}{x-4}$ 39. $\dfrac{1}{y+5}$

17.4 Exercises, page 423

1. 400 3. 480 5. $42x^2y^5$ 7. $60a^2bc^3$

9. $1200a^2b^3x^3y^2z^8$ 11. $1452wx^5y^3z^7$ 13. $60(x+y)$

15. $20(3x+2y)$ 17. $63(2a-3b+c)$

19. $5(x+y)(x-y)$ 21. $18x(x+y)$

23. $(x+1)(x+6)(x-6)$ 25. $12xy(x-2)(x+7)$

27. $36(2y+1)(2y-1)(4y^2+2y+1)$

29. $5x(5x-7y)^2(5x+7y)$

17.5 Exercises, page 426

1. $\dfrac{5b + 7a}{ab}$ 3. $\dfrac{29}{18w}$ 5. $\dfrac{5y - 6x}{x^2y^2}$ 7. $\dfrac{4c + 3a - b}{abc}$

9. $\dfrac{67x}{240}$ 11. $\dfrac{a - 6b}{105}$ 13. $\dfrac{t + w}{10}$ 15. $\dfrac{-b^2}{(a - b)(a + b)}$

17. $\dfrac{2x^2 + xy + 3y^2}{(3y - 2x)(3y + 2x)}$ 19. $\dfrac{3x^2 - 8xy - 4y^2}{(x + 2y)(x - 2y)}$

21. $\dfrac{1}{77}$ 23. $\dfrac{2}{a - 4b}$ 25. $\dfrac{6}{x - 7}$

27. $\dfrac{12(a - b)}{a(5a - 7b)(7a - 5b)}$ 29. $\dfrac{60xy}{(2x - 5y)(2x + 5y)}$

31. $\dfrac{4x + 3y}{x^2 + xy + y^2}$ 33. $\dfrac{a - b}{a^2 - ab + b^2}$

35. $\dfrac{y}{(x - y)(x - 3y)}$ 37. $\dfrac{1}{w - 3}$ 39. $\dfrac{y}{x^2 - xy + y^2}$

17.6 Exercises, page 431

1. $\dfrac{1}{6}$ 3. $\dfrac{5}{4}$ 5. $\dfrac{a}{8}$ 7. $\dfrac{3}{2}$ 9. $\dfrac{24}{a}$ 11. $\dfrac{3}{2}$ 13. -2

15. $\dfrac{15}{8}$ 17. $\dfrac{2}{xy(w + z)}$ 19. $\dfrac{9x(x - 7)}{10y^2(x + 2)}$

21. $\dfrac{x}{2x + 1}$ 23. $\dfrac{a^2 - 1}{a^2 + 1}$ 25. $\dfrac{x - 1}{x}$ 27. $\dfrac{x - 3}{x^2}$

29. $\dfrac{2x - 1}{2x + 3}$ 31. $\dfrac{1}{a - b}$ 33. $\dfrac{-4aw}{w^2 + a^2}$

35. $\dfrac{(b + c)(b - c)}{bc}$ 37. $\dfrac{10x}{21}$ 39. $\dfrac{5x - 3y}{4xy}$

41. $3a^2b^2(a^2 - ab + b^2)$ 43. $4(5w + z)$ 45. $\dfrac{4}{x + 3}$

17.7 Exercises, page 435

1. $x = 1$ 3. $x = 1$ 5. $x = 3, x = 2$ 7. $x = 2$

9. $x = 4, x = -1$ 11. $x = 2$ 13. $x = 2$ 15. $x = \dfrac{4}{3}$

17. $x = -5$ 19. no solution   21. $x = 0$ 23. $x = 8$
25. $x = -16$

Chapter 17 Review Exercises, page 435

1. $\dfrac{3}{4}$ 2. $\dfrac{x - y}{5}$ 3. $\dfrac{2a - 3b}{2a + 3b}$ 4. $\dfrac{s - t}{9}$ 5. $25xy$

6. $3(x^2 - y^2)$ 7. $4x(x - y)$ 8. $2(x + 5)(2x + 1)$

9. 5 10. $\dfrac{5y}{7x}$ 11. 3 12. $\dfrac{(x + 1)(x - 2)}{(x + 7)(x^2 + 2x + 4)}$

13. $\dfrac{3yz + 4xz - 5xy}{xyz}$ 14. 1 15. $\dfrac{11x}{15(x - 2)}$

16. $\dfrac{x^2 - xy + y^2}{xy}$ 17. $\dfrac{-x + 71y}{105(x + y)^2(x - y)}$ 18. $\dfrac{-3}{5}$

19. $\dfrac{30}{x(x + 3y)}$ 20. $\dfrac{35y^2z}{12wx}$ 21. $\dfrac{xy(x - 5y)}{2(2x - y)(x - 2y)}$

22. $\dfrac{1}{3}$ 23. $\dfrac{3y(x - y)}{x}$ 24. $\dfrac{3}{(2x - y)(2x - y)}$

25. $\dfrac{6}{7xy(y - x)}$ 26. $x = 3$ 27. no solution

28. $x = 5$ 29. $x = 3$ 30. $x = -\dfrac{7}{4}$

Chapter 18

18.1 Exercises, page 444

1. 6 3. 3 5. -1 7. 10 9. -12 11. a

13. y^{20} 15. x^{12} 17. $-u^7$ 19. n^2 21. $2x$

23. $4a^4$ 25. $5y^2$ 27. $10a$ 29. $9y^5$ 31. 10

33. x 35. $3a$ 37. x^2 39. $2x^3$ 41. $2\sqrt{2}$

43. $5\sqrt{2}$ 45. $4\sqrt{2}$ 47. $3\sqrt{3}$ 49. $7\sqrt{2}$

51. $x\sqrt{x}$ 53. $a^4\sqrt{a}$ 55. $b^2\sqrt{b}$ 57. $x^3\sqrt{x}$

59. $y^7\sqrt{y}$ 61. $2x\sqrt{x}$ 63. $a\sqrt{5}$ 65. $2x^2\sqrt{2x}$

67. $4xy\sqrt{y}$ 69. $2y\sqrt{6x}$ 71. $6\sqrt{10}$ 73. $10\sqrt{6}$

75. $4\sqrt{30}$ 77. $3xy^2\sqrt{30}$ 79. $a^7b^{10}\sqrt{ac}$ 81. 2

83. $\sqrt{11}$ 85. $\sqrt{13}$ 87. $\sqrt{2}$ 89. $\sqrt{5}$ 91. $\dfrac{2}{3}$

93. $\dfrac{\sqrt{3}}{8}$ 95. $\dfrac{11}{10}$ 97. $\dfrac{\sqrt{10}}{9}$ 99. $\dfrac{2}{x}$ 101. $\dfrac{3\sqrt{2}}{2}$

103. $\dfrac{\sqrt{15}}{3}$ 105. $\dfrac{\sqrt{7}}{7}$ 107. $\dfrac{\sqrt{35}}{5}$ 109. $\dfrac{\sqrt{15}}{15}$

111. $\dfrac{4\sqrt{x}}{x}$ 113. $\dfrac{\sqrt{3y}}{y}$ 115. $\dfrac{\sqrt{a}}{a}$ 117. $\dfrac{\sqrt{2x}}{2x}$

119. $\dfrac{4\sqrt{3x}}{3x^3}$ 121. $8\sqrt{2}$ 123. $4\sqrt{7}$ 125. $-\sqrt{19}$

127. $3\sqrt{5}$ 129. $\sqrt{x}$ 131. $3\sqrt{3}$ 133. $5\sqrt{2}$

135. $\dfrac{7}{2}\sqrt{2}$ 137. $8\sqrt{2}$ 139. $x\sqrt{x}$ 141. $9\sqrt{6}$

143. $\dfrac{\sqrt{5x}}{x}$ 145. $\dfrac{\sqrt{3xy}}{y}$ 147. $\dfrac{\sqrt{y}}{y^2}$ 149. $5a^5b^3c\sqrt{7bc}$

18.2 Exercises, page 447

1. quadratic   3. not quadratic   5. quadratic
7. $2x^2 - 4x + 8 = 0$ 9. $x^2 - 3x + 2 = 0$
11. $x = -2, x = -1$ 13. $x = 6, x = -3$

15. $x = 0, x = -3$ 17. $x = -\dfrac{3}{2}, x = 5$

19. $x = -\dfrac{2}{3}, x = -\dfrac{5}{2}$ 21. $x = 4, x = -4$

23. $x = 5, x = -5$ 25. $x = -6, x = -7$

27. $x = -\dfrac{3}{7}, x = \dfrac{1}{2}$ 29. $x = \dfrac{9}{11}, x = -\dfrac{9}{11}$

18.3 Exercises, page 450

1. $x^2 - 5x - 6 = 0; a = 1, b = -5, c = -6$
3. $7x^2 - 2x + 3 = 0; a = 7, b = -2, c = 3$
5. $x^2 + 2x - 8 = 0; a = 1, b = 2, c = -8$
7. $x^2 + x - 1 = 0; a = 1, b = 1, c = -1$
9. $x^2 - 6x + 9 = 0; a = 1, b = -6, c = 9$
11. $x^2 - 2x = 0; a = 1, b = -2, c = 0$
13. $6x^2 - 5 = 0; a = 6, b = 0, c = -5$
15. $10x^2 + 3x - 1 = 0; a = 10, b = 3, c = -1$
17. $x^2 + 3x - 2 = 0; a = 1, b = 3, c = -2$
19. $x^2 - 14x + 49 = 0; a = 1, b = -14, c = 49$
21. $x^2 - 3x = 0; a = 1, b = -3, c = 0$

23. $x = 6$, $x = -1$; real roots

25. $x = \dfrac{2 \pm \sqrt{-80}}{14}$; nonreal roots

27. $x = +2\sqrt{2}$, $x = -2\sqrt{2}$; real roots

29. $x = \dfrac{-1 \pm \sqrt{5}}{2}$; real roots   31. $x = 3$; real root

33. $x = 2$, $x = 0$; real roots

35. $x = \dfrac{+\sqrt{30}}{6}$, $x = \dfrac{-\sqrt{30}}{6}$; real roots

37. $x = -\dfrac{1}{2}$, $x = \dfrac{1}{5}$; real roots

39. $x = \dfrac{-3 + \sqrt{17}}{2}$, $x = \dfrac{-3 - \sqrt{17}}{2}$; real roots

41. $x = 7$; real root   43. $x = 6$, $x = -6$; real roots

45. $x = 2$, $x = -2$ 47. $x = -4$, $x = -2$

49. $x = \dfrac{1}{3}$, $x = -2$

18.4 Exercises, page 453

1. -10 and -13 or 10 and 13 3. 16 feet $\times$ 31 feet

5. 5 meters   7. 13 and 14   9. 6 centimeters

11. $x = \dfrac{3}{2}$, $x = 1$ 13. $x = \dfrac{1 + \sqrt{-7}}{2}$, $x = \dfrac{1 - \sqrt{-7}}{2}$

15. $x = 2$, $x = \dfrac{19}{5}$ 17. $\dfrac{+\sqrt{30}}{3}$, $\dfrac{-\sqrt{30}}{3}$ 19. no solution

Chapter 18 Review Exercises, page 453

1. $11\sqrt{7}$ 2. $\sqrt{3}$ 3. $10\sqrt{3}$ 4. $5\sqrt{6}$ 5. $-7\sqrt{5}$

6. $2\sqrt{22}$ 7. $\dfrac{\sqrt{5ab}}{b}$ 8. $5ab^2$ 9. $\dfrac{\sqrt{y}}{y^2}$

10. $2xy\sqrt{31y}$ 11. quadratic   12. not quadratic

13. not quadratic   14. quadratic   15. not quadratic

16. not quadratic

17. $2x^2 - 4x + 6 = 0$; $a = 2$; $b = -4$; $c = 6$

18. $3x^2 - 27 = 0$; $a = 3$; $b = 0$; $c = -27$

19. $x^2 - 25 = 0$; $a = 1$; $b = 0$; $c = -25$

20. $x^2 + 3x = 0$; $a = 1$; $b = 3$; $c = 0$

21. $x^2 - 3x + 6 = 0$; $a = 1$; $b = -3$; $c = 6$

22. $x^2 + 3x - 2 = 0$; $a = 1$; $b = 3$; $c = -2$ 23. $6, -2$

24. $9, -9$ 25. $\dfrac{1}{3}, -4$ 26. $\dfrac{-2}{5}, 3$

27. $\dfrac{-3 + 3\sqrt{5}}{2}, \dfrac{-3 - 3\sqrt{5}}{2}$ 28. $2.6, -1.1$

29. $\dfrac{3 + \sqrt{57}}{12}, \dfrac{3 - \sqrt{57}}{12}$ 30. $\dfrac{-3 + \sqrt{33}}{2}, \dfrac{-3 - \sqrt{33}}{2}$

31. $5\sqrt{2}, -5\sqrt{2}$ 32. $\dfrac{1}{2}, 1$ 33. $-8, -2$

34. $x = 4$, $x = 35$ 35. $x = 7$, $x = 1$

36. 0 and -5 or 10 and 5   37. 10 and 12

38. 20 meters wide $\times$ 35 meters long

39. 15 and 17 or -15 and -17 40. 8 and 12

41. 14 ft $\times$ 20 ft

Appendix A

A.1 Exercises, page 459

1. $6 \in S$ 3. $4 \notin P$ 5. $D = \{5, 15, 1\}$

7. $S = \{4, 19, 105\}$ 9. One-half is not an element of N.

11. One hundred seventeen is an element of W.

13. Set S is not equal to set T.

15. A is the set containing two, fourteen, and thirty-three.

17. F   19. F   21. T   23. F   25. T

27. $F = \{5, 10, 15, 20, 25, 30, 35, 40\}$ 29. no

31. no   33. yes   35. no

A.2 Exercises, page 460

1. $I \subseteq Q$ 3. $N \subseteq W$ 5. I is a subset of R.

7. W is not equal to N.   9. E is a subset of N.   11. F

13. F   15. F   17. T   19. T

A.3 Exercises, page 461

1. $\dfrac{5}{10}$ 3. $\dfrac{-13}{3}$ 5. $\dfrac{-33}{1}$ 7. $\dfrac{605}{1}$ 9. $\dfrac{-33\frac{1}{3}}{100} = \dfrac{-1}{3}$

11. $\dfrac{-407}{100}$ 13. $\dfrac{407}{10}$ 15. $\dfrac{-407}{10000}$ 17. F   19. T

21. T   23. F   25. T   27. F   29. F   31. T

33. F   35. F   37. F   39. F   41. T   43. F

45. T   47. T   49. T

A.4 Exercises, page 463

1. $\{1, 2, 4, 6, 8\}$ 3. $\{4, 5\}$ 5. $\{1, 2, 6, 7, 8, 10\}$

7. $\{10\}$ 9. $\{2, 4\}$ 11. $\varnothing$ 13. $\{2, 4\}$

15. $\{5, 10, 15, 20, 25, ...\}$ 17. W   19. N

21. $\{1000, 2\}$ 23. R   25. I   27. $\varnothing$ 29. yes

A.5 Exercises, page 465

1. T   3. T   5. T   7. T   9. F   11. T   13. F

15. T   17. T   19. T   21. N, W, I, Q, R   23. T, R

25. W, I, Q, R   27. I, Q, R   29. N, W, I, Q, R

Appendix A Review Exercises, page 466

1. $\{1, 2, 3, 4, 5, 6, 7\}$ 2. $\{5\}$ 3. $\varnothing$

4. $\{1, 2, 3, 4, 5, 10, 15, 20, ...\}$ 5. -12 6. $5 \in A$

7. $X \nsubseteq Y$ 8. $M \cap N = \varnothing$

9. One and two-tenths is an element of I.

10. T is a subset of R.

11. A is the set which contains one, three, five, and seven.

12. The intersection of the set containing four and eight and the set containing four is the set containing four.

13. $\dfrac{13}{100}$ 14. $\dfrac{21}{5}$ 15. $-\dfrac{9}{40}$ 16. $\dfrac{-24}{1000}$ or $\dfrac{-3}{125}$

17. $\dfrac{-15}{1}$ 18. F   19. F   20. T   21. T   22. F

23. T   24. $\varnothing$ 25. W   26. I   27. R   28. W

29. T

Appendix B

B.1 Exercises, page 472

1. mm **3.** m **5.** dg **7.** km **9.** g **11.** 1
13. hg **15.** cm **17.** dal **19.** dm
21. meter, kilogram, liter **23.** milli—0.001 of,
kilo—1000 times, centi—0.01 of **25.** 0.001 km
27. 100 cg **29.** 1000 ml **31.** 10 dl **33.** 0.1 dam
35. 0.01 hg **37.** 5.261 m **39.** 13400000 mm
41. 13400 m **43.** 520 mm **45.** 0.00215 km
47. 1.520 km **49.** 1200 ml **51.** 0.001 kg
53. 5040000 mg **55.** 24 g **57.** 350 cm
59. 600 ml **61.** 0.02519 kg **63.** 4.28 m **65.** >
67. < **69.** > **71.** < **73.** < **75.** < **77.** >
79. > **81.** < **83.** < **85.** > **87.** > **89.** =
91. < **93.** > **95.** > **97.** −40° **99.** 15°
101. 100° **103.** 22° **105.** −12° **107.** 0.001 liter
109. 2 ml **111.** 24 ml

B.2 Exercises, page 475

5. > **7.** < **9.** < **11.** < **13.** > **15.** >
17. > **19.** > **21.** > **23.** ≐ **25.** about 1 mm
27. about 20 cm **29.** about 23.5 cm × 30.5 cm
31. 180 cm **33.** 75 cm **41.** more
43. 1 kg, 2 kg, 5 kg

B.3 Exercises, page 482

1. 22.5 m **3.** 36 cm^2 **5.** 60.5 mi **7.** 15
9. 56 g **11.** 5 cm **13.** 255.6 g **15.** 3.498 qts
17. 90 m **19.** 7600 ml **21.** 430000 cl
23. 0.000043 km **25.** 43 mg **27.** 3 hr 25 min
29. 22 hr 10 min **31.** 4 hr 38 min
33. 5 hr 3 min 25 sec **35.** 4 hr 5 min
37. 40 hr 7 min **39.** 17 hr 56 min 3 sec
41. 9 hr 16 min 28 sec **43.** 61 hr 17 min
45. 44 min 40 sec **47.** 4 hr 30 min **49.** 13 hr 30 min
51. 101 hr 47 min 30 sec **53.** 1 hr 15 min
55. 3 hr 8 min 8 sec **57.** 1 hr 50 min
59. 114 hr 5 min 26 sec
61. 1 hr 18 min 30 sec; 5 hr 14 min **63.** 7 hr 53 min
65. 3:58 p.m.

Appendix B Review Exercises, page 484

1. F **2.** F **3.** T **4.** T **5.** T **6.** T **7.** F
8. F **9.** F **10.** T **11.** 90 m **12.** 30 m
13. 4000000 m **14.** 400 cm **15.** 40000 mm
16. 40 liter **17.** 4 km **18.** 400000 cl
19. 1968.5 in. **20.** 1.201 g **21.** 0.1201 g
22. 0.1201 kg **23.** 120100 g **24.** 1201 ml
25. 2 oz **26.** 95 liters **27.** about $2.18
28. 1.3 kg **29.** 1680 cm **30.** 3.33 m

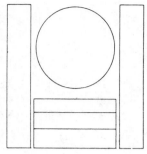

INDEX